Springer Monographs in Mathematics

Springer-Verlag Berlin Heidelberg GmbH

Edward R. Fadell • Sufian Y. Husseini

Geometry and Topology of Configuration Spaces

Springer

Edward R. Fadell
University of Wisconsin-Madison
Department of Mathematics
Madison, WI 53706
USA
e-mail: fadell@math.wisc.edu

Sufian Y. Husseini
University of Wisconsin-Madison
Department of Mathematics
Madison, WI 53706
USA
e-mail: husseini@math@wisc.edu

Library of Congress Cataloging-in-Publication Data

Fadell, Edward R., 1926-
Geometry and topology of configuration spaces / Edward R. Fadell, Sufian Husseini.
p. cm. -- (Springer monographs in mathematics, ISSN 1439-7382)
Includes bibliographical references and index.
ISBN 978-3-540-66669-1 ISBN 978-3-642-56446-8 (eBook)
DOI 10.1007/978-3-642-56446-8
1. Configuration space. 2. Algebraic topology. I. Husseini, S. Y. II. Title. III. Series.

QA607 .F34 2000
516.3'5--dc21

00-049221

Mathematics Subject Classification (2000): 20F36, 55M30, 55R10, 55R20, 55T10, 55T20, 58B05, 58E05, 70G25

ISSN 1439-7382
ISBN 978-3-540-66669-1

Originally published by Springer-Verlag Berlin Heidelberg New York in 2001

Cover design: *Erich Kirchner*, Heidelberg
Typesetting by the authors using a Springer LaTeX macro package
Printed on acid-free paper SPIN 10724127 41/3142ck-5 4 3 2 1 0

To Patricia and Barbara

Preface

The configuration space of k particles in the smooth manifold M is the space

$$\mathbb{F}_k(M) = \{(x_1, x_2, \cdots, x_k) \mid x_i \neq x_j\} \subset M^{k\times}.$$

These spaces and the associated free and based loop spaces, $\Omega\mathbb{F}_k(M)$ and $\Lambda\mathbb{F}_k(M)$, respectively, play an important role in topology and geometry and related areas. For example, the space $\mathbb{F}_k(M)$ provides additional topological invariants for the mainfold M and, more generally, for an imbedding $f : M \to M'$ of one manifold into another (see, for example [11, Bott], [13, Bott-Taubes], [69, Kohno], [108, Vasiliev], [51, Haefliger], [52, Haefliger], [99, Shapiro], [111, Wu], and [112, Wu]). Also, $\mathbb{F}_k(\mathbb{R}^{n+1})$ is the space of k noncolliding particles $x_1, \cdots, x_k$ in $\mathbb{R}^{n+1}$, and its free loop space $\Lambda\mathbb{F}_k(\mathbb{R}^{n+1})$ is the space of its k closed curves (orbits) in $\mathbb{R}^{n+1}$. The existence of periodic solutions to a Hamiltonian system of the k-body type is deduced from the study of the Lusternik-Schnirelman category and the Poincaré series of $\Lambda\mathbb{F}_k(\mathbb{R}^{n+1})$ (see, for example [8, Bahri-Rabinowitz], [37, Fadell-Husseini], [39, Fadell-Husseini], [40, Fadell-Husseini], [75, Majer-Terracini], [76, Majer-Terracini], and [91, Riahi]).

The special case of $\mathbb{F}_k(\mathbb{R}^2)$ is the underlying space of the pure braid group on k strands. As it is aspherical, it is also the classifying space of said group. There is an extensive literature on this case ([9, Birman], [54, Hansen]), and there is also a fairly extensive literature regarding stable homotopy questions concerning $\mathbb{F}_k(\mathbb{R}^{n+1})$ ([10, Bödigheimer et al.], [17, Cohen], [19, Cohen-Taylor]). However, some fundamental and interesting questions of a different sort, arising naturally in other contexts, have received little or no attention in the literature. All in all, the literature on configuration spaces is scattered and, in many cases, incomplete.

The aim of this monograph is to give a systematic and self-contained account of the main ideas and methods in configuration spaces and their loop spaces. It is our hope that it will be accessible not only topologists but also to readers whose chief interest is in related areas. In order to keep the book within reasonable bounds, we limit ourselves to the cases when $M = \mathbb{R}^{n+1}$, S^{n+1}, although many of the methods employed here apply more generally (cf. [48, Fulton-Macpherson], [18, Cohen-Gitler] and [113, Xicoténcatl Merino]).

The book is in three parts. In Part One, we study $\mathbb{F}_k(M)$ with $M = \mathbb{R}^{n+1}, S^{n+1}$ from the point of view of classical homotopy theory, as most of the problems that one encounters concerning $\mathbb{F}_k(M)$ reduce to questions in homotopy theory. The upshot is that, when $n > 1$, the space $\mathbb{F}_k(M)$ is simply connected and the graded Lie algebra of homotopy groups $\pi_*(\mathbb{F}_k(M))$ is, in effect, multilayered split extensions of the analogous algebras of certain bouquets of spheres. These split extensions are reflected in relations that we call the *Yang-Baxter relations.* These play a fundamental role in the development of the theory. In fact, in Part Two we use them to determine the homology and cohomology of $\mathbb{F}_k(M)$ as well as of certain corresponding cellular structures on the configuration spaces. These structures, in turn, provide us with the means to understand the structure of the associated spaces $\Omega\mathbb{F}_k(M)$ and $\Lambda\mathbb{F}_k(M)$. In Part Three, having the topological problems that arise in connection with some variational problems in mind, we take up the study of the loop spaces of $\mathbb{F}_k(M)$. Such problems require, typically, finding the Poincaré series and compact subsets of arbitrarily high Lusternik-Schnirelman category of the relevant loop space. We also give, by way of illustration, applications to some problems of the k-body type.

To explain how the various topics in the monograph are related to each other, let us begin with a brief outline of its main features. Suppose that $M = \mathbb{R}^{n+1}$. Denote the basic vector $(1, 0, \cdots, 0)$ by e, and put

$$q_i = 4(i-1)e, \;\; Q_i = \{q_1, q_2, \cdots, q_i | 1 \leq i \leq k\}, \; Q_0 = \emptyset.$$

For the sake of brevity, let $\mathbb{F}_{k-r,r} = \mathbb{F}_{k-r}(\mathbb{R}^{n+1} - Q_r)$. Note that

$$\mathbb{F}_{k-r}(\mathbb{R}^{n+1} - Q_r) = \{(x_1, \cdots, x_k) \in \mathbb{F}_k(M) \mid x_i = q_i, 1 \leq i \leq r\}.$$

These spaces lead to the natural filtration

$$\mathcal{F}_k(\mathbb{R}^{n+1}) : \left\{ \begin{array}{ccccccc} \mathbb{F}_{k,0} & \leftarrow \mathbb{F}_{k-1,1} & \leftarrow & \cdots & \leftarrow \mathbb{F}_{2,k-2} & \leftarrow & \mathbb{R}^{n+1}_{k-1} \\ \downarrow & \downarrow & & \downarrow & \downarrow & & \\ \mathbb{R}^{n+1} & \mathbb{R}^{n+1}_1 & & \cdots & \mathbb{R}^{n+1}_{k-2}, & & \end{array} \right.$$

where the vertical maps are projections on the first entry different from the q_i's, and $\mathbb{R}^{n+1}_r = \mathbb{R}^{n+1} - Q_r$. The vertical maps in $\mathcal{F}_k$ are fibrations that have sections. We think of such a fibration as a *sectioned twisted product* of base by fiber. For example, the notation $\mathbb{F}_{k-r,r} = \mathbb{R}^{n+1}_r \rtimes \mathbb{F}_{k-r-1,r+1}$ suggests that the vertical map $\mathbb{F}_{k-r} \to \mathbb{R}^{n+1}_r$ has a section and is a fibration with $\mathbb{F}_{k-r-1,r+1}$ as fiber. This filtration shows that $\mathbb{F}_k(\mathbb{R}^{n+1})$ is homotopy equivalent to a multifold twisted product of the spaces $\mathbb{R}^{n+1}_r$, a fact of great value in the study of the loop spaces, both based and free; for, passing to based loop spaces, one sees readily that $\Omega\mathbb{F}_k(\mathbb{R}^{n+1})$ is, as a space, homotopy equivalent to the product of loop spaces of the spaces $\mathbb{R}^{n+1}_r$.

The maps $\alpha'_{rs} : S^n \to \mathbb{F}_k(\mathbb{R}^{n+1})$ for all $1 \leq s < r \leq k$, where

$$\xi \mapsto (q_1, q_2, \cdots, , q_{r-1}, q_s + \xi, q_r, \cdots),$$

are imbeddings of S^n in $\mathbb{F}_k(\mathbb{R}^{n+1})$. Denote by $S_{rs} \subset \mathbb{R}_r^{n+1}$ the image of the composite of α'_{rs}, with the projection $\mathbb{F}_{k-r,r} \to \mathbb{R}_r^{n+1}$. Then, we obtain a homotopy equivalence $S_{r1} \vee \cdots \vee S_{rr-1} \to \mathbb{R}_r^{n+1}$, and the maps $\alpha'_{r1}, \cdots, \alpha'_{rr-1}$ together induce a section of the pull-back of $\mathbb{F}_{k-r,r} \to \mathbb{R}_r^{n+1}$ to $S_{r1} \vee \cdots \vee S_{rr-1} \to \mathbb{R}_r^{n+1}$. Suppose that $n > 1$, and let

$$\alpha_{rs} = [\alpha'_{rs}], \ 1 \leq s < r \leq k$$

denote the corresponding elements of $\pi_n(\mathbb{F}_k(\mathbb{R}^{n+1}))$. Then, using the diagram $\mathcal{F}_k(\mathbb{R}^{n+1})$, one shows that these elements generate $\pi_n(\mathbb{F}_k(\mathbb{R}^{n+1}))$.

The structure of $\pi_*(\mathbb{F}_k(\mathbb{R}^{n+1}))$ as a graded Lie algebra, with the Whitehead product providing the multiplication ([110, Whitehead]), is a good measure of the twisting of the fibrations in the filtration $\mathcal{F}_k(\mathbb{R}^{n+1})$. The elements α_{rs} satisfy the *Yang-Baxter relations*, which we call the Y-B relations (see §3 of Chapter II). These relations can be viewed as the higher dimensional version of similar relations that appear in the group algebra of the braid group on k strands in [67, Kohno] and [68, Kohno]. There, they are called the *infinitesimal braid relations*. The Y-B relations turn out to be of great importance to us: they imply that the splitting $\Omega\mathbb{F}_k(\mathbb{R}^{n+1}) \simeq \prod_r \Omega(\mathbb{R}_r^{n+1})$ is one of spaces and not H-spaces and that the natural projections $\mathbb{F}_k(\mathbb{R}^{n+1}) \to \mathbb{F}_r(\mathbb{R}^{n+1})$ are not fiber homotopically trivial as fiber spaces, except for special cases. The Y-B relations also figure prominently in the calculations involving the space of free loops $\Lambda\mathbb{F}_k(\mathbb{R}^{n+1})$.

The case of $\mathbb{F}_k(\mathbb{R}^2)$ is complicated by the lack of simple connectivity. Nevertheless, the program outlined for the case $n+1 > 2$ can be adapted successfully. Moreover, thinking of $\mathbb{F}_k(\mathbb{R}^2)$ as a multifold twisted product of bouquets of circles, with the twisting being given by the analogue of the Y-B relations, we are able to present the classical theory of pure braids in a novel fashion and obtain, as a consequence, a new and concise presentation of the pure braid group.

The case of $\mathbb{F}_{k+1}(S^{n+1})$ is closely related to the case of $\mathbb{F}_k(\mathbb{R}^{n+1})$. There is a filtration similar to that of $\mathcal{F}_k(\mathbb{R}^{n+1})$. However, because $\mathbb{F}_3(S^{n+1})$ is homotopy equivalent to the Stiefel manifold $O_{n+2,2}$ of orthonormal 2-frames in $\mathbb{R}^{n+2}$, the similarities begin only beyond $k = 3$. There are Y-B relations coming from the space $\mathbb{F}_k(\mathbb{R}^{n+1})$ and others coming from the twisting of the fibration $O_{n+2,2} \to S^{n+1}$. These relations show that the projections $\mathbb{F}_{k+1}(S^{n+1}) \to \mathbb{F}_r(S^{n+1})$ are not trivial as fiber spaces, except for a few special cases.

Also, the space of free loops, $\Lambda\mathbb{F}_k(\mathbb{R}^{n+1})$ is the homotopy theoretic analogue of the adjoint bundle $EG \times_G G \to BG$, if one thinks of the loop space $\Omega\mathbb{F}_k(\mathbb{R}^{n+1})$ and the path fibration $\mathcal{P}\mathbb{F}_k(\mathbb{R}^{n+1}) \to \mathbb{F}_k(\mathbb{R}^{n+1})$ as the analogues, respectively, of the group G and its universal bundle $EG \to BG$. The Y-B relations show that the splitting of $\Omega\mathbb{F}_k(\mathbb{R}^{n+1})$ as a product of loop spaces of bouquets of spheres is not one of H-spaces. Therefore, the adjoint action of $\Omega\mathbb{F}_k(\mathbb{R}^{n+1})$ on itself is not equivalent to the product of the adjoint actions of the constituent components, and, hence, $\Lambda\mathbb{F}_k(\mathbb{R}^{n+1})$ does not split in the manner of based loops. The questions concerning $H_*(\Lambda\mathbb{F}_k(\mathbb{R}^{n+1});\mathbb{K})$ are, as one would expect, much harder to settle than analogous questions regarding $H_*(\Omega\mathbb{F}_k(\mathbb{R}^{n+1});\mathbb{K})$ and the answers depend very much on $\mathbb{K}$. For example, as is well known, $H_*(\Lambda(S^2);\mathbb{Q})$ need not have any nontrivial coproducts, in spite of the fact that $H_*(\Omega(S^2);\mathbb{Q})$ is rich in them (see [105, Sullivan and Vigué-Poirrier]). On the other hand, $H_*(\Lambda(S^2);\mathbb{Z}_2)$ behaves like $H_*(S^2;\mathbb{Z}_2)\otimes H_*(\Omega(S^2);\mathbb{Z}_2)$ [37, Fadell-Husseini].

Our main tool in solving these is the Serre spectral sequence of the evaluation map $\Lambda\mathbb{F}_k(\mathbb{R}^{n+1}) \to \mathbb{F}_k(\mathbb{R}^{n+1})$. We observe that the mod 2 Serre spectral sequence of the canonical fibration $\mathcal{P}\mathbb{F}_3(\mathbb{R}^{n+1}) \to \mathbb{F}_3(\mathbb{R}^{n+1})$ stabilizes after the nth stage. That is to say, $\mathrm{E}^{n+1} \cong \mathrm{E}^{\infty}$. We prove this in general. Therefore, the E^n term, of the mod 2 Serre spectral sequence of $\mathcal{P}\mathbb{F}_k(\mathbb{R}^{n+1}) \to \mathbb{F}_k(\mathbb{R}^{n+1})$ can be viewed as an acyclic free resolution of $\mathbb{Z}_2$ over $H_*(\Omega\mathbb{F}_k(\mathbb{R}^{n+1}))$. Hence, for the fibration $\Lambda\mathbb{F}_k(\mathbb{R}^{n+1}) \to \mathbb{F}_k(\mathbb{R}^{n+1})$, the E^{n+1} term can be viewed as

$$\mathrm{Tor}^{H_*(\Omega\mathbb{F}_k(\mathbb{R}^{n+1});\mathbb{Z}_2)}(H_*(\Omega\mathbb{F}_k(\mathbb{R}^{n+1}),\mathbb{Z}_2),$$

which is just the E^2-term of the Eilenberg-Moore spectral sequence of the evaluation map

$$\mathrm{eval} : \Lambda\mathbb{F}_k(\mathbb{R}^{n+1}) \to \mathbb{F}_k(\mathbb{R}^{n+1}),\ \omega \mapsto \omega(1).$$

We ask then whether the Eilenberg-Moore spectral sequence of the evaluation map eval collapses. To answer this question, we proceed as follows. First, we define a minimal finite CW-complex X homotopy equivalent to $\mathbb{F}_k(\mathbb{R}^{n+1})$. The Y-B relations turn out to be the key. They give us the cells and their attaching maps. The complex X also gives, in turn, a cellular model $M(X)$ of the reduced product type for the based loops. The corresponding cellular chain complex $C_*(M(X))$ admits a homomorphism

$$C_*(M(X)) \to CS_*(\Omega\mathbb{F}_k(\mathbb{R}^{n+1}))$$

of DG-algebras, where $CS_*(\Omega\mathbb{F}_k(\mathbb{R}^{n+1}))$ is the singular chain complex of the loop space. The fact that the attaching cells and their attaching maps are defined by the Y-B relations allows one to define a homomorphism

$$\phi : C_*(M(X)) \to H_*(\Omega\mathbb{F}_k(\mathbb{R}^{n+1}))$$

that induces a homotopy equivalence on homology. Using this, we prove that the mod 2 Serre spectral sequence of the free loop fibration stablizes at the $(n+1)$ stage, and the calculation of the homology of the space $\Lambda\mathbb{F}_k(\mathbb{R}^{n+1})$ is thereby greatly simplified.

More detailed description, chapter by chapter, introduces each part.

We take this opportunity to express our appreciation to the referees for their careful reading and useful comments. We offer our thanks to colleagues and students in Madison, Pisa, Milan, and Heidelberg who attended our lectures on the subject of this monograph and offered valuable advice. We also wish to express our sincere thanks to Fred Cohen for many contacts and conversations. Our special thanks to Paul Rabinowitz, Abbas Bahri, Hasna Riahi, Pietro Majer, and Susannah Terracini whose questions on the topology of the free loop space of the configuration spaces of Euclidean spaces provided the original motive for the project that resulted in this work.

Contents

Part III. Homology and Cohomology of Loop Spaces

Part I

The Homotopy Theory of Configuration Spaces

Introduction

Our aim in Part One is to develop basic methods for Parts Two and Three, where we study the associated spaces of the based and free loops of, respectively, $\mathbb{F}_k(M)$, with M being the Euclidean space $\mathbb{R}^m$, or the sphere S^m. Part One is devoted to the homotopy theory of configuration spaces.

In Chapter I we consider the natural projections $\mathrm{proj}_{k,r} : \mathbb{F}_k(M) \to \mathbb{F}_r(M)$, where $r < k$. We prove that these projections are fiber bundles in the sense of Steenrod ([103, Steenrod]). These fibrations are basic for the work to follow. They provide one with an inductive procedure for studying the relevant topological questions.

The local Euclidean nature of M makes $\mathbb{F}_k(\mathbb{R}^m)$ part of every configuration space. In Chapter II we assume that $n > 1$. The configuration spaces $\mathbb{F}_k(\mathbb{R}^{n+1})$ are, therefore, simply connected, and the general pattern of the theory is simplified. The Whitehead product turns $\pi_n(\mathbb{F}_{k+1}(\mathbb{R}^{n+1}))$ into a graded Lie algebra. This algebra plays a prominent role in our study. We introduce elements α_{rs}, $1 \leq s < r \leq k$ in $\pi_n(\mathbb{F}_{k+1}(\mathbb{R}^{n+1}))$, and study their behaviour under the natural action of the symmetric group. The filtration of $\mathbb{F}_k(\mathbb{R}^{n+1})$ by the ascending sequence of subspaces described in the diagram $\mathcal{F}_k(\mathbb{R}^{n+1})$ given in the preface leads to naturally to the filtration of $\pi_*(\mathbb{F}_k(\mathbb{R}^{n+1}))$

$$L_1 \subset \cdots \subset L_{k-r} \subset \cdots \subset L_k = \pi_*(\mathbb{F}_k(\mathbb{R}^{n+1}))$$

by ideals, where $L_{k-r} \cong L_{k-r-1} \rtimes L_{k-r,k-r-1}$, $L_{k-r,k-r-1}$ is the Lie algebra of a bouquet of r spheres of dimension n, and L_{k-r-1} consists of multifold split extensions defined by the relevant Y-B relations (see §2 of Chapter II).

We seek the relations satisfied by their Whitehead products. Thinking of $\mathbb{F}_k(\mathbb{R}^m)$ as the space underlying a problem of the k-body type, one obtains the following simple relations. They are of two kinds: the first, in a manner of speaking, describes triplets, where two bodies circle around a third in different orbits; the second describes two different pairs moving about independently of each other, with each pair consisting of bodies where one moves around the other. Other relations are obtained from the preceding relations by applying to them the symmetries of the full symmetric group Σ_k. We call these basic relations the *Yang-Baxter relations* (Y-B for short) because they resemble differential equations that go by the same name (see [68, Kohno], [100, Shnider-Sternberg]).More complicated relations are obtained in Part Two, when we take up the question of finding a minimal CW-complex for $\mathbb{F}_k(\mathbb{R}^m)$.

Finally, we show in Chapter II that the twisted-product structure on $\mathbb{F}_k(\mathbb{R}^m)$, except for special cases, is not naturally trivial.The multifold twisted-product structure of $\mathbb{F}_k(\mathbb{R}^{n+1})$ leads one naturally to ask whether the fiber bundle $\mathrm{proj}_{k,r} : \mathbb{F}_k(\mathbb{R}^{n+1}) \to \mathbb{F}_r(\mathbb{R}^{n+1})$ is, in fact, fiber homotopically equivalent to the trivial bundle. As an interesting application of the Y-B relations, we show that, except for a few exceptional cases, this is never the case (see §5 of Chapter II).

In Chapter III we study the configuration space $\mathbb{F}_{k+1}(S^{n+1})$. The fiber of the projection $\mathrm{proj}_{k+1,1} : \mathbb{F}_{k+1}(S^{n+1}) \to S^{n+1}$ on the last factor can be identified by means of the appropriate stereographic projection with $\mathbb{F}_k(\mathbb{R}^{n+1})$. There is an analogue of diagram $\mathcal{F}_k$, and, in addition to the Y-B relations that come from the fiber, there are others that come from the Stiefel manifold $O_{n+2,2}$ of orthonormal 2-frames in $\mathbb{R}^{n+2}$. For, here, $O_{n+2,2}$ can be naturally identified, up to homotopy, with $\mathbb{F}_3(S^{n+1})$. Consequently, the parity of $(n+1)$ plays a role.

It is worthwhile to note that the constructions described in Chapter III for S^m when m is odd require only the existence of a nonvanishing vector field and are possible for any odd- dimensional closed manifold.

In Chapter IV we take up the study of the case $\mathbb{F}_k(\mathbb{R}^2)$. This and its orbit space $\mathbb{F}_k(\mathbb{R}^2)/\Sigma_k$, where Σ_k is the group of symmetries on k letters, are the underlying spaces of the classical braid groups. The pattern, established in the preceding chapters, is modified to take care of the difficulties due to the lack of simple connectivity. A noteworthy result is a presentation of the pure braid group $\pi_1(\mathbb{F}_k(\mathbb{R}^2), q)$ based on the Y-B relations. The presentation is both concise and easy to visualize. It is different from the more familiar presentation given in [9, Birman], where the point of view is more algebraic.

I. Basic Fibrations

Consider now the natural projection $\mathrm{proj}_{k,r} : \mathbb{F}_k(M) \to \mathbb{F}_r(M), \quad r < k$ that sends $(x_1, \cdots, x_k)$ to $(x_1, \cdots, x_r)$, where M is a connected manifold of dimension m. With the configuration spaces being regarded as the space of imbeddings of the sets $\mathbf{k}$ and $\mathbf{r}$, respectively, one sees that these projections are just the restriction maps induced by the injection $\mathbf{r} = \{1, \cdots, r\} \subset \mathbf{k} = \{1, \cdots, k\}$. Hence, according to [107, Thom] (see also [87, Palais]), these projections are locally trivial fibrations. As the fibration of $\mathbb{F}_k(M)$ over $\mathbb{F}_r(M)$ plays a central role in the study of the geometry and topology of these configuration spaces ([43, Fadell-Neuwirth], [17, Cohen]), a simple and direct proof of the fact that $\mathrm{proj}_{k,r} : \mathbb{F}_k(M) \to \mathbb{F}_r(M)$ is locally trivial, independent of [107, Thom] (or [87, Palais]), is in order. In §1 below, we give such a proof. The structure group will be a subgroup of $\mathrm{Top}(M)$, and its fiber the configuration space $\mathbb{F}_{k-r}(M - Q_r)$, where $Q_r = \{q_1, \cdots, q_r\}$ and $q = (q_1, \cdots, q_r)$, is the basepoint of $\mathbb{F}_r(M)$.

In applications, one often has a compact Lie group $G \subset \mathrm{Top}(M)$ acting on M, and one is interested in the restriction of the fibration $\mathrm{proj}_{k,r} : \mathbb{F}_k(M) \to \mathbb{F}_r(M)$ to the orbit Gq. With $H \subset G$ being the isotropy group of a tuple $(q_1, q_2, \cdots, q_r)$, we show in §2 that the restriction to G/H is, in fact, the bundle, associated with the principal bundle $G \to G/H$, the fiber of which is the subspace of all tuples in $\mathbb{F}_k(M)$ that start with the tuple $(q_1, \cdots, q_r)$. As the tangent bundle of G/H is also associated with the same principal bundle, we see that the restriction of the configuration fibration to G/H and the tangent bundle of G/H are also associated with the same principal bundle. This result will prove useful in Chapters II and III, where we discuss the triviality, or not, of the configuration fibrations of $\mathbb{R}^m$ and S^m.

Stiefel manifolds $O_{n+1,r}$ appear as homogeneous spaces imbedded in the configuration spaces, with the orthogonal group O_{n+1} acting linearly on $\mathbb{R}^{n+1}$ and S^n. In §§2 and 3 we discuss the restrictions of the configuration fibrations to these homogeneous spaces. In particular, the restriction of the bundle $\mathrm{proj}_{k,2} : \mathbb{F}_k(\mathbb{R}^{n+1}) \to \mathbb{F}_2(\mathbb{R}^{n+1})$ to $S^n \subset \mathbb{F}_2(\mathbb{R}^{n+1})$, $x \mapsto (x, -x)$ is shown to be associated with the principal tangent bundle of S^n.

1 The Projection $\text{proj}_{k,r} : \mathbb{F}_k(M) \to \mathbb{F}_r(M)$

Denote by Top(M) the group of homeomorphisms $\phi : M \to M$ with the compact-open topology. A homeomorphism ϕ is said to be *stable* if, and only if, ϕ is a homeomorphism of M fixed outside some proper closed subset of M. The stable homeomorphisms generate a subgroup of Top(M), say, $\text{Top}_s(M)$. There are other subgroups of Top(M) that occur in the study of configuration spaces: e.g., the group of diffeomorphisms of M, as well as the subgroup $\text{Top}_{ss}(M)$ of Top(M) generated by the *strongly stable* homeomorphisms $\phi \in$ Top(M), where ϕ is said to be strongly stable if, and only if, for some m-cell $D^m \subset M$, $\phi | M - \text{int}\, \text{D}^\text{m}$ is fixed. Note that, according to the classical result ([6, Alexander]), a strongly stable homoeomorphism ϕ is isotopic to the identity through an isotopy fixed outside of some m-cell. This fact implies easily that $\text{Top}_{ss}(M)$ is a path-connected subgroup of $\text{Top}_s(M)$.

For simplicity, here we shall only consider $\text{Top}_s(M)$. By definition, for $x = (x_1, \cdots, x_k) \in \mathbb{F}_k(M)$, let $Q_i^x = \{x_1, \cdots, x_i\}$, where $1 \le i \le k$. When x is the basepoint $q = (q_1, \cdots, q_k)$, Q_r^q is denoted simply by Q_r. Let us put

$$\mathbb{F}_{k-r,r}(M) = \mathbb{F}_{k-r}(M - Q_r).$$

Note that $\text{Top}_s(M)$ acts naturally on the associated configuration space $\mathbb{F}_k(M)$. Thus $\text{Top}_s(M)$ is naturally a subgroup of the group of homeomorphisms of $\mathbb{F}_k(M)$, for all k. Also put

$$\text{Top}_{sr}(M) = \{\phi \in \text{Top}_s(M) \mid \phi(q_i) = q_i,\ 1 \le i \le r\}.$$

Theorem 1.1 *The projection* $\text{proj}_{k,r} : \mathbb{F}_k(M) \to \mathbb{F}_r(M)$, $r < k$, *fibers* $\mathbb{F}_k(M)$, *in the sense of Steenrod ([103, Steenrod]; also [101, Spanier]), over* $\mathbb{F}_r(M)$. *The fiber and structure group are* $\mathbb{F}_{k-r,r}(M)$ *and* $\text{Top}_{sr}(M)$, *respectively.*

The proof depends on the following lemma. Let D^m be the disk in $\mathbb{R}^m$ of radius 1 and centered at the origin, and denote by V^m its interior. Observe that the maps $g : V^m \to \mathbb{R}^m$ and $h : \mathbb{R}^m \to V^m$ defined by the rules

$$\begin{aligned} g(y) &= \tfrac{y}{1-|y|}, \quad y \in V^m \text{ and} \\ h(x) &= \tfrac{x}{1+|x|}, \quad x \in \mathbb{R}^{n+1}, \end{aligned}$$

are homeomorphisms and that $g^{-1} = h$. Also denote by $G_0(D^m)$ the group of homeomorphisms of D^m to itself that leave the boundary pointwise fixed, with the usual compact-open topology.

Lemma 1.1 *There is a map* $\gamma : V^m \to G_0(D^m)$ *such that,*

$$\begin{aligned} &(i)\gamma(x)(x) = 0, \quad \forall\, x \in V^m \text{ and} \\ &(ii)\gamma(x)(y) = y, \quad \forall\, y \in \partial D^m. \end{aligned}$$

Proof. First, define $\gamma' : V^m \times D^m \to D^m$ to be the map such that

$$\gamma'(x,y) = \begin{cases} h(g(y) - g(x)), \forall\, y \in V^m \\ y, \qquad\qquad\qquad \forall\, y \in \partial D^m. \end{cases}$$

Let $\gamma : D^m \to G_0(D^m)$ be the map that takes $x \in V^m$ to the homoeomorphism $y \mapsto \gamma'(x,y)$. □

Let U be a proper open subset of M the closure of which, $\bar{U}$, is homeomorphic to D^m in such a way that V^m corresponds to U, and 0, to $x_0 \in U$. Lemma 1.1 implies the following immediately.

Corollary 1.1 *There is a map $\gamma : U \to \mathrm{Top}_s(M)$ with the following properties.*

$$(i)\gamma(x)(x){=}x_0, \forall\, x \in U; \text{ and}$$
$$(ii)\gamma(x)(y){=}y, \ \forall\, x \in U, \text{ and } y \in M - U.$$

Proof of Theorem 1.1
Step 1.

Let us first prove that the projection $\mathrm{proj}_{k,r} : \mathbb{F}_k(M) \to \mathbb{F}_r(M)$, $1 \le r < k$, is locally trivial. Let $(x_1, \cdots, x_r) \in \mathbb{F}_r(M)$ be a fixed r-tuple, and choose mutually disjoint disk neighborhoods $\bar{U}_1, \cdots, \bar{U}_r$ in M, with $x_i \in U_i$, the interior of $\bar{U}_i$, for $1 \le r \le r$. For each i, let $\gamma_i : U_i \to \mathrm{Top}_s(M)$ be the map of Corollary 1.1. Identify the fiber of $\mathrm{proj}_{k,r} : \mathbb{F}_k(M) \to \mathbb{F}_r(M)$ at $(x_1, \cdots, x_r)$ with the space $\mathbb{F}_{k-r}(M - Q_r^x)$, where $Q_r^x = \{x_1, \cdots, x_r\}$. Now let $U = U_1 \times \cdots \times U_r$, and define the local trivialization

$$\begin{array}{ccc} U \times \mathbb{F}_{k-r}(M - Q_r^x) & \xrightarrow{\phi'_U} & \mathrm{proj}^{-1}(U) \\ \downarrow & & \downarrow \\ U & \xrightarrow{\mathrm{id}} & U, \end{array}$$

where the vertical maps are the natural projections such that

$$(x'_1, \cdots, x'_r; y) \mapsto (x'_1, \cdots, x'_r, (\gamma_1^{-1}(x'_1) \circ \cdots \circ \gamma_r^{-1}(x'_r))(y)).$$

Note that ϕ'_U depends on U and on the choice of $(x = (x_1, \cdots, x_r)$.

Step 2

By Corollary 1.1 of Lemma 1.1 and the connectivity of M, there is a $\alpha_x \in \mathrm{Top}_s(M)$ that induces a stable homeomorphism $\alpha'_x : \mathbb{F}_{k-r}(M - Q_r) \to \mathbb{F}_{k-r}(M - Q_r^x)$, where $Q_r = \{q_1, \cdots, q_r\}$, and $q = (q_1, \cdots, q_r)$ is a chosen basepoint in $\mathbb{F}_r(M)$. Now, by definition, put $\phi_U = \phi'_U \circ (1 \times \alpha'_x)$. Then, ϕ_U is a trivialization

$$\begin{array}{ccc} U \times \mathbb{F}_{k-r}(M - Q_r) & \xrightarrow{\phi'_U} & \mathrm{proj}^{-1}(U) \\ \downarrow & & \downarrow \\ U & \xrightarrow{\mathrm{id}} & U \end{array}$$

over U, where the fiber is now fixed, i.e., independent of x.

Next we exhibit the structure group of the bundle.

Step 3

Let U and V denote the product neighborhoods as in Step 2. Then the map

$$\phi_V^{-1}\phi_U : (U \cap V) \times \mathbb{F}_{k-r}(M - Q_r) \to (U \cap V) \times \mathbb{F}_{k-r}(M - Q_r)$$

defines a map $U \cap V \to \mathrm{Top}_{sr}(M)$, as the group of the bundle as defined in [103, Steenrod]. □

Consider the map $\mathrm{p} : \mathrm{Top}_s(M) \to \mathbb{F}_k(M)$, $g \mapsto (gq_1, \cdots, gq_r)$, and put $H = \mathrm{p}^{-1}(q_1, \cdots, q_r) = \mathrm{Top}_{sr}(M)$. Corollary 1.1 readily implies that $(\mathrm{Top}_s(M), p, \mathbb{F}_r(M))$ is a principal bundle, with H as group and fiber. Let H act, on the left, on $\mathrm{Top}_s(M) \times \mathbb{F}_{k-r}(M)$, according to the rule $h(g, y) = (gh^{-1}, hy)$.

Proposition 1.1 *The bundle* $(\mathrm{Top}_s(M), \mathrm{p}, \mathbb{F}_r(M))$ *is the principal bundle associated with the bundle* $\mathrm{proj}_{k,r} : \mathbb{F}_k(M) \to \mathbb{F}_r(M)$, *the fiber of which is* $\mathbb{F}_{k-r}(M - Q_r)$.

Proof. Define a map $\psi : \mathrm{Top}_s(M) \times \mathbb{F}_{k-r}(M - Q_r) \to \mathbb{F}_k(M)$ by the rule $(g, (y_{r+1}, \cdots, y_k)) = g(q_1, \cdots, q_r, y_{r+1}, \cdots, y_k)$. Note that

$$\psi(gh^{-1}, h(y_{k+1}, \cdots, y_k)) = g(q_1, \cdots, q_r, y_{r+1}, \cdots, y_k).$$

Hence, ψ induces the homeomorphism

$$\begin{array}{ccc} \mathrm{Top}_s(M) \times \mathbb{F}_{k-r}(M - Q_r) & \xrightarrow{\bar{\psi}} & \mathbb{F}'_k(M) \\ \bar{\mathrm{p}} \downarrow & & \downarrow \mathrm{proj}_{k,r} \\ \mathbb{F}_r(M) & \xrightarrow{\mathrm{id}} & \mathbb{F}_r(M) \end{array}$$

over $\mathbb{F}_r(M)$, where $\bar{\mathrm{p}}$ is induced by the map $\mathrm{p} : \mathrm{Top}_s(M) \to \mathbb{F}_r(M)$. This proves the assertion. □

It is worthwhile to note here that it is possible to reduce the group of the bundle $\mathrm{proj}_{k,1} : \mathbb{F}_k(M) \to M$ from $\mathrm{Top}_{sk}(M)$ to

$$\mathrm{Top}_{ssk}(M) = \mathrm{Top}_{ss} \cap \mathrm{Top}_{sk}(M),$$

where $\mathrm{Top}_{ss}(M)$ is the group of strongly stable homeomorphisms. To see this, first, as above, fix points $q_1, \cdots, q_k$ of M, and define $\mathrm{p} : \mathrm{Top}_{ss} \to M$ according to the rule $\mathrm{p}(g) = (gq_1, \cdots, gq_k)$. Now, proceed as in the proof of Theorem 1.1 to show that the projection $\mathrm{p} : \mathrm{Top}_{ss}(M) \to M$ is a principal bundle with group and fiber equal to $\mathrm{Top}_{ssk}(M) = \mathrm{p}^{-1}(q_1, \cdots, q_k)$. The latter is also associated with the bundle $\mathrm{proj}_{k,1} : \mathbb{F}_k(M) \to M$.

2 Relations to Homogeneous Spaces G/H

The preceding fibrations are related to some classical fibrations associated with homogeneous spaces G/H and their tangent spaces when M is either the Euclidean space or the unit sphere S^n. To see this, it is convenient to consider the more general case when G is a compact Lie group acting on M on the left. The given action of G, in turn, induces an action

$$G \times \mathbb{F}_k(M) \to \mathbb{F}_k(M),$$

where $(g,(x_1,\cdots,x_k)) \mapsto (gx_1,\cdots,gx_k)$. Denote by H the isotropy subgroup of the basepoint $q=(q_1,\cdots,q_r)$, $r<k$. Then, the map $G \to \mathbb{F}_r(M)$, where $g \mapsto gq$ induces an imbedding $\rho : G/H \to \mathbb{F}_r(M)$. By definition, let

$$\mathbb{F}'_k(M) = \{(gH, x_1, x_2, \cdots, x_k) \mid (x_1,\ldots,x_r) = gq = \rho(gH)\}$$

be the pull-back of the bundle $\mathrm{proj}_{k,r} : \mathbb{F}_k(M) \to \mathbb{F}_r(M)$ by the imbedding $\rho : G/H \to \mathbb{F}_r(M)$ defined above. Denote by $\mathrm{proj}'_{k,r} : \mathbb{F}'_k(M) \to G/H$ the natural projection, where $(\rho(gH), x_1, \cdots, x_k) \mapsto gH$.

Theorem 2.1 *The bundle* $\mathrm{proj}'_{k,r} : \mathbb{F}'_k(M) \to G/H$ *is equivalent to the bundle with fiber* $\mathbb{F}_{k-r,r}(M)$ *associated with the principal bundle* $G \to G/H$.

Proof. To prove the theorem, consider the map

$$\phi' : G \times \mathbb{F}_{k-r,r}(M) \to \mathbb{F}'_k(M)$$

such that $(g,x) \mapsto (gq, gx)$, where $q=(q_1,\cdots,q_r)$, and $x=(x_{r+1},\cdots,x_k)$. Next, on identifying $\mathbb{F}_{k-r,r}(M)$ with the subspace

$$\{(x_1,\cdots,x_k) \mid x_1 = q_1, \cdots, x_r = q_r\}$$

of $\mathbb{F}_k(M)$, we see that $\mathbb{F}_{k-r,r}(M)$ is invariant under H. The map ϕ' induces a homeomorphism

$$\begin{array}{ccc} G \times_H \mathbb{F}_{k-r,r}(M) & \xrightarrow{\phi} & \mathbb{F}'_k(M) \\ \mathrm{p} \downarrow & & \downarrow \mathrm{proj}_{k,r} \\ G/H & \xrightarrow{\mathrm{id}} & G/H \end{array}$$

over G/H, proving the theorem. □

Observe here that the tangent bundle of G/H is also associated with the principal bundle $(G,p,G/H)$, with H as group and fiber. In fact, if $L(G)$ denotes the Lie algebra of G, then

$$L(G) \cong L(H) \oplus L(H)^{\perp}$$

as H-modules, since the adjoint action of H on $L(G)$ is reductive. Proceeding as above, with H acting diagonally on $G \times L(H)^{\perp}$, we obtain the equivalence

$$\begin{array}{ccc} G \times_H L(H)^{\perp} & \xrightarrow{\phi} & T(G/H) \\ \mathrm{p} \downarrow & & \downarrow \mathrm{proj}_{k,r} \\ G/H & \xrightarrow{\mathrm{id}} & G/H \end{array}$$

of vector bundles, where $\mathrm{p} : T(G/H) \to G/H$ is the tangent bundle of G/H. Thus, in some sense, the restriction of the configuration space fibration to G/H, described in Theorem 1.1 above, and the tangent bundle of G/H are *associates*, that is, they arise from the same principal bundle.

When M is the Euclidean space $\mathbb{R}^{n+1}$ or the unit sphere $S^n \subset \mathbb{R}^{n+1}$ and the group G is the orthogonal group O_{n+1} acting linearly, then the resulting homogeneous spaces G/H are the Stiefel manifolds $O_{n+1,r}$ of orthonormal r-frames in $\mathbb{R}^{n+1}$.

3 The Pull-back to $O_{n+1,r}$

Let the orthogonal group O_{n+1} act on $\mathbb{R}^{n+1}$ in the usual fashion. Then O_{n+1} leaves S^n invariant, and the induced action on S^n induces, in turn, an action on the spaces $\mathbb{F}_r(S^n)$ for all $r \geq 1$. Now, take $q = (q_1, \cdots, q_r)$ to be the basepoint of $\mathbb{F}_r(S^n)$, where $r \leq (n+1)$, and let q_i be the ith basic vector of $\mathbb{R}^{n+1}$. Note that the isotropy subgroup of q is exactly the subgroup O_{n+1-r} of O_{n+1}. Therefore, the map that takes an r-frame to its end points induces an imbedding

$$\rho : O_{n+1,r} = O_{n+1}/O_{n+1-r} \to \mathbb{F}_r(S^n).$$

Denote the pull-back of $\mathrm{proj}_{k-r,r} : \mathbb{F}_k(S^n) \to \mathbb{F}_r(S^n)$ to $O_{n+1,r}$ by

$$\mathrm{proj}'_{k,r} : \mathbb{F}'_k(S^n) \to O_{n+1,r}.$$

Theorem 1.1 above implies the following corollary.

Corollary 3.1 *The bundle* $\mathrm{proj}'_{k-r,r} : \mathbb{F}'_k(S^n) \to O_{n+1,r}$*, with* $\mathbb{F}_{k-r,r}(S^n)$ *as fiber, is associated with the principal bundle* $O_{n+1} \to O_{n+1,r}$.

There is another interesting imbedding of the Stiefel manifolds in the configuration spaces: namely, the imbedding

$$\lambda : O_{n+1,r} \to \mathbb{F}_{2r}(S^n), \; r \leq (n+1)$$

that sends an r-frame $x = (x_1, \cdots, x_r)$ to the $2r$-tuple

$$(x, -x) = (x_1, \cdots, x_r, -x_1, \cdots, -x_r)$$

in $\mathbb{F}_{2r}(S^n)$. Let $\mathrm{proj}''_{k,2r} : \mathbb{F}''_k(S^n) \to O_{n+1,r}$, $k > 2r$, be the pull-back to $O_{n+1,r}$ by λ.

Corollary 3.2 *The bundle* $\mathrm{proj}''_{k,2r} : \mathbb{F}''_k(S^n) \to O_{n+1,r}$*, with* $\mathbb{F}_{k-2r,2r}(S^n)$ *as fiber, is associated with the principal bundle* $O_{n+1} \to O_{n+1,r}$*.*

Observe that $\mathbb{F}_2(S^n) \to S^n$ admits a section that sends x to $(x, -x)$. As the fiber is contractible, we see that the imbedding given by the section is a homotopy equivalence. This implies the following special case (cf. [29, Fadell]). Identify S^n with its image in $\mathbb{F}_2(S^n)$ under the imbedding $x \mapsto (x, -x)$, and $\mathbb{R}^n - \{0\} = \mathbb{R}^n_0$ with $S^n - \{e, -e\}$ by means of the stereographic projection of $S^n - \{-e\}$ onto $\mathbb{R}^n$, where e is the basic vector $(0, 0, \cdots, 0, 1) \in S^n \subset \mathbb{R}^{n+1}$.

Corollary 3.3 *The restriction of the bundle* $\mathrm{proj}_{3,2} : \mathbb{F}_3(S^n) \to \mathbb{F}_2(S^n)$ *to* $S^n \subset \mathbb{F}_2(S^n)$ *is equivalent to the bundle* $O_{n+1} \times_{O_n} (\mathbb{R}^n)_0 \to O_{n+1,1} = S^n$*, with fiber* $\mathbb{R}^n_0$ *associated with the sphere tangent bundle of* S^n*. In particular* $\mathrm{proj}_{3,2} : \mathbb{F}_3(S^n) \to \mathbb{F}_2(S^n)$ *contains the sphere tangent bundle* $O_{n+1,2} \to S^n$ *as a sub-bundle and a deformation retract.*

4 $\mathbb{F}_{k-1,1}(\mathbb{R}^{n+1})$ Restricted to $O_{n+1,r}$

Let us consider next the case of $\mathbb{R}^{n+1}$. First, we need to introduce some suitable notation. Put q_1 equal to the origin in $\mathbb{R}^{n+1}$, and, for $1 \le i \le (n+1)$ and $r \le (n+2)$, let q_{i+1} be the ith basic vector. Choose $(q_1, \cdots, q_r)$ to be the basepoint of $\mathbb{F}_k(\mathbb{R}^{n+1})$. Put $Q_r = \{q_1, \cdots, q_r\}$. Now, according to Theorem 1.1 above, for $k > r$, the projection $\mathrm{proj}_{k,r} : \mathbb{F}_k(\mathbb{R}^{n+1}) \to \mathbb{F}_r(\mathbb{R}^{n+1})$ that sends a k-tuple to the r-tuple consisting of the first r entries is a bundle with fiber $\mathbb{F}_{k-r,r}(\mathbb{R}^{n+1})$ and group $\mathrm{Top}_{sr}(\mathbb{R}^{n+1})$.

Consider next the imbedding $\rho : O_{n+1,r-1} \to \mathbb{F}_r(\mathbb{R}^{n+1})$ such that $g \mapsto (gq_1, \cdots, gq_r)$, for $g \in O_{n+1}$. Observe that $(x_1, \cdots, x_{r-1}) \mapsto (0, x_1, \cdots, x_{r-1})$. Thus we obtain the commutative diagram

$$\begin{array}{ccccc}
O_{n+1,r-1} & \to \cdots \to & O_{n+1,2} & \to & S^n \\
\downarrow & & \downarrow & & \downarrow \\
\mathbb{F}_r(\mathbb{R}^{n+1}) & \to \cdots \to & \mathbb{F}_3(\mathbb{R}^{n+1}) & \to & \mathbb{F}_2(\mathbb{R}^{n+1}),
\end{array}$$

where the projections in the first row $\mathrm{p}_{k-1s} : O_{n+1,s-1} \to O_{n+1,s-2}$ are those that take an $(s-1)$-frame to the subframe consisting of the first $(s-2)$-vectors. Let $\mathbb{F}'_k(\mathbb{R}^{n+1}) = \{(0, x, x_{r+1}, \cdots, x_k) \in \mathbb{F}_k(\mathbb{R}^{n+1}) | x \in O_{n+1,r-1}\}$ be the subspace $\mathrm{proj}^{-1}_{k,r}(O_{n+1,r-1})) \subset \mathbb{F}_k(\mathbb{R}^{n+1})$. The following is an immediate consequence of Theorem 2.1 above.

Theorem 4.1 *The projection* $\mathrm{proj}'_{k,r} : \mathbb{F}'_k(\mathbb{R}^{n+1}) \to O_{n+1,r-1}$ *that maps a* k*-tuple to its first* r *coordinates is a fiber bundle, with fiber* $\mathbb{F}_{k-r,r}(\mathbb{R}^{n+1})$*, associated with the principal bundle* $\mathrm{p}_{n+1,r-1} : O_{n+1} \to O_{n+1,r-1}$*.*

Let us consider next the special case $\text{proj}_{k,2} : \mathbb{F}_k(\mathbb{R}^{n+1}) \to \mathbb{F}_2(\mathbb{R}^{n+1})$, where $\text{proj}_{k,2}$ sends $(x_1, \cdots, x_k)$ to (x_1, x_2).

Corollary 4.1 *The bundle* $\text{proj}'_{k,2} : \mathbb{F}'_k(\mathbb{R}^{n+1}) \to S^n$ *is equivalent to the bundle, with fiber* $\mathbb{F}_{k-2,2}(\mathbb{R}^{n+1})$, *associated with the principal tangent bundle of* S^n.

Proof. It is easy to see that the last map $\rho_2 : S^n \to \mathbb{F}_2(\mathbb{R}^{n+1})$ of the commutative diagram given above is, in fact, a homotopy equivalence. Theorem 2.1 readily implies that the pull-back $\text{proj}'_{k,2} : \mathbb{F}'_k(\mathbb{R}^{n+1}) \to S^n$ by ρ is the bundle associated with the principal bundle $\text{p}_{n+11} : O_{n+1} \to S^n$, where p_{n+11} sends a matrix to its first column. But the latter is exactly the principal tangent bundle of S^n. □

5 Historical Remarks

Configuration spaces of $\mathbb{R}^{n+1}$ and S^{n+1} are those mostly studied in the literature, particularly the case when $n = 1$ in connection with braid theory ([43, Fadell-Neuwirth], [44, Fadell-Van Buskirk]). For $n > 1$, see [17, Cohen]. Configuration spaces of more general manifolds are studied in [19, Cohen-Tayler] and various other places. Configuration spaces of fibrations are introduced in [22, Crabb-James]. In [113, Xicotèncatl Merino] one finds a notion of configuration spaces with an equivariant flavor.

The fibration theorem, Theorem 1.1, is a special case of a theorem due to Thom on the differentiable imbeddings of manifolds (see [107, Thom], [87, Palais]). The result for topological manifolds appears without proof in [43, Fadell-Neuwirth], and a more general result, with proof, appears in [30, Fadell]. A proof based on a variant of Lemma 1.1 and its Corollary 1.1, for the case of $\mathbb{F}_2(M) \to M$, where M is any connected topological manifold, is given in [101, Spanier]. All these proofs are essentially based on the parallelizability of $\mathbb{R}^m$ and the fact that a manifold of dimension m is covered by open sets homeomorphic to $\mathbb{R}^m$. In fact, the map $\mathbb{R}^m \times \mathbb{R}^m \to \mathbb{R}^m$ such that $(x, y) \mapsto (x, y - x)$ induces a trivialization of fiber-pairs

$$\begin{array}{ccc} \mathbb{R}^m \times (\mathbb{R}^m, \mathbb{R}^m_0) & \xrightarrow{\phi} & (\mathbb{R}^m \times \mathbb{R}^m, \mathbb{R}^\times\mathbb{R}^m - \Delta) \\ \text{proj} \downarrow & & \downarrow \text{proj} \\ \mathbb{R}^m & \xrightarrow{\text{id}} & \mathbb{R}^m, \end{array}$$

where Δ is the diagonal, and $\mathbb{R}^m_0 = \mathbb{R}^m - \{0\}$. Thus, one can think of this trivialization as a map $\mathbb{R}^m \to \text{Top}(\mathbb{R}^m)$. Since a point on a manifold has neighborhoods homeomorphic to $\mathbb{R}^m$, this provided the basis for the proofs of the Lemma 1.1 and Corollary 1.1 given above.

In the special case when $k = 3$, Massey proves a result similar to Corollary 4.1 above (see [79, Massey]).

II. Configuration Space of $\mathbb{R}^{n+1}$, $n > 1$

In this chapter we shall consider the configuration space $\mathbb{F}_k(\mathbb{R}^{n+1})$ and $n > 1$. The space is simply connected. The case when $n = 1$ will be taken up in Chapter IV.

In §1 we choose a suitable basepoint $(q_1, \cdots, q_k) \in \mathbb{F}_k(\mathbb{R}^{n+1})$ and consider the natural filtration

$$\mathbb{F}_{1,k-1}(\mathbb{R}^{n+1}) \subset \cdots \subset \mathbb{F}_{k-r,r}(\mathbb{R}^{n+1}) \subset \cdots \subset \mathbb{F}_{k,0}(\mathbb{R}^{n+1}) = \mathbb{F}_k(\mathbb{R}^{n+1}),$$

where $\mathbb{F}_{k-r,r}(\mathbb{R}^{n+1}) = \{(x_1, \cdots, x_k) \mid x_i = q_i, 1 \leq i \leq r\}$. We show that each $\mathbb{F}_{k-r,r}(\mathbb{R}^{n+1})$ fibers over $\mathbb{R}^{n+1} - Q_r$, where $Q_r = \{q_1, \cdots, q_r\}$, and that the fiber is the preceding space $\mathbb{F}_{k-r-1,r+1}(\mathbb{R}^{n+1})$. Moreover, each of these fibrations has a section, and, therefore,

$$\pi_*(\mathbb{F}_k(\mathbb{R}^{n+1})) \cong \bigoplus_{r=1}^{k-1} \pi_*(\mathbb{R}^{n+1} - Q_r).$$

Also, each $(\mathbb{R}^{n+1} - Q_r)$ for all $1 \leq r \leq (k-1)$ is homotopy equivalent to a bouquet of r spheres of dimension n.

The Whitehead product turns $\pi_*(\mathbb{F}_k(\mathbb{R}^{n+1}))$ into a graded Lie algebra. A natural question is whether the preceding splitting of $\pi_*(\mathbb{F}_k(\mathbb{R}^{n+1}))$ in terms of the homotopy groups of bouquets of spheres is one of graded Lie algebras. To study this question, we first recall the definition of the homotopy classes $\{\alpha_{rs} | 1 \leq s < r \leq k\}$. We show that these elements satisfy quadratic relations of two different kinds in the algebra $\pi_*(\mathbb{F}_k(\mathbb{R}^{n+1}))$: those that involve three indices, and those that involve four. These are the Y-B relations.

To facilitate the description of the Y-B relations, we consider, in §2, the natural action of the full symmetric group Σ_k on $\mathbb{F}_k(\mathbb{R}^{n+1})$, induced by permuting the coordinates. This action is obviously free, and, therefore, one does not have a fixed basepoint here. But the space concerned is simply connected, and, hence, any path joining q to $\sigma(q)$, for any σ in Σ_k, defines a unique isomorphism between the homotopy groups based at q and $\sigma(q)$. So, the action of the symmetric group induces an action on the graded Lie algebra $\pi_*(\mathbb{F}_k(\mathbb{R}^{n+1}))$. In §2, we show that the effect of an element $\sigma \in \Sigma_k$ on α_{rs} is to permute the indices: σ takes α_{rs} to $\alpha_{\sigma r \sigma s}$. This is a key result in our study.

Having determined the effect of Σ_k on $\pi_*(\mathbb{F}_k(\mathbb{R}^{n+1}))$, we show, by applying the group Σ_k, that the Y-B relations are

$$(i)\ [\alpha_{21}, \alpha_{31} + \alpha_{32}] = 0,$$
$$(ii)\qquad [\alpha_{21}, \alpha_{43}] = 0,$$

and all those that are obtained therefrom. The Relation (i) describes a cluster of three distinct particles bound together in motion, where one is at the center, while the other two circle around it in two different orbits. Relation (ii), on the other hand, describes a formation of two pairs of distinct particles; the particles of each pair form a 2-body system, but the two systems move independently of each other.

The filtration of $\mathbb{F}_k(\mathbb{R}^{n+1})$ given in §1 leads to a filtration of the Lie algebra $\pi_*(\mathbb{F}_k(\mathbb{R}^{n+1}))$, which is subject matter of §3. We show that it leads to a filtration

$$L_1 \subset L_2 \subset \cdots \subset L_{k-r} \subset \cdots \subset L_k = \pi_*(\mathbb{F}_k(\mathbb{R}^{n+1})),$$

where each L_{k-r} is an ideal in L_k and is a split product $L_{k-r-1} \ltimes L_{k-rr}$, with L_{k-rr} being the Lie algebra of a bouquet of r spheres. The structure of the latter is given by the well-known Hilton theorem ([55, Hilton]; and see also [110, Whitehead])). This characterization of the Lie algebra $\pi_*(\mathbb{F}_k(\mathbb{R}^{n+1}))$ is useful for the study of the rational homotopy type of the space. However, we shall not be concerned with that question in this work.

In Parts Two and Three we study, respectively, the based and free loop spaces $\Omega\mathbb{F}_k(\mathbb{R}^{n+1})$, $\Lambda\mathbb{F}_k(\mathbb{R}^{n+1})$ of $\mathbb{F}_k(\mathbb{R}^{n+1})$ and show that, by virtue of the Y-B relations, the filtration of $\mathbb{F}_k(\mathbb{R}^{n+1})$ described above induces a splitting $\Omega\mathbb{F}_k(\mathbb{R}^{n+1}) \simeq \prod_{r=1}^{k-1} \Omega(\mathbb{R}_r^{n+1})$ as spaces, but not as H-spaces. Even so, the problem of computing the homology of the space of based loops is simplified greatly by this splitting. The nontriviality of the canonical fibrations for $r > 2$, which we establish in §4, indicates that a splitting in a similar fashion is not possible for the space of free loops. This is one of the reasons why the task of determining the homology and cohomology of the space of free loops of $\mathbb{F}_k(\mathbb{R}^{n+1})$ is considerably harder. This matter will be taken up in detail in Part Three of this work.

In §5 we consider the question of when the natural fibrations

$$\text{proj}_{k,r} : \mathbb{F}_k(\mathbb{R}^{n+1}) \to \mathbb{F}_r(\mathbb{R}^{n+1}),\ r \geq 2$$

are fiber homotopically trivial. We show that this is the case if, and only if, $r = 2$ and $n = 3$ or 7.

1 Filtration of $\mathbb{F}_k(\mathbb{R}^{n+1})$

To begin, we introduce the following notation. Denote by e the unit vector $(1, 0, \cdots, 0) \in \mathbb{R}^{n+1}$, and put

$$q_1 = (0, \cdots, 0),\ q_i = q_1 + 4(i-1)e,\ Q_i = \{q_1, \cdots, q_i\},\ i \geq 1,\ Q_0 = \emptyset$$

for $1 \leq i \leq k$.

The Whitehead product (see §7 of Chapter X of [110, Whitehead])

$$\pi_r(\mathbb{F}_k(\mathbb{R}^{n+1})) \times \pi_s(\mathbb{F}_k(\mathbb{R}^{n+1})) \to \pi_{r+s-1}(\mathbb{F}_k(\mathbb{R}^{n+1}))$$

turns $\pi_*(\mathbb{F}_k(M)) = \sum_r \pi_r(\mathbb{F}_k(M))$ into a graded Lie algebra. For the sake of brevity, put

$$\mathbb{F}_{k-r,r} = \mathbb{F}_{k-r,r}(\mathbb{R}^{n+1}) = \mathbb{F}_{k-r}(\mathbb{R}^{n+1} - Q_r),\ \ \mathbb{R}^{n+1}_r = (\mathbb{R}^{n+1} - Q_r).$$

Note now that the map

$$\phi : \mathbb{R}^{n+1} \times \mathbb{F}_{k-1,1} \to \mathbb{F}_k(\mathbb{R}^{n+1}),$$

where $(x_1, (q_1, x_2, \cdots, x_k))$ is mapped to $(x_1, x_1 + x_2, \cdots, x_1 + x_k)$, induces a trivialization of the bundle

$$\begin{array}{ccc} \mathbb{R}^{n+1} \times \mathbb{F}_{k-1,1}(\mathbb{R}^{n+1}) & \xrightarrow{\phi} & \mathbb{F}_k(\mathbb{R}^{n+1}) \\ \downarrow & & \downarrow \\ \mathbb{R}^{n+1} & \xrightarrow{\text{id}} & \mathbb{R}^{n+1} \end{array}$$

over $\mathbb{R}^{n+1}$. As the space $\mathbb{R}^{n+1}$ is contractible, the map

$$i_{k-1,k} : \mathbb{F}_{k-1,1}(\mathbb{R}^{n+1}) \to \mathbb{F}_k(\mathbb{R}^{n+1}),$$

where $(x_2, \cdots, x_k) \mapsto (q_1, x_2, \cdots, x_k)$, is, in fact, a homotopy equivalence.

To describe $\mathbb{F}_{k-1}(\mathbb{R}^{n+1} - Q_1)$, let us consider the diagram

$$\mathcal{F}_k(\mathbb{R}^{n+1}) : \left\{ \begin{array}{ccccccccc} \mathbb{F}_{k,0} & \leftarrow & \cdots & \leftarrow & \mathbb{F}_{k-r,r} & \leftarrow & \mathbb{F}_{k-r-1,r+1} & \leftarrow & \cdots \\ \downarrow & & \downarrow & & \downarrow & & \downarrow & & \downarrow \\ \mathbb{R}^{n+1} & & \cdots & & \mathbb{R}^{n+1}_r & & \mathbb{R}^{n+1}_{r+1} & & \cdots, \end{array} \right.$$

with $1 \leq r < k$, where the vertical maps $\mathrm{p}_r : \mathbb{F}_{k-r,r}(\mathbb{R}^{n+1}) \to \mathbb{R}^{n+1}_r$ are the projections such that $(q_1, \ldots, q_r, x_{r+1}, \cdots, x_k) \mapsto x_{r+1}$. Note here that $\mathbb{F}_{k,0} = \mathbb{F}_k(\mathbb{R}^{n+1})$, since $Q_0 = \emptyset$. We shall call $\mathcal{F}_k(\mathbb{R}^{n+1})$ the *fundamental fiber sequence* of $\mathbb{F}_k(\mathbb{R}^{n+1})$ (cf. [43, Fadell-Neuwirth]). Observe that for each r we have $\mathbb{R}^{n+1}_r \simeq (S^n \vee \cdots \vee S^n) = (S^n)^{\vee r}$, where $\simeq$ indicates homotopy equivalence. Then, the vertical maps are fibrations, which, according to elementary obstruction theory, admit sections. In fact, one can readily construct sections directly, using the following simple lemma.

Lemma 1.1 *Let M denote a connected n-manifold, with $n \geq 2$, and $Q_r = \{q_1, \cdots, q_r\}$, $r \geq 1$, a set of r distinct points. Then, the projection*

$$\mathrm{proj}_{k,1} : \mathbb{F}_k(M - Q_r) \to M - Q_r,\ (x_1, \cdots, x_k) \mapsto x_1$$

admits a section.

Proof. Choose any $a \in M$, and let L denote a simple arc from a to q_1, avoiding q_i for all $i \geq 2$. Then, $L - Q_1$ is a closed subset of $M - Q_r$ homeomorphic to the half-line $[0, \infty) \subset \mathbb{R}$. Let $g_i : L - Q_1 \to L - Q_1$ denote a translation along $L - Q_1$, so that $(x, g_1(x), \cdots, g_{k-1}(x))$ is in $\mathbb{F}_k(L - Q_1)$. Since $L - Q_1$ is an absolute retract, there is a retraction $\rho : M - Q_r \to L - Q_1$. Then, the map

$$\sigma : M - Q_r \to \mathbb{F}_k(M - Q_r),\ x \mapsto (x, \rho(g_1(x)), \cdots, \rho(g_{k-1}(x)))$$

is a section. □

An immediate consequence is the following.

Corollary 1.1 *Suppose that $n > 1$. Then $\mathbb{F}_{k-r,r}(\mathbb{R}^{n+1})$ is simply connected.*

Using the homotopy long exact sequence of a fibration, one easily obtains the following theorem.

Theorem 1.1 *There is an isomorphism*

$$\pi_*(\mathbb{F}_k(\mathbb{R}^{n+1})) \cong \bigoplus_{r=2}^{k-1} \pi_*(\mathbb{R}_r^{n+1}) \cong \bigoplus_{r=2}^{k-1} \pi_*(S^{n \vee r})$$

of modules.

Note that the preceding direct sum decomposition of the homotopy groups $\pi_*\mathbb{F}_k(\mathbb{R}^{n+1})$ in terms the homotopy groups of r-bouquets of S^n, where r ranges from 1 to $(r-1)$, is just an additive isomorphism that depends on the choice of sections of the fibrations in Diagram $\mathcal{F}_k(\mathbb{R}^{n+1})$. One would like to find a decomposition, and corresponding generators, that would lead to a simple description of the Whitehead product

$$\pi_*(\mathbb{F}_k(\mathbb{R}^{n+1})) \times \pi_*(\mathbb{F}_k(\mathbb{R}^{n+1})) \to \pi_*(\mathbb{F}_k(\mathbb{R}^{n+1})).$$

This is how we shall proceed.

Definition 1.1 *By definition, for $1 \leq s \neq r \leq k$, let*

$$\alpha'_{rs} : S^n \to \mathbb{F}_{k-r}(\mathbb{R}_r^{n+1})$$

be the map where

$$\xi \in S^n \mapsto (q_1, \cdots, q_{r-1}, q_s + \xi, q_r, \dots, q_{k-1}),$$

with S^n being the unit sphere in $\mathbb{R}^{n+1}$. Denote the homotopy class $[\alpha'_{rs}] \in \pi_n(\mathbb{F}_k(\mathbb{R}^{n+1})$ by α_{rs}.

Observe that the elements are the analogues of the classical braids. In order to proceed further, we need the following result.

Proposition 1.1 *The elements* $\{\alpha_{rs} | 1 \leq s < r \leq k\}$ *generate the group* $\pi_n(\mathbb{F}_k(\mathbb{R}^{n+1}))$.

Proof. Observe that the composite maps

$$S^n \xrightarrow{\alpha'_{rs}} \mathbb{F}_{k-r+1,r-1}(\mathbb{R}^{n+1}) \stackrel{\text{proj}_{k-r+1,1}}{\longrightarrow} \mathbb{R}^{n+1}_{r-1}$$

are imbeddings, and put $S_{rs} = (\alpha'_{rs} \circ \text{proj}_{k-r+1,1})(S^n)$. Note that, for each pair r, s such that $1 \leq s < r \leq k$, the restriction of the vertical fibration

$$\mathrm{p}_r : \mathbb{F}_{k-r+1,r-1} \to \mathbb{R}^{n+1}_{r-1}$$

in Diagram $\mathcal{F}_k(\mathbb{R}^{n+1})$ to the imbedded sphere $S_{rs} \subset \mathbb{R}^{n+1}_r$ admits a section induced by the map α'_{rs}. Moreover, the spheres $\{S_{rs} | 1 \leq s < r\}$ together yield a map $S_{r1} \vee \cdots \vee S_{rr-1} \stackrel{\simeq}{\to} \mathbb{R}^{n+1}_{r-1}$, that can easily be seen to be a homotopy equivalence. Thus we obtain a decomposition

$$\pi_n(\mathbb{F}_k(\mathbb{R}^{n+1})) \cong \bigoplus_{r=2}^{k} \pi_n(S_{r1} \vee \cdots \vee S_{rr-1}).$$

As $\pi_n(S_{r1} \vee \cdots \vee S_{rr-1})$ is generated by the elements α_{rs} with $1 \leq s < r$ (see §8, Chapter XI of [110, Whitehead]), it follows that the group $\pi_n(\mathbb{F}_k(\mathbb{R}^{n+1}))$ is generated by the elements listed in the proposition. □

The following result shows that the elements α_{rs}, for $r > s$, determine those when $r < s$.

Proposition 1.2 *If* $1 \leq s < r \leq k$, *then*

$$\alpha'_{sr} = (-1)^{n+1} \alpha_{rs}.$$

Proof. By definition, put $\mathrm{p}_{rs} : \mathbb{F}_k(\mathbb{R}^{n+1}) \to \mathbb{F}_2(\mathbb{R}^{n+1})$ equal to the map where $(x_1, x_2, \cdots, x_k) \mapsto (x_s, x_r)$. It is easy to verify that

$$\pi_n(\mathrm{p}_{ij})(\alpha_{rs}) = \begin{cases} \alpha_{21} & (r,s) = (i,j) \\ 0 & (r,s) \neq (i,j). \end{cases}$$

Since $\{\alpha_{rs} \mid 1 \leq s < r \leq k\}$ generate the group $\pi_n(\mathbb{F}_k(\mathbb{R}^{n+1}))$, we see that

$$\alpha_{sr} = \sum c_{ij} \alpha_{ij}, \ 1 \leq j < i \leq k,$$

where $c_{ij} \in \mathbb{Z}$. Applying the maps p_{ij} for all $j < i$, one sees immediately that the left side of the equation goes to zero unless $(i,j) = (r,s)$ or (s,r). Hence, the same is true of right hand side of the equation, and we see that $\alpha_{sr} = \pm \alpha_{rs}$, $s < r$. To determine the sign, consider the projection ϕ : $\mathbb{F}_2(\mathbb{R}^{n+1}) \to S^n$, where $(x_1, x_2) \mapsto (x_2 - x_1)/\|x_2 - x_1\|$. It is easy to see that ϕ is a homotopy equivalence, and the morphism it induces on the homotopy groups takes α_{21} to the class $\iota_n \in \pi_n(S^n)$ of the identity map of S^n. On the other hand, α_{12} is taken to the class of the antipodal map, which has degree $(-1)^{n+1}$. □

2 Action of Σ_k

A property that is extremely useful for the study of the action induced on the homotopy groups of $\mathbb{F}_k(\mathbb{R}^{n+1})$ by the natural action of the symmetric group Σ_k on $\mathbb{K}$ letters is the invariance of the set of generators $\{\alpha_{st}\}$ under the action of the group Σ_k.

Theorem 2.1 *The elements $\{\alpha_{rs} \mid 1 \le s \ne r \le k\}$ satisfy the relations*

$$\sigma_* \alpha_{rs} = \alpha_{\sigma r \sigma s}$$

for all $\sigma \in \Sigma_k$ and admissible $1 \le s \ne t \le k$.

Proof. As the transpositions $\tau_t = (t, t+1)$ generate the symmetric group Σ_k on k letters, it suffices to prove the theorem when $\sigma = \tau_t$. There are four cases to consider:

$$\begin{aligned}
&(i)\ \tau_t(r) \ne r\ ,\ \ \tau_t(s) = s,\\
&(ii)\ \tau_t(r) = r\ ,\ \ \tau_t(s) \ne s,\\
&(iii)\ \tau_t(r) \ne r\ ,\ \ \tau_t(s) \ne s, \text{ and}\\
&(iv)\ \tau_t(r) = r\ ,\ \ \tau_t(s) = s.
\end{aligned}$$

In Case(i) one can easily see that $r \in \{t, t+1\}$ and $s \notin \{t, t+1\}$. So either $t = r$ or $r = t+1$. Thus, the following assertion takes care of this case.

Assertion 2.1 *For $s < t$, we have the relations*

$$\begin{aligned}
&(i)\ (\tau_t)_*(\alpha_{ts}) &&= \alpha_{t+1s} \text{ and}\\
&(ii)\ (\tau_t)_*(\alpha_{t+1s}) &&= \alpha_{ts}.
\end{aligned}$$

As τ_t is an involution, (ii) is implied by (i). To prove (i), consider the following equations:

$$\begin{aligned}
\tau_t \alpha'_{ts} &= \tau_t(q_1, \cdots, q_s, \cdots, q_{t-1}, q_s + \xi, q_s, \ldots)\\
&= (q_1, \cdots, q_{t-1}, q_t, q_s + \xi, q_{t+1}, \cdots).
\end{aligned}$$

It is easy to see that, as required, $(\tau_t)_*(\alpha_{ts}) = \alpha_{t+1s}$. If Case (ii) holds, then $r \notin \{t, t+1\}$ and $s \in \{t, t+1\}$. If, moreover, $r > t$, then $r > t+1$, as well. The following Assertion takes care of this case.

Assertion 2.2 *For $r > t+1 > t$, the relations*

$$\begin{aligned}
&(i)\ (\tau_t)_*(\alpha_{rt+1}) &&= \alpha_{rt} \text{ and}\\
&(ii)\ (\tau_t)_*(\alpha_{rt}) &&= \alpha_{rt+1}
\end{aligned}$$

hold.

Again observe that (ii) follows from (i). To prove (i) note that

$$\begin{aligned}\tau_t\alpha'_{rt+1}(\xi) &= \tau_t(\cdots, q_t, q_{t+1}, \cdots, q_{r-1}, q_{t+1}+\xi, q_r, \cdots)\\ &= (\cdots, q_{t+1}, q_t, \cdots, q_r, q_{t+1}+\xi, \cdots).\end{aligned}$$

Hence, using a homotopy that permutes q_t with q_{t+1} and takes the unit sphere centered at q_{t+1} to the unit sphere centered at q_t, we see that, as required, $\tau_t\alpha'_{r+1t+1} \simeq \alpha'_{r+1t}$.

Next, in Case (iii), we have $r \in \{t, t+1\}$ and $s \in \{t, t+1\}$. Hence, if $r > s$, it follows that $r = t+1$ and $s = t$; thus, the following assertion takes care of this case.

Assertion 2.3 $\tau_t\alpha_{t+1t} = (-1)^{n+1}\alpha_{t+1t}$.

To prove this, consider the expression

$$\begin{aligned}\tau_t\alpha'_{t+1t}(\xi) &= \tau_t(q_1, \cdots, q_t, q_t+\xi, q_{t+1}, \cdots, q_{k-1})\\ &= (q_1, \cdots, q_t+\xi, q_t, q_{t+1}, \cdots).\end{aligned}$$

The homotopy $(\xi, u) \mapsto (\cdots, q_{t-1}, q_t + (1-u)\xi, q_t - u\xi, q_{t+1}, \cdots)$ shows that $\tau_t\alpha'_{t+1t} \simeq \alpha'_{t+1t} \circ a$, where $a : S^n \to S^n$ is the antipodal map. The assertion now follows.

Finally, the following assertion takes care of Case (iv).

Assertion 2.4 $\tau_t\alpha_{r+1s} = \alpha_{r+1s}$ *for the other cases.*

To prove the assertion, note that

$$\begin{aligned}\tau_t\alpha'_{r+1s}(\xi) &= \tau_t(q_1, \cdots, q_s, \cdots, q_r, q_s+\xi, q_{r+1}, \cdots)\\ &= (q_1, \cdots, q_r, q_s+\xi, q_{r+1}, \cdots),\end{aligned}$$

since neither r nor s is equal to t or $t+1$. Hence we have $\tau_t\alpha'_{r+1s} \simeq \alpha'_{r+1s}$, and the assertion follows.

The theorem is easily seen to follow from these four assertions □

The preceding theorem can also be proved using the method of proof for Proposition 1.2 of the previous section. In fact, the effect of $\tau \in \Sigma_k$ on α_{rs} is to send it to some element in $\pi_n(\mathbb{F}_k(\mathbb{R}^{n+1}))$. Hence, one can express $\pi_n(\tau)(\alpha_{rs})$ as a linear combination $\sum c_{ij}\alpha_{ij}$, and, making use of the projections

$$\mathrm{p}_{ij} : \mathbb{F}_k(\mathbb{R}^{n+1}) \to \mathbb{F}_2(\mathbb{R}^{n+1}),$$

one can show that the values of c_{ij} are either 0 or ± 1, as stated in the four cases of the previous proof. (When $n = 1$, the basepoint cannot be ignored, and one needs the homotopies used in the preceding proof. The effect on the basepoint can be described precisely; see §2 of Chapter IV below.)

Thus, the elements $\{\alpha_{rs} \in \pi_n(\mathbb{F}_k(\mathbb{R}^{n+1})) | 1 \leq s \neq r \leq k\}$ constitute a set of generators of the group $\pi_n(\mathbb{F}_k(\mathbb{R}^{n+1}))$, invariant set-wise under the action of Σ_k induced by the natural action on $\mathbb{F}_k(\mathbb{R}^{n+1})$.

3 The Y-B Relations

The next result is the key to the computation of the Whitehead products.

Theorem 3.1 *For all $\sigma \in \Sigma_k$, the following identities hold:*

$$\begin{aligned} &(i)\ [\alpha_{\sigma 2\sigma 1}, \alpha_{\sigma 3\sigma 1} + \alpha_{\sigma 3\sigma 2}] = 0, \quad for\ k \geq 3, \quad and \\ &(ii)\ [\alpha_{\sigma 2\sigma 1}, \alpha_{\sigma 4\sigma 3}] = 0, \quad for\ k \geq 4, \end{aligned}$$

in $\pi_*(\mathbb{F}_k(\mathbb{R}^{n+1})$.

Proof. Note that $\sigma[\alpha, \beta] = [\sigma\alpha, \sigma\beta]$. This, together with Theorem 2.1, shows that it is enough to prove that the statements

$$\begin{aligned} &(i)\ [\alpha_{21}, \alpha_{31} + \alpha_{32}] = 0, \text{ for } k \geq 3 \text{ and} \\ &(ii)\ [\alpha_{21}, \alpha_{34}] = 0, \text{ for } k \geq 4 \end{aligned}$$

hold in $\pi_*(\mathbb{F}_k(\mathbb{R}^{n+1}))$. Now, to prove Statement (i), consider the map

$$\phi : S^n \times S^n \to \mathbb{F}_k(\mathbb{R}^{n+1})$$

such that $(\xi_1, \xi_2) \mapsto (q_1, q_1 + \xi_1, q_1 + 5\xi_2, q_3, \cdots)$. It is easy to verify that

$$[\phi | S^n \times \{*\}] = \alpha_{21}, \quad [\phi | \{*\} \times S^n] = \alpha_{31} + \alpha_{32},$$

using the method of proof of Proposition 1.2, or drawing directly on the definitions. This proves Statement(i). To prove Statement(ii), consider the map

$$\psi : S^n \times S^n \to \mathbb{F}_k(\mathbb{R}^{n+1}),$$

under which $(\xi_1, \xi_2) \mapsto (q_1, q_1 + \xi_1, q_2, q_3 + \xi_2, q_4, \cdots)$. Again, as in the preceding case, it is easy to verify that

$$[\psi | S^n \times \{*\}] = \alpha_{21}, \quad [\psi | \{*\} \times S^n] = \alpha_{43}.$$

Hence, Statement (ii) follows. □

The following special cases are very useful.

Corollary 3.1 *The following relations*

$$\begin{aligned} &(i)\ [\alpha_{32}, \alpha_{21} + \alpha_{31}] = 0 \quad and \\ &(ii)\ [\alpha_{31}, \alpha_{21} + (-1)^{n+1}\alpha_{32}] = 0 \end{aligned}$$

hold.

Proof. Both are consequences of (i) of Theorem 3.1. To prove (i), let σ be the transposition $(1, 3)$; to prove (ii), let σ be the transposition $(2, 3)$. □

Relations similar to the Y-B relations of Theorem 3.1 appear in [67, 68, Kohno] in connection with monodromy representations of the pure braid group on n strings. They also arise in the classical Yang-Baxter relations ([67, 68, Kohno]; and see also [100, Shnider-Sternberg]).

4 Filtration of $\pi_*(\mathbb{F}_k(\mathbb{R}^{n+1}))$

The following concept provides a helpful way to think of the fibrations of Diagram $\mathcal{F}_k(\mathbb{R}^{n+1})$.

Definition 4.1 *Let* p : $E \to B$ *be a fibration with a section* $\sigma : B \to E$. *Suppose that* b_0 *is a basepoint in* B, *and denote the fiber at* b_0 *by* F. *We shall say that* E *is a* sectioned twisted product *of* B *by* F *and write* $E = B \ltimes F$ *for the quadruple* (E, p, B, F).

Note that the section σ is not displayed in the notation $E = B \ltimes F$. We emphasize here that the concept just defined is only a convenient and intuitive way of thinking of these fibrations and will only be used as such. However, it will induce *split extensions*, when appropriate algebraic invariants are introduced later in this work.

The fibrations in Diagram $\mathcal{F}_k$ can be thought of as a tower of twisted products. A natural question to ask is: how does the following multilayered sectioned twisted-product structure

$$\mathbb{F}_k(\mathbb{R}^{n+1}) \simeq S_{21} \ltimes (\cdots(\cdots \ltimes (S_{k1} \vee \cdots \vee S_{kk-1})))$$

determine the Lie algebra structure of $\pi_*(\mathbb{F}_k(\mathbb{R}^{n+1}))$? We shall see in this section that the homotopy groups $\pi_*(\mathbb{F}_k(\mathbb{R}^{n+1})$ are constructed out of the groups

$$\pi_*(\mathbb{R}^{n+1}_r) \cong \pi_*(S_{r1} \vee \cdots \vee S_{rr-1})$$

as multilayered split extensions of Lie algebras.

So let us consider first the groups $\pi_*(\mathbb{R}^{n+1}_r)$. Observe that

$$\mathbb{R}^{n+1}_r = \mathbb{R}^{n+1} - Q_r \simeq (S_{r+11} \vee \cdots \vee S_{r+1r}),$$

and, therefore,

$$\pi_*(\mathbb{R}^{n+1}_r) \cong \pi_*(S_{r+11} \vee \cdots \vee S_{r+1r}).$$

For simplicity, put

$$\begin{aligned} L_{r+1,r} &= \pi_*(\mathbb{R}^{n+1}_r) \\ &= \pi_*(S_{r+11} \vee \cdots \vee S_{r+1r}). \end{aligned}$$

Also, let $\mathcal{W}_r$ denote the set of basic Whitehead products on r symbols (see Chapter XI, §6 of [110, Whitehead]). Let S_w denote the unit sphere the dimension of which is $(n-1)\|w\| + 1$, where $\|w\|$ is the weight of w. Thus $S_w = S^n$, if w is a single letter, and $S_w = S^{2n-1}$, if w is a word of weight two.

Regard $L_{r+1,r}$ as a graded Lie algebra, where the Lie product, is defined by the Whitehead product (see Chapters X and XI of [110, Whitehead]).

Theorem 4.1 *The following assertions hold.*

(*i*) *For each $w \in \mathcal{W}$, the Lie algebra $\pi_* S_w$ is a subalgebra of $L_{r+1,r}$;*
(*ii*) *$L_{r+1,r} \cong \bigoplus \pi_*(S_w)$ is an isomorphism of groups, but not of Lie algebras, where w ranges over the set $\mathcal{W}_r$; and*
(*iii*) *$L_{r+1,r} \otimes \mathbb{Q}$ is the free Lie algebra over the rationals on r letters, where $\mathbb{Q}$ stands for the rational numbers.*

Proof. The first assertion follows from the naturality of the Whitehead product. The second is an immediate consequences of Hilton's decomposition theorem (see Chapter XI, §6 of [110, Whitehead]). To prove the third, recall that one of steps in Hilton's proof is that $\Omega(S_{r+11} \vee \cdots \vee S_{r+1r}) \simeq \prod_w \Omega S_w$ as spaces but not as H-spaces. Next recall that the Pontryagin algebra

$$H_*(\Omega(\vee_{j=1}^r S_{r+1j}); \mathbb{Q})$$

is the free associative algebra on r symbols that are in correspondence with the spheres in the bouquet. Observe also that the preceding algebra is, in fact, a Hopf algebra. Therefore, its primitive elements themselves form a Lie algebra, which is the free Lie algebra on these symbols (see [14, Bourbaki], Chapter 2). But the set of primitive elements of the Pontryagin algebra $H_*(\Omega(\vee_{j=1}^r S_{r+1j}); \mathbb{Q})$ coincides with the rational homotopy groups

$$\pi_*(S_{r+11} \vee \cdots S_{r+1r}) \otimes \mathbb{Q},$$

according to [83, Milnor-Moore]. This proves the third assertion. □

Next, let us consider the problem of determining the structure of the Lie algebra of the homotopy groups of the configuration space $\mathbb{F}_k(\mathbb{R}^{n+1})$. Put

$$L_{k-r} = \pi_*(\mathbb{F}_{k-r,r}(\mathbb{R}^{n+1})) = \pi_*(\mathbb{F}_{k-r}(\mathbb{R}^{n+1} - Q_r)).$$

Theorem 4.2 *The Lie algebra L_k contains the descending chain*

$$L_k = L_{k-1} \supset \cdots \supset L_{k-r} \supset \cdots L_1$$

of subalgebras. Moreover, each L_{k-r} is an ideal in L_k.

Proof. Consider the sequence

$$\mathbb{F}_k(\mathbb{R}^{n+1}) \leftarrow \mathbb{F}_{k-1,1} \leftarrow \mathbb{F}_{k-r,r} \leftarrow \cdots \leftarrow \mathbb{F}_{1,r} = \mathbb{R}^{n+1}_{k-1}$$

of imbeddings in Diagram $\mathcal{F}_k(\mathbb{R}^{n+1})$, where $\mathbb{F}_{k-r,r} = \mathbb{F}_{k-r,r}(\mathbb{R}^{n+1})$, and note that $\mathbb{F}_{k-r}$ is the fiber of the projection $\text{proj}_{k,r} : \mathbb{F}_k(\mathbb{R}^{n+1}) \to \mathbb{F}_r(\mathbb{R}^{n+1})$, where $k > r$. But, according to Lemma 1.1, the latter admits a section. Hence, the natural injection $\mathbb{F}_{k-r,r} \to \mathbb{F}_k$ induces an injection $L_{k-r} \to L_k$. Thus, the preceding sequence of spaces induces the ascending sequence of Lie algebras

$$L_1 \subset L_2 \subset \cdots \subset L_{k-r} \subset \cdots \subset L_{k-1} = L_k.$$

Observe that L_{k-r} is an ideal in L_k, since L_{k-r} is the kernel of $\pi_*(\text{proj}_{k,r})$. Thus, Theorem 4.2 is proved. □

Recall that $L = L' \ltimes L''$, where L, L' and L'' are Lie algebras, means that L is a split extension of L'' by L'.

Theorem 4.3 *For each r such that $1 \leq k - r \leq k$,*

$$L_{k-r} \cong L_{k-r-1} \ltimes L_{r+1,r},$$

and the split extension is not trivial.

Proof. Consider the fibration $p_r : \mathbb{F}_{k-r,r} \to \mathbb{R}_r^{n+1}$ the fiber of which is $\mathbb{F}_{k-r-1,r+1}$. Now note that $\mathbb{R}_{r-1}^{n+1} \simeq (S_{r+11} \vee \cdots \vee S_{r+1r})$. We may assume, without loss of generality, that the fibration above admits a section $\rho : \mathbb{R}_r^{n+1} \to \mathbb{F}_{k-r,r}$ and that the restriction of ρ to the sphere $S_{r+1,s}$ represents α_{r+1s}, for $1 \leq s < r+1$. Therefore, the fibration is a sectioned twisted product, and we have $\mathbb{F}_{k-r+1,r-1} \simeq (S_{r1} \vee \cdots \vee S_{rr-1}) \ltimes \mathbb{F}_{k-r,r}$. The fact that $L_{k-r} \cong L_{k-r-1} \ltimes L_{r+1,r}$ is now an immediate consequence.

To show that the split extension is not trivial, note that, for $t > r+1 > s$, the Y-B relations imply that

$$\text{(I)} \qquad [\alpha_{r+1s}, \alpha_{ts}] = [\alpha_{tr+1}, \alpha_{ts}]$$

in L_{k-r} (see Theorem 3.1). Now, $\alpha_{tr+1}, \alpha_{ts}$ belong to L_{k-r-1}; in fact, they are in the summand defined by $\mathbb{R}_{t-1}^{n+1}$. Hence, the right-hand side of (I) is nontrivial according to Hilton's decomposition theorem (see Theorem 4.1). On the other hand, α_{r+1s} is in the image of L_{r+11} under $\pi_*(\rho)$, and the term on the left-hand side of (I) expresses the action of α_{r+1s} induced by the action of L_{k-r} on the ideal L_{k-r-1}. Hence, $L_{k-r} \cong L_{k-r-1} \ltimes L_{r+1,r}$ is not an isomorphism of Lie algebras, and the split extension is not trivial. This finishes the proof of the theorem. □

Summing up, note that Theorems 4.1, 4.2, and 4.3 imply the following.

Corollary 4.1 *L_k has an ascending sequence*

$$L_1 \subset L_2 \subset \cdots \subset L_{k-r} \subset \cdots \subset L_{k-1} \subset L_k$$

of ideals, where

(*i*)$L_{r+1,r} \cong L_{k-r}/L_{k-r-1}$;
(*ii*)$L_{r+1,r} \otimes \mathbb{Q}$ *is the free graded Lie algebra over $\mathbb{Q}$ on the elements $\{\alpha_{r+11}, \cdots, \alpha_{r+1r}\}$; and*
(*iii*)$L_{k-r} \cong L_{k-r-1} \ltimes L_{r+1,r}$ *is not an isomorphism of algebras.*

The groups $\pi_{n+1}(\mathbb{F}_k(\mathbb{R}^{n+1}))$ are also useful, as we shall see in Chapter III. Put

$$\tilde{\gamma}_{rs} = \alpha_{rs} \circ \eta_n,\ 1 \leq s \neq r \leq k,\ \eta_n = E^{n-2}(\eta),$$

where $\eta \in \pi_3(S^2)$ is the homotopy class of the Hopf map $\eta' : S^3 \to S^2$ and $E^{n-2} : \pi_3(S^2) \to \pi_{n+1}(S^n)$ is the $(n-2)$-fold Freudenthal suspension. Here $\omega \mapsto \omega \circ \eta_n$ is the composition product (see Chapter X, §8.1 of [110, Whitehead]).

To determine the generators of $\pi_{n+1}(\mathbb{F}_k(\mathbb{R}^{n+1}))$, we first consider the case when $n > 2$.

Proposition 4.1 *The set* $\{\tilde{\gamma}_{rs} | 1 \leq s < r \leq k\}$ *generates the group* $\pi_{n+1}(\mathbb{F}_k(\mathbb{R}^{n+1}))$, *if* $n > 2$.

Proof. Consider the multilayered twisted-product structure

$$\mathbb{F}_k(\mathbb{R}^{n+1}) \simeq S_{21} \ltimes (\cdots(\cdots \ltimes (S_{k1} \vee \cdots \vee S_{kk-1}))).$$

It readily implies that $\pi_m(\mathbb{F}_k(\mathbb{R}^{n+1})) \cong \bigoplus_{r=2}^k \pi_m(S_{r1} \vee \cdots \vee S_{rr-1})$. However, according to Hilton's decomposition theorem, $\pi_m(S_{r1} \vee \cdots \vee S_{rr-1}) \cong \sum_w \pi_m(S_w)$, where w ranges over the set $\mathcal{W}_r$ of basic Whitehead products on r letters ([110, Whitehead]). Now, since $n > 2$ by assumption, it follows that $\dim S_w = (n-1)\|w\| + 1 > n+1$, if $\|w\| \geq 2$, where S_w is the sphere defined by the basic product $w \in \mathcal{W}_r$. Hence,

$$\pi_{n+1}(S_{r1} \vee \cdots \vee S_{rr-1}) \cong \bigoplus_{j=1}^{r-1} \pi_{n+1}(S_{rj}).$$

Since $\pi_{n+1}(S^n)$ is generated by η_n, this proves the assertion. □

The case when $n = 2$ is special.

Proposition 4.2 *The set* $\{\tilde{\gamma}_{rs} | 1 \leq s < r \leq k\} \cup \{[\alpha_{rs}, \alpha_{tu}]\}$, *where* $1 \leq u < t \leq k$, *generates the group* $\pi_3(\mathbb{F}_k(\mathbb{R}^3))$.

Proof. Consider the isomorphism $\pi_3(\mathbb{F}_k(\mathbb{R}^3)) \cong \bigoplus_{r=2}^k \pi_3(S_{r1} \vee \cdots \vee S_{rr-1})$. Order the elements $\{\alpha_{rs} | 1 \leq s < r \leq k\}$ according to the rule $\alpha_{rs} < \alpha_{uv}$, if $r < u$, and $\alpha_{rs} < \alpha_{rv}$, if $s < v$. Hilton's Theorem implies the result. □

5 When Are the Canonical Fibrations Trivial?

The aim of this section is to consider the problem of when

$$\text{proj}_{k,r} : \mathbb{F}_k(\mathbb{R}^{n+1}) \to \mathbb{F}_r(\mathbb{R}^{n+1})$$

is fiber homotopically trivial.

If $r = 1$, then $\mathbb{F}_1(\mathbb{R}^{n+1})$ is just $\mathbb{R}^{n+1}$, and the preceding fibration is trivial, as a bundle, because $\mathbb{R}^{n+1}$ is contractible. But the situation is different when $r > 1$. The following theorem describes the state of affairs. The Y-B relations play a major role in the proof.

We assume that $n + 1 \geq 3$. (For $n = 1$, see §6 of Chapter IV.)

Theorem 5.1 *The fibration* $\mathrm{proj}_{k,r} : \mathbb{F}_k(\mathbb{R}^{n+1}) \to \mathbb{F}_r(\mathbb{R}^{n+1})$, $r \geq 2$, *with fiber* $\mathbb{F}_{k-r,r}(\mathbb{R}^{n+1})$, *where a* k*-tuple goes to its first* r *entries, is fiber homotopically trivial if, and only if,* $r = 2$ *and* $n = 3$ *or* 7.

Proof of ($\Leftarrow$)

We prove the stronger statement that, if $r = 2$ and 3 or 7, then the fibration in the theorem is actually trivial, i.e., the fibration is a product bundle.

Consider the mapping

$$\rho_2 : O_{n+1,1} \to \mathbb{F}_2(\mathbb{R}^{n+1}),\ e_1 \mapsto (0, e_1)$$

of §4 of Chapter I, where $e_1 \in S^n = O_{n+1,1}$, and denote by

$$\mathrm{proj}'_{k,2} : \mathbb{F}'_k(\mathbb{R}^{n+1}) \to S^n$$

the pull-back of $\mathrm{proj}_{k,2} : \mathbb{F}_k(\mathbb{R}^{n+1}) \to \mathbb{F}_2(\mathbb{R}^{n+1})$. Note that the fiber is $\mathbb{F}_{k-2,2} = \mathbb{F}_{k-2}(\mathbb{R}^{n+1} - Q_2)$. Now, according to Theorem 4.1 of Chapter I, the principal bundle $\mathrm{p}_1 : O_{n+1} \to S^n$, $(x_1, \cdots, x_{n+1}) \mapsto x_1$, is the associated principal bundle of $\mathbb{F}'_k(\mathbb{R}^{n+1}) \to S^n$. But, $\mathrm{p}_1 : O_{n+1} \to S^n$ is the principal bundle of the tangent bundle of S^n. Therefore, if $n = 3$ or 7, the principal tangent bundle and, hence, $\mathbb{F}'_k(\mathbb{R}^{n+1}) \to S^n$ is a product bundle.

To prove the converse, assume that $\mathrm{proj}_{k,r} : \mathbb{F}_k(\mathbb{R}^{n+1}) \to \mathbb{F}_r(\mathbb{R}^{n+1})$ is fiber homotopically trivial. We need to show that $n = 3$ or 7, and $r = 2$. We shall do so in two steps. First, we prove that $n = 3$ or 7.

Proof of ($\Rightarrow$), Step 1

Let $\{\alpha_{st} \mid 1 \leq t < s \leq k\}$ denote the generators of $\pi_n(\mathbb{F}_k(\mathbb{R}^{n+1}))$. Regard the subset $\{\alpha_{st} \mid 1 \leq t < s \leq r\}$ as the generators of $\pi_n(\mathbb{F}_r(\mathbb{R}^{n+1}))$. Let $\alpha'_{21} : S^n \to \mathbb{F}_r(\mathbb{R}^{n+1})$ be the representative of α_{21} (see Definition 1.1), and denote by S_{21} its image. Put

$$\mathrm{proj}'_{k,2} : \mathbb{F}'_k(\mathbb{R}^{n+1}) \to S_{21}$$

equal to the restriction of $\mathrm{proj}_{k,r} : \mathbb{F}_k(\mathbb{R}^{n+1}) \to \mathbb{F}_r(\mathbb{R}^{n+1})$ to S_{21}. It is a fibration with $\mathbb{F}_{k-r,r}(\mathbb{R}^{n+1})$ as fiber. Suppose that

$$\begin{array}{ccc} S_{21} \times \mathbb{F}_{k-r,r}(\mathbb{R}^{n+1}) & \xrightarrow{\phi} & \mathbb{F}'_k(\mathbb{R}^{n+1}) \\ \mathrm{p} \downarrow & & \downarrow \mathrm{proj}_{k,r} \\ S_{21} & \xrightarrow{\mathrm{id}} & S_{21} \end{array}$$

is a homotopy equivalence over S_{21}. Note that ϕ can be adjusted, if necessary, so that $\phi|(\mathbb{F}_{k-r,r}(\mathbb{R}^{n+1})) : \mathbb{F}_{k-r,r}(\mathbb{R}^{n+1}) \to \mathbb{F}_{k-r,r}(\mathbb{R}^{n+1})$ is the identity. As the configuration spaces here are all simply connected, we have the direct sum decomposition

$$\text{(I)}\ \pi_n(\mathbb{F}'_k(\mathbb{R}^{n+1})) \cong \pi_n(S_{21}) \oplus (\textstyle\bigoplus_{s=r+1}^k \pi_n(\vee_{t=1}^{s-1} S_{st}))$$

induced by the sectioned split product representations

$$\text{(I*)} \begin{cases} \mathbb{F}'_k(\mathbb{R}^{n+1}) & \simeq S_{21} \ltimes \mathbb{F}_{k-r,r}(\mathbb{R}^{n+1}) \\ \mathbb{F}_{k-r,r}(\mathbb{R}^{n+1}) & \simeq \ltimes_{s=r+1}^k (\vee_{t=1}^{s-1} S_{st}). \end{cases}$$

The morphism $\pi_n(\phi)$ takes α_{21} to $\alpha_{21} + \sum_{s=r+1}^k \beta_{s*}$, where each β_{s*} is in $\pi_n(S_{s1} \vee \cdots \vee S_{ss-1})$. Since $[\alpha_{21}, \alpha_{r+11}] = 0$ in $\pi_{2n-1}(S_{21} \times \mathbb{F}_{k-r,r}(\mathbb{R}^{n+1}))$, and because $\pi_*(\phi)$ preserves Whitehead products, it follows that

$$\text{(II)} \begin{cases} \phi_*[\alpha_{r+11}, \alpha_{21}] = [\alpha_{r+11}, \alpha_{21}] + \sum_{s=r+1}^k [\alpha_{r+11}, \beta_{s*}] \\ \qquad\qquad\qquad = 0. \end{cases}$$

The Y-B relations imply that the elements $[\alpha_{r+11}, \alpha_{21}]$ and $[\alpha_{r+11}, \beta_{r+1*}]$ are in $\pi_{2n-1}(S_{r+11} \vee \cdots \vee S_{r+1r})$, whereas, for $s > (r+1)$, $[\alpha_{r+11}, \beta_{s*}]$ is in $\pi_{2n-1}(S_{s1} \vee \cdots \vee S_{ss-1})$. As these are two distinct summands, (II) implies that

$$\text{(III)}\quad [\alpha_{r+11}, \alpha_{21}] + [\alpha_{r+11}, \beta_{r+1*}] = 0,$$

and the second term of (III) can be expanded in the form

$$[\alpha_{r+11}, \beta_{r+1*}] = \sum_{s=1}^{r} c_{r+1s}[\alpha_{r+11}, \alpha_{r+1s}],$$

with $c_{r+1s} \in \mathbb{Z}$, where $\beta_{r+1*} = \sum_{t=1}^r c_{r+1t}\alpha_{r+1t}$. So, (III) becomes

$$\text{(III)*}\ [\alpha_{r+11}, \alpha_{21}] + c_{r+11}[\alpha_{r+11}, \alpha_{r+11}] + \cdots + c_{r+1r}[\alpha_{r+11}, \alpha_{r+1r}] = 0.$$

On replacing $[\alpha_{r+11}, \alpha_{21}]$ by its value in terms of the Whitehead product $[\alpha_{r+11}, \alpha_{r+12}]$ and using the Y-B relation $[\alpha_{r+11}, \alpha_{21} + (-1)^{n+1}\alpha_{r+12}] = 0$, we obtain the following version of (III)*

$$\text{(III)**} \begin{cases} c_{r+11}[\alpha_{r+11}, \alpha_{r+11}] + \\ (c_{r+12} - (-1)^{n+1})[\alpha_{r+11}, \alpha_{r+12}] + \\ \sum_{t>2}^r c_{r+1t}[\alpha_{r+11}, \alpha_{r+1t}] \qquad = 0. \end{cases}$$

Next, recall that Hilton's Theorem gives, for each s such that $(r+1) \leq s \leq k$, the direct sum decomposition

$$\pi_{2n-1}(S_{s1} \vee \cdots \vee S_{ss-1}) \cong (\bigoplus_{i=1}^{s-1} \pi_{2n-1}(S_{si})) \oplus (\bigoplus_{w} \pi_{2n-1}(S_w)),$$

where w ranges over all Whitehead products of weight 2 on the set of symbols $\{\alpha_{st} | 1 \leq t < s\}$ ([110, Whitehead]). Observe that the various Whitehead products in (III)** belong to different summands in the Hilton formula. Thus we obtain the equations

$$\text{(IV)}\begin{cases} (i) \quad c_{r+11}[\alpha_{r+11}, \alpha_{r+11}] & = 0 \\ (ii) \quad (c_{r+12} - (-1)^{n+1})[\alpha_{r+11}, \alpha_{r+12}] & = 0 \\ (iii) \; c_{r+1t}[\alpha_{r+11}, \alpha_{r+1t}] & = 0. \end{cases}$$

Now, note that $w = [\alpha_{r+11}, \alpha_{r+12}] \in \pi_{2n-1}(S^w)$ is a basic product. Therefore, it defines a summand in the Hilton Theorem. Also note that it is of infinite order. This clearly implies that $c_{r+12} = (-1)^{n+1}$. Next, starting with the fact $[\alpha_{21}, \alpha_{r+12}] = 0 \in \pi_{2n-1}(S^n_{21} \times \mathbb{F}_{k-r,r}(\mathbb{R}^{n+1}))$, apply the same argument as that given above to prove that $c_{r+11} = 1$. Hence, (i) of (IV) implies that $[\alpha_{r+11}, \alpha_{r+11}] = 0$. However, $[\iota_n, \iota_n] = [\alpha_{r+11}, \alpha_{r+11}]$. Therefore, $[\iota_n, \iota_n] = 0$ and, hence, S^n is an H-space. This proves that $n = 3$ or 7, as required (see[2, Adams], [3, Adams], [4, Adams]).

Step 2

Assume now that $n = 3$ or 7. Consider (ii) of (IV). The basic product $w = [\alpha_{r+11}, \alpha_{r+1t}]$ generates the infinite cyclic group $\pi_{2n-1}(S_w)$. This implies that its coefficient c_{r+1t} is zero and, therefore,

$$\beta_{r+1*} = \alpha_{21} + \alpha_{r+11} + \alpha_{r+12}.$$

Suppose now that $r > 2$. Observe that, for all γ in $\{\alpha_{r+1t} | 2 < t \leq r\}$, we have $[\beta_{r+1*}, \gamma] = 0$. For each such t, we have

$$\text{(V)}\begin{cases} 0 = [\beta_{r+1*}, \alpha_{r+1,t}] \\ \quad = [\alpha_{21} + \alpha_{r+11} + \alpha_{r+12}, \alpha_{r+1t}] \\ \quad = [\alpha_{r+11}, \alpha_{r+1t}] + [\alpha_{r+12}, \alpha_{r+1t}], \end{cases}$$

since $[\alpha_{21}, \alpha_{r+1,t}] = 0$ for any t such that $2 < t \leq r$. Notice now that

$$\{[\alpha_{r+11}, \alpha_{r+1t}], [\alpha_{r+12}, \alpha_{r+1t}]\}$$

for any t such that $3 \leq 3 \leq r$ consists of distinct basic products. Therefore, the corresponding summands in Hilton's description,

$$\pi_{2n-1}(S_{s1} \vee \cdots \vee S_{ss-1}) \cong (\bigoplus_{i=1}^{i=s-1} \pi_{2n-1}(S_{si})) \oplus (\bigoplus_{w} \pi_{2n-1}(S_w)),$$

are distinct summands, and

$$[\alpha_{r+11}, \alpha_{r+1t}] = 0 = [\alpha_{r+12}, \alpha_{r+1t}],$$

since (V) implies that their sum vanishes. This is a contradiction. Thus $r = 2$ when $n = 3$ or 7. □

6 Historical Remarks

As we mentioned in §1, the homotopy classes α_{rs} are the higher dimensional version of the usual braid generators. Analogous spherical homology classes are defined in ([17, Cohen]) and were modified later in ([19, Cohen-Taylor]).

The induced action of Σ_k on $H_n(\mathbb{F}_k(\mathbb{R}^{n+1}); \mathbb{Z})$ was investigated by Cohen in [17, Cohen]. Our Theorem 2.1 implies his result and our proof is basically the same as his. We needed, however, to reformulate his proof, in order to be able to deal with the problems that arise in Chapter IV in the case $n = 1$, when $\mathbb{F}_k(\mathbb{R}^{n+1})$ is no longer simply connected.

As diagram $\mathcal{F}_k(\mathbb{R}^{n+1})$ described in §1 above is an important tool for the study of configuration spaces, it is worthwhile to note that it is a special case of a more general diagram $\mathcal{F}_k(M)$ (see [43, Fadell-Neuwirth]). In Chapter III below we define the analogous diagram for spheres. To describe it in general, let M be an arbitrary manifold and choose a base point $(q_1, \cdots, q_k) \in \mathbb{F}_k(M)$. Put $Q_r = \{q_1, \cdots, q_r\}$, for $1 \leq r \leq k$. Let $M_r = M - Q_r$. The diagram is

$$\mathcal{F}_k(M): \begin{cases} \mathbb{F}_{k,0} \leftarrow \cdots \leftarrow \mathbb{F}_{k-r,r} \leftarrow \mathbb{F}_{k-r-1,r+1} \leftarrow \cdots \\ \downarrow \qquad \downarrow \qquad \downarrow \qquad \downarrow \qquad \downarrow \\ M \qquad \cdots \qquad M_r \qquad M_{r+1} \qquad \cdots, \end{cases}$$

with $1 \leq r < k$, where $\mathbb{F}_{k-r,r}$ stands for $\mathbb{F}_{k-r,r}(M) = \mathbb{F}_{k-r}(M_r)$ and the vertical maps are the projections $\text{proj}_{k-r,1} : \mathbb{F}_{k-r,r}(M) \to M_r$ such that $(q_1, \ldots, q_r, x_{r+1}, \cdots, x_k) \mapsto x_{r+1}$. The diagram is called the *fundamental fiber sequence of* $\mathbb{F}_k(M)$. Note that, in general, except for the first one, the vertical projections admit sections. However, for manifolds M, which admit a nonzero tangent vector field, all of the vertical projections admit sections.

Results similar to those of §4 have been independently arrived at (see [18, Cohen-Gitler]). Using different methods, Massey ([79, Massey]) studied a problem concerning the canonical fibration, when $k = 3$ and $r = 2$, similar in nature to that of Theorem 5.1 above. Theorem 5.1 appears in [41, Fadell-Husseini].

III. Configuration Spaces of S^{n+1}, $n > 1$

The configuration spaces of spheres, as to be expected, are intimately related to those of the Euclidean spaces of the same dimension. Nevertheless, they present important novel features. The primary difference is due to the fact that the tangent bundle of the sphere is nontrivial except for the cases when the sphere S^m is S^1, S^3, or S^7. In this chapter we consider the case $m > 2$ only, so the relevant configuration spaces are simply connected. The case S^2 presents a new kind of difficulty, as the corresponding configuration spaces are no longer simply connected. It will be taken up in Chapter IV.

In §1 we show how the first nontrivial homotopy groups of $\mathbb{F}_{k+1}(S^{n+1})$ are determined by those of $\mathbb{F}_k(\mathbb{R}^{n+1})$. The fiber of the projection

$$\text{proj}_{k+1,1} : \mathbb{F}_{k+1}(S^{n+1}) \to S^{n+1}$$

is homeomorphic to $\mathbb{F}_k(\mathbb{R}^{n+1})$, and the natural imbedding

$$i_k : \mathbb{F}_k(\mathbb{R}^{n+1}) \to \mathbb{F}_{k+1}(S^{n+1})$$

takes the elements $\alpha_{rs} \in \pi_n(\mathbb{F}_k(\mathbb{R}^{n+1}))$ to $\pi_n(i_k)(\alpha_{rs})$ of $\pi_n(\mathbb{F}_{k+1}(S^{n+1}))$, with $1 \leq s < r \leq k$. It is not hard to see that these elements generate $\pi_n(\mathbb{F}_{k+1}(S^{n+1}))$.

However, the fibration used in the definition of these elements is not invariant under all permutations, and, hence, the induced action of the full symmetric group on these generators is not immediately obvious. So, in §2 we use the stereographic map that sends the south pole to the point at infinity to introduce elements $\{\beta_{rs} \mid 1 \leq s < r \leq (k+1)\}$. These behave exactly like their counterparts under the action of the full symmetric group Σ_{k+1}. We show that $\beta_{rs} = \pi_n(i_k)(\alpha_{rs})$, for $r \leq k$, and that $\sum_{s=1}^{k} \beta_{k+1s} = 0$. The latter formula allows one to express the elements β_{k+1s} in terms of those β_{rs} where $r \leq k$.

There are elements γ_{rs} in $\pi_{n+1}(\mathbb{F}_{k+1}(S^{n+1}))$ that come from the Hopf map $S^3 \to S^2$, suspended an appropriate number of times if necessary.

In §§3 and 4, we consider the case when $(n+1)$ is odd. Here, the sphere tangent bundle of S^{n+1} admits a section, and the section, in turn, defines an element, say, δ_{k+1}, of $\pi_{n+1}(\mathbb{F}_{k+1}(S^{n+1}))$. The element δ_{k+1} and the set

$\{\gamma_{rs} \mid 1 \leq s < r \leq k\}$ generate together the group $\pi_{n+1}(\mathbb{F}_{k+1}(S^{n+1}))$. We demonstrate how δ_{k+1} behaves under the action of the symmetric group Σ_{k+1}: if $(n+2) = 0 \bmod 4$, we prove that δ_{k+1} is invariant. The reason is that, under the preceding condition, the ambient space $\mathbb{R}^{n+2}$ is a quaternionic space, and one has two mutually orthogonal vector fields that come, respectively, from multiplication by i and j. Having these two vector fields available, one is then able to perform the necessary deformations between the element δ_{k+1} and its image under a symmetry. But, if $(n+2)$ is not congruent to 0 modulo 4, then δ_{k+1} is no longer invariant under Σ_{k+1}. For example, if $\tau_r = (r, r+1)$, then its effect on δ_{k+1} is to change the latter by γ_{r+1r}.

In §5, with $(n+1)$ being odd, we look for the Y-B relations in the Lie algebra $\pi_*(\mathbb{F}_{k+1}(S^{n+1}))$. We show that, in addition to the those induced by the Y-B relations of $\pi_*(\mathbb{F}_k(\mathbb{R}^{n+1}))$, there are new relations involving δ_{k+1} and β_{rs}. Here the fiber homotopy type of the tangent sphere bundle of S^{n+1} is relevant: the product $[\delta_{k+1}, \beta_{rs}]$ is in $\pi_{2n}(\mathbb{F}_{k+1}(S^{n+1}))$ and is determined by a certain element in the image of the J-homomorphism.

In §6 the case when $(n+1)$ is even is considered. Here, as in the case when $(n+1)$ is odd, there is a map $\psi_{k+1} : O_{n+2,2} \to \mathbb{F}_{k+1}(S^{n+1})$ such that $\psi_3 : O_{n+2,2} \to \mathbb{F}_3(S^{n+1})$ is a homotopy equivalence, and ψ_{k+1} is a section of the pull-back $\mathrm{proj}_{k+1,3} : \mathbb{F}'_{k+1}(S^{n+1}) \to O_{n+2,2}$ by ψ_3 of the natural projection $\mathbb{F}_{k+1}(S^{n+1}) \to \mathbb{F}_r(S^{n+1})$. Thus, we see that $\pi_*(\mathbb{F}_{k+1}(S^{n+1})) \cong \pi_*(O_{n+2,2}) \oplus \pi_*(\mathbb{F}_{k-2}(\mathbb{R}^{n+1}))$. Looking at $\mathbb{F}_{k+1}(S^{n+1})$ differently, we also show that $\pi_n(i_k) : \pi_n(\mathbb{F}_k(\mathbb{R}^{n+1})) \to \pi_n(\mathbb{F}_{k+1}(S^{n+1}))$ is surjective. Therefore, the set $\{\beta_{rs} \mid 1 \leq s < r \leq (k+1)\}$ generates the group $\pi_n(\mathbb{F}_{k+1}(S^{n+1}))$. This group has a cyclic subgroup of order two that comes from the Stiefel manifold $O_{n+2,2}$. Moreover, the generator of this subgroup is the image of a class Δ_k, which we call the *Dirac class.* This class is represented by $(k-1)$ concentric spheres in $\mathbb{F}_k(\mathbb{R}^{n+1})$. The next group, $\pi_{n+1}(\mathbb{F}_{k+1}(S^{n+1}))$, is generated by the set $\{\gamma_{rs} \mid 1 \leq s < r \leq k\}$.

In §7 we consider the Lie algebra $\pi_*(\mathbb{F}_{k+1}(S^{n+1}))$. It is best described in terms of the algebras $\pi_*(\mathbb{F}_{k-2,2}(\mathbb{R}^{n+1}))$, $\pi_*(O_{n+2,2})$. For $(k+1) \geq 3$, one has a section $\psi_{k+1} : O_{n+2,2} \to \mathbb{F}_{k+1}(S^{n+1})$. Furthermore, ψ_3 is a homotopy equivalence. The algebra $\pi_*(\mathbb{F}_{k+1}(S^{n+1}))$ is a split extension of the ideal $\pi_*(\mathbb{F}_{k-2,2}(\mathbb{R}^{n+1}))$ by $\pi_*(O_{n+2,2})$.

Finally, in §8, we address the question of when the natural projection

$$\mathrm{proj}_{k+1,r} : \mathbb{F}_{k+1}(S^{n+1}) \to \mathbb{F}_r(S^{n+1})$$

is fiber homotopically trivial. Under the assumption that $(n+1)$ is odd, we show that it is fiber homotopically trivial if, and only if, $r \leq 2$ and $(n+1)$ is 3 or 7. Although the situation here bears resemblance to the Euclidean case, there is a new ingredient: namely, the fiber homotopy type of the tangent bundle of the sphere. Our answer to the case when $(n+1)$ is even is incomplete. We show that if the projection is fiber homotopically trivial, then $r = 3$ and n is 3 or 7. The converse remains an open question.

1 Filtration of $\pi_*(\mathbb{F}_{k+1}(S^{n+1}))$, $n > 1$

The filtration we wish to obtain is induced by the fundamental sequence of $\mathbb{F}_{k+1}(S^{n+1})$. To begin, regard $\mathbb{R}^{n+2}$ as $\mathbb{R} \times \mathbb{R}^n \times \mathbb{R}$, and denote the subspace $\{x \in \mathbb{R} \times \mathbb{R}^n \times \mathbb{R} | x_{n+2} = 0\}$ by V^e. Let $\phi^e : S^{n+1} - -e \to V^e = \mathbb{R} \times \mathbb{R}^n$ be the stereographic projection from $-e$ to V^e, where $e = (0, \cdots, 0, 1)$. Define q_i^e by the rule

$$q_1^e = e,\ q_2^e = e_1,\ \phi^e(q_i^e) = q_i = 4(i-1)e_1, \quad 3 \leq i \leq k,$$

where $e_1 = (1, 0, \cdots, 0) \in \mathbb{R} \times \mathbb{R}^n$, and put

$$Q_0^e = \{-e\},\quad Q_i^e = \{q_1^e, \cdots, q_i^e, -e\},\ 1 \leq i < k.$$

We choose $(q_1^e, \cdots, q_k^e, -e)$ as a basepoint. Note that the points q_i^e lie on the great circle in S^{n+1} cut by the plane spanned by the first and last basic vectors in $\mathbb{R}^{n+2}$. Consider next the projection $\text{proj}_{k+1,1} : \mathbb{F}_{k+1}(S^{n+1}) \to S^{n+1}$ where

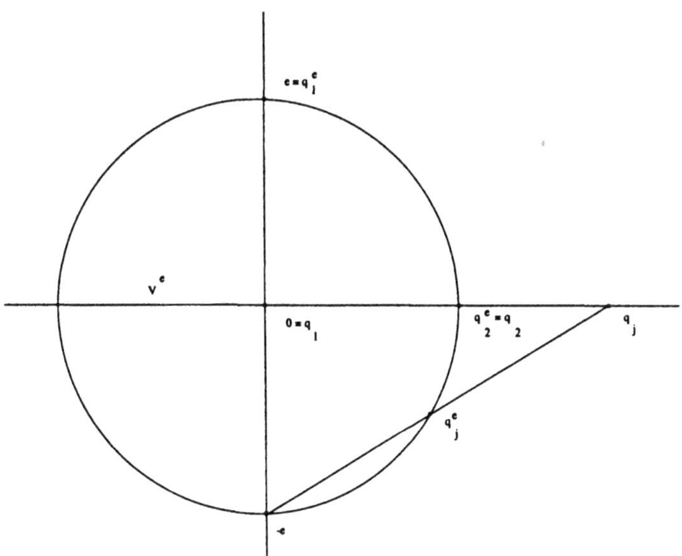

Fig. III.1. Stereographic Projection

$(x_1, \cdots, x_{k+1}) \mapsto x_{k+1}$. Note that, on taking $-e$ as the basepoint of S^{n+1}, we see that the fiber of the $\text{proj}_{k+1,1}$ can be identified with $\mathbb{F}_k(\mathbb{R}^{n+1})$ by means of the map $\phi^e : S^{n+1} - \{-e\} \to V^e = \mathbb{R}^{n+1}$. Define $i_k : \mathbb{F}_k(\mathbb{R}^{n+1}) \to \mathbb{F}_{k+1}(S^{n+1})$ to be the map where $(x_1, \cdots, x_k) \mapsto ((\phi^e)^{-1}(x_1), \cdots, (\phi^e)^{-1}(x_k), -e)$. As in Chapter II, §2, and with $\mathbb{R}^{n+1}_r$ standing for $S^{n+1} - Q^e_r$, the preceding fibration fits into the diagram

$$\mathcal{F}_{k+1}(S^{n+1}) : \begin{cases} \mathbb{F}_{k+1}(S^{n+1}) \leftarrow \mathbb{F}_{k,0} \leftarrow \cdots \mathbb{F}_{k-r,r} \leftarrow \\ \quad\downarrow \qquad\qquad \downarrow \qquad \downarrow \quad \downarrow \\ \quad S^{n+1} \qquad \mathbb{R}^{n+1} \quad \cdots \ \mathbb{R}^{n+1}_r, \end{cases}$$

where the spaces above are

$$\begin{aligned} \mathbb{F}_{k-r,r} &= \mathbb{F}_{(k+1)-(r+1),r+1}(S^{n+1}) = \mathbb{F}_{k-r}(S^{n+1} - Q^e_r) \\ &\cong \mathbb{F}_{k-r,r}(\mathbb{R}^{n+1}) \qquad\qquad = \mathbb{F}_{k-r}(\mathbb{R}^{n+1} - Q_r) \\ &= \mathbb{F}_{k-r}(\mathbb{R}^{n+1}_r) \end{aligned}$$

and the vertical maps are the projections $\text{p}_r : \mathbb{F}_{k-r,r}(S^{n+1}) \to \mathbb{R}^{n+1}_r$, with the property that $(q_1, \cdots, q_r, x_{r+1}, \cdots, x_k) \in \mathbb{F}_{k-r,r}(\mathbb{R}^{n+1}) \mapsto x_{r+1} \in \mathbb{R}^{n+1}$.

Proposition 1.1 *The map $\psi_3 : O_{n+2,2} \to \mathbb{F}_3(S^{n+1})$, $(x, y) \mapsto (x, y, -x)$ is a homotopy equivalence.*

Proof. The map ψ_3 induces a map of fibrations

$$\begin{array}{ccc} O_{n+2,2} & \xrightarrow{\psi_3} & \mathbb{F}_3(S^{n+1}) \\ \downarrow & & \downarrow \\ S^{n+1} & \xrightarrow{\psi_2} & \mathbb{F}_2(S^{n+1}), \end{array}$$

where the first vertical projection takes (x, y) to x, while the second sends (x_1, x_2, x_3) to (x_1, x_3), and $\psi_2(x) = (x, -x)$. The map ψ_2 is a homotopy equivalence, and ψ_3 restricted to the fiber is the identity. This is enough to prove the proposition. □

Corollary 1.1 *The map $\psi_3 : O_{n+2,2} \to \mathbb{F}_3(S^{n+1})$ induces a fiber homotopy equivalence*

$$\begin{array}{ccc} O_{n+2,2} & \xrightarrow{\psi_3} & \mathbb{F}_3(S^{n+1}) \\ \downarrow & & \downarrow \\ S^{n+1} & \xrightarrow{\text{id}} & S^{n+1}, \end{array}$$

where the vertical maps are the projections on the first factor.

The fibrations $\text{p}_r : \mathbb{F}_{k-r,r}(\mathbb{R}^{n+1}) \to \mathbb{R}^{n+1}_r$ in diagram $\mathcal{F}_k(S^{n+1})$ admit sections, as we saw in Chapter II.

Proposition 1.2 *There is a map $\psi_{k+1} : O_{n+2,2} \to \mathbb{F}_{k+1}(S^{n+1})$ such that $\text{proj}_{k+1,3} \circ \psi_{k+1} = \psi_3$, where $\text{proj}_{k+1,3}$, is the projection that takes $(x_1, \cdots, x_{k+1})$ to (x_1, x_2, x_{k+1}).*

Proof. Let $(x,y) \in O_{n+2,2}$, and denote by V^x the subspace in $\mathbb{R}^{n+2}$ orthogonal to x. The stereographic projection $\phi^x : S^{n+1} - \{-x\} \to V^x$, $-x \mapsto \infty$ takes the great circle on S^{n+1}, determined by the pair (x,y), to a ray $\mathbb{R}^{xy} \subset V^x$. Choose the unit vector in $\mathbb{R}^{xy}$ to be y, and put $e^{xy} = y$. By definition,

$$q_1^x = x,\ q_2^x = y,\ q_{k+1}^x = -x \text{ and } q_i^{xy} = (\phi^x)^{-1}(4(i-1)e^{xy}), \quad i = 3, \cdots, k.$$

One can easily verify that the map $\psi_{k+1} : O_{n+2,2} \to \mathbb{F}_{k+1}(S^{n+1})$, defined by the rule, where $(x,y) \mapsto (x, q_2^{xy}, q_3^{xy}, \cdots, q_{k-1}^{xy}, -x)$, satisfies the required conditions. □

Going back to diagram $\mathcal{F}_k(S^{n+1})$, we see that $\mathbb{F}_{k-2,2}$ is the fiber of the projection $\mathrm{proj}_{k+1,3} : \mathbb{F}_{k+1}(S^{n+1}) \to \mathbb{F}_3(S^{n+1})$ and that this fibration has a section, by virtue of the previous propositions. So we get the representation

$$\mathbb{F}_{k+1}(S^{n+1}) \simeq O_{n+2,2} \ltimes (\ltimes_{r=3}^{k-1}(S_{r1} \vee \cdots \vee S_{rr-1}))$$

of $\mathbb{F}_{k+1}(S^{n+1})$ as a multifold twisted product, with the notation being the same as that of Chapter II. As an immediate corollary, we have the following analogue of Theorem 1.1 of Chapter II.

Theorem 1.1 *There is an isomorpism*

$$\pi_*(\mathbb{F}_{k+1}(S^{n+1})) \cong \pi_*(O_{n+2,2}) \oplus \oplus_{r=3}^{k-1}\pi_*(S_{r1} \vee \cdots \vee S_{rr-1})$$

of groups.

The first component, $\pi_*(O_{n+2,2})$, depends on the parity of $(n+1)$. We shall see later that, in addition, the isomorphism in the theorem is only one of groups. To describe the Lie algebra structure, we need to find suitable generators and the relations among them. This is our aim in the following sections.

2 Relation with $\mathbb{F}_k(\mathbb{R}^{n+1})$

According to Theorem 1.1 of Chapter II, $\pi_m(\mathbb{F}_k(\mathbb{R}^{n+1})) \cong \sum_{r=1}^{k-1} \pi_m(S^{\vee r})$, where $S^{\vee r} = S_{r+11} \vee \cdots \vee S_{r+1r}$. For $m = n$, we found the generators $\{\alpha_{rs} | 1 \leq s < r \leq k\}$, for $\pi_n(\mathbb{F}_k(\mathbb{R}^{n+1}))$, and $\{\gamma_{rs} | 1 \leq s < r \leq k\}$, for $\pi_{n+1}(\mathbb{F}_k(\mathbb{R}^{n+1}))$. The latter did not play a special role in the study of $\mathbb{F}_k(\mathbb{R}^{n+1})$ carried out in Chapter II. In the study of $\mathbb{F}_{k+1}(S^{n+1})$, we shall define similar generators $\beta_{rs} \in \pi_n(\mathbb{F}_{k+1}(S^{n+1}))$ and $\gamma_{rs} \in \pi_{n+1}(\mathbb{F}_{k+1}(S^{n+1}))$ and, then, study their transformation under the action of the symmetric group Σ_{k+1}. Generators for $\pi_n(\mathbb{F}_{k+1}(S^{n+1}))$ can be found in the image of the nth homotopy group of $\mathbb{F}_k(\mathbb{R}^{n+1})$, regarded as the fiber at $-e$. However, as this fiber is not invariant under the action of Σ_{k+1}, the task of determining the manner in which these elements behave under the action of Σ_{k+1} is made difficult. Therefore, we need elements that arise from a suitable invariant imbedding of $\mathbb{F}_{k+1}(\mathbb{R}^{n+1})$ in $\mathbb{F}_{k+1}(S^{n+1})$. We shall define these elements, describe the effect of the symmetric group Σ_{k+1} on them, and then show how they are related to the images of $\{\alpha_{rs} \mid 1 \leq s < r \leq k\}$ in the fiber. We shall also derive the Yang-Baxter relations among them.

Consider the stereographic projection $\phi^e : S^{n+1} - \{-e\} \to V^e$, where e is the basic vector $(0,\cdots,0,1) \in \mathbb{R}^{n+2}$. As $V^e = \mathbb{R}\times\mathbb{R}^n$, the map ϕ^e induces a homeomorphism $\mathbb{F}_{k+1}(S^{n+1}-\{-e\}) \to \mathbb{F}_{k+1}(\mathbb{R}^{n+1})$. Put $\beta'_{rs} = (\phi^e)^{-1}\circ\alpha'_{rs}$ for all r,s such that $1 \leq r,s \leq (k+1)$ and $r \neq s$. Explicitly, the map

$$\beta'_{rs} : S^n \to \mathbb{F}_{k+1}(S^{n+1}), \quad \{r,s \mid 1 \leq s < r \leq (k+1)\}$$

is that where $\xi \mapsto (e, q^e_2, \cdots, q^e_s, \cdots, q^e_{r-1}, u_s(\xi), q^e_r, \cdots, q^e_k)$, with $u_s(\xi)$ being defined by the rule $\phi^e(u_s(\xi)) = (\phi^e)(q^e_s) + \xi$, $\|\xi\| = 1$. Denote by β_{rs} the homotopy class of β'_{rs}.

The natural action of the symmetric group Σ_{k+1} on $\mathbb{F}_{k+1}(S^{n+1})$ induces an action on the homotopy group $\pi_n(\mathbb{F}_{k+1}(S^{n+1}))$. Observe that, under this action, the subspace $\mathbb{F}_{k+1}(S^{n+1}-\{-e\})$ of $\mathbb{F}_{k+1}(S^{n+1})$ is invariant. Thus, the assertions of the following theorem are immediate consequences, respectively, of Proposition 1.2 and Theorem 2.1 of Chapter II.

Theorem 2.1 *For all $1 \leq s < r \leq (k+1)$ and $\sigma \in \Sigma_{k+1}$, it follows that*

$$\begin{array}{ll} (i) & \beta_{sr} = (-1)^{n+1}\beta_{rs} \quad \textit{and} \\ (ii) & \pi_n(\sigma)(\beta_{rs}) = \beta_{\sigma r \sigma s} \end{array}$$

hold in $\pi_n(\mathbb{F}_{k+1}(S^{n+1}))$.

The elements $\beta_{rs} \in \pi_n(\mathbb{F}_{k+1}(S^{n+1}))$ are not independent, but they constitute an invariant set of generators of the group. To see how they are related to each other, consider the imbedding $i_k : \mathbb{F}_k(\mathbb{R}^{n+1}) \to \mathbb{F}_{k+1}(S^{n+1})$, where $(x_1, x_2, \cdots, x_k) \mapsto (x'_1, x'_2, \cdots, x'_k, -e)$, with $x'_i = (\phi^e)^{-1}(x_i)$, for $1 \leq i \leq k$.

Theorem 2.2 *The elements $\{\beta_{rs} | 1 \leq s < r \leq (k+1)\}$ satisfy the following relations*

$$\begin{array}{ll} (i) & \pi_n(i_k)(\alpha_{rs}) = \beta_{rs} \quad \textit{and} \\ (ii) & 0 = \sum_{s=1}^{k} \beta_{k+1 s} \end{array}$$

in $\pi_n(\mathbb{F}_{k+1}(S^{n+1}))$.

Proof. To prove Statement (i), one can easily check that $\beta'_{rs} \simeq i_k \circ \alpha'_{rs}$ for all r,s such that $1 \leq s < r \leq k$.

To prove Statement (ii), we observe first that an argument along the lines of Proposition 1.2 of Chapter II yields the result that $\sum_{s=1}^{k} \beta_{k+1s}$ is represented by a sphere S of dimension n in $S^{n+1} - \{-e\}$ surrounding $\{q^e_1, \cdots, q^e_k\} \subset S^{n+1}$. On the other hand, the sphere S also bounds the disk containing $\{-e\}$. Hence, it is null-homotopic, and Statement (ii) is thus proved. □

Corollary 2.1 *We have the additional relations*

$$\begin{aligned} (iii)\ \beta_{k+1k} &= (-1)^n \textstyle\sum_{s=1}^{k-1} \beta_{ks} \quad \text{and} \\ (iv)\ \beta_{k+1t} &= (-1)^n \textstyle\sum_{s=1}^{k-1} \beta_{\sigma k \sigma s}, \end{aligned}$$

where $1 \leq t < k$ *and* $\sigma = (k,t)$.

Proof. To prove Statement (iii), observe first that Statement (ii) implies the relation $\beta_{k+1k} = (-1)\sum_{s=1}^{k-1} \beta_{k+1s}$. Now, apply the transposition $(k, k+1)$ to obtain $\beta_{k+1k} = (-1)^n \sum_{s=1}^{k-1} \beta_{ks}$. Finally, to obtain Statement (iv) apply the transposition (k,t) to Statement (iii). □

Following the usual convention, let $\eta_n = E^{n-2}(\eta)$, with $\eta' : S^3 \to S^2$ being the Hopf map and $E : \pi_3(S^2) \to \pi_4(S^3)$, the Freudenthal suspension. In analogy with the definition of the elements $\tilde{\gamma}_{rs}$ given in Chapter II §5, let

$$\gamma_{rs} = \beta_{rs} \circ \eta_n \in \pi_{n+1}(\mathbb{F}_{k+1}(S^{n+1})),$$

where $\omega \circ \eta_n$ is the composition product (see Theorem 8.1 of Chapter X in [110, Whitehead]). Note that $\pi_{n+1}(\mathbb{F}_{k+1}(\mathbb{R}^{n+1})) \to \pi_{n+1}(\mathbb{F}_{k+1}(S^{n+1}))$ takes $\tilde{\gamma}_{rs}$ to γ_{rs}.

Theorem 2.3 *For all* $\sigma \in \Sigma_{k+1}$, *it follows that*

$$\pi_{n+1}(\sigma)(\gamma_{rs}) = \gamma_{\sigma r \sigma s}.$$

Proof. The proof is straightforward. Consider the diagram

$$\begin{array}{ccccc} S^{n+1} & \xrightarrow{\eta'_n} & S^n & \xrightarrow{\beta'_{rs}} & \mathbb{F}_{k+1}(S^{n+1}) \\ \downarrow = & & \downarrow = & & \downarrow \sigma_\# \\ S^{n+1} & \xrightarrow{\eta'_n} & S^n & \xrightarrow{\beta_{\sigma r \sigma s}} & \mathbb{F}_{k+1}(S^{n+1}), \end{array}$$

where η'_n is the $(n-2)$-fold suspension of η'. Note first that $\sigma_\# \circ \beta'_{rs} \circ \eta'_n$ represents the class $\pi_{n+1}(\sigma)(\gamma_{rs})$. Also, the composite $\beta'_{\sigma r \sigma s} \circ \eta'_n$ represents $\gamma_{\sigma r \sigma s}$. Now, observe that the square to the right commutes up to homotopy, according to Theorem 2.2 of Chapter II. Hence, the diagram commutes up to homotopy. This implies that the composites above are indeed homotopic, and the theorem follows. □

The elements $\{\gamma_{rs} | 1 \leq s < r \leq (k+1)\}$ are related among themselves in the manner of the elements $\{\beta_{rs} | 1 \leq s < r \leq (k+1)\}$. There are two cases to consider.

Theorem 2.4 *Suppose that* $n > 2$. *Then*

$$\begin{aligned} (i)\ \gamma_{sr} &= (-1)^{n+1}\gamma_{rs} \quad \text{and} \\ (ii)\ 0 &= \textstyle\sum_{s=1}^{k} \gamma_{k+1s} \end{aligned}$$

hold in $\pi_{n+1}(\mathbb{F}_{k+1}(S^{n+1}))$.

Proof. By virtue of Theorem 2.2 above, it is enough to show that η_{n-2} is additive. To do so, observe that $n > 2$ implies that η_{n-2} is a suspension. Hence, the map $\eta' : S^{n+1} \to S^n$ is a map of co-H-spaces, which implies that the composition product $\pi_n(X) \to \pi_{n+1}(X)$ by η_{n-2} is additive (cf. Chapter X, Theorem 8.2 of [110, Whitehead]). □

Similarly one proves the following additional relations.

Corollary 2.2

$$\begin{aligned} (iii)\ \gamma_{k+1k} &= (-1)^n \textstyle\sum_{s=1}^{k-1} \gamma_{ks} \quad and \\ (iv)\ \gamma_{k+1t} &= (-1)^n \textstyle\sum_{s=1}^{k-1} \gamma_{\sigma k \sigma s}, \end{aligned}$$

where $\sigma = (k, t) \in \Sigma_{k+1}$

On the other hand, the case $n = 2$ is quite different and more complicated, as the composition product here is not additive.

Theorem 2.5 *Suppose that* $n = 2$. *Then*

$$\begin{aligned} (i)\ \gamma_{sr} &= - \gamma_{rs} + [\beta_{rs}, \beta_{rs}], \\ (ii)\ \gamma_{k+1k} &= \textstyle\sum_{s=1}^{k-1} \gamma_{ks} + \sum_{s<t}^{k-1} [\beta_{ks}, \beta_{kt}], \quad and \\ &\quad + \textstyle\sum_{k=1}^{k-1} [\beta_{kt}, \beta_{kt}] \\ (iii)\ \gamma_{k+1u} &= \textstyle\sum_{s=1}^{k-1} \gamma_{\sigma k \sigma s} + \sum_{s,t=1} [\beta_{\sigma k \sigma s}, \beta_{\sigma k \sigma t}] \\ &\quad + \textstyle\sum_{t=1}^{k-1} [\beta_{\sigma k \sigma t}, \beta_{\sigma k \sigma t}], \end{aligned}$$

where $\sigma = (u, k) \in \Sigma_{k+1}$, *hold in* $\pi_3(\mathbb{F}_{k+1}(S^3))$.

First, we need the following calculation.

Assertion 2.1 *For all* $\omega_1, \cdots, \omega_p \in \pi_2(X)$, *we have*

$$(\sum_{i=1}^{p} \omega_i) \circ \eta = \sum_{i=1}^{p} \omega_i \circ \eta + \sum_{1 \le i < j \le p} [\omega_i, \omega_j].$$

Proof. To start, we note that Formula (1.10) of [110, Whitehead] implies that

$$(\omega_1 + \omega_2) \circ \eta = \omega_1 \circ \eta + \omega_2 \circ \eta + [\omega_1, \omega_2],$$

for all $\omega_1, \omega_2 \in \pi_2(S^2)$, because η is a map of Hopf invariant one (see §1, Chapter XI of [110, Whitehead]). Proceeding inductively, we obtain the desired formula. □

To prove Statement (i) of the theorem, note first that Assertion 2.1 readily implies that

$$(*) \qquad (-\omega) \circ \eta = -(\omega) \circ \eta + [\omega, \omega].$$

Next, apply $(*)$ inductively to obtain Statement (ii), starting with Statement (iii) of Corollary 2.1. Now Statement (iii) follows from Statement (ii) and the invariance properties under Σ_{k+1} of the elements γ_{rs}. □

3 The Groups π_n, π_{n+1}, $(n+1)$ Odd

When $(n+1)$ is odd, the ambient space $\mathbb{R}^{n+2}$ is the underlying space of a complex space; if $(n+2)$ is divisible by 4, then $\mathbb{R}^{n+2}$ is actually the underlying space of a quaternionic space. These two facts will be of great value in this and the succeeding section.

So, assume that $(n+1)$ is odd, and let $\mathfrak{J} : \mathbb{R}^{n+2} \to \mathbb{R}^{n+2}$ be a real linear morphism such that $\mathfrak{J}^2 = -\mathrm{id}$. Observe that $\mathfrak{J}$ defines the complex structure on $\mathbb{R}^{n+2}$. Put $h : S^{n+1} \to O_{n+2,2}$, $x \mapsto (x, h^x)$, where, for $x = (x_1, x_2, \cdots, x_{n+1}, x_{n+2}) \in S^{n+1}$,

$$h^x = \mathfrak{J}(x) = (-x_2, x_1, \cdots, \cdots, -x_{n+2}, x_{n+1}).$$

Next, observe that the tuple $(x, h^x, -x)$ spans a plane in $\mathbb{R}^{n+2}$ that intersects the unit sphere S^{n+1} in a great circle. The stereographic projection ϕ^x from $-x$ takes this great circle to a straight line in V^x through the origin. Denote this line by $\mathbb{R}^{h^x}$. Observe here that the point $\phi^x(x)$ is the origin in $\mathbb{R}^x$, while $\phi^x(h^x)$ is a unit vector in $\mathbb{R}^x$. Let $q_i^x \in S^{n+1}$ be the point defined by the rule that

$$\phi^x(q_i^x) = 4(i-1)\phi^x(h^x) \in \mathbb{R}^x \subset V^x,$$

where $i = 1, 2, \cdots, k$. Note that $q_1^x = x$ and $q_2^x = 4h^x$. Let $\delta'_{k+1} : S^{n+1} \to \mathbb{F}_{k+1}(S^{n+1})$ be the map such that $x \mapsto (x, q_2^x, \cdots, q_k^x, -x) = (x, h^x, q_3^x, \cdots, q_k^x, -x)$, where $h : S^{n+1} \to O_{n+2,2}$ is the section defined in the preceding paragraph.

Note that there is an alternative section , $\delta''_{k+1} : S^{n+1} \to \mathbb{F}_{k+1}(S^{n+1})$, given by $x \mapsto (x, q_2^x, q_3^x, \cdots, q_{k+1}^x)$. The sections δ''_{k+1} and δ'_{k+1} are homotopic to each other.

Consider now the projection $\mathrm{proj}_{k+1,1} : \mathbb{F}_{k+1}(S^{n+1}) \to S^{n+1}$, where the tuple $(x_1, x_2, \cdots, x_{k+1})$ is mapped to x_{k+1}. Identify the fiber at $-e$ with the space $\mathbb{F}_k(\mathbb{R}^{n+1})$ by the map $i_k : \mathbb{F}_k(\mathbb{R}^{n+1}) \to \mathbb{F}_{k+1}(S^{n+1})$ such that $(x_1, x_2, \cdots, x_k) \mapsto ((\phi^e)^{-1}(x_1), \cdots, (\phi^e)^{-1}(x_k), -e)$, where e is the last standard basic vector $(0, 0, \cdots, 1)$.

Theorem 3.1 *The homotopy exact sequence of the fibration*

$$\mathrm{proj}_{k+1,1} : \mathbb{F}_{k+1}(S^{n+1}) \to S^{n+1}$$

reduces to the short exact sequence

$$0 \to \pi_*(\mathbb{F}_k(\mathbb{R}^{n+1})) \overset{\pi_*(i_k)}{\to} \pi_*(\mathbb{F}_{k+1}(S^{n+1})) \overset{\pi_*(\mathrm{proj}_{k+1,1})}{\to} \pi_*(S^{n+1}) \to 0.$$

Proof. Note that $\mathrm{proj}_{k+1,1} \circ \delta'_{k+1} = \lambda$, where $\lambda : S^{n+1} \to S^{n+1}$ is the antipodal map. As $(n+1)$ is odd, λ is homotopic to the identity. Hence, δ'_{k+1} is a homotopy section of the fibration $\mathrm{proj}_{k+1,1} : \mathbb{F}_{k+1}(S^{n+1}) \to S^{n+1}$ described in the theorem. This implies the theorem. □

As a corollary we have the following.

Corollary 3.1 *There are isomorphisms*

$$\begin{aligned} \pi_n(\mathbb{F}_k(\mathbb{R}^{n+1})) &\cong \pi_n(\mathbb{F}_{k+1}(S^{n+1})) \\ \pi_{n+1}(\mathbb{F}_k(\mathbb{R}^{n+1})) \oplus \pi_{n+1}(S^{n+1}) &\cong \pi_{n+1}(\mathbb{F}_{k+1}(S^{n+1})). \end{aligned}$$

In particular, it follows that the set $\{\beta_{rs} \mid 1 \leq s < r \leq k\}$ generates $\pi_n(\mathbb{F}_{k+1}(S^{n+1}))$ (cf. Theorem 2.1 of Chapter II).

The description of the generators of $\pi_{n+1}(\mathbb{F}_{k+1}(S^{n+1}))$ is more complicated. There are two cases to consider: $n > 2$ and $n = 2$.

Theorem 3.2 *Suppose that $n > 2$. Then, the group $\pi_{n+1}(\mathbb{F}_{k+1}(S^{n+1}))$ is generated by the set $\{\delta_{k+1}, \gamma_{rs} \mid 1 \leq s < r \leq k\}$, where δ_{k+1} is the class of the homotopy section $\delta'_{k+1} : S^{n+1} \to \mathbb{F}_{k+1}(S^{n+1})$ defined above.*

Proof. Recall that, as in Theorem 3.1 above, the homotopy section δ'_{k+1} leads to the split-exact sequence,

$$0 \to \pi_{n+1}(\mathbb{F}_k(\mathbb{R}^{n+1})) \to \pi_{n+1}(\mathbb{F}_{k+1}(S^{n+1})) \to \pi_{n+1}(S^{n+1}) \to 0.$$

Also recall that, as a consequence of the multifold sectioned twisted-product structure $\mathbb{F}_k(\mathbb{R}^{n+1}) = \rtimes_{r=2}^{k}(\vee_{s=1}^{r-1} S_{rs}))$, we know that $\pi_n(\mathbb{F}_k(\mathbb{R}^{n+1})) \cong \bigoplus_{r=2}^{k} \pi_n(S_{r1} \vee \cdots \vee S_{rr-1})$. However, according to Hilton's Theorem, $\{\beta_{rs} \mid 1 \leq s < r\}$, $\{\gamma_{rs} \mid 1 \leq s < r\}$ generate, respectively, the groups π_n and π_{n+1} of $(S_{r1} \vee \cdots \vee S_{rr-1})$. Here we use the fact that each $\pi_{n+1}(S_{rs})$ is generated by the η_n (see Corollary (2.7) of Chapter XII of [110, Whitehead]). Finally, note that δ_{k+1} generates the summand corresponding to $\pi_{n+1}(S^{n+1})$. □

The generators for the case when $n = 2$ are determined in a similar fashion.

Theorem 3.3 *The group $\pi_3(\mathbb{F}_{k+1}(S^3))$ is generated by the set*

$$\{\delta_{k+1}, \gamma_{rs}, [\beta_{rs}, \beta_{tu}]\},$$

where $1 \leq s < r \leq k$, $1 \leq u < t \leq k$, and $r < t$.

Proof. Consider the isomorphism $\pi_3(\mathbb{F}_k(\mathbb{R}^3)) \cong \bigoplus_{r=2}^{k} \pi_3(S_{r1} \vee \cdots \vee S_{rr-1})$. Order the elements $\{\beta_{rs} \mid 1 \leq s < r \leq k\}$ so that $\beta_{rs} < \beta_{uv}$ if $r < u$ and $\beta_{rs} < \beta_{rv}$ if $s < v$. The theorem now follows immediately from Hilton's Theorem. □

4 Symmetry Invariance of δ_{k+1}

The effect of Σ_{k+1} on the elements $\{\gamma_{rs} \mid 1 \leq s < r \leq (k+1)\}$ is given in the Theorem 2.3 above. Here, we wish to describe the effect of Σ_{k+1} on δ_{k+1}. Recall that $(n+1)$ is odd. There are two cases to consider.

Theorem 4.1 *If* $n+2 = 0 \bmod 4$, *then* $\pi_{n+1}(\sigma)(\delta_{k+1}) = \delta_{k+1}$, *for all* $\sigma \in \Sigma_{k+1}$.

Proof. The assumption that $(n+2) = 0 \bmod 4$ implies that $\mathbb{R}^{n+2}$ is isomorphic to $\mathbb{H}^m$, as a real vector space, where $\mathbb{H}$ is the quaternion algebra and $4m = (n+2)$. Hence, the morphism $\mathfrak{J} : \mathbb{R}^{n+2} \to \mathbb{R}^{n+2}$, defined by multiplication by j in each factor, is a real morphism.

To prove the theorem it is enough to show that $\pi_{n+1}(\tau_r)(\delta_{k+1}) = \delta_{k+1}$, where $\tau_r = (r, r+1)$. Choose the map $\delta''_{k+1} : S^{n+1} \to \mathbb{F}_{k+1}(S^{n+1})$, defined in §4 above, as a representative of δ_{k+1}. Recall that $\delta''_{k+1}(x) = (q_1^x, \cdots, q_{k+1}^x)$, and consider the expression

$$(\tau_r \circ \delta'_{k+1})(x) = (\cdots, q_{r+1}^x, q_r^x, \cdots)$$

for $x \in S^{n+1}$. Denote by W^x the space spanned by the vectors x, $\mathfrak{J}(x)$ and $\mathfrak{J}(x)$. Following the notation of the previous section, put

$$\phi^x(q_1^x, \cdots, q_{k+1}^x) = (u_1^x, \cdots, u_{k+1}^x) \in V^x,$$

and let

$$\begin{aligned} \xi_{(x,t)} &= 2(\cos \pi\, t\, \mathfrak{J}(x) + \sin \pi\ t\, \mathfrak{J}(x)), \quad \text{for } 0 \le t \le 1 \\ u_r^{x,t} &= (u_r^x + u_{r+1}^x)/2 - \xi_{(x,t)} \\ u_{r+1}^{x,t} &= ((u_r^x + u_{r+1}^x)/2 + \xi_{(x,t)}, \quad \text{and} \\ u_j^{x,t} &= u_j^x, \quad \text{for } j \neq r, r+1. \end{aligned}$$

Next, let $q_j^{x,t} = (\phi^x)^{-1}(u_j^{x,t})$. Finally, define $H : S^{n+1} \times I \to \mathbb{F}_k(S^{n+1})$ by the rule

$$(x,t) \to (q_1^x, \cdots, q_r^{x,t}, q_{r+1}^{x,t}, \cdots, q_{k+1}^x).$$

It is easy to verify that H is a homotopy from δ''_{k+1} to $\tau_r \circ \delta''_{k+1}$. This finishes the proof. □

If $(n+2) \neq 0 \bmod 4$, then, obviously, $\mathbb{R}^{n+2}$ does not admit the structure of an $\mathbb{H}$-module. Otherwise $\mathbb{R}^{n+2}$ would be a free module over $\mathbb{H}$ and, hence, $(n+2) = 0 \bmod 4$. So the preceding argument does not apply, and we need to proceed in a different manner.

The Stiefel manifold $O_{n+2,2}$ of orthonormal 2-frames in $\mathbb{R}^{n+2}$ plays an important role. Let $\mathrm{p}_{n+2,1} : O_{n+2,2} \to S^{n+1}$ be the projection onto the unit sphere S^{n+1}, where $(x, y) \mapsto x$.

Theorem 4.2 *If* $(n+2) \neq 0 \bmod 4$, *then*

$$\pi_{n+1}(\tau_r)(\delta_{k+1}) = \delta_{k+1} + \gamma_{r+1r},$$

where $\tau_r = (r, r+1)$.

Proof of Theorem 4.2 when $(k+1) = 3$

Observe that the map $\delta_3' : S^{n+1} \to \mathbb{F}_3(S^{n+1})$, $x \mapsto (x, h^x, -x)$, defined above, lies in the image of the map ψ_3, defined in §2 of Chapter II. Hence, there is a section

$$u' : S^{n+1} \to O_{n+2,2} = \{(x,y)|x \perp y\} \subset S^{n+1} \times S^{n+1}$$

of the fibration of the Stiefel manifold, where $u'(x) = (x, h^x)$. Recall that

$$h^x = (-x_2, x_1, \cdots, -x_{2n+2}, x_{2n+1}).$$

There is another section,

$$\rho_3' : S^{n+1} \to O_{n+2,2} \subset \mathbb{F}_3(S^{n+1}),$$

where $x \mapsto (x, -h^x, -x)$. Note that, according to Theorem 1.10 of [64, James], these two sections are not in the same homotopy class when $(n+2) \neq 0 \bmod 4$. So $\delta_3 - \rho_3 \neq 0 \in \pi_{n+1}(O_{n+2,2}) \cong \pi_{n+1}(\mathbb{F}_3(S^{n+1}))$. Now, note that $n > 3$, since $(n+2) \neq 0 \bmod 4$ and $(n+1)$ is odd. Therefore, in this case, we have $\pi_{n+1}(\mathbb{F}_3(\mathbb{R}^{n+1})) \cong \mathbb{Z}/2\mathbb{Z} \oplus \mathbb{Z}$, and, consequently, $\delta_3 - \rho_3 \in \pi_{n+1}(\mathbb{F}_3(S^{n+1}))$ generates the summand $\mathbb{Z}/2\mathbb{Z}$.

Observe that, according to Theorem 2.4, we have $\gamma_{32} = (-1)^n \gamma_{21}$, $\gamma_{31} = \gamma_{21}$. Therefore, $\gamma_{32} = \gamma_{31} = \gamma_{21}$, since the elements belong to a cyclic group of order two. This means that, if $r = 2$ or 1, then γ_{r+1r} generates the torsion subgroup $\mathbb{Z}/2\mathbb{Z}$ of $\pi_{n+1}(\mathbb{F}_3(S^{n+1}))$. Now, one proves directly that $\tau_r \circ \delta_3'$ is homotopic to the other section ρ_3', if $r = 1$ or 2. This proves that

$$\pi_{n+1}(\tau_r)(\delta_3) = \delta_3 + \gamma_{r+1r},\ r = 1, 2,$$

as required.

To prove the general case we use the projections

$$\mathrm{proj}_{t,s,r} : \mathbb{F}_{k+1}(S^{n+1}) \to \mathbb{F}_3(S^{n+1}),\ (x_1, \cdots, x_{k+1}) \mapsto (x_t, x_s, x_r)$$

for suitable t, s, and r such that $1 \leq t < s < r \leq (k+1)$. Consider the induced morphism $\pi_{n+1}(\mathrm{proj}_{t,s,r}) : \pi_{n+1}(\mathbb{F}_{k+1}(S^{n+1})) \to \pi_{n+1}(\mathbb{F}_3(S^{n+1}))$.

Assertion 4.1 *Suppose that $1 \leq v < u \leq k$. Then $\pi_{n+1}(\mathrm{proj}_{t,s,r})$ is such that*

$$\gamma_{uv} \mapsto \begin{cases} \gamma_{21}, \text{ if} & t = v, s = u, r \geq (u+1) \\ \gamma_{31}, \text{ if} & t = v, v < s < u, r = u \\ \gamma_{32}, \text{ if} & t < v, s + v, r = u \\ 0, & \text{otherwise.} \end{cases}$$

Proof. To prove the assertion, note that γ_{uv} is given by a sphere around the point q_v in the uth position. One easily verifies that the projection that selects the tth, s^{th}, and r^{th} positions takes, respectively, the sphere of γ_{uv} to a sphere around the first in the second and third positions in the first and second cases, while in the third case it takes it to a sphere around the second point in the third position. Finally, in other cases one verifies that the sphere of γ_{uv} is projected either to a point or to a sphere that can be contracted to a point. □

Proof of Theorem 4.2 when $(k+1) > 3$

Proof. As $\{\gamma_{k+1s} \mid 1 \leq s \leq k\}$ can be expressed in terms of the others,

$$(*) \qquad \pi_{n+1}(\tau_r)(\delta_{k+1}) - \delta_{k+1} = \sum c_{st}\gamma_{st},$$

where $c_{st} \in \mathbb{Z}/2\mathbb{Z}$ and the sum is over all pairs s, t such that $1 < s \leq k$ and $t < s$. To compute the coefficients c_{st}, choose $\delta'_{k+1}(x)$ as a representative of $\delta_{k+1}(x)$; recall that

$$(**) \qquad \tau_r \circ \delta'_{k+1}(x) = (x, h^x, q_3^x, \cdots, q_{r+1}^x, q_r^x, \cdots, q_k^x, q_{k+1}^x),$$

and use the projections defined by triples $(r\, r+1\, k+1)$ and $(s\, k\, k+1)$. Note that, according to Assertion 4.1 for $r < k$, the induced morphism

$$\pi_{n+1}(\text{proj}_{r,r+1,k+1}) : \pi_{n+1}(\mathbb{F}_{k+1}(S^{n+1})) \to \pi_{n+1}(\mathbb{F}_3(S^{n+1}))$$

takes γ_{st} to γ_{21}, if $(s,t) = (r, r+1)$, and to 0, otherwise, while it takes γ_{tu} to γ_{21}, if $(t,u) = (k,s)$, and to 0, otherwise. One shows that $c_{rr+1} = 1$ and $c_{st} = 0$ when $(st) \neq (r\, r+1)$. Hence, $\pi_{n+1}(\tau_r)(\delta_{k+1}) - \delta_{k+1} = \gamma_{r+1r}$. This proves the general case of Theorem 4.2. □

5 The Y-B Relations, $(n+1)$ Odd

We wish to find the Whitehead products of the generators of the groups $\pi_n(\mathbb{F}_{k+1}(S^{n+1}))$ and $\pi_{n+1}(\mathbb{F}_{k+1}(S^{n+1}))$ described in the previous sections. To achieve this we need a new homotopy-theoretic invariant associated with the fiber homotopy type of $O_{n+2,2}$ as given in [65, James-Whitehead].

Recall that the principal bundle $\mathrm{p}_{n+2,1} : O_{n+2,n+2} = O_{n+2} \to O_{n+2,1} = S^{n+1}$, where an $(n+2)$-frame is projected to its last member, is the principal bundle of sphere tangent bundle of S^{n+1}, $\mathrm{p}_{n+2,1} : O_{n+2,2} \to O_{n+2,1} = S^{n+1}$ The section $u' : S^{n+1} \to O_{n+2,2}$, $x \mapsto (x, h^x)$ leads to a splitting $\tau(S^{n+1}) \cong \xi \oplus o^1$ of the tangent bundle $\tau(S^{n+1})$ of S^{n+1}. Here o^1 is the trivial bundle of dimension 1. By abuse of language, denote the characteristic map ([103, Steenrod]) of the bundle ξ by the same symbol $\xi : S^n \to O_n$. Observe that the natural imbedding, $i : O_n \to O_{n+1}$, takes $[\xi] \in \pi_n(O_n)$ to the class of the characteristic map of $\tau(S^{n+1})$, since $\tau(S^{n+1})$ is the bundle obtained from ξ by extending the group. By definition, let $J(\xi) \in \pi_{2n}(S^n)$ be the image of $[\xi]$ under the J-homomorphism $J : \pi_*(O_n) \to \pi_{*+n}(S^n)$ (see Chapter XI, §4 of [110, Whitehead]). By definition, let $\zeta_{\sigma 2,\sigma 1} = \pi_{2n}(\beta_{\sigma 2 \sigma 1})(J(\xi))$, where $\sigma \in \Sigma_{k+1}$.

Theorem 5.1 *Assume that $n > 2$. Then, for all $\sigma \in \Sigma_{k+1}$, the relations*

$$\begin{aligned}
(i)\ & [\beta_{\sigma 2 \sigma 1}, \beta_{\sigma 3 \sigma 1} + \beta_{\sigma 3 \sigma 2}] &= 0,\\
(ii)\ & [\beta_{\sigma 2 \sigma 1}, \beta_{\sigma 4 \sigma 3}] &= 0,\\
(iii)\ & [\gamma_{\sigma 2 \sigma 1}, \gamma_{\sigma 3 \sigma 1} + \gamma_{\sigma 3 \sigma 2}] &= 0,\\
(iv)\ & [\gamma_{\sigma 2 \sigma 1}, \gamma_{\sigma 4 \sigma 3}] &= 0,\\
(v)\ & [\beta_{\sigma 2 \sigma 1}, \gamma_{\sigma 3 \sigma 1} + \gamma_{\sigma 3 \sigma 2}] &= 0,\\
(vi)\ & [\beta_{\sigma 2 \sigma 1}, \gamma_{\sigma 4 \sigma 3}] &= 0, \text{ and}\\
(vii)\ & [\sigma \circ \delta_{k+1}, \beta_{\sigma 2 \sigma 1}] + \zeta_{\sigma 2 \sigma 1} &= 0
\end{aligned}$$

hold in the Lie algebra $\pi_*(\mathbb{F}_{k+1}(S^{n+1}))$.

Proof. Observe, first of all, that Relations (i) and (ii) are easy consequences of the analogous relations that hold in $\pi_*(\mathbb{F}_k(\mathbb{R}^{n+1}))$ (see Theorem 3.1 of Chapter II).

To prove Relation (iii), it is enough to show that $[\gamma_{21}, \gamma_{31} + \gamma_{32}] = 0$, because the invariance properties of $\{\gamma_{rs}\}$ under Σ_{k+1} would then imply Relation (iii). So consider

$$\begin{aligned}
[\gamma_{21}, \gamma_{31} + \gamma_{32}] &= [\beta_{21} \circ \eta_n, \beta_{31} \circ \eta_n + \beta_{32} \circ \eta_n]\\
&= [\beta_{21} \circ \eta_n, (\beta_{31} + \beta_{32}) \circ \eta_n]\\
&= 0,
\end{aligned}$$

as required (see Chapter X, Theorem(8.14) of [110, Whitehead]).

Then Relations (iv), (v), and (vi) are proved along the lines of the proofs of Relations (i) and (ii) (cf. Theorem 3.1 of Chapter II). To prove Relation (vii), it is enough to show that

$$(\mathrm{I})_{k+1} \quad [\delta_{k+1}, \beta_{21}] + \zeta_{21} = 0.$$

Let us consider the following special case first.

Case of $(k+1) = 3$

Let $S^{n+1} \subset O_{n+2,2}$ be the image of the section δ_3' and $S^n \subset O_{n+2,2}$, the fiber of the projection $O_{n+2,2} \to S^{n+1}$ at $e \in S^{n+1}$, and consider the homotopy equivalence $\psi_3 : O_{n+2,2} \to \mathbb{F}_3(S^{n+1})$, where $(x, y) \mapsto (x, y, -x)$. As δ_3 and β_{21} are, respectively, the homotopy classes of

$$\psi_3 | S^{n+1} : S^{n+1} \to O_{n+2,2},\ \psi_3 | \{e\} \times S^n : S^n \to O_{n+2,2},$$

we see that it is enough to prove that the relation

$$(*)_3 \quad [\delta_3, \iota_n] + \zeta = 0$$

holds in $\pi_*(O_{n+2,2})$. Here, $\zeta = J(\xi)$, and ι_n is the homotopy class of the identity map of S^n. Note that $(*)_3$ follows directly from the following proposition (see §3 of [65, James-Whitehead] and §5 of [64, James]). It gives the cellular structure of $O_{n+2,2}$.

Proposition 5.1 *Assume that $n+1$ is odd, and identify S^{n+1} with its image in $O_{n+2,2}$ under a section u'. Then, as a cell complex,*

$$O_{n+2,2} \simeq (S^{n+1} \vee S^n) \cup_\mu D^{2n+1},$$

where the homotopy class $[\mu]$ of the attaching map is

$$\zeta + [\iota_{n+1}, \iota_n], \quad \zeta = J(\xi),$$

where ι_{n+1} is the homotopy class of the identity map of S^{n+1}.

For the proof we need the following well-known facts. Let $p_\xi : V_\xi \to S^{n+1}$ be the n-dimensional vector bundle associated with ξ. Consider the Thom-space $B^\xi = D(V_\xi)/S(V_\xi)$ of the vector bundle $p_\xi : V_\xi \to B_\xi$. Here,

$$D(V_\xi) \to B_\xi, \quad S(V_\xi) \to B_\xi$$

are, respectively, the associated disk and sphere bundles, and the space $D(V_\xi)/S(V_\xi)$ is that obtained by collapsing $S(V_\xi)$ to a point.

Let $D_\pm(V_{\xi\oplus o^1}) \subset S(V_{\xi\oplus o^1})$ denote, respectively, the upper and lower disk subbundle of the sphere bundle $p_\xi : S(V_{\xi\oplus o^1}) \to B_\xi$. The following is easily verifiable (cf. [64, James], p.35).

Assertion 5.1 *The imbedding of $D(V_\xi) \to S(V_{\xi\oplus o^1})$ as the upper disk bundle $D_+(V_{\xi\oplus o^1})$ induces a homotopy equivalence $(B_\xi)^\xi \to (S(V_{\xi\oplus o^1})/B_\xi)$, where B_ξ is identified with $\{(b,(0,-1) \mid b \in B_\xi\}$.*

Next, let us describe more generally the homotopy type of the Thom spaces of bundles over spheres. Let $p_\xi : S(V_\xi) \to S^{n+1}$ be the sphere bundle associated with the r-dimensional vector bundle, $p_\xi : V_\xi \to S^{n+1}$. Denote by $\xi \in \pi_n(O_r)$ its characteristic map ([103, Steenrod]). Then, ξ defines naturally an action

$$\mu_\xi : S^n \times (D^r, S^{r-1}) \to (D^r, S^{r-1}), \quad (x,y) \mapsto \xi(x)y.$$

Assertion 5.2 *The Thom space $(S^{n+1})^\xi$ is the mapping cone of the map*

$$J(\mu_\xi) : S^{n+r} \to S^r$$

defined by the Hopf construction.

The proof is given in [64, James], p. 37. The associated sphere and disk bundles are, respectively, of the form,

$$\begin{aligned} S(V_\xi) &= D^{n+1} \times S^{r-1} \cup_{\mu_\xi} S^{r-1} \quad \text{and} \\ D(V_\xi) &= \quad D^{n+1} \times D^r \cup_{\mu_\xi} D^r. \end{aligned}$$

Therefore, the bundle pair $(D(V_\xi), S(V_\xi)) \to S^{r+1}$ can be expressed in the form

$$(D(V_\xi), S(V_\xi)) = D^{n+1} \times (D^r, S^{r-1}) \cup_{\mu_\xi} (D^r, S^{r-1}).$$

Hence, the Thom space is the quotient space obtained from the disjoint union

$$(S^{r-1} \times I) \coprod (S^{r-1} \times I) \times D^{n+1},$$

with obvious identifications. It is easy to verify that $(S^{n+1})^\xi$ is the space obtained from the disjoint union, $S^r \coprod D^{n+1} * S^{r-1}$ by identifying

$$S^n * S^{r-1} \cong S^{n+r}$$

with its image under the map $J(\mu_\xi)$. This proves the assertion. □

Proof of (I)$_3$

Consider the attaching map

$$\nu : S^{2n} \to S^{n+1} \vee S^n, \quad [\nu] = [\iota_{n+1}, \iota_n] + \lambda.$$

Observe that $O_{n+2,2}$ is the sphere bundle of the tangent bundle $\tau(S^{n+1})$ of S^{n+1}. The section $u' : S^{n+1} \to O_{n+2,2}$ leads to a splitting $\tau(S^{n+1}) \cong \xi \oplus o^1$. Hence, by Assertion 5.2, $O_{n+2,2}/S^{n+1} = (S^{n+1})^\xi$, where S^{n+1} is identified with its image under u'. Now, note that

$$O_{n+2,2}/S^{n+1} = S^n \cup_{p\circ\nu} D^{2n+1},$$

where $p : S^{n+1} \vee S^n \to S^n$ is the projection on the second factor. But, by Assertion 5.2, the Thom space $(S^{n+1})^\xi = S^n \cup_{J(\xi)} D^{2n+1}$. Hence, $\lambda = J(\xi)$, as required. □

The special case of (I)$_3$ follows easily.

Proof of (I)$_{k+1}$, $(k+1) > 3$

To start, let $(x, y) \in O_{n+2,2}$, and consider the stereographic projection

$$\phi^x : S^{n+1} - \{x\} \to V^x$$

onto the plane perpendicular to the line spanned by x. The plane spanned by the pair $(x, h^x) = u'(x)) \in O_{n+2,2}$ in $\mathbb{R}^{n+2}$ intersects S^{n+1} in a great circle. Put

$$q_1^x = x,\ \phi^x(q_2^{xy}) = y,\ \phi^x(q_i^x) = 4(i-1)\phi^x(h^x),\ i = 3, \cdots, (k+1).$$

Observe here that q^{xy} is the intersection of the plane spanned by the pair (x, y) with V^x. Consider now the map $\psi_{k+1} : O_{n+2,2} \to \mathbb{F}_{k+1}(S^{n+1})$, where $(x, y) \mapsto (q_1^x, q_2^{xy}, q_3^x, \cdots, q_{k+1}^x)$. It is clear that the homotopy class of

$$\psi_{k+1}|S^n : S^n \to \mathbb{F}_{k+1}(S^{n+1}),$$

where $S^n = (\{(e, y) \mid e \perp y\}) \subset O_{n+2,2}$, is just β_{21}. Moreover,

$$\pi_{n+1}(\psi_{k+1})(u) = \delta_{k+1} \in \pi_{n+1}(\mathbb{F}_{k+1}(S^{n+1})$$

and $\zeta \mapsto \zeta_{21}$ under ψ_{k+1}. These facts, along with the naturality of the Whitehead products, enable one to deduce the general case of the theorem from the special case when $k = 3$. □

The Y-B relations for the case $n = 2$ are somewhat different, because the composition product by η_2 is not additive (cf. Theorem 2.5 above).

Theorem 5.2 *Assume that $n = 2$. Then, for all $\sigma_{k+1} \in \Sigma_{k+1}$, the relations*

$$\begin{aligned}
(i)\ & [\beta_{\sigma 2\sigma 1}, \beta_{\sigma 3\sigma 1} + \beta_{\sigma 3\sigma 2}] &= 0,\\
(ii)\ & [\beta_{\sigma 2\sigma 1}, \beta_{\sigma 4\sigma 3}] &= 0,\\
(iii)\ & [\gamma_{\sigma 2\sigma 1}, \gamma_{\sigma 3\sigma 1} + \gamma_{\sigma 3\sigma 2}] + [\gamma_{\sigma 2\sigma 1}, [\beta_{\sigma 3\sigma 1}, \beta_{\sigma 3,\sigma 2}]] &= 0,\\
(iv)\ & [\gamma_{\sigma 2\sigma 1}, \gamma_{\sigma 4\sigma 3}] &= 0,\\
(v)\ & [\beta_{\sigma 2\sigma 1}, \gamma_{\sigma 3\sigma 1} + \gamma_{\sigma 3\sigma 2}] + [\beta_{\sigma 2\sigma 1}, [\beta_{\sigma 3\sigma 1}, \beta_{\sigma 3\sigma 2}]] &= 0,\\
(vi)\ & [\beta_{\sigma 2\sigma 1}, \gamma_{\sigma 4\sigma 3}] &= 0, \quad \textit{and}\\
(vii)\ & [\delta_{k+1}, \beta_{\sigma 2\sigma 1}] &= 0
\end{aligned}$$

hold in the Lie algebra $\pi_*(\mathbb{F}_{k+1}(S^3))$.

Proof. One proves Statements (i), (ii), (iv), (v), and (vi) in the same manner as Statements (i) and (iv) of Theorem 5.1. To prove Statement (iii,) one applies Assertion 2.1 above to Statement (i). To prove Statement (vii), one proceeds as in the proof of Theorem 4.5: note that $O_{n+2,2} \cong S^3 \times S^2$, as S^3 is parallelizable. □

6 The Dirac Class Δ_{k+1}

Consider the map

$$\psi_{k+1} : O_{n+2,2} \to \mathbb{F}_{k+1}(S^{n+1}),\ k+1 \geq 3,\ \psi_{k+1}(x, y) = (x, y, q_3^x, \cdots, q_k^x, -x),$$

defined in §1 above. By definition, let $\Delta'_{k+1}(S^{n+1}) = \psi_{k+1}|S^n$, where S^n is identified with the set $\{y|(e, y) \in O_{n+2,2}$. The *Dirac class* $\Delta_{k+1}(S^{n+1})$ is the homotopy class $[\Delta'_{k+1}(S^{n+1})] \in \pi_n(\mathbb{F}_{k+1}(S^{n+1}))$.

There is an analogous class in $\pi_n(\mathbb{F}_{k+1}(\mathbb{R}^{n+1}))$. To see this, note that the image of $\Delta'_{k+1}(S^{n+1})$ consists of $k+1$ spheres of dimension n, with center on the e-axis and with the first and last being degenerate corresponding to the points e, and $-e$, respectively. Define the Dirac $\Delta(\mathbb{R}^{n+1}) \in \pi_n(\mathbb{F}_k(\mathbb{R}^{n+1}))$ to be the homotopy class of the map

$$\phi^e \circ \mathrm{proj}_{k+1,k} \circ \Delta'_{k+1}(S^{n+1}) : S^n \to \mathbb{F}_k(\mathbb{R}^{n+1}),$$

where $\mathrm{proj}_{k+1,k} : \mathbb{F}_{k+1}(S^{n+1}) \to \mathbb{F}_k(S^{n+1})$ sends a $k+1$ tuple to its first k entries. Here we identify the space $V^e \subset \mathbb{R}^{n+2}$ orthogonal to e with $\mathbb{R}^{n+1}$ by means of the stereographic map ϕ^e. Note that the map $\phi^e \circ \mathrm{proj}_{k+1,k} \circ \Delta'_{k+1}(S^{n+1})$ omits the degenerate point $-e$ and keeps k concentric n-spheres in $\mathbb{R}^{n+1}$ centered at the origin, with the first being the degenerate sphere at the origin.

The following is a simple consequence of this description of $\Delta_k(\mathbb{R}^{n+1})$.

Proposition 6.1 *The relation* $\Delta_k(\mathbb{R}^{n+1}) = \sum_{r=2}^{k} \omega_r$, *where* $\omega_r = \sum_{s=1}^{r-1} \alpha_{rs}$, *holds in* $\pi_n(\mathbb{F}_{k+1}(\mathbb{R}^{n+1}))$.

Consider next the map

$$i_k : \mathbb{F}_k(\mathbb{R}^{n+1}) \to \mathbb{F}_k(S^{n+1}),$$

where $(x_1, \cdots, x_k) \mapsto ((\phi^e)^{-1}(x_1), \cdots, (\phi^e)^{-1}(x_k), -e)$. The following is an immediate consequence of Proposition 6.1.

Corollary 6.1 *The morphism* $\pi_n(i_k) : \pi_n(\mathbb{F}_k(\mathbb{R}^{n+1})) \to \pi_n(\mathbb{F}_k(S^{n+1}))$ *takes* $\Delta_k(\mathbb{R}^{n+1})$ *to* $\Delta_{k+1}(S^{n+1})$. *In particular,* $\Delta_{k+1}(S^{n+1}) = \sum_{r=2}^{k} \sum_{s=1}^{r-1} \beta_{rs}$.

The class $\Delta_k(\mathbb{R}^{n+1})$ is the higher dimensional analogue of the Dirac braid when $n = 1$ ([29, Fadell], and see IV.7 below). The order of the Dirac class $\Delta_{k+1}(S^{n+1})$ depends on the parity of $n+1$.

Proposition 6.2 *Assume that* $k+1 \geq 3$ *and* $n+1$ *odd. Then the Dirac class* $\Delta_{k+1}(S^{n+1})$ *has infinite order.*

Proof. Recall that the morphism induced by i_k takes α_{rs} to β_{rs}, for $1 \leq s < r \leq k$, by Theorem 2.2 above. To finish the proof, note that the fibration $\mathrm{proj}_{k+1,1} : \mathbb{F}_{k+1}(S^{n+1}) \to S^{n+1}$ admits a section, and, hence, the homotopy exact sequence of the fibration yields the short exact sequence

$$0 \to \pi_n(\mathbb{F}_k(\mathbb{R}^{n+1})) \to \pi_n(\mathbb{F}_k(S^{n+1})) \to \pi_n(S^{n+1}) \to 0.$$

Now, observe that the image of $\Delta_k(\mathbb{R}^{n+1})$ is nontrivial because $\pi_n(i_k)$ is injective, according to Theorem 3.1 above. As $\pi_n(\mathbb{F}_k(\mathbb{R}^{n+1}))$ is a finitely generated free abelian group, we see that $\Delta_k(\mathbb{R}^{n+1})$ has infinite order. Hence, $\Delta_k(S^{n+1})$ also has infinite order. □

To determine the order of $\Delta_{k+1}(S^{n+1}))$ when $(n+1)$ is even, we need the following well-known description of the space $O_{n+2,2}$. Consider $O_{n+2,2}$ as the homogeneous space O_{n+2}/O_n. Denote by e_1 the standard first basic vector $(1, 0, \cdots, 0)$ in $\mathbb{R}^{n+2}$. Define $\mu_x : S^n \to O_{n+1}$ to be the map $x \mapsto \tau_x \tau_{e_1} \in O_{n+1}$, where τ_{e_1}, τ_x are the reflections in the hyperplanes perpendicular to the lines spanned by e_1 and x, respectively ([56, Hopf], [104, Steenrod-Whitehead]). Then, $\partial(\iota_{n+1}) = [\mu_x]$, where ∂ is the boundary homomorphism of the homotopy exact sequence of the fibration $p_{21} : O_{n+2} \to S^{n+1}$, and $\iota_{n+1} \in \pi_{n+1}(S^{n+1})$ is the homotopy class of the identity map. The map μ_e determines the map $\mu : S^n \times S^n \to S^n$, where $\mu(x, y) = \mu_x{\cdot}y$. It is easy to check that μ has bidegree $(2, 1)$.

Proposition 6.3 *There is a homotopy equivalence*

$$S^n \cup_\mu D^{n+1} \times S^n \to O_{n+2,2}.$$

Proof. Let $\chi_{t,x} \in O_{n+2}$ be the rotation in the plane spanned by x, e that takes x to e. Then the map

$$H : S^n \times I \to O_{n+1},\ (x, t) \mapsto \chi_{x,t} \tau_x \chi_{x,t}^{-1} \tau_{e_1}$$

is a homotopy in O_{n+2}, and $\chi_{x,t} \tau_x \chi_{x,t}^{-1}$ is the reflection in the plane in $\mathbb{R}^{n+2}$ orthogonal to $\chi_{x,t}(e) \in S^{n+1}$. The homotopy H induces, in turn, an extension $\mu'_x : D^{n+1} \to O_{n+2}$ of μ_x. Next, one verifies that the composite

$$D^{n+1} \overset{\mu'_x}{\to} O_{n+2} \overset{p_{n+2}}{\to} S^{n+1}$$

takes the boundary $S^n = \partial(D^{n+1})$ to a point. Thus, one obtains, over D^{n+1}, a trivialization of the pull-back of $p_{n+2,2} : O_{n+2,2} \to S^{n+1}$ by the natural projection $(D^{n+1}, S^n) \to (S^{n+1}, *)$. The rest of the proof follows easily now. □

Summing up, we have the following result.

Theorem 6.1 *Let $i_k : \mathbb{F}_k(\mathbb{R}^{n+1}) \to \mathbb{F}_{k+1}(S^{n+1})$ be the natural imbedding of $\mathbb{F}_k(\mathbb{R}^{n+1})$ as the fiber at $-e$. Suppose that $n+1$ is even. Then,*

(*i*) *The set $\{\beta_{rs} \mid 1 \le s < r \le k)\}$ generates the group $\pi_n(\mathbb{F}_{k+1}(S^{n+1}))$;*
(*ii*) $\ker(\pi_n(\mathbb{F}_k(\mathbb{R}^{n+1})) \overset{\pi_n(i_k)}{\to} \pi_n(\mathbb{F}_{k+1}(S^{n+1})))$ *is infinite cyclic; and*
(*iii*) $\pi_n(i_k)(\Delta_k(\mathbb{R}^{n+1})) = \Delta_{k+1}(S^{n+1})$*; and $\Delta_{k+1}(S^{n+1})$ has order two in $\pi_n(\mathbb{F}_{k+1}(S^{n+1}))$.*

Proof. Put $\bar{\psi}_{k+1} = \psi_{k+1}|S^n$. Observe that $\bar{\psi}_{k+1}$ defines $\Delta_k(\mathbb{R}^{n+1})$. Since $\psi_{k+1} : O_{n+2,2} \to \mathbb{F}_{k+1}(S^{n+1})$ is a map of fibrations, we obtain the commutative diagram

$$\begin{array}{ccccc}
\pi_{n+1}(S^{n+1}) \overset{\partial}{\to} & \pi_n(S^n) & \to & \pi_n(O_{n+2,2}) & \to 0 \\
\downarrow = & \downarrow \bar{\psi}_k & & \downarrow \psi_{k+1} & \\
\pi_{n+1}(S^{n+1}) \overset{\partial}{\to} & \pi_n(\mathbb{F}_k(\mathbb{R}^{n+1})) & \to & \pi_n(\mathbb{F}_{k+1}(S^{n+1})) & \to 0
\end{array}$$

with exact rows. Here, the upper and lower horizontal sequences are the tail of the homotopy sequences of, respectively, the projections $O_{n+2,2} \to S^{n+1}$ and $\mathbb{F}_{k+1}(S^{n+1}) \to S^{n+1}$. Each of these projections sends a tuple to its first entry (see Propositions 1.1 an 1.2).

According to Proposition 1.1 of Chapter II, $\{\alpha_{rs} | 1 \le s < r \le k\}$ generates $\pi_n(\mathbb{F}_k(\mathbb{R}^{n+1}))$. Moreover, Theorem 2.2 above asserts that $\alpha_{rs} \mapsto \beta_{rs}$. Thus, we see that the set $\{\beta_{rs} | 1 \le s < r \le k\}$ generates $\pi_n(\mathbb{F}_k(S^{n+1}))$. This proves statement (i).

Next, note that (ii) is an immediate consequence of the fact that $\pi_n(\mathbb{F}_k(\mathbb{R}^{n+1}))$ is a free abelian group (cf. Proposition 6.2).

The first assertion of (iii) is obvious (see Corollary 6.1). To finish the proof of (iii), note first that the third morphisms of both sequences are surjective, since $\pi_n(S^{n+1})$ is trivial. Also, as we pointed at the beginning of the proof, $\pi_n(\bar{\psi}_3)(\iota_n) = \Delta_2(\mathbb{R}^{n+1})$. Observe also that the Dirac class $\Delta_2(\mathbb{R}^{n+1})$ goes to the Dirac class $\Delta_k(\mathbb{R}^{n+1})$. Hence, $\Delta_3(S^{n+1})$ goes to $\Delta_{k+1}(S^{n+1})$. Now, as a consequence of Proposition 6.3, we conclude that the order of $\Delta_3(S^{n+1})$ in $\pi_n(O_{n+2,2}) \cong \pi_n(\mathbb{F}_3(S^{n+1}))$ is exactly two. Hence the order of $\Delta_{k+1}(S^{n+1})$ is also two. This proves (iii) and the theorem. □

The preceding result has some bearing on $\pi_{n+1}(\mathbb{F}_{k+1}(S^{n+1}))$.

Theorem 6.2 *Suppose that $(n+1)$ is even. Then $\pi_{n+1}(\mathbb{F}_{k+1}(S^{n+1}))$ is generated by $\{\gamma_{rs} \mid 1 \le s < r \le k\}$. In particular, $\pi_{n+1}(\mathbb{F}_{k+1}(S^{n+1}))$ is a torsion group where every element is of order 2.*

Proof. Consider the following section of the homotopy exact sequence

$$\begin{array}{rll}
\pi_{n+2}(S^{n+1}) & \overset{\partial}{\to} & \pi_{n+1}(\mathbb{F}_k(\mathbb{R}^{n+1})) \\
\to \pi_{n+1}(\mathbb{F}_{k+1}(S^{n+1})) \to & & \pi_{n+1}(S^{n+1}) \\
\overset{\partial}{\to} \quad \pi_n((\mathbb{F}_k(\mathbb{R}^{n+1})) & & \to \pi_n(\mathbb{F}_{k+1}(S^{n+1})) \to 0
\end{array}$$

of the fibration $\mathrm{proj}_{k+1,1} : \mathbb{F}_{k+1}(S^{n+1}) \to S^{n+1}$, where a tuple is projected to its last entry. As we have seen in the proof of the previous theorem, $\pi_{n+1}(S^{n+1}) \overset{\partial}{\to} \pi_n(\mathbb{F}_k(\mathbb{R}^{n+1}))$ takes the generator ι_{n+1} of $\pi_{n+1}(S^{n+1})$ to twice the Dirac class $\Delta_k(\mathbb{R}^{n+1})$ and is, therefore, injective. Hence, the morphism

$$\pi_{n+1}(i_k) : \pi_{n+1}(\mathbb{F}_k(\mathbb{R}^{n+1})) \to \pi_{n+1}(\mathbb{F}_{k+1}(S^{n+1}))$$

is surjective.

To continue the proof, recall that the set $\{\tilde{\gamma}_{rs} | 1 \leq s < r \leq k\}$ generates the group $\pi_{n+1}(\mathbb{F}_k(\mathbb{R}^{n+1})$ (see Proposition 4.1 of Chapter II). But $\bar{\gamma}_{rs} \mapsto \gamma_{rs}$. Hence, the set $\{\gamma_{rs} | 1 \leq s < r \leq k\}$ generates $\pi_{n+1}(\mathbb{F}_{k+1}(S^{n+1}))$, and this group is finite with every element being of order 2. □

7 The Lie Algebra $\pi_*(\mathbb{F}_r(S^{n+1}))$, $n > 1$

To start, consider the fundamental fiber sequence

$$\mathcal{F}_{k+1}(S^{n+1}) : \left\{ \begin{array}{cccc} \mathbb{F}_{k+1}(S^{n+1}) \leftarrow & \mathbb{F}_{k,0} & \leftarrow \cdots & \mathbb{F}_{k-r,r} \leftarrow \\ \downarrow & \downarrow & \downarrow & \downarrow \\ S^{n+1} & \mathbb{R}^{n+1} & \cdots & \mathbb{R}_r^{n+1} \end{array} \right.$$

introduced in §1 above. Identify the spaces $\mathbb{F}_{r,k-r}(\mathbb{R}^{n+1})$ with their images in $\mathbb{F}_{k+1}(S^{n+1})$. Then, the filtration

$$\mathbb{F}_{1,k-1} \subset \cdots \mathbb{F}_{k-r,r} \subset \cdots \subset \mathbb{F}_{k-2,2} \subset \mathbb{F}_{k-1,1} \subset \mathbb{F}_k(\mathbb{R}^{n+1}) \subset \mathbb{F}_{k+1}(S^{n+1})$$

of §1 leads naturally to a filtration of the Lie algebra $\pi_*(\mathbb{F}_{k+1}(S^{n+1}))$, as follows. First, put

$$\begin{aligned} L_{k+1} &= \pi_{k+1}(\mathbb{F}_{k+1}(S^{n+1})), \\ L_k &= \pi_*(\mathbb{F}_{k,0}), \\ L_{k-r} &= \pi_*(\mathbb{F}_{k-r,r}), \text{ and} \\ L_{rr-1} &= \pi_*(\mathbb{R}_r^{n+1}), \end{aligned}$$

with $\mathbb{R}_r^{n+1} = \mathbb{R}^{n+1} - Q_r$. Each L_{rr-1} is the graded Lie algebra of a bouquet of r spheres of dimension n. The space $\mathbb{F}_{k-2,2}$ can be identified with the fiber of the projection $\mathrm{proj}_{k+1,3} : \mathbb{F}_{k+1}(S^{n+1}) \to \mathbb{F}_3(S^{n+1})$ that sends a tuple to its last three entries. Since this fibration admits a section, the morphism $\pi_*(i_{k-2}) : L_{k-2} \to L_{k+1}$ is injective. Moreover, $\mathbb{F}_{k-r,r}$ can be identified with the fiber of $\mathrm{proj}_{k,k-r} : \mathbb{F}_k(\mathbb{R}^{n+1}) \to \mathbb{F}_r(\mathbb{R}^{n+1})$, and this fibration has a section. Hence, the induced morphism $\pi_*(i_{k-r}) : L_{k-r} \to \pi_*(\mathbb{F}_{k+1}(S^{n+1}))$ is injective for all $2 \leq r \leq (k-2)$. For simplicity, let us identify L_{k-r} with its image.

Theorem 7.1 *There is an ascending chain of ideals*

$$L_1 \subset \cdots \subset L_{k-r} \subset \cdots \subset L_{k-2} \subset L_{k+1}$$

such that

(*i*) L_{rr-1} *is isomorphic to the Lie algebra of a bouquet of* r *spheres of dim* n;
(*ii*) $L_{k-r} \cong L_{k-r-1} \rtimes L_{r+1r}$, *for* $r \leq (k-2)$; *and*
(*iii*) $L_k \cong L_{k-2} \rtimes M_{n+2,2}$,

where $M_{n+2,2} = \pi_*(O_{n+2,2})$.

Proof. To see that L_{k-r}, for $r \geq 2$, is an ideal in L_{k+1}, just observe that

$$\mathbb{F}_{k-r,r}(S^{n+1}) \cong \mathbb{F}_{k-r}(S^{n+1} - Q^e_{r+1}) \cong \mathbb{F}_{k-r}(\mathbb{R}^{n+1} - Q_r)$$

can be identified with the fiber of $\text{proj}_{k+1,r+1} : \mathbb{F}_{k+1}(S^{n+1}) \to \mathbb{F}_{r+1}(S^{n+1})$ that takes a tuple to its first r entries. The first assertion follows from the fact that $\mathbb{R}^{n+1} - Q_r$ is homotopy equivalent to a bouquet of r spheres of dimension n. The other assertions follow easily from the fact the appropriate fibrations have sections. Moreover, the various split extensions are determined by the Y-B relations. □

The previous result uses the filtration up to $\mathbb{F}_{k-2,2}$. How the remainder

$$\mathbb{F}_{k-2,2}(\mathbb{R}^{n+1}) \subset \mathbb{F}_{k-1,1}(\mathbb{R}^{n+1}) \subset \mathbb{F}_{k}(\mathbb{R}^{n+1}) \subset \mathbb{F}_{k+1}(S^{n+1})$$

of the filtration behaves depends on the parity of $(n+1)$.

Theorem 7.2 *Suppose that $(n+1)$ is odd. Then the ascending chain of ideals given above can be extended by the chain $L_{k-2} \subset L_{k-1} = L_k \subset L_{k+1}$ such that*

(i) $L_{k-1} \cong L_{k-2} \rtimes L(n)$, and $L(n)$ is the Lie algebra of a single sphere of dimension n; and
(ii) $L_{k+1} \cong L_k \rtimes L(n+1)$, and $L(n+1)$ is the Lie algebra of a single sphere of dimension $(n+1)$.

Proof. The proof is similar to that of the previous theorem. The fact that L_k is an ideal in L_{k+1} is a consequence of the fact that $\mathbb{F}_k(\mathbb{R}^{n+1})$ is the fiber of the projection $\text{proj}_{k+1,1} : \mathbb{F}_{k+1}(S^{n+1}) \to S^{n+1}$, which sends a tuple to its last entry: the fiber has a section, since $(n+1)$ is odd. Furthermore, $L_{k+1,k}$ contains δ_{k+1}, and the Y-B relations of Theorems 5.1 and 5.2 describe its Lie product with the elements $\{\beta_{rs}, \gamma_{rs}\}$.

Similarly, L_{k-2} is an ideal in $L_{k-1} = L_k$ because, by definition, $L_{k-2} = \pi_n(\mathbb{F}_{k-2,2})$, and $\mathbb{F}_{k-2,2}$ is the fiber of the map $\text{p}_1 : \mathbb{F}_{k-1,1} \to \mathbb{R}^{n+1}_1$ that sends a tuple to its last nonconstant entry, x_k. The algebra $L(n)$ contains β_{21}, and Theorems 6.1 and 6.2 describe its Lie product with the elements $\{\beta_{rs}, \gamma_{rs}\}$. □

The case when $(n+1)$ is even is different, as would be expected. Ignoring torsion, the elements $\{\beta_{rs} \mid 1 \leq s < r \leq t\}$ generate the algebra L_t, for all $1 \leq t \leq (k-2)$, and the split extension $L_r \cong L_{r-1} \rtimes L_{r,r-1}$ is described by the Y-B relations among the generators. The latter are induced by the Y-B relations among the elements $\{\alpha_{rs}\}$.

The Stiefel manifold $O_{n+2,2}$, when $(n+2)$ is even, is obtained by adjoining a $(2n+1)$-cell to the Moore space $X_2 = S^n \cup_2 D^{n+1}$. So,

$$O_{n+2,2} \cong X_2 \cup_\nu D^{2n+1}.$$

As the homotopy groups of X_2 are all 2-power finite groups, it follows that the homotopy class $[\nu] \in \pi_{2n}(S^n)$ is annihilated by some power 2^t. Therefore, the fundamental class $[O_{n+2,2}]$ becomes spherical on multiplication by some 2^t. Define Ψ_{2n+1} to be the class in $\pi_{2n+1}(O_{n+2,2})$ such that $\Psi_{2n+1} = 2^t[O_{n+2,2}]$. Note that the subalgebra

$$\pi_*(\mathbb{F}_{n-2,2}(\mathbb{R}^{n+1})) \otimes \mathbb{Q} \subset \pi_*(\mathbb{F}_{k+1}(S^{n+1})) \otimes \mathbb{Q}$$

is generated by $\{\beta_{rs} \mid 1 < r \leq (k-2),\ 1 \leq s < r \leq (k-2)\}$ (see §4 of Chapter II).

Theorem 7.3 *Suppose that $(n+1)$ is even. Then,*

$$\pi_*(\mathbb{F}_{k+1}(S^{n+2})) \otimes \mathbb{Q} \cong \pi_*(O_{n+2,2}) \otimes \mathbb{Q} \otimes \pi_*(\mathbb{F}_{k-2,2}(\mathbb{R}^{n+1}) \otimes \mathbb{Q}$$

as graded Lie algebras.

Proof. The map $\Psi_{2n+1} : S^{2n+1} \to O_{n+2,2}$ induces an isomorphism

$$\pi_*(S^{2n+1}) \otimes \mathbb{Q} \to \pi_*(O_{n+2,2}) \otimes \mathbb{Q}$$

of Lie algebras. Also, $\pi_*(S^{2n+1}) \otimes \mathbb{Q}$ is the exterior algebra generated by ψ_{2n+1}, since the groups $\pi_m(S^{2n+1})$ are finite for $m \neq (2n+1)$. To prove the theorem, it suffices to show that $[\Psi_{2n+1}, \beta_{rs}] = 0$ for all admissible r, s. Now, the algebra $\pi_*(\mathbb{F}_{k-2,2}(\mathbb{R}^{n+1}))$ is only nontrivial in the dimensions where the basic Whitehead products occur. The p-fold products are of dimension $p(n-1)+1$. Hence, $2n+1 = p(n-1)+1$ implies that $n = 3$ and $p = 3$; and, if $n \neq 3$, then $[\Psi_{2n+1}, \beta_{rs}] = 0$, as desired.

As for the case when $n = 3$, we first recall that, according to §3 of Chapter I, the pull-back of $\mathrm{proj}_{k+1,3} : \mathbb{F}_{k+1}(S^{n+1}) \to \mathbb{F}_3(S^{n+1})$ by the map $\psi_3 : O_{5.2} \to \mathbb{F}_3(S^{n+1}$ is the bundle $O_5 \times_{O_3} \mathbb{F}_{k-2,2}$ obtained from $O_5 \to O_{5,2}$. So, consider the natural projection

$$O_5 \times \mathbb{F}_{k-2,2} \to O_5 \times_{O_3} \mathbb{F}_{k-2,2}.$$

It is a fibration, with O_3 as a fiber. As $\pi_*(O_3) \otimes \mathbb{Q}$ is trivial except in dimension 3, we conclude that the induced morphism

$$\pi_7(O_5) \otimes \mathbb{Q} \oplus \pi_7(\mathbb{F}_{k-2,2}) \otimes \mathbb{Q} \to \pi_7(O_{5,2}) \otimes \mathbb{Q} \oplus \pi_7(\mathbb{F}_{k-2,2}) \otimes \mathbb{Q}$$

is an isomorphism. On considering the homotopy exact sequence of the fiber bundle $O_5 \to O_{5,2}$ with O_3 as fiber, it is easy to see that

$$\pi_*(O_5) \otimes \mathbb{Q} \cong \pi_*(O_3) \otimes \mathbb{Q} \oplus \pi_*(O_{5,2}) \otimes \mathbb{Q},$$

and that there is a class Ψ_7' in $\pi_7(O_5) \otimes \mathbb{Q}$ that maps to Ψ_7 in $\pi_7(O_{5,2}) \otimes \mathbb{Q}$. Now the $[\Psi_7', \beta_{rs}] = 0$ in $\pi_*(O_5 \times \mathbb{F}_{k-2,2}) \otimes \mathbb{Q}$. This implies that $[\Psi_7, \beta_{rs}] = 0$ in $\pi_*(\mathbb{F}_{k+1}(S^4)) \otimes \mathbb{Q}$, as required. □

8 Are The Canonical Fibrations Trivial?

In this section we shall consider the fiber homotopy triviality, or lack of it, of the canonical projections $\mathrm{proj}_{k+1,r} : \mathbb{F}_{k+1}(S^{n+1}) \to \mathbb{F}_r(S^{n+1})$. The situation here is somewhat different from that of $\mathbb{R}^{n+1}$ because the tangent bundle of S^{n+1} is nontrivial, except when $(n+1) = 3$ or 7. This fact and its implications in the Y-B relations add a new ingredient to the problem.

There are two cases depending on the parity of $n+1$. We shall consider them separately.

Theorem 8.1 *Suppose that $(n+1)$ is odd. Then, the fibration*

$$\mathrm{proj}_{k+1,r} : \mathbb{F}_{k+1}(S^{n+1}) \to \mathbb{F}_r(S^{n+1}), \quad r \geq 1$$

induced by projection on the first r coordinates is fiber homotopically trivial if, and only if, $r \leq 2$ and $(n+1)$ is 3 or 7.

In preparation for the proof, consider first the commutative diagram

$$\begin{array}{ccc} \mathbb{F}_{k+1}(S^{n+1}) & \stackrel{\mathrm{proj}_{k+1,2}}{\longrightarrow} & \mathbb{F}_2(S^{n+1}) \\ \mathrm{id} \downarrow & & \mathrm{proj}_{3,2} \downarrow \\ \mathbb{F}_{k+1}(S^{n+1}) & \stackrel{\mathrm{proj}_{k+1,1}}{\longrightarrow} & S^{n+1}. \end{array}$$

Note that the fiber of first horizontal projection is $\mathbb{F}_{k-1,1}(\mathbb{R}^{n+1})$, while that of the second is $\mathbb{F}_k(\mathbb{R}^{n+1})$. But the maps $\mathbb{F}_{k-1,1}(\mathbb{R}^{n+1}) \to \mathbb{F}_k(\mathbb{R}^{n+1})$ and $\mathbb{F}_2(S^{n+1}) \to S^{n+1}$ are homotopy equivalences. Hence, the first fibration, $\mathrm{proj}_{k+1,2} : \mathbb{F}_{k+1}(S^{n+1}) \to \mathbb{F}_2(S^{n+1})$, is fiber homotopically equivalent to the pull-back of $\mathrm{proj}_{k+1,1} : \mathbb{F}_{k+1}(S^{n+1}) \to S^{n+1}$ by the map $\mathrm{proj}_{2,1} : \mathbb{F}_2(S^{n+1}) \to S^{n+1}$. Thus it is enough to consider the case when $r \geq 2$ in order to establish the theorem.

Proof of ($\Leftarrow$), Step 1

Assume that $r \leq 2$ and $(n+1) = 3$ or 7. Consider commutative diagram

$$\begin{array}{ccc} O_{n+2,2} & \stackrel{\psi_3}{\to} & \mathbb{F}_3(S^{n+1}) \\ \downarrow & & \downarrow \\ S^{n+1} & \stackrel{\psi_2}{\to} & \mathbb{F}_2(S^{n+1}), \end{array}$$

where ψ_2 takes x to $(x, -x)$ and ψ_3 takes (x, y) to $(x, y, -x)$. According to Corollary 3.2 of Chapter I, the pull-back of the fibration $\mathrm{proj}_{k+1,2} : \mathbb{F}_{k+1}(S^{n+1}) \to \mathbb{F}_2(S^{n+1})$ by ψ_2 is associated with the principal tangent bundle $O_{n+2} \to S^{n+1}$ of S^{n+1}. But $(n+1) = 3$, or 7. Hence, the latter is trivial. Finally, observe that ψ_2 is a homotopy equivalence. Therefore, the fibration $\mathrm{proj}_{k+1,2} : \mathbb{F}_{k+1}(S^{n+1}) \to \mathbb{F}_2(S^{n+1})$ is itself trivial. □

Next, let us consider the converse. Assume that

$$(\mathrm{I})_{k+1,r} \quad \mathrm{proj}_{k+1,r} : \mathbb{F}_{k+1}(S^{n+1}) \to \mathbb{F}_r(S^{n+1})$$

is fiber homotopically trivial. We shall proceed in two steps.

Proof of (⇒), Step 1

First we wish to show that $r \leq 2$. So suppose, by way of contradiction, that $r > 2$. Identify the fiber of the projection $\mathrm{proj}_{r,1} : \mathbb{F}_r(S^{n+1}) \to S^{n+1}$ at the point $-e$, with $\mathbb{F}_{r-1}(\mathbb{R}^{n+1})$, as in §1 above. Observe that the preimage of $\mathbb{F}_{r-1}(\mathbb{R}^{n+1})$ under $\mathrm{proj}_{k,r-1}(\mathbb{R}^{n+1})$ is $\mathbb{F}_k(\mathbb{R}^{n+1})$. Note that the restriction

$$\mathrm{proj}_{k,r-1}|\mathbb{F}_k(\mathbb{R}^{n+1}) : \mathbb{F}_k(\mathbb{R}^{n+1}) \to \mathbb{F}_{r-1}(\mathbb{R}^{n+1})$$

of the fibration $(\mathrm{I})_{k+1,r}$ to $\mathbb{F}_{r-1}(\mathbb{R}^{n+1})$ is also trivial. Therefore, according to Theorem 1.6 of Chapter II, $(r-1) \leq 2$ and $n = 3$ or 7. In particular, this implies that $(n+1) = 4$ or 8. But this is impossible, since $(n+1)$ is odd by assumption. Hence, $r \leq 2$, as required. □

Proof of (⇒), Step 2

Next, we wish to show that if the fibration

$$(\mathrm{I})_{k+1,2} \quad \mathrm{proj}_{k+1,2} : \mathbb{F}_{k+1}(S^{n+1}) \to \mathbb{F}_2(S^{n+1})$$

is fiber homotopically trivial, then $(n+1) = 3$ or 7. It is enough to prove that if $(n+1) \neq 3$, then $(n+1) = 7$. Proceeding along the lines of the proof of Theorem 1.6 of Chapter II, let

$$\begin{array}{ccc} S^{n+1} \times \mathbb{F}_{k-,1}(\mathbb{R}^{n+1}) & \overset{\phi}{\to} & \mathbb{F}_{k+1}(S^{n+1}) \\ \downarrow & & \downarrow \\ S^{n+1} & \overset{\simeq}{\to} & S^{n+1} \end{array}$$

be a trivialization of $\mathrm{proj}_{k+1,2}; \mathbb{F}_{k+1,2}(S^{n+1}) \to \mathbb{F}_2(S^{n+1})$. Then,

$$\pi_{n+1}(\phi)(\iota_{n+1}) = \delta_{k+1} + c_{21}\gamma_{21} + \sum_{s=3}^{k} \omega_s,$$

where $\omega_s = \sum_{t=2}^{s-1} c_{st}\gamma_{st}$ and $c_{st} \in \mathbb{Z}/2\mathbb{Z}$, since $(n+1) \neq 3$. Assuming, without loss of generality, that $\phi|\mathbb{F}_{k-1,1}(\mathbb{R}^{n+1}) : \mathbb{F}_{k-1,1}(\mathbb{R}^{n+1}) \to \mathbb{F}_{k-1,1}(\mathbb{R}^{n+1})$ is homotopic to the identity, we see that

$$\pi_{n+1}(\phi)(\gamma_{rs}) = \gamma_{rs}, \; \pi_n(\phi)(\alpha_{rs}) = \beta_{rs}, \; 1 \leq s < r \leq k.$$

Note that in $S^{n+1} \times \mathbb{F}_{k-1,1}(\mathbb{R}^{n+1})$ the Whitehead products $[\iota_{n+1}, \lambda] = 0$, for all λ in $\pi_*(\mathbb{F}_{k-1,1}(\mathbb{R}^{n+1}))$. Hence, it follows that

$$[\delta_{k+1} + c_{21}\gamma_{21} + \sum_{s=3}^{k} \omega_s, \beta_{21}] = 0$$

in $\mathbb{F}_{k+1}(S^{n+1})$. Observe that

$$\pi_*(\mathbb{F}_{k-1,1})(\mathbb{R}^{n+1})) \cong \bigoplus_{r=2}^{k} \pi_*(\vee_{s=1}^{r-1} S_{rs})$$
$$\cong \pi_*(S_{21}) \oplus \pi_*(S_{31} \vee S_{32}) \oplus \cdots \quad .$$

Now, simplify the preceding formula according to the Y-B relations (see Theorem 5.1 above), and note that the term in $\pi_*(S_{21})$ is

$$\text{(II)} \quad c_{21}[\gamma_{21}, \beta_{21}] + [\delta_{k+1}, \beta_{21}] = 0.$$

Using (vii) of Theorem 5.1, we see that the preceding equation can be written as

$$c_{21}[\gamma_{21}.\beta_{21}] - \zeta_{21} = 0.$$

If $c_{21} = 0$, then $\zeta_{21} = 0$, and this in turn implies that there is a retraction $O_{n+2,2} \to S^n$ (cf. Theorem (1.11) of [64, James]). If $c_{21} = 1$, consider the map $g : S^{n+1} \vee S^n_{21} \to S^{n+1} \vee S^n_{21}$, where $g|S^n_{21}$ is the identity, and $g|S^{n+1}$ is in the homotopy class $\gamma_{21} + \delta_{k+1}$. Observe now that

$$[\iota_{n+1}, \beta_{21}] \mapsto [\gamma_{21} + \delta_{k+1}, \beta_{21}].$$

However, $[\delta_{k+1} + \gamma_{21}, \beta_{21}] = 0$, according to Formula(II) above. Hence, g can be extended to a map, also denoted by g, $g : S^{n+1} \times S^n \to O_{n+2,2}$ that can easily be seen to be a homotopy equivalence (cf. §3 of [65, James-Whitehead]). Hence, there is a retraction of $O_{n+2,2}$ onto the fiber, and this means that $O_{n+2,2} \to S^{n+1}$ is fiber homotopically trivial. Theorem (1.11) of [64, James] implies that $(n+1) = 7$, as required. □

Theorem 8.2 *Suppose that $(n+1)$ is even, and assume that the projection* $\text{proj}_{k+1,r} : \mathbb{F}_{k+1}(S^{n+1}) \to \mathbb{F}_r(S^{n+1})$ *is fiber homotopically trivial. Then, $r = 3$, and $n = 3$ or 7.*

Proof. Note that the restriction of $\text{proj}_{k+1,r} : \mathbb{F}_{k+1}(S^{n+1}) \to \mathbb{F}_r(S^{n+1})$ to the subspace $\mathbb{F}_{r-1}(\mathbb{R}^{n+1}) \subset \mathbb{F}_r(S^{n+1})$ is equivalent to the fibration

$$\text{proj}_{k,r-1} : \mathbb{F}_k(\mathbb{R}^{n+1}) \to \mathbb{F}_{r-1}(\mathbb{R}^{n+1}).$$

This restriction is also fiber homotopically trivial. Applying Theorem 5.1 of Chapter II, we see that $(r-1) \leq 2$ and n is 3 or 7.

If $(r-1) < 2$ or, equivalently, $r < 3$, then the assumption in the theorem would imply that $\mathbb{F}_3(S^{n+1}) \to S^{n+1}$ has a section. But this fibration is homotopy equivalent to $O_{n+2,2} \to S^{n+1}$ and, hence, the latter has a section. However, this is impossible, since $(n+1)$ is even. Hence, $r = 3$, as desired. □

The question of whether the fibration

$$\text{proj}_{k+1,3} : \mathbb{F}_{k+1}(S^{n+1}) \to \mathbb{F}_3(S^{n+1})$$

is fiber homotopically trivial when $n = 3$ or 7 remains open.

9 Historical Remarks

The relation of $\mathbb{F}_3(S^{n+1})$ to Stiefel manifolds was first recognized in [29, Fadell], where the Dirac braid is also defined. Its relation to the Stiefel manifold $O_{n+2,2}$ is also known to F.Cohen, and is used extensively in his work.

The invariance under the action of the symmetric group Σ_k is new, as are results on the fiber homotopy triviality of the canonical configuration space bundles.

IV. The Two Dimensional Case

As the spaces $\mathbb{F}_k(\mathbb{R}^2)$ and $\mathbb{F}_k(S^2)$ are not simply connected, the methods in the previous chapters need to be adapted accordingly. In particular, the choice of the basepoint $q = (q_1, \cdots, q_k)$ must always be considered.

In §1, using a filtration of $\mathbb{F}_k(\mathbb{R}^2)$ similar to that of $\mathbb{F}_k(\mathbb{R}^{n+1})$ with $n > 1$ given in §1 of Chapter II, we prove that $\mathbb{F}_k(\mathbb{R}^2)$ is aspherical. This filtration also yields an ascending chain of normal subgroups of $\pi_1(\mathbb{F}_k(\mathbb{R}^2), q)$ such that each subgroup is a split-extension of the preceding subgroup by a free subgroup. This result is analogous to the results of §5 of Chapter II.

In §2 we turn to the problem of finding the elements in the fundamental group $\pi_1(\mathbb{F}_k(\mathbb{R}^2), q)$ analogous to the elements α_{rs} of $\pi_n(\mathbb{F}_k(\mathbb{R}^{n+1}))$ when $n > 1$. We first observe that the maps $\alpha'_{s+1s} : S^1 \to \mathbb{F}_k(\mathbb{R}^2)$, as defined in Chapter II, are in effect basepoint preserving, but α'_{rs}, with $r > s+1$, are not. In fact, we see that α'_{rs} is obtained from α'_{s+1s} by post-composing with $\tau_{r-1} \circ \cdots \tau_{s+1}$. Hence, on conjugating with the composite path

$$\mu_\sigma = \sigma'_{r-1} * \cdots * \sigma'_{s+1},$$

where σ'_i is the path given the homotopy of Assertion 2.2 of Chapter II that interchanges q_i and q_{i+1}, we obtain the usual elements $\alpha_{rs} \in \pi_1(\mathbb{F}_k(\mathbb{R}^2), q)$ represented by braids, where the rth strand circles around the sth particle. The natural projection $\mathbb{F}_k(\mathbb{R}^2)$ onto the braid space $\mathbb{B}_k(\mathbb{R}^2)$ takes $[\sigma'_i]$ to the standard braid permutation $\sigma_i \in \pi_1(\mathbb{B}_k(\mathbb{R}^2), [q])$. Also, we obtain the well-known expressions of the braids α_{rs} in terms of the σ_i.

in §§3 and 4 we consider the effect of the symmetric group Σ_k on the elements α_{rs}. The formula $\sigma_*(\alpha_{rs}) = \alpha_{\sigma r \sigma s}$, which is valid when $n > 1$, needs to be modified for the case $n = 1$, as $\mathbb{F}_k(\mathbb{R}^2)$ is no longer simply connected, and $\sigma : \mathbb{F}_k(\mathbb{R}^2) \to \mathbb{F}_k(\mathbb{R}^2)$ changes basepoints, certainly if $\sigma \neq 1 \in \Sigma_k$. However, we know by virtue of §2 of Chapter II that, given σ in Σ_k, there is a path μ_σ in $\mathbb{F}_k(\mathbb{R}^2)$ from q to $\sigma(q)$ such that $\mu_\sigma * \sigma \circ \alpha'_{rs} * \mu_\sigma^{-1}$ is in the class of $\alpha_{\sigma r \sigma s}$. The path μ_σ depends on σ and is by no means canonical. If one uses the paths constructed by stringing together appropriate permutation braids, one is led to Theorem 3.1 and Corollary 4.1 below. The latter points to a class of permutations for which the formula $\sigma_*(\alpha_{rs}) = \alpha_{\sigma r \sigma s}$ holds. In §4 we seek the Y-B relations among the braids of the previous section. They turn out to be, as in the simply connected case but with changes made necessary by the lack of simple connectivity, of two basic kinds: one among the elements α_{21}, α_{31}, and α_{32}, and another involving α_{21} and α_{43}, as well all those obtained from the basic relations by use of certain symmetries, in the manner of §3, of the symmetric group Σ_k.

In §5 we show that the Y-B relations lead to a concise presentation of the pure braid group $\pi_1(\mathbb{F}_k(\mathbb{R}^2), q)$. The idea is to construct the total space of a quasi fibration as a complex made up of the 1-skeletons of $\mathbb{F}_{k-1}(\mathbb{R}^2)$ and $\mathbb{R}^2 - Q_{k-1}$. The 2-cells of the complex correspond to the Y-B relations of the first and second kind that hold among the α_{ks}, on the one hand, and α_{rt} with $r < k$, on the other. The union of this complex with the 2-skeleton of $\mathbb{F}_{k-1}(\mathbb{R}^2)$ gives us a complex, the fundamental group of which is shown to be the pure braid group on k strands.

In §6 we prove that the natural projection $\mathrm{proj}_{k,r} : \mathbb{F}_k(\mathbb{R}^2) \to \mathbb{F}_r(\mathbb{R}^2)$ is fiber-homotopically trivial if, and only if, $r \leq 2$. The argument is similar to that of Theorem 5.1 of Chapter II, but it is complicated by lack of commutativity. To get around this difficulty, we use the lower central series of the pure braid group to carry out the necessary calculations.

In §7 we consider the group $\pi_1(\mathbb{F}_k(S^2), q)$. As in the simply connected case, $\mathbb{F}_2(S^2)$ is homotopy equivalent to S^2, and $\mathbb{F}_3(S^2)$ is of the homotopy type of the Stiefel manifold $O_{3,2}$. The latter fact allows us to prove that the pull-back $\mathbb{F}'_k(S^2) \to O_{3,2}$ of the configuration space fibration is trivial. Next, as in §§6 and 7 of Chapter III, we describe how the space $\mathbb{F}_{k+1}(S^2)$ is related to $\mathbb{F}_k(\mathbb{R}^2)$. Then we introduce the *Dirac class* Δ_k and exploit the presentation of $\pi_1(\mathbb{F}_k(\mathbb{R}^2), q)$ to obtain a presentation for the group $\pi_1(\mathbb{F}_k(S^2), q)$.

1 Asphericity of $\mathbb{F}_k(\mathbb{R}^2)$

Following the notation of §1 of Chapter II, put $\mathbb{F}_{k-r,r}(\mathbb{R}^2) = \mathbb{F}_{k-r}(\mathbb{R}^2 - Q_r)$, and let $i_{k-r} : \mathbb{F}_{k-r,r}(\mathbb{R}^2) \to \mathbb{F}_k(\mathbb{R}^2)$ be the map where $(x_{r+1}, \cdots, x_k)$ is mapped to $(q_1, \cdots, q_r, x_{r+1}, \cdots, x_k)$. Unlike the case when $n > 1$, the space $\mathbb{F}_{k-r,r}(\mathbb{R}^2)$ has no higher homotopy groups.

Theorem 1.1 *The space* $\mathbb{F}_{k-r,r}(\mathbb{R}^2)$ *is aspherical.*

Proof. We shall prove inductively that each of the spaces $\mathbb{F}_{r,k-r}(\mathbb{R}^2)$ is aspherical. To start, note that $\mathbb{R}^2_{k-1} = \mathbb{R}^2 - Q_{k-1}$ is homotopy equivalent to a bouquet of $(k-1)$ circles and, therefore, is aspherical. Suppose now that $\mathbb{F}_{k-r-1,r+1}(\mathbb{R}^2)$ is aspherical, and consider the fibration $\mathrm{p}_r : \mathbb{F}_{k-r,r}(\mathbb{R}^2) \to \mathbb{R}^2_r$, where the p_r takes a tuple to its first entry. Note that its fiber, $\mathbb{F}_{k-r-1,r+1}(\mathbb{R}^2)$, is aspherical by the induction assumption. The base space is aspherical, as it is homotopy equivalent to a bouquet of circles. Hence the total space $\mathbb{F}_{k-r,r}(\mathbb{R}^2)$ is also aspherical. This proves the induction step, and, thus, the theorem is proved. □

The map i_{k-r} induces the morphism $\pi_1(i_{k-r}) : \pi_1(\mathbb{F}_{k-r,r}(\mathbb{R}^2), q^{k-r}) \to \pi_1(\mathbb{F}_k(\mathbb{R}^2, q^k)$, where $q^{k-r} = (q_{r+1}, \cdots, q_k)$ and $q^k = (q_1, \cdots, q_k)$ are the basepoints of $\mathbb{F}_{k-r,r}(\mathbb{R}^2)$ and $\mathbb{F}_k(\mathbb{R}^2)$, respectively. Let L_{k-r} denote $\pi_1(\mathbb{F}_{k-r,r}(\mathbb{R}^2, q^{k-r})$. Note that L_k is just $\pi_1(\mathbb{F}_k(\mathbb{R}^2), q)$. The next theorem is the analogue of Theorem 1.4 of Chapter II.

Theorem 1.2 *The morphisms $\pi_1(i_{k-r}) : L_{k-r} \to L_k$ are all injective. They induce an ascending chain of normal subgroups*

$$L_1 \subset L_2 \subset \cdots \subset L_k$$

such that L_{k-r}/L_{k-r-1} is a free group on r generators. Moreover, L_{k-r} is isomorphic to $L_{k-r-1} \rtimes (L_{k-r}/L_{k-r-1})$.

Proof. Note that $\mathbb{F}_{k-r,r}(\mathbb{R}^2)$ is the fiber of the projection $\mathrm{proj}_{k,r} : \mathbb{F}_k(\mathbb{R}^2) \to \mathbb{F}_r(\mathbb{R}^2)$ that sends $(x_1, x_2, \cdots, x_k)$ to $(x_1, x_2, \cdots, x_r)$ and that this fibration admits a section. Hence, the morphism $\pi_1(i_{k-r})$ is injective, and L_{k-r} is a normal subgroup of L_k for each $1 \leq r \leq k$.

Next, note that the projection $\mathrm{p}_r : \mathbb{F}_{k-r,r}(\mathbb{R}^2) \to \mathbb{R}^2_r$ that sends the tuple $(q_1, \cdots, q_r, x_{r+1}, \cdots, x_k)$ to x_{r+1} is a fibration with a section of which the fiber and base space are, respectively, $\mathbb{F}_{k-r-1,r+1}(\mathbb{R}^2)$ and $\mathbb{R}^2_r = \mathbb{R}^2 - Q_r$. This implies that L_{k-r}/L_{k-r-1} is isomorphic to the fundamental group of the base space. But the base space is homotopy equivalent to a bouquet of r circles. This proves that L_{k-r}/L_{k-r-1} is a free group on r generators, and $L_{k-r} \cong L_{k-r-1} \rtimes (L_{k-r}/L_{k-r-1})$, where $L_{k-r}/L_{k-r-1} \cong \pi_1(\mathbb{R}^2, q^r)$. □

2 Generators for $\pi_1(\mathbb{F}_k(\mathbb{R}^2), q)$

We wish to define generators for $\pi_1(\mathbb{F}_k(\mathbb{R}^2), q^k)$ analogous to $\{\alpha_{rs} \mid 1 \leq s < r \leq k\}$, defined in Chapter II, where $n > 1$. As the space $\mathbb{R}^2$ is not simply connected, we need to modify the definitions to take care of the basepoint, which we choose to be $q^k = (q_1, \cdots, q_k)$. However, for simplicity, from here on we shall denote it by q.

Note that the group of permutations Σ_k acts freely on $\mathbb{F}_k(\mathbb{R}^2)$ by permuting the factors. The quotient space $\mathbb{F}_k(\mathbb{R}^2)/\Sigma_k$ is called the *braid space on k strands* and is denoted by $\mathcal{B}_k(\mathbb{R}^2)$. Let $\mathrm{p} : \mathbb{F}_k(\mathbb{R}^2) \to \mathbb{B}_k(\mathbb{R}^k)$ be the natural projection. The basepoint of the braid space $\mathbb{B}_k(\mathbb{R}^k)$ is, by definition, the image of the basepoint $q = (q_1, \cdots, q_k)$ of $\mathbb{F}_k(\mathbb{R}^k)$ and will be denoted by $[q] - \mathrm{p}(q)$.

Consider now the maps $\alpha'_{rs} : S \to \mathbb{F}_k(\mathbb{R}^2)$, defined in Chapter II by the formula $\alpha'_{rs}(\xi) = (q_1, \cdots, q_s, \cdots, q_{r-1}, q_s + \xi, \cdots)$, where $S \subset \mathbb{R}^2$ is the unit circle with $1 = (1, 0)$ as basepoint. The map α'_{rs} is not basepoint preserving, for it takes $1 \in S$ to $(q_1, q_2, \cdots, q_s, \cdots, q_{r-1}, q'_s, \cdots)$, with q'_s being a point one unit to the right of q_s. It is easy to see that

$$\alpha'_{rs} = \tau_{r-1} \circ \cdots \circ \tau_{s+1} \circ \alpha'_{s+1s},$$

where τ_i is the transposition $(i, i+1)$ in Σ_k. After sliding q_s to the right along the line on which the points $q_1, q_2, \cdots$ lie, we can arrange it so that $\alpha'_{s+1s}(1) = (q_1, \cdots, q_s, q_{s+1}, \cdots)$, where $1 \in S$ is the basepoint of S. In this manner we make α'_{s+1s} basepoint preserving. However, to make the maps α'_{rs} basepoint preserving, we shall slide $\alpha'_{rs}(1)$ to $(q_1, q_2 \cdot \cdots)$ along suitably chosen paths. These paths are, in fact, the tracks of the basepoint under the relevant free homotopies of Theorem 1.2 of Chapter II.

Definition 2.1 *Let $\sigma'_i : I \to \mathbb{F}_k(\mathbb{R}^2)$ be the map such that*

$$t \mapsto (q_1, \cdots, q_{i-1}, (q_i + q_{i+1})/2 - 2\xi_t, (q_i + q_{i+1})/2 + 2\xi_t, \cdots),$$

where $\xi_t = (-\cos \pi t)e_1 + (\sin \pi t)e_2$. Here, e_1 and e_2 are the basic vectors $(1,0)$ and $(0,1)$, respectively.

The homotopy class $[p \circ \sigma'_i]$ is denoted by $\sigma_i \in \pi_1(\mathbb{B}_k(\mathbb{R}^2), [q])$ and is called the *braid permutation.* Observe that σ'_i is a path from the point $(q_1, \cdots, q_i, q_{i+1}, \cdots, q_k) \in \mathbb{F}_k(\mathbb{R}^2)$ to the point $(q_1, \cdots, q_{i+1}, q_i, \cdots, q_k) \in \mathbb{F}_2(\mathbb{R}^2)$, as in Fig. IV.1. (Here we follow the usual convention: the figures represent the graphs of the maps in $\mathbb{R}^2 \times \mathbb{R}$, with time t running from 0(top) to 1(bottom).) Observe that σ'_i is the path traversed by the basepoint under the homotopy described in the proof of (ii) of Assertion 2 of Theorem 2.1 in Chapter II.

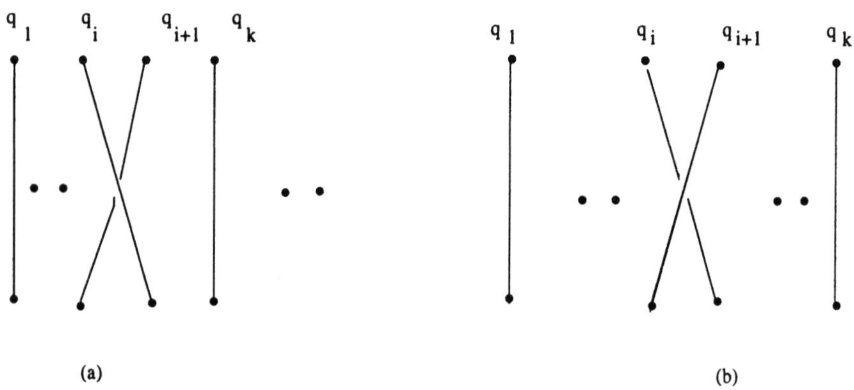

Fig. IV.1. (a) the braid σ_i, (b) the braid σ_i^{-1}

Definition 2.2 *Let $\alpha''_{rs} : S \to \mathbb{F}_k(\mathbb{R}^2)$ be the loop*

$$(\sigma'_{r-1} * \cdots * \sigma'_{s+1}) * \alpha'_{rs} * (\sigma'_{r-1} * \cdots * \sigma'_{s+1})^{-1},$$

where $1 \leq s < r \leq k$. Here, we compose paths from left to right.

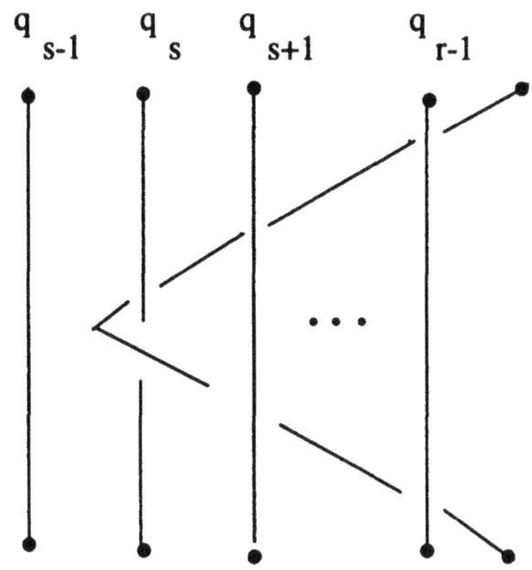

Fig. IV.2. Definition of α_{rs}

The loops $\alpha''_{rs} : S \to \mathbb{F}_k(\mathbb{R}^2)$ are based at $q \in \mathbb{F}_k(\mathbb{R}^2)$ and thus define elements α_{rs} in the fundamental group $\pi_1(\mathbb{F}_k(\mathbb{R}^2), q)$ (see Fig.IV.2). The paths correcting the change of basepoint are projected to loops in $\mathbb{B}_k(\mathbb{R}^2)$ by the natural projection. They define the following elements of $\pi_1(\mathbb{B}_k(\mathbb{R}^2), [q])$.

Definition 2.3 *Put, for $1 < s < r \leq k$,*

$$\mu_{rs} = \sigma_{r-1} \cdots \sigma_s$$

and $\mu_{rr} = 1$, for all $1 < r \leq k$.

Observe here that the image of μ_{rs} in Σ_k under the natural homomorphism $\pi_1(\mathbb{B}_k(\mathbb{R}^2), [q]) \to \Sigma_k$ is, in fact, the cycle $(s, s+1, \cdots, r-1, r)$, and the expression defining μ_{rs} mimics the factorization

$$(s, s+1, \cdots, r-1, r) = (r-1, r) \cdots (s+1, s+2)(s, s+1)$$

of the cycle as a product of transpositions that involve successive indices, where the product is from left to right.

The following result expresses α_{s+1s} in terms of the braid σ_i and α_{rs} as a conjugate of α_{s+1s} (cf. [9, Birman, p. 27], and [16, Chow]).

Proposition 2.1 *In $\pi_1(\mathbb{B}_k(\mathbb{R}^2), [q])$, we have $\pi_1(p)(\alpha_{rs}) = \mu_{rs+1}\alpha_{s+1s}\mu_{rs+1}^{-1}$. In particular, $\alpha_{s+1s} = \sigma_s^2$.*

Proof. Let us prove first that $\alpha_{s+1s} = \sigma_s^2$. To do this, consider the homotopy $H : S \times I \to \mathbb{F}_k(\mathbb{R}^2)$, where (t, u) is mapped to

$$(q_1, \cdots, q_{s-1}, (q_s + q_{s+1})/2 - 2(1-u)\xi_t, (q_s + q_{s+1})/2 + 2\xi_t, q_{s+2}, \cdots)$$

with $\xi_t = (-\cos \pi t e_1 + \sin \pi t e_2)$. The homotopy H moves the entry in the sth place linearly to the midpoint $(q_s + q_{s+1})/2$, while that in the $(s+1)^{\text{st}}$ place remains on the circle of radius 2 centered at the midpoint. This homotopy H can be corrected to make it basepoint preserving by translating along the path traversed by the basepoint (see Fig.IV.3). Observe that H is a homotopy from the loop $\gamma' = \sigma'_s * \sigma'_s : S \to \mathbb{F}_k(\mathbb{R}^2)$, where the entries in the s-th and $(s+1)$-st places constitute a pair of antipodal points on the circle of radius 2 centered at the midpoint $(q_s + q_{s+1})/2$, to the loop $S \to \mathbb{F}_k(\mathbb{R}^2)$, where

$$t \mapsto (q_1, \cdots, q_{s-1}, (q_s + q_{s+1})/2, (q_s + q_{s+1})/2 - 2\xi_t, q_{s+2}, \cdots).$$

The latter loop is certainly α'_{s+1s}, and the former is taken to a loop in $\mathbb{B}_k(\mathbb{R}^2)$ representing σ_s^2 in $\pi_1(\mathbb{B}_k(\mathbb{R}^2), [q])$. Hence, $\pi_1(p)(\alpha_{s+1s}) = \sigma_s^2$, as required (see Fig.IV.3).

The first statement can be proved similarly. Observe that the projection $\mathbb{F}_k(\mathbb{R}^2) \to \mathbb{B}_k(\mathbb{R}^2)$, takes the path $[\sigma']$ to the loop σ'_i representing $\sigma_i \in \pi(\mathbb{B}_k(\mathbb{R}^2), [q])$. As composition of paths goes to the multiplication in the group $\pi_1(\mathbb{B}_k(\mathbb{R}^2), [q]))$, the result follows immediately. □

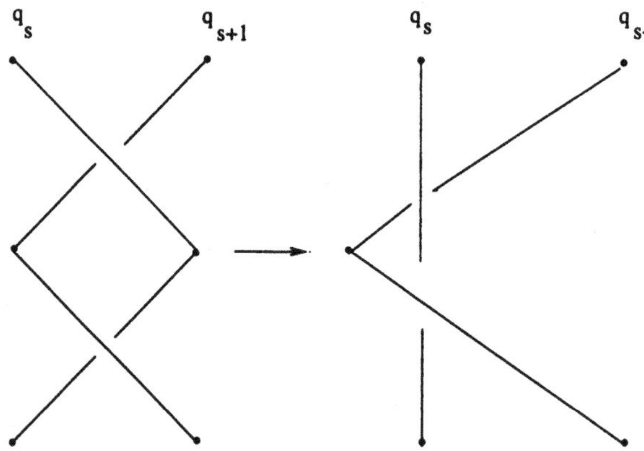

Fig. IV.3. the relation $\alpha_{s+1s} = \sigma_s^2$

Let us remark here that homotopy used in the proof of the assertion that $\alpha_{s+1s} = \sigma_s^2$ is just the simple homotopy indicated in Fig.IV.3. As to the general statement, the proof we have given can also be replaced by using a figure similar to that above.

The following result, which gives the classic braid relations in the group $\pi_1(\mathbb{B}_k(\mathbb{R}^2), [q])$, is the key to many calculations; these are easily proved by considering Fig.IV.4 (cf. [9, Birman]).

Lemma 2.1 *The relations*

$$\begin{array}{lccc} (i) & \sigma_{s+1}\sigma_s\sigma_{s+1} & = & \sigma_s\sigma_{s+1}\sigma_s \quad and \\ (ii) & \sigma_s\sigma_t & = & \sigma_t\sigma_s, \end{array}$$

where $|t-s| \geq 2$ *and* $1 \leq s \leq (k-1)$, *hold in* $\pi(\mathbb{B}_k(\mathbb{R}^2), [q])$.

It is a classical result that the braid relations are complete. However, our objective is to obtain a complete presentation of $\pi_1(\mathbb{F}_k(\mathbb{R}^2), q)$ independently of this fact.

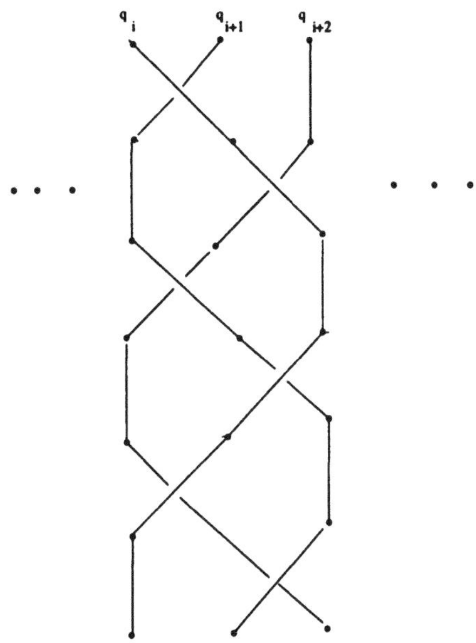

Fig. IV.4. the relation $\sigma_{s+1}\sigma_s\sigma_{s+1}\sigma_s^{-1}\sigma_{s+1}^{-1}\sigma_s^{-1} = 1$

3 The Action of Σ_k on $\mathbb{F}_k(\mathbb{R}^2)$

The natural action of the symmetric group Σ_k on k letters on $\mathbb{F}_k(\mathbb{R}^{n+1})$ changes basepoints. One can translate along appropriate paths to go back to the basepoint. If $n > 1$, we showed that the effect of a permutation on the elements α_{rs} is to permute the indices. The proof given there depends on homotopies that are not basepoint preserving. Therefore, here, where $n = 1$, the effect of a permutation on the elements α_{rs} is, up to conjugacy, the same as in the simply connected case, and the relevant conjugacies are defined by the paths traversed by these homotopies. But these paths define the braids σ_i. Therefore the relevant conjugacies affecting the generators $\{\alpha_{rs} \mid 1 \leq s < r \leq k\}$ are provided by these braids.

Given $\sigma \in \Sigma_k$, we wish to describe the effect of the induced morphism

$$\pi_1(\sigma) : \pi_1(\mathbb{F}_k(\mathbb{R}^2), q) \to \pi_1(\mathbb{F}_2(\mathbb{R}^2), \sigma(q))$$

on the generators $\{\alpha_{rs} \mid 1 \leq s < r \leq k\}$. To do so, we need suitable paths in $\mathbb{F}_k(\mathbb{R}^2)$ from q to $\sigma(q)$. First, for simplicity, let us identify the group $\pi_1(\mathbb{F}_k(\mathbb{R}^2), q)$ with its image in $\pi_1(\mathbb{B}_k(\mathbb{R}^2), [q])$ under the morphism induced by the natural projection $\mathrm{p} : \mathbb{F}_k(\mathbb{R}^2) \to \mathbb{B}_k(\mathbb{R}^2)$. Now, put

$$\mathcal{M} = \{\mu_{2i_2} \cdots \mu_{ki_k} | 1 \leq i_1 \leq 2, \cdots 1 \leq i_k \leq k\},$$

where $\mu_{rs} = \sigma_{r-1} \cdots \sigma_s$, as in Definition 2.3 above. Note that $\mathcal{M}$ is a set of right coset representatives of $\pi_1(\mathbb{B}_k(\mathbb{R}^2), [q])$ modulo $\pi_1(\mathbb{F}_k(\mathbb{R}^2), q)$. Observe that $\mathcal{M}$ is such that any initial segment of a right coset representative is again a coset representative. That is to say, $\mathcal{M}$ is a *Schreier set* (or a *transversal*, as it is called in [71, Lyndon-Schupp]). The elements in $\mathcal{M}$ are represented by paths in $\mathbb{F}_k(\mathbb{R}^2)$ that are composites of the paths σ_i'.

Since $\mathcal{M}$ is a Schreier set, every $\sigma \in \Sigma_k$ can be expressed in the form $\sigma = \prod_{m=2}^k \sigma_{m,i_m}$, where $\sigma_{m,i_m} = \mathrm{id}$, if $i_m = m$, and

$$\begin{aligned} \sigma_{m,i_m} &= (i_m, i_m + 1, \cdots, m-1, m) \\ &= (m, m-1) \cdots (i_m + 1, i_m), \end{aligned}$$

if $1 \leq i_m < m$, where the permutations are applied from left to right.

Lemma 3.1 *The natural projection*

$$\phi : \pi_1(\mathbb{B}_k(\mathbb{R}^2), [q]) \to \Sigma_k, \ \sigma_i \mapsto (i, i+1)$$

takes $\mu_\sigma = \mu_{2i_2} \cdots \mu_{ki_k}$ *to* σ.

Proof. First, note that

$$(i_m, i_m + 1, \cdots, m-1, m) = (m, m-1) \cdots (i_m + 1, i_m).$$

Hence, it is enough to show that μ_{mi_m} maps to the cycle σ_{mi_m}. To do this, note that the element $\pi(p)(\mu_{mi_m})$ is represented by the image under p of the composite path $\mu'_{mi_m} = \sigma'_{m-1} * \cdots * \sigma'_{i_m}$, where σ'_i is the braid that interchanges the points q_i and q_{i+1}, as defined in the previous section. Observe now that $[p \circ \mu'_{mi_m}] = \sigma_{m-1} \cdots \sigma_{i_m}$, and the expression on the right hand side is taken by ϕ to the cycle σ_{m-1,i_m} in Σ_k. □

Now let us return to the problem of describing the effect of a permutation $\sigma \in \Sigma_k$ on the homotopy groups. We showed in §2 of Chapter II, when $n > 1$, that, as the space $\mathbb{F}_k(\mathbb{R}^{n+1})$ is simply connected, its homotopy groups are independent of the basepoint and $\pi_1(\sigma)(\alpha_{rs}) = \alpha_{\sigma r \sigma s}$. As we pointed out above, the case when $n = 1$ is more complicated because the space

$\mathbb{F}_k(\mathbb{R}^2)$ is not simply connected. For one thing, a permutation σ takes the fundamental group based at q to that based at $\sigma(q)$. However, using the paths μ'_{mi_m} constructed by composition out of the paths described above, one can translate back to the original basepoint q and determine where α_{rs} goes under σ. This procedure requires a careful analysis of conjugations in the group $\pi_1(\mathbb{B}(\mathbb{R}^2),[q])$ by the elements μ_{mi_m} described above. Precisely, the problem is to describe the morphism $\pi_1(\sigma) : \pi_1(\mathbb{F}_k(\mathbb{R}^2), q) \to \pi_1(\mathbb{F}_k(\mathbb{R}^2), \sigma(q))$ and to characterize permutations σ where the effect, after conjugation, is just the action of the permutation on the indices. So, consider the loops α''_{rs}, defined in Theorem 2.2. Note that the class $\pi_1(\sigma)(\alpha_{rs})$ in the group $\pi_1(\mathbb{F}_k(\mathbb{R}^2), \sigma(q))$ is represented by the element $\sigma \circ \alpha''_{rs}$. The element $\sigma \in \Sigma_k$ leads to a composite path $\mu_\sigma = \mu_{2i_2} \cdots \mu_{ki_k}$, as described above. The loop

$$\mu_{2i_2} * \cdots * \mu_{ki_k} * \sigma \circ \alpha''_{rs} * (\mu_{2i_2} * \cdots * \mu_{ki_k})^{-1}$$

is based at $q \in \mathbb{F}_2(\mathbb{R}^2)$, and its projection into the braid space $\mathbb{B}_k(\mathbb{R}^2)$ represents the homotopy class

$$(\mu_{2i_2} \cdots \mu_{ki_k})\pi_1(p)(\alpha_{rs})(\mu_{2i_2} \cdots \mu_{ki_k})^{-1}.$$

Thus, the problem of describing the morphism $\pi_1(\sigma) : \pi_1(\mathbb{F}_k(\mathbb{R}^2), q) \to \pi_1(\mathbb{F}_k(\mathbb{R}^2), \sigma(q))$ is reduced to the task of computing the effect of conjugation of the image of α_{rs} in $\pi_1(\mathbb{B}_k(\mathbb{R}^2),[q])$ by the elements μ_{tu} of $\mathcal{M}$.

Theorem 3.1 *For all $1 \le s < r \le k$ and $1 \le u < t \le k$, the relations*

$$\mu_{tu}\alpha_{rs}\mu_{tu}^{-1} = \begin{cases} \alpha_{rs} & 1 \le u < t < s < r \le k \\ \alpha_{ts-1}\alpha_{rs-1}\alpha_{ts-1}^{-1} & 1 \le u < s \le t < r \le k \\ \alpha_{r-1s-1} & 1 \le u < s < r \le t \le k \\ \alpha_{rt} & 1 \le s = u < t < r \le k \\ \alpha_{tr-1} & 1 \le s = u < r \le t \le k \\ \alpha_{rs} & 1 \le s < u < t < r \le k \\ \alpha_{tr-1}\alpha_{r-1s}\alpha_{tr-1}^{-1} & 1 \le s < u < r \le t \le k \\ \alpha_{ts} & 1 \le s < r = u < t \le k \\ \alpha_{rs} & 1 \le s < r < u < t \le k \end{cases}$$

hold in $\pi_1(\mathbb{F}_k(\mathbb{R}^2), q) \subset \pi_1(\mathbb{B}_k(\mathbb{R}^2),[q])$.

To prove the theorem, we begin by proving the following basic lemma.

Lemma 3.2 *In $\pi_1(\mathbb{F}_k(\mathbb{R}^2), q)$, the relations*

$$\sigma_t\alpha_{rs}\sigma_t^{-1} = \begin{cases} \alpha_{r+1s}, & \text{if } 1 \le s < t = r < k \\ \alpha_{rs+1}, & \text{if } 1 \le t = s < r-1 \le k-1 \\ \alpha_{rr-1}\alpha_{r-1s}\alpha_{rr-1}^{-1}, & \text{if } 1 \le s < t = r-1 \le k-1 \\ \alpha_{ss-1}\alpha_{rs-1}\alpha_{ss-1}^{-1}, & \text{if } 1 \le t = s-1 < r-1 \le k-1 \\ \alpha_{rs}, & \text{otherwise} \end{cases}$$

are valid.

It is sometimes convenient to restate the lemma as follows.

Lemma 3.3 *The following relations in* $\pi_1(\mathbb{F}_k(\mathbb{R}^2), q)$ *are valid.*

$$\begin{array}{llll}
(i)\ \sigma_r\alpha_{rs}\sigma_r^{-1} & = & \alpha_{r+1s}, & if\ 1 \le s < r < k,\\
(ii)\ \sigma_s\alpha_{rs}\sigma_s^{-1} & = & \alpha_{rs+1}, & if\ 1 \le s < r-1 \le k-1,\\
(iii)\ \sigma_{r-1}\alpha_{rs}\sigma_{r-1}^{-1} & = & \alpha_{rr-1}\alpha_{r-1s}\alpha_{rr-1}^{-1}, & if\ 1 \le s < r-1 \le k-1,\\
(iv)\ \sigma_{s-1}\alpha_{rs}\sigma_{s-1}^{-1} & = & \alpha_{ss-1}\alpha_{rs-1}\alpha_{ss-1}^{-1}, & if\ 1 \le s-1 < r-1 \le k-1,\\
(v)\ \sigma_t\alpha_{rs}\sigma_t^{-1} & = & \alpha_{rs}, & otherwise.
\end{array}$$

Proof. We shall give the proof in stages.

Proof of (i) and (iii)

According to the preceding proposition,

$$\alpha_{r+1s} = \mu_{r+1s+1}\alpha_{s+1s}\mu_{r+1s+1}^{-1}.$$

Now, the left-hand side of the preceding equation equals $\sigma_r(\alpha_{rs})\sigma_r^{-1}$. Hence,

$$\alpha_{r+1s} = \sigma_r(\alpha_{rs})\sigma_r^{-1},$$

which proves (i), as required. To prove (iii), note that (i) implies that

$$\sigma_{r-1}\alpha_{r-1s}\sigma_{r-1}^{-1} = \alpha_{rs}.$$

Conjugate the preceding equality by σ_r to obtain (iii). □

Proof of (ii) and (iv)

Note, to start, that $\sigma_s\alpha_{s+2s}\sigma_s^{-1} = \sigma_s\sigma_{s+1}\sigma_s^2\sigma_{s+1}^{-1}\sigma_s^{-1}$. But Lemma 2.1 implies that $\sigma_s\sigma_{s+1}\sigma_s = \sigma_{s+1}\sigma_s\sigma_{s+1}$. Hence,

$$\begin{aligned}
\sigma_s\alpha_{s+2s}\sigma_s^{-1} &= \sigma_{s+1}^2\\
&= \alpha_{s+2s+1}.
\end{aligned}$$

This special case implies the general case for $r > s+2$. To see this, first note that $\sigma_s(\mu_{rs+2})\sigma_s^{-1}) = \mu_{rs+2}$. Hence,

$$\begin{aligned}
\sigma_s\alpha_{rs}\sigma_s^{-1} &= \sigma_s(\mu_{rs+2}\alpha_{s+2s}\mu_{rs+2}^{-1})\sigma_s^{-1}\\
&= \mu_{rs+2}(\sigma_s\alpha_{s+2s}\sigma_s^{-1})\mu_{rs+2}^{-1}\\
&= \mu_{rs+2}\sigma_{s+1}^2\mu_{rs+2}^{-1}\\
&= \mu_{rs+2}\alpha_{s+2s+1}\mu_{rs+2}^{-1}\\
&= \alpha_{rs+1}.
\end{aligned}$$

□

Proof of (v)

Consider the expression

$$\alpha_{rs} = \sigma_{r-1}\cdots\sigma_{s+1}\sigma_s^2\sigma_{s+1}^{-1}\cdots\sigma_{r-1}^{-1}.$$

On repeatedly conjugating by σ_t, we see that σ_t commutes with the braids σ_i for all $s \le i \le (t-1)$, since $t \neq (s-1), s, (r-1)$, and r. This clearly proves (v). □

A similar argument based on the analogous expression of α_{rs} as a product of braids proves (ii). □

Now let us begin the proof of the theorem. It involves simple applications of Lemma 3.1 and is included here for the benefit of the reader.

Proof of Theorem 3.1, Statement (i)

Since $1 \le u < t < s < r \le k$ by assumption, Statement (v) of the previous lemma implies that σ_i for $u \le i \le (t-1)$ commutes with α_{rs}. Therefore,

$$\begin{aligned}\mu_{tu}\alpha_{rs}\mu_{tu}^{-1} &= (\sigma_{t-1}\cdots\sigma_u)\alpha_{rs}(\sigma_{t-1}\cdots\sigma_u)^{-1},\\ &= \alpha_{rs},\end{aligned}$$

as required.

Proof of Theorem 3.1, Statement (ii)

As $1 \le u < s \le t < r \le k$ by assumption, $\mu_{tu} = \mu_{ts-1}\mu_{s-1u}$. But μ_{s-1u} commutes with α_{rs} by Statement (i). Hence,

$$\begin{aligned}\mu_{tu}\alpha_{rs}\mu_{tu}^{-1} &= \mu_{ts-1}\alpha_{rs}\mu_{ts-1}^{-1},\\ &= \mu_{ts}\sigma_{s-1}\alpha_{rs}\sigma_{s-1}^{-1}\mu_{ts}^{-1}.\end{aligned}$$

Observe that

$$\sigma_{s-1}\alpha_{rs}\sigma_{s-1}^{-1} = \alpha_{ss-1}\alpha_{rs-1}\alpha_{ss-1}^{-1}$$

by Statement (iv) of Lemma 3.2. Hence,

$$\begin{aligned}\mu_{ts-1}\alpha_{rs}\mu_{ts-1}^{-1} &= \mu_{ts}\alpha_{ss-1}(\alpha_{rs-1})\alpha_{ss-1}^{-1}\mu_{ts-1}^{-1},\\ &= \alpha_{ts-1}\alpha_{rs-1}\alpha_{ts-1}^{-1}\end{aligned}$$

by Proposition 2.1. This proves the assertion.

Proof of Theorem 3.1, Statement (iii)

Arguing as before, we obtain the equality

$$\mu_{tu}\alpha_{rs}\mu_{tu}^{-1} = \alpha_{ts-1}\mu_{ts}\alpha_{rs}\mu_{ts}^{-1}\alpha_{ts-1}^{-1}.$$

Again arguing in a similar manner, we see that

$$\begin{aligned}\mu_{ts}\alpha_{rs-1}\mu_{ts}^{-1} &= \mu_{tr-1}\alpha_{rs-1}\mu_{tr-1}^{-1}\\ &= \mu_{tr}\alpha_{rr-1}\alpha_{r-1s-1}\alpha_{rr-1}^{-1}\mu_{tr}^{-1}\\ &= \alpha_{tr-1}\alpha_{r-1s-1}\alpha_{tr-1}^{-1},\end{aligned}$$

since μ_{tr} commutes with α_{rs}. Hence,

$$\mu_{tu}\alpha_{rs}\mu_{tu}^{-1} = (\alpha_{ts-1}\alpha_{tr-1})\alpha_{t-1s-1}(\alpha_{ts-1}\alpha_{tr-1})^{-1}.$$

Assertion 3.1 *The elements* $(\alpha_{ts-1}\alpha_{tr-1})$ *and* α_{r-1s-1} *commute with each other.*

Consider the map

$$\phi_1 : S \times S \to \mathbb{F}_k(\mathbb{R}^2)$$

under which

$$(\xi_1, \xi_2) \mapsto (q_1, q_1 + \xi_1, q_1 + 2\xi_2, \cdots).$$

This implies that $[\alpha_{21}, \alpha_{31}\alpha_{32}] = 1$. The general statement of the Assertion is proved by conjugating successively with $\sigma_3, \cdots, \sigma_{t-2}$ and then with $\sigma_2, \cdots, \sigma_{s-2}$.

Proof of Theorem 3.1, Statement (iv)

Here $1 \le s = u < t < r \le k$. Arguing as in the previous cases, we see that

$$\begin{aligned}\mu_{tu}\alpha_{rs}\mu_{tu}^{-1} &= \mu_{ts}\alpha_{rs}\mu_{ts}^{-1}\\ \mu_{ts}\alpha_{rs}\mu_{ts}^{-1} &= \alpha_{rs}\\ &= \mu_{ts+1}(\sigma_s\alpha_{rs}\sigma_s^{-1})\mu_{ts+1}^{-1}\\ &= \mu_{ts+1}\alpha_{rs}\mu_{ts}^{-1}\\ &= \mu_{rt}.\end{aligned}$$

This proves the assertion. □

Proof of Theorem 3.1, Statement (v)

Here $1 \le u = s < r \le t \le k$. Again we have

$$\begin{aligned}\mu_{tu}\alpha_{rs}\mu_{tu}^{-1} &= \mu_{ts}\alpha_{rs}\mu_{tu}^{-1}\\ &= \mu_{tr}(\mu_{rs}\alpha_{rs}\mu_{rs}^{-1})\mu_{tr}^{-1}\\ &= \mu_{tr}\alpha_{rr-1}\mu_{tr}^{-1}\\ &= \alpha_{tr-1}.\end{aligned}$$

Proof of Theorem 3.1, Statement (vi)

Here the conditions on the indices are such that μ_{tu} and α_{rs} commute with each other.

Proof of Theorem 3.1, Statement (vii)

Here $1 \leq s < u < r \leq t \leq k$. Therefore, we have

$$\begin{aligned} \mu_{tu}\alpha_{rs}\mu_{tu}^{-1} &= \mu_{tr-1}(\mu_{r-1u}\alpha_{rs}\mu_{tr-1}^{-1})\mu_{tr-1}^{-1} \\ &= \mu_{tr-1}\alpha_{rs}\mu_{tr-1}^{-1} \\ &= \mu_{tr}(\sigma_{r-1}\alpha_{rs}\sigma_{r-1}^{-1})\mu_{tr} \\ &= \mu_{tr}(\alpha_{rr-1}\alpha_{r-1s}\alpha_{rr-1}^{-1})\mu_{tr}^{-1} \\ &= \alpha_{tr-1}\alpha_{r-1s}\alpha_{tr-1}^{-1}. \end{aligned}$$

Proof of Theorem 3.1, Statement (viii)

Here $1 \leq s < r = u < t \leq k$. Hence,

$$\begin{aligned} \mu_{tu}\alpha_{rs}\mu_{tu}^{-1} &= \mu_{tr}\alpha_{rs}\mu_{tr}^{-1} \quad \text{and} \\ &= \alpha_{ts}. \end{aligned}$$

Finally, to prove Statement (ix), note that the indices here are such that the terms μ_{tu} and α_{rs} commute with each other. This finishes the proof of Theorem 3.1. □

Theorem 3.1 shows that conjugating the elements α_{rs} by the elements of $\mathcal{M}$ is equivalent to conjugating the said elements by suitable elements in $\pi_1(\mathbb{F}_k(\mathbb{R}^2), q)$.

4 The Y-B Relations

The Y-B relations in the simply connected case describe the commutators among the generators $\{\alpha_{rs} \mid 1 \leq s < r\}$. As in the simply connected case, when $n > 2$ the Y-B relations are of two kinds: those that involve three indices, and those that involve four. Our plan is to find some simple basic relations from which the rest are obtained by applying appropriate symmetries.

The next result can be viewed as a corollary of Theorem 3.1 above. It also points to the type of restriction that one must impose on a permutation σ so that effect of conjugation by the braid corresponding to σ is to yield an element the indices of which are the result of applying σ to the indices.

Proposition 4.1 *Let $\mu_\sigma = \mu_{2i_2} \cdots \mu_{ki_k} \in \mathcal{M}$, and assume that $i_m \geq r$, for $m \geq r$, and $i_m \geq s$, for $m \geq s$. Then, $\mu_\sigma\alpha_{rs}\mu_\sigma^{-1} = \alpha_{\sigma r\sigma s}$, where $\sigma \in \Sigma_k$ is the image of μ_σ under the natural projection $\pi_1(\mathbb{B}_k(\mathbb{R}^2), [q]) \to \Sigma_k$.*

Proof. Consider, first, the expression

$$\mu_\sigma = \mu_{2i_2} \cdots \mu_{ki_k}, \; 1 < i_m \leq m, \, 1 < m \leq k$$

of μ_σ as the product of μ_{mi_m}.

Assume, without loss of generality, that $\sigma(r) \neq r$. Next, suppose that $r < n \leq k$ is the first integer such that $i_n = r$. Then Statement (ix) of Theorem 3.1 implies that $\mu_{mi_m}\alpha_{rs}\mu_{mi_m}^{-1} = \alpha_{rs}$ for all $n < m \leq k$. On the other hand, Statement (vii) implies that $\mu_{ni_n}\alpha_{rs}\mu_{ni_n}^{-1} = \alpha_{ns}$. Now, conjugating by μ_{mi_m} for all $m < n$ leaves the index n unchanged.

Again assume, without loss of generality, that $\sigma(s) \neq s$. Next, let t be the first index such that $i_t = s$. Then, as above, Statement (ix) of Theorem 3.1, implies that conjugation by μ_{uw} for all $u \leq n$ and $w \geq t$ leaves α_{rs} unchanged. Moreover, conjugating by μ_{ut} takes $\alpha_{n,s}$ to $\alpha_{n,t}$. But $\sigma = \prod_{m=2}^{k}(m, m-1) \cdots (i_m+1, i_m)$ takes (r, s) to (n, t). This proves the proposition. □

The preceding proposition suggests the following as appropriate concepts for dealing with the effect of conjugation on the indices.

Definition 4.1 *The permutation $\rho \in \Sigma_k$ is said to be of the* first kind *if, and only if, $\rho(1) < \rho(2) < \rho(3)$. It is said to be of the* second kind *if, and only if, $\rho(1) < \rho(2) < \rho(3) < \rho(4)$.*

As the state of affairs here presents us with new features not encountered in that of the simply connected case, we find it convenient to write ρ_m for the cycle

$$\begin{aligned}\sigma_{m,i_m} &= (i_m, i_m+1, \cdots, m-1, m) \\ &= (m, m-1) \cdots (i_m+1, i_m).\end{aligned}$$

Proposition 4.2 *Given natural numbers n_1, n_2, and n_3 such that $1 \leq n_1 < n_2 < n_3 \leq k$, put $\rho = \rho_3\rho_2\rho_1$, where $\rho_i = (n_i, \cdots, i)$ for $i = 1, 2$, and 3. Then ρ is a permutation of the first kind and*

$$\mu_{n_11}\mu_{n_22}\mu_{n_33}\alpha_{rs}\mu_{n_33}^{-1}\mu_{n_22}^{-1}\mu_{n_11}^{-1} = \alpha_{\rho r \rho s}$$

for $1 \leq s < r \leq 3$.

The proof is an easy consequence of Theorem 3.1 (cf. Proposition 4.1). Similarly, one proves the following analogous result for permutations of the second kind.

Proposition 4.3 *Given $n_1, n_2, n_3, n_4 \in \mathbb{N}$ such that $1 \leq n_1 < n_2 < n_3 < n_4$, put $\rho = \rho_4\rho_3\rho_2\rho_1$, where $\rho_i = (n_i, \cdots, i)$ for $i = 1, 2, 3$, and 4. Then $\rho \in \Sigma_k$ is of the second kind, and*

$$\mu_{n_11}\mu_{n_22}\mu_{n_33}\mu_{n_44}\alpha_{rs}\mu_{n_44}^{-1}\mu_{n_33}^{-1}\mu_{n_22}^{-1}\mu_{n_11}^{-1} = \alpha_{\rho r \rho s},$$

where $1 \leq r < s \leq 4$.

The proof is again a consequence of Proposition 4.1.

Definition 4.2 *The Y-B Relations of the* first kind *involve three indices and are as follows:*

$$\begin{cases} (i) & [\alpha_{\rho 2 \rho 1}\,,\, \alpha_{\rho 3 \rho 1}\alpha_{\rho 3 \rho 2}] = 1 \\ (ii) & [\alpha_{\rho 3 \rho 1}\,,\, \alpha_{\rho 3 \rho 2}\alpha_{\rho 2 \rho 1}] = 1 \\ (iii) & [\alpha_{\rho 3 \rho 2}\,,\, \alpha_{\rho 2 \rho 1}\alpha_{\rho 3 \rho 1}] = 1, \end{cases}$$

where $\rho \in \Sigma_k$ is of the first kind.

One defines those of the second kind similarly.

Definition 4.3 *The Y-B Relations of the* second kind *involve four indices and are as follows:*

$$\begin{cases} (iv) & [\alpha_{\rho 4 \rho 3}\,,\, \alpha_{\rho 2 \rho 1}] = 1 \\ (v) & [\alpha_{\rho 4 \rho 2}\,,\, \alpha_{\rho 3 \rho 2}^{-1}\alpha_{\rho 3 \rho 1}\alpha_{\rho 3 \rho 2}] = 1 \\ (vi) & [\alpha_{\rho 4 \rho 1}\,,\, \alpha_{\rho 3 \rho 2}] = 1 \\ (vii) & [\alpha_{\rho 4 \rho 2}, \alpha_{\rho 4 \rho 3}\alpha_{\rho 3 \rho 1}\alpha_{\rho 4 \rho 3}^{-1}] = 1, \end{cases}$$

where $\rho \in \Sigma_k$ is of the second kind.

Theorem 4.1 *The Y-B Relations of the first and second kind are valid in the fundamental group $\pi_1(\mathbb{F}_k, q)$.*

Proof. Consider the maps $\phi_1 : S \times S \to \mathbb{F}_k(\mathbb{R}^2)$, where (ξ_1, ξ_2) is mapped to $(q_1, q_1 + \xi_1, q_1 + 2\xi_2, \cdots)$ (see the Fig.IV.5).

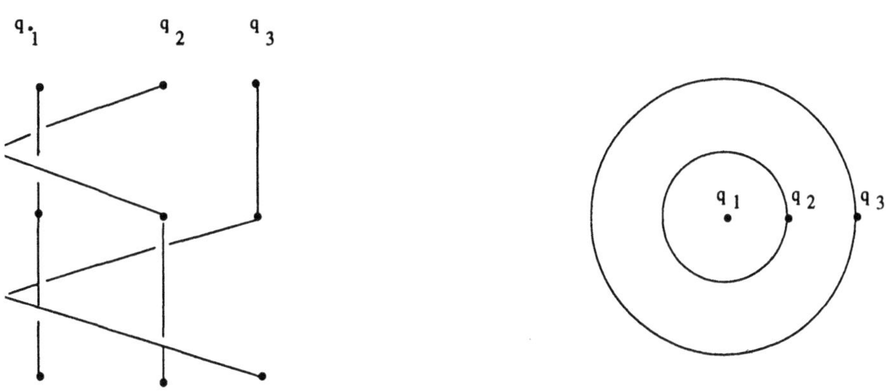

Fig. IV.5. Y-B Relations, 1st Kind

They lead to the relation

$$(i)\ [\alpha_{21}\,,\, \alpha_{31}\alpha_{32}] = 1$$

in the group $\pi_1(\mathbb{F}_k(\mathbb{R}^2), q)$. Now, conjugate Relation (i), first by σ_2 and then by $\sigma_1\sigma_2$, using the appropriate conjugation formulas of Lemma 3.1, to get

$$(ii)\ [\alpha_{31}\,,\,\alpha_{32}\alpha_{21}] = 1 \text{ and}$$
$$(iii)\ [\alpha_{32}\,,\,\alpha_{21}\alpha_{31}] = 1.$$

(Note that, in fact, (i) and (ii) imply (iii).)
To finish the proof, suppose that $\rho \in \Sigma_k$ of the first kind. Then $\rho(1) = n_1 < \rho(2) = n_2 < \rho(3) = n_3$. Choose the permutation $\rho' = \rho_3\rho_2\rho_1$, where ρ_1, ρ_2, and ρ_3 are the cycles defined in Proposition 4.2. Now, conjugate the relations $(i), (ii)$, and (iii) above by $\mu_{n_1 1}, \mu_{n_2 2}$, and $\mu_{n_3 3}$ to establish Statement (i) of the theorem.

To verify the Y-B Relations of the second kind, let us start with the map $\phi_2 : S \times S \to \mathbb{F}_k(\mathbb{R}^2)$, where $(\xi_1, \xi_2) \to (q_1, q_1 + \xi_1, q_2, q_2 + \xi_2, q_3, \cdots)$. as in Fig.IV.6.

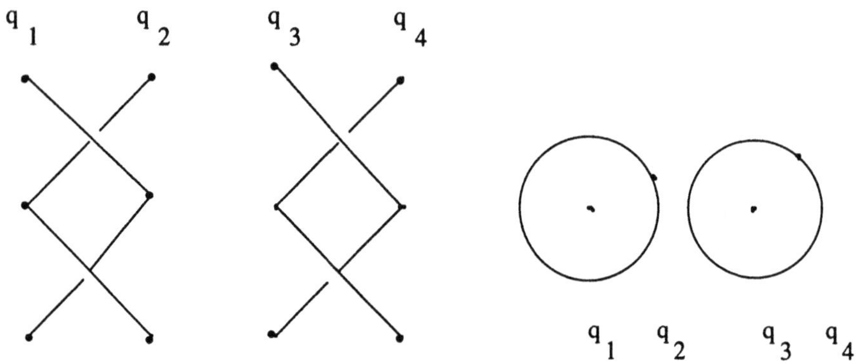

Fig. IV.6. Y-B Relations, 2nd Kind

The map ϕ_2 implies that α_{21}, α_{43} commute with each other, and this, in turn, leads to the following relation

$$(iv)\ [\alpha_{43}, \alpha_{21}] = 1.$$

Conjugate by σ_2, σ_1 to obtain the relation $[\alpha_{32}\alpha_{42}\alpha_{32}^{-1}, \alpha_{31}] = 1$, which implies

$$(v)\ [\alpha_{42}\,,\,\alpha_{32}^{-1}\alpha_{31}\alpha_{32}] = 1.$$

Next, conjugate Relation (v) by σ_1 and, using Lemma 3.2, obtain

$$[\alpha_{21}\alpha_{41}\alpha_{21}^{-1}, (\alpha_{21}\alpha_{31}\alpha_{21}^{-1})^{-1}\alpha_{32}(\alpha_{21}\alpha_{31}\alpha_{21}^{-1})] = 1.$$

The latter can be simplified to yield

$$[\alpha_{41}, (\alpha_{21}\alpha_{31})^{-1}\alpha_{32}(\alpha_{21}\alpha_{31})] = 1.$$

However, $\alpha_{21}\alpha_{31}$ commutes with α_{32}, by Relation (iii), and thus we obtain

$$(vi)\ [\alpha_{41}, \alpha_{32}] = 1.$$

To prove Relation (vii), conjugate Relation (vi) by σ_3 to obtain

$$(vii)\ [\alpha_{43}\alpha_{31}\alpha_{43}^{-1}, \alpha_{42}] = 1.$$

Now, let $\rho \in \Sigma_k$ of the second kind. We may assume that $\rho = \rho_4\rho_3\rho_2\rho_1$, as in Proposition 4.3. With $\mu_{n_1}, \mu_{n_2}, \mu_{n_3}$ and μ_{n_4} being as in Proposition 4.3, conjugate the relations $(iv), (v), (vi)$, and (vii) by the element $\mu = \mu_{n_1}\mu_{n_2}\mu_{n_3}\mu_{n_4}$ to prove that the Y-B relations of the second kind are satisfied in $\pi_1(\mathbb{F}_k, q)$. □

To prove the basic identity $[\alpha_{43}, \alpha_{21}] = 1$, one may, alternatively, appeal to the following assertion.

Assertion 4.1 *Suppose that $\alpha', \beta' \in \Omega(\mathbb{F}_k(\mathbb{R}^2))$ are of the form*

$$\alpha' = (q_1, \cdots, \alpha'_i, q_{i+1}, \cdots, q_k) \text{ and}$$
$$\beta' = (q_1, \cdots, \beta'_j, q_{j+1}, \cdots, q_k),$$

where $i \neq j$. Assume further that $\alpha'_i(s) \neq \beta'_j(t)$ for all $s, t \in S$. Then $[\alpha, \beta] = 0$, where α, β are the homotopy classes in $\pi_1(\mathbb{F}_k(\mathbb{R}^2), q)$ of, respectively, α' and β'.

Proof. Consider the map

$$\phi : S \times S \to \mathbb{F}_k(\mathbb{R}^2),\ (s,t) \mapsto (q_1, \alpha(s), \beta(s), ..., q_k) \in \mathbb{F}_k(\mathbb{R}^2).$$

The map ϕ provides one with the required deformation to establish the relation $\alpha\beta = \beta\alpha$. □

5 A Presentation of $\pi_1(\mathbb{F}_k(\mathbb{R}^2), q)$

Denote by $\mathcal{F}_k$ the free group on the set $\{\alpha_{ks} | 1 \leq s < k\}$, and define the *pure braid group on k strands* $\mathcal{P}_k$ to be the free group generated by the symbols $\{\alpha_{rs} \mid 1 \leq s < r \leq k\}$ modulo the normal closure of the set of Y-B Relations of the first and second kind (see Definitions 4.2 and 4.3 above). We wish to show that $\pi_1(\mathbb{F}_k(\mathbb{R}^2), q)$ is isomorphic to $\mathcal{P}_k$. We give two proofs: both are by induction on k using the fibration $\text{proj}_{k,k-1} : \mathbb{F}_k(\mathbb{R}^2) \to \mathbb{F}_{k-1}(\mathbb{R}^2)$ with $\mathbb{R}^2_{k-1} = \mathbb{R}^2 - Q_{k-1}$ as fiber. The first is more algebraic. The idea in the second is to construct a finite cell complex X_k of dimension two, of which the fundamental group is an extension of $\mathcal{P}_k$ and which contains $\pi(\mathbb{R}^2_{k-1})$ as a normal subgroup. The complex X_k is a subcomplex of a complex homotopy equivalent to $\mathbb{F}_2(\mathbb{R}^2)$ that we construct in Chapter VI. The complex X_k is regarded as a model for the twisted product of $B_{k-1} = \bigvee_{r=2}^{r=k-1}(S_{r1} \vee \cdots \vee S_{rr-1})$ by $(S_{k1} \vee \cdots \vee S_{kk-1})$. The space B_{k-1} is regarded as a 1-skeleton for $\mathbb{F}_{k-1}(\mathbb{R}^2)$, and the twisted product, as the restriction of the fibration $\mathbb{F}_k(\mathbb{R}^2) \to \mathbb{F}_{k-1}(\mathbb{R}^2)$ to B_{k-1}. The attaching maps for the 2-cells are induced by the Y-B relations that describe the conjugation of the generators $\{\alpha_{ks} | 1 \le s < k\}$ by the other generators.

To proceed, consider a more general case of a fibration $\mathrm{p} : E \to B,\ \mathrm{p}^{-1}(b_0) = F$, where $b_0 \in B$ is the basepoint. Recall that a lifting function for the fibration $\mathrm{p} : E \to B$ is a map

$$\lambda : E^* \to E^I,\ \ E^* = \{(\omega, e) \mid \omega(0) = \mathrm{p}(e)\} \subset B^I \times E$$

such that $\tilde{\mathrm{p}} \circ \lambda = \mathrm{id}$, where $\tilde{\mathrm{p}} : E^I \to E^*$ is the map that sends ω to $(\mathrm{p}^I(\omega), \omega(0))$, with p^I being the map induced by p. Let $s : B \to E$ be a section. Then, because we are in the category of ANR(metric) spaces, λ can be chosen so that $\lambda(\omega, e_0)(t) \in s(B)$ for all $t \in I$, where $s(b_0) = e_0 \in F \subset E$. The lifting function λ induces, in the usual manner, an action

$$\phi_\lambda : \Omega(B) \times F \to F,\ \ (\omega, e) \mapsto \lambda(\omega, e)(1).$$

Since λ keeps the section invariant, ϕ_λ induces an action, also denoted by ϕ_λ,

$$\phi_\lambda : \Omega(B) \times \Omega(F) \to \Omega(F).$$

Observe here that a section $s : \mathrm{B} \to \mathrm{E}$ induces a weak homotopy equivalence

$$\Omega(B) \times \Omega(\mathrm{F}) \to \Omega(\mathrm{F}),\ (\alpha, \beta) \mapsto \Omega(\mathrm{s})(\alpha) * \beta,$$

where $\Omega(s) : \Omega(\mathrm{B}) \to \Omega(\mathrm{E})$ is the map induced by s. We shall identify $\pi_1(\mathrm{B}, \mathrm{b}_0)$ with its image in $\pi_1(\mathrm{E}, e_0)$.

Proposition 5.1 *Assume that* $s : \mathrm{B} \to \mathrm{E}$ *is a section. Then, the action* ϕ_λ *induces the map* $\pi_1(B, b_0) \times \pi_1(F, e_0) \to \pi_1(F, e_0)$, *where* $[\phi_\lambda(\alpha^{-1}, \beta)] = \alpha\beta\alpha^{-1}$ *in* $\pi_1(E, e_0)$.

Proof. To give the proof, note that the lifting function ϕ_λ induces a map $\psi : S \times I \to E$, where $\psi | S \times \{0\} = \beta$, while $\psi | \{1\} \times I = \alpha$ and $\psi \mid \{1\} = \phi_\lambda(\alpha^{-1}, \beta)$. Now the proposition follows easily. □

In the case at hand, $E = \mathbb{F}_k(\mathbb{R}^2)$, $B = \mathbb{F}_{k-1}(\mathbb{R}^2)$, and p is the natural projection on the first $(k-1)$ factors. The fiber is $\mathbb{R}^2_{k-1}$, and the section $s : \mathbb{F}_{k-1}(\mathbb{R}^2) \to \mathbb{F}_k(\mathbb{R}^2)$ takes the elements α_{rs}, with $r < k$, to the corresponding elements in the total space.

In order to define the complex X_k, we need to describe the conjugation of the generators α_{ks} by α_{rs}, with $r < k$. The following identities in the group $\mathcal{P}_k$, which we shall call, respectively, identities of the first and second kind, address this problem.

Proposition 5.2 *Let $\alpha_{rs} \in \mathcal{P}_k$ be a generator, and suppose that $r < k$. Then*

$$(i)\ \alpha_{rs}^{-1}\alpha_{ks}\alpha_{rs} = \alpha_{kr}\alpha_{ks}\alpha_{kr}^{-1} \text{ and}$$
$$(ii)\ \alpha_{rs}^{-1}\alpha_{kr}\alpha_{rs} = \alpha_{ks}\alpha_{kr}\alpha_{ks}^{-1},$$

for all generators α_{ks}, $\alpha_{kr} \in \mathcal{P}_k$ such that $1 \le s < r < k$.

Proof. Let $\rho \in \Sigma_k$ be a permutation of the first kind such that $\rho(3) = k$, $\rho(2) = r$, $\rho(1) = s$. To prove the first statement, note that the second Y-B relation is equivalent to the relation

$$\alpha_{\rho 3\rho 1}\alpha_{\rho 2\rho 1}\alpha_{\rho 3\rho 2}\alpha_{\rho 3\rho 1}^{-1} = \alpha_{\rho 2\rho 1}\alpha_{\rho 3\rho 2}.$$

This implies immediately that

$$\alpha_{\rho 2\rho 1}^{-1}\alpha_{\rho 3\rho 1}\alpha_{\rho 2\rho 1}\alpha_{\rho 3\rho 2}\alpha_{\rho 3\rho 1}^{-1}. = \alpha_{\rho 3\rho 2}.$$

Multiplying on the right, first by $\alpha_{\rho 3\rho 1}$ and then by $(\alpha_{\rho 3\rho 2}^{-1})$, we obtain the statement

$$\alpha_{\rho 2\rho 1}^{-1}\alpha_{\rho 3\rho 1}\alpha_{\rho 2\rho 1} = \alpha_{\rho 3\rho 2}\alpha_{\rho 3\rho 1}\alpha_{\rho 3\rho 2}^{-1}$$

of (i), as required.

The proof of (ii) is similar. The third Y-B relation is equivalent to

$$\alpha_{\rho 3\rho 2}\alpha_{\rho 2\rho 1}\alpha_{\rho 3\rho 1}\alpha_{\rho 3\rho 2}^{-1} = \alpha_{\rho 2\rho 1}\alpha_{\rho 3\rho 1},$$

which, in turn, leads immediately to

$$\alpha_{\rho 2\rho 1}^{-1}\alpha_{\rho 3\rho 2}\alpha_{\rho 2\rho 1} = \alpha_{\rho 3\rho 1}\alpha_{\rho 3\rho 2}\alpha_{\rho 3\rho 1}^{-1}.$$

The latter is the assertion of (ii). This proves the proposition. □

The next result is an immediate corollary.

Corollary 5.1 *The following identities of the first kind are valid in $\mathcal{P}_k$.*

$$(i)\ [\alpha_{\rho 3\rho 2}, \alpha_{\rho 3\rho 1}] = [\alpha_{\rho 2\rho 1}^{-1}, \alpha_{\rho 3\rho 1}] \text{ and}$$
$$(ii)\ [\alpha_{\rho 3\rho 1}, \alpha_{\rho 3\rho 2}] = [\alpha_{\rho 2\rho 1}^{-1}, \alpha_{\rho 3\rho 2}],$$

In particular, taking ρ to be such that $\rho(3) = k, \rho(2) = r$, and $\rho(1) = s$, we obtain the relations

$$(i)\ [\alpha_{rs}^{-1}, \alpha_{ks}] = [\alpha_{kr}, \alpha_{ks}] \text{ and}$$
$$(ii)\ [\alpha_{rs}^{-1}, \alpha_{kr}] = [\alpha_{ks}, \alpha_{kr}]$$

for $1 \le s < r \le k$.

Observe that only the Y-B relation (ii) is used in the proof of the first statement of Proposition 5.2 and only the Y-B relation (iii), in the proof of the second statement. Moreover, the arguments are reversible. Therefore, the two statements of Proposition 5.2 are equivalent to the Y-B relations (ii) and (iii), respectively.

The next result describes the conjugation of an element α_{rs} by α_{tu} when the indices intertwine in the sense that $1 \leq u < s < t < r \leq k$. It involves the Y-B relations of the second kind.

Proposition 5.3 *Let $\alpha_{rs} \in \mathcal{P}_k$ be a generator such that $r < k$. Then, given α_{kt}, we have $\alpha_{rs}^{-1}\alpha_{kt}\alpha_{rs} = \omega_{krst}\alpha_{kt}\omega_{krst}^{-1}$ in $\mathcal{P}_k$, where ω_{krst} equals $[\alpha_{ks}, \alpha_{kr}]$, if $1 \leq s < t < r < k$, and equals 1, if $1 \leq s < r < t < k$ or $1 \leq t < s < r < k$.*

Proof of Case $1 \leq s < r < t < k$

Choose a permutation $\rho \in \Sigma_k$ of the second kind such that $\rho(4) = k$, $\rho(3) = t$, $\rho(2) = r$, $\rho(1) = s$. Then, the Y-B relation (iv) implies that α_{rs}, α_{kt} commute with each other. Hence, $\alpha_{rs}\alpha_{kt}\alpha_{rs}^{-1} = \alpha_{kt}$ and, therefore, $\omega_{krst} = 1$, as required.

Proof of Case $1 \leq t < s < r < k$

Choose a $\rho \in \Sigma_k$ such that $\rho(4) = k$, $\rho(3) = r$, $\rho(2) = s$, and $\rho(1) = t$. Then the Y-B relation (vi) implies that α_{rs}, α_{kt} commute with each other. Take $\omega_{krst} = 1$ to prove the assertion.

Proof of Case $1 \leq s < t < r < k$

Let $\rho \in \Sigma_k$ be such that $\rho(4) = k$, $\rho(3) = r$, $\rho(2) = t$, $\rho(1) = s$. Now, using the Y-B relation (vii), we obtain

$$\alpha_{\rho4\rho3}\alpha_{\rho3\rho1}\alpha_{\rho4\rho3}^{-1}.\alpha_{\rho4\rho2}.\alpha_{\rho4\rho3}\alpha_{\rho3\rho1}^{-1}\alpha_{\rho4\rho3}^{-1} = \alpha_{\rho4\rho2}.$$

On conjugating both sides by $\alpha_{\rho3\rho1}^{-1}$, we readily obtain the relation

$$[\alpha_{\rho3\rho1}^{-1}, \alpha_{\rho4\rho3}].\alpha_{\rho4\rho2}.[\alpha_{\rho3\rho1}^{-1}, \alpha_{\rho4\rho3}]^{-1} = \alpha_{\rho3\rho1}^{-1}.\alpha_{\rho4\rho2}.\alpha_{\rho3\rho1}.$$

According to the previous corollary, we have

$$[\alpha_{\rho3\rho1}^{-1}, \alpha_{\rho4\rho3}] = [\alpha_{\rho4\rho1}, \alpha_{\rho4\rho3}].$$

Hence, taking $\omega_{krst} = [\alpha_{\rho4\rho1}, \alpha_{\rho4\rho3}]$, one proves the required statement. This finishes the proof of the proposition. □

Note that the Y-B relations involved in the proof of the preceding relations are equivalent to the respective assertions of the proposition. The following result is an easy consequence of the preceding proposition.

Corollary 5.2 *Suppose that* $1 \leq s < t < r < k$. *Then,*

$$[\alpha_{\rho 3 \rho 1}^{-1}, \alpha_{\rho 4 \rho 2}] = [[\alpha_{\rho 4 \rho 1}, \alpha_{\rho 4 \rho 3}], \alpha_{\rho 4 \rho 2}]$$

holds in $\mathcal{P}_k$ *for all* $\rho \in \Sigma_k$ *of the second kind.*

Consider next the projection

$$\mathrm{proj}_k : \mathcal{P}_k \to \mathcal{P}_{k-1}$$

that takes each α_{rs} for $r < k$ to itself and each of the remaining elements α_{ku} to 1. This morphism is the analogue of the morphism of the fundamental groups induced by the natural projection $\mathbb{F}_k(\mathbb{R}^2) \to \mathbb{F}_{k-1}(\mathbb{R}^2)$.

Following the notation of the previous chapters, put, for simplicity,

$$\mathbb{R}^2_{k-1} = \mathbb{R}^2 - Q_{k-1} \ , \ \mathbb{F}_k = \mathbb{F}_k(\mathbb{R}^2).$$

Theorem 5.1 *There is a morphism* $\phi_k : \mathcal{P}_k \to \pi_1(\mathbb{F}_k(\mathbb{R}^2), q)$ *that makes the diagram*

$$\begin{array}{lccccccc} & 1 \to & \mathcal{F}_k & \to & \mathcal{P}_k & \to & \mathcal{P}_{k-1} & \to 1 \\ \mathcal{D}_k : & & \downarrow & & \downarrow & & \downarrow & \\ & 1 \to & \pi_1(\mathbb{R}^2_{k-1}, q) & \to & \pi_1(\mathbb{F}_k, q) & \to & \pi_1(\mathbb{F}_{k-1}, q^{k-1}) & \to 1 \end{array}$$

commute, and the vertical morphisms are isomorphisms.

We offer two proofs: the first is more algebraic and is based on the (normality) identities of Propositions 5.2, 5.3, and Corollary 5.2; the second is more geometric and regards $\pi_1(\mathbb{F}_{k-1}(\mathbb{R}^2), q)$ as a group of automorphisms of $\mathcal{F}_k$. It follows the outline mentioned earlier in the section.

First Proof of Theorem 5.1

Let $\phi_k : \mathcal{P}_k \to \pi_1(\mathbb{F}_k(\mathbb{R}^2), q)$ be the homomorphism that takes $\alpha_{rs} \in \mathcal{P}_k$ to $\alpha_{rs} \in \pi_1(\mathbb{F}_k(\mathbb{R}^2), q)$. (The use of the same notation for elements in $\mathcal{P}_k$ and $\pi_1(\mathbb{F}_k(\mathbb{R}^2), q)$ should cause no difficulty.) The morphism ϕ_k is well-defined and surjective.

Proceed by induction on k. The theorem is obviously true when $k = 2$. Next, define $\mathcal{U}_k$ to be the subgroup of $\mathcal{P}_k$ generated by $\{\alpha_{k1}, \cdots, \alpha_{kk-1}\}$. One sees easily that $\mathcal{U}_k$ is normal by virtue of Propositions 5.2, 5.3, and Corollary 5.2. This implies that the upper sequence in $\mathcal{D}_k$, with $\mathcal{U}_k$ replacing $\mathcal{F}_k$, is exact (see [23, Crowell-Fox], p.43). The induction assumption is that $\phi_{k-1} : \mathcal{P}_{k-1} \to \pi_1(\mathbb{F}_{k-1}(\mathbb{R}^2), q)$ is an isomorphism. As $\phi_k : \mathcal{P}_k \to \pi_1(\mathbb{F}_k(\mathbb{R}^2), q)$ is surjective, it follows that $\phi_k : \mathcal{P}_k \to \pi_1(\mathbb{F}_k(\mathbb{R}^2), q)$ takes $\mathcal{U}_k$ to the image of $\pi_1(\mathbb{R}^2_{k-1}, q) \to \pi_1(\mathbb{F}_k(\mathbb{R}^2), q)$. But this induced morphism is clearly surjective, and $\pi_1(\mathbb{R}^2_{k-1}, q)$ is the free group on the $\{\alpha_{k1}, \cdots, \alpha_{kk-1}\}$. Consequently, $\mathcal{U}_k$ is free, and $\phi_k|\mathcal{U}_k$ is an isomorphism of $\mathcal{U}_k$ onto the image of $\pi_1(\mathbb{R}^2_{k-1}, q) \to \pi_1(\mathbb{F}_k(\mathbb{R}^2), q)$. By the five-lemma, ϕ_k is itself an isomorphism. Thus the theorem is established. □

Second Proof of Theorem 5.1

To construct the desired 2-dimensional complex X_k, proceed as follows. For every (r, s) such that $1 \leq s < r \leq (k-1)$, denote by ϕ_{rs} the automorphism

$$\phi_{rs} : \mathcal{F}_k \to \mathcal{F}_k,\ \alpha_{kt} \mapsto \gamma_{ksrt}\alpha_{kt}\gamma_{ksrt}^{-1},$$

where

$$\gamma_{ksrs} = \begin{cases} \alpha_{kr}, \text{ when } t = s \\ \omega_{ksrt}, \text{ otherwise,} \end{cases}$$

with ω_{ksrt} being the element of Proposition 5.3 (cf. Proposition 5.2). Put $\mathrm{F_k} = \mathrm{S_{k1}} \vee \cdots \vee \mathrm{S_{kk-1}}$. Note that the elements α_{ks}, $1 \leq s < k$ induce a homotopy equivalence $\mathrm{F_k} \to \mathbb{R}^2_{k-1}$. Since $\mathrm{F_k}$ is aspherical and $\mathcal{F}_k \cong \pi_1(\mathbb{R}^2_{k-1}, q)$, there is a homotopy equivalence $\phi'_{rs} : \mathrm{F_k} \to \mathrm{F_k}$ such that $\pi_1(\phi'_{rs})$ is the automorphism $\phi_{rs} : \mathcal{F}_k \to \mathcal{F}_k$.

Definition 5.1 *Let*

$$X_k = \cup_{rs} D_{rs} \times \mathrm{F_k} \cup_{\phi_{rs}} \mathrm{F_k},$$

where D_{rs} is a one dimensional disk, and $1 \leq s < r < k$.

Observe that X_k is the union of the mapping tori of the maps ϕ'_{rs} along $\mathrm{F_k}$.

Assertion 5.1 *Let* $\mathrm{B_{k-1}} = \bigvee_{\mathrm{r=2}}^{\mathrm{k-1}}(\mathrm{S_{r1}} \vee \cdots \vee \mathrm{S_{rr-1}})$, *and denote by*

$$\mathrm{p}_k : X_k \to \mathrm{B_{k-1}}$$

the projection on the first factor. Then p_k is the projection of a quasi fibration the fiber of which is $\mathrm{F_k} \simeq \mathbb{R}^2_{k-1}$.

The proof follows easily from the fact that the attaching maps are homotopy equivalences, since the maps ϕ'_{rs} are isomorphisms (see [26, Dold-Lashof]).

Note that the complex $\mathrm{B_{k-1}}$ can be regarded as the 1-dimensional skeleton of $\mathbb{F}_{k-1}(\mathbb{R}^2)$, since $\{\alpha_{rs} | 1 \leq s \leq r < k-1\}$ generates $\pi_1(\mathbb{F}_{k-1}(\mathbb{R}^2), q^{k-1})$. Consequently, X_k can be viewed as the new part of the 2-dimensional skeleton, the other part coming from $\mathbb{F}_{k-1}(\mathbb{R}^2)$.

To prove the theorem, proceed, as in the first proof, inductively on k. The theorem is trivially true for the case $k = 2$. Assuming that it is true for $k-1$, we consider the fibration $\mathrm{proj}_{k,k-1} : \mathbb{F}_k(\mathbb{R}^2) \to \mathbb{F}_{k-1}(\mathbb{R}^2)$, the fiber of which at the basepoint is $\mathbb{R}^2_{k-1} \simeq (S_{k1} \vee \cdots \vee S_{kk-1})$, where, as in earlier chapters, S_{kr} stands for the generator α_{kr}. Now, there is a fiber-preserving map $\psi_k : X_k \to \mathbb{F}_k(\mathbb{R}^2)$ such that the diagram

$$\begin{array}{ccccc} \mathbb{R}^2_{k-1} & \to & X_k & \to & \mathrm{B_{k-1}} \\ \downarrow & & \downarrow & & \downarrow \\ \mathbb{R}^2_{k-1} & \to & \mathbb{F}_k & \to & \mathbb{F}_{k-1}(\mathbb{R}^2) \end{array}$$

commutes. So, by virtue of the Van Kampen-Seifert Theorem, it is easy to see that X_k is a complex with fundamental group generated by $\{\alpha_{rs} \mid 1 \leq s < r \leq k\}$ modulo the normal closure of the relations of Propositions 5.2 and 5.3 above. Therefore, the morphism $\pi_1(\psi_k) : \pi_1(X_k, q) \to \mathcal{P}_k$ is surjective. Moreover, we have the commutative diagram

$$\begin{array}{ccccccccc} 1 & \to & \pi_1(\mathbb{R}^2_{k-1}, q) & \to & \pi_1(X_k, q) & \to & \pi_1(\mathrm{B}_{k-1}, q) & \to & 1 \\ & & \downarrow & & \downarrow & & \downarrow & & \\ 1 & \overset{?}{\to} & \mathcal{F}_k & \to & \mathcal{P}_k & \to & \mathcal{P}_{k-1} & \to & 1 \\ & & \downarrow & & \downarrow & & \downarrow & & \\ 1 & \to & \pi_1(\mathbb{R}^2_{k-1}, q) & \to & \pi_1(\mathbb{F}_k, q) & \to & \pi_1(\mathbb{F}_{k-1}, q^{k-1}) & \to & 1, \end{array}$$

in which the first and third horizontal sequences are exact, because they are part of the homotopy exact sequences of fibrations with sections. As the image of $(\pi_1(\mathbb{R}^2_{k-1}, q)$ in $\pi_1(X_k, q))$ is normal, and $\pi_1(\mathbb{R}^2_{k-1}, q) \to \mathcal{F}_k$ is an isomorphism, it follows that the image of $\mathcal{F}_k \to \mathcal{P}_k$ is also normal. Note that $\mathcal{F}_k \to \pi_1(\mathbb{R}^2_{k-1}, q)$ is injective, because $\pi_1(\mathbb{R}^2_{k-1}, q) \cong \mathcal{F}_k$, and $\pi_1(\mathbb{R}^2_{k-1}, q) \to \pi_1(\mathbb{F}_k, q)$ is injective. Hence, $\mathcal{F}_k$ is mapped onto $\mathcal{U}_k \subset \pi_1(\mathbb{R}^2_{k-1}, q)$. This implies that $\mathcal{F}_k$ is isomorphic to the kernel of the projection $\mathcal{P}_k \to \mathcal{P}_{k-1}$ (see [23, Crowell-Fox], p.43 or [101, Spanier], Theorem 3.10, p. 147). Hence, the middle sequence is exact. Also, the first and third vertical morphisms going from the second row to the third are isomorphisms. This finishes the proof of the theorem. □

As $\mathcal{F}_k$ is normal in $\pi_1(\mathbb{F}_k(\mathbb{R}^2), q)$, the action of $\pi_1(\mathbb{F}_k(\mathbb{R}^2), q)$ on itself by inner automorphisms induces a morphism $\iota : \pi_1(\mathbb{F}_k(\mathbb{R}^2), q) \to \mathrm{Aut}(\mathcal{F}_\mathrm{k})$, $\alpha_{\mathrm{rs}} \mapsto \iota_{\mathrm{rs}}$, where $\iota_{rs}(\beta) = \alpha_{rs}\beta\alpha_{rs}^{-1}$ (cf.[9, Birman], Corollary 1.8.3).

Corollary 5.3 *The morphism*

$$\iota_k : \pi_1(\mathbb{F}_{k-1}(\mathbb{R}^2), q) \to \mathrm{Aut}(\mathcal{F}_\mathrm{k}),\ \alpha_{\mathrm{rs}} \mapsto \phi_{\mathrm{rs}}$$

is a monomorphism that takes α_{rs} to ϕ_{rs}.

Proof. According to Proposition 5.1, the fibration $\mathrm{proj}_{k,k-1} : \mathbb{F}_k(\mathbb{R}^2) \to \mathbb{F}_{k-1}(\mathbb{R}^2)$ leads to an action of

$$\pi_1(\mathbb{F}_{k-1}(\mathbb{R}^2), q) \times \pi_1(\mathbb{R}^2_{k-1}, q) \to \pi_1(\mathbb{R}^2_{k-1}, q).$$

If $\pi_1(\mathbb{R}^2_{k-1}, q)$ is regarded as a subgroup of $\pi_1(\mathbb{F}_k(\mathbb{R}^2_k), q)$, then the action is that where $(\alpha, \beta) \mapsto \alpha\beta\alpha^{-1}$. In other words, $\pi_1(\mathbb{F}_{k-1}(\mathbb{R}^2), q)$, regarded as a subgroup of $\pi_1(\mathbb{F}_k(\mathbb{R}^2_k), q)$, acts on $\pi_1(\mathbb{R}^2_{k-1}, q)$ by inner automorphisms. But Propositions 5.2 and 5.3 imply that the action is that induced by ι. □

The following result is of some interest (cf. Theorem 4.2 of [9, Birman]).

Proposition 5.4 *There is a natural isomorphism*

$$\pi_1(\mathbb{F}_k(\mathbb{R}^2), q)/\mathrm{Cent}(\pi_1(\mathbb{F}_\mathrm{k}(\mathbb{R}^2), \mathrm{q})) \to \pi_0(\mathrm{Top}_{\mathrm{sk}}(\mathbb{R}^2)).$$

Proof. Note that $\mathrm{Top}_s(\mathbb{R}^2)$ acts transitively on $\mathbb{F}_k(\mathbb{R}^2)$ and that $\mathrm{Top}_{sk}(\mathbb{R}^2)$ is the isotropy subgroup of $q = (q_1, \cdots, q_k)$. Hence, $\mathrm{Top}_s(\mathbb{R}^2)/\mathrm{Top}_{sk}(\mathbb{R}^2) \cong \mathbb{F}_k(\mathbb{R}^2)$ (see §2 of Chapter I). As $\mathrm{Top}_s(\mathbb{R}^2) \simeq SO_1$, the long homotopy exact sequence of $\mathrm{Top}_s(\mathbb{R}^2) \to \mathbb{F}_k(\mathbb{R}^2)$ leads immediately to the exact sequence

$$\begin{aligned} \cdots \to \pi_1(\mathrm{Top}_s(\mathbb{R}^2), \mathrm{id}) \to \\ \pi_1(\mathbb{F}_k(\mathbb{R}^2), q) \to \pi_0(\mathrm{Top}_{sk}(\mathbb{R}^2), \mathrm{id}) \to \\ \pi_0(\mathrm{Top}_s(\mathbb{R}^2), \mathrm{id}) \to 1. \end{aligned}$$

However, $\pi_0(\mathrm{Top}_s(\mathbb{R}^2), \mathrm{id}) = 1$, and

$$\mathrm{image}(\pi_1(\mathrm{Top}_s(\mathbb{R}^2), \mathrm{id}) \to \pi_1(\mathbb{F}_k(\mathbb{R}^2), q)) = \mathrm{Cent}(\pi_1(\mathbb{F}_\mathrm{k}(\mathbb{R}^2), \mathrm{q})).$$

The result now follows. □

The characterization of the pure braid group of $\mathbb{R}^2$ modulo its center given in Proposition 5.4 is similar to a characterization of the pure braid group of a 0-connected manifold M based on the group $\mathrm{Top}_{ss}(M)$ of strongly stable homeomorphisms (see §2 of Chapter I). Note that the group $\mathrm{Top}_{ss}(M)$ is path-connected, because each $g \in \mathrm{Top}_{ss}(M)$ is isotopic through strongly stable homeomorphisms to the identity. Also, the evaluation map

$$\mathrm{eval} : \mathrm{Top}_{ss}(M) \to \mathbb{F}_k(M),\ g \in \mathrm{Top}_{ss}(M) \mapsto (g.q_1, \cdots, g.q_k)$$

yields a principal bundle with $\mathrm{Top}_{ssk}(M)$ as fiber, where the latter is the isotropy subgroup of $(q_1, \cdots, q_k) \in \mathbb{F}_k(M)$. Observe that the tail of the homotopy exact sequence is

$$\cdots \to \pi_1(\mathrm{Top}_{ss}(M), \mathrm{id}) \to \pi_1(\mathbb{F}_k(M), q) \to \pi_0(\mathrm{Top}_{ssk}(M), \mathrm{id}) \to 0,$$

since $\pi_0(\mathrm{Top}_{ss}(M), \mathrm{id})$ is trivial. With a simple argument, valid for a general M, one sees that

$$\ker(\pi_1(\mathbb{F}_k(M), q) \to \pi_0(\mathrm{Top}_{ssk}(M), \mathrm{id}) \cong \mathrm{Cent}(\pi_1(\mathbb{F}_\mathrm{k}(\mathrm{M}), \mathrm{q})).$$

Thus we obtain the following result.

Proposition 5.5 *The homotopy exact sequence of the evaluation map* eval : $\mathrm{Top}_{ss}(M) \to \mathbb{F}_k(M)$ *induces an isomorphism*

$$\pi_1(\mathbb{F}_k(M), q)/\mathrm{Cent}(\pi_1(\mathbb{F}_\mathrm{k}(\mathrm{M}), \mathrm{q})) \to \pi_0(\mathrm{Top}_{\mathrm{ssk}}(\mathrm{M}), \mathrm{id}).$$

Proposition 5.5 is clearly the analogue of Proposition 5.4. There is a similar result concerning the braid group $\mathcal{B}_{k-1}$. As above, $\mathrm{Top}_s(\mathbb{R}^2)$ acts transitively on $\mathbb{B}_k(\mathbb{R}^2)$. Denote the isotropy subgroup of $\{q_1, \cdots, q_k\}$ by $\widetilde{\mathrm{Top}}_{sk}$. On considering the long exact sequence of the evaluation map $\mathrm{Top}_s(\mathbb{R}^2) \to \mathbb{B}_k(\mathbb{R}^2)$, one obtains directly an isomorphism

$$\pi_1(\mathbb{B}(\mathbb{R}^2),[q])/\mathrm{Cent}(\pi_1(\mathbb{B}(\mathbb{R}^2),[q])) \to \pi_0(\widetilde{\mathrm{Top}_{\mathrm{sk}}},\mathrm{id}).$$

Consider next the fibration $p \circ \mathrm{proj}_{k,k-1} : \mathbb{F}_k(\mathbb{R}^2) \to \mathbb{B}_{k-1}(\mathbb{R}^2)$, and let $\phi_\lambda : \Omega(\mathbb{B}_{k-1}) \times \mathbb{R}^2_{k-1} \to \mathbb{R}^2_{k-1}$ be the action defined by a lifting function λ. The action ϕ_λ induces an action $\pi_1(\mathbb{B}_{k-1},[q]) \times \pi_1(\mathbb{R}^2_{k-1},q) \to \pi_1(\mathbb{R}^2_{k-1},q)$, which, in turn, induces a homomorphism $\iota'_{k-1} : \mathcal{B}_{k-1} \to \mathrm{Aut}(\mathcal{F}_k)$. Using Lemma 3.3 and Propositions 5.2 and 5.3, one verifies that $\iota'_{k-1}(\sigma_s)$ is such that $\alpha_{ks} \mapsto \alpha_{ks}\alpha_{ks+1}\alpha_{ks}^{-1}$, $\alpha_{ks+1} \mapsto \alpha_{ks}$ and is the identity otherwise (cf. [9, Birman], p. 26).

There is another way of describing the preceding presentation. Let $\phi : \mathcal{B}_k \to \Sigma_k$ be the morphism of Lemma 3.1. Recall that ϕ takes $\mu_\sigma \in \mathcal{B}_k$ to $\sigma \in \Sigma_k$. Put $\mathcal{B}'_{k-1} = \phi^{-1}(\Sigma_{k-1})$. Observe that $\mathcal{B}'_{k-1} \cap \mathcal{B}_{k-1} = \mathcal{F}_k$ and $\mathcal{B}'_{k-1} = \mathcal{B}_{k-1}\mathcal{F}_k$. The following result, due to [16, Chow], follows easily.

Proposition 5.6 $\mathcal{B}'_{k-1} = \mathcal{B}_{k-1} \rtimes \mathcal{F}_k$. *In particular,* $\mathcal{B}_{k-1}$ *acts on* $\mathcal{F}_k$.

6 When Are the Canonical Fibrations Trivial?

Consider the natural projection

$$\mathrm{proj}_{k,k-1} : \mathbb{F}_k(\mathbb{R}^2) \to \mathbb{F}_{k-1}(\mathbb{R}^2)$$

that sends a tuple $(x_1, \cdots, x_k)$ to $(x_1, \cdots, x_{k-1})$. We showed in Chapter II that the pull-back to a circle by any of the maps α'_{rs} is trivial. We present here a stronger theorem.

Theorem 6.1 *The bundle* $\mathrm{proj}_{k,r} : \mathbb{F}_k(\mathbb{R}^2) \to \mathbb{F}_r(\mathbb{R}^2)$ *is fiber-homotopically trivial if, and only if,* $r \leq 2$.

Proof of ($\Leftarrow$)

If $r = 1$, then $\mathbb{F}_1(\mathbb{R}^2) = \mathbb{R}^2$, which is contractible, and the theorem is true. Suppose now that $r = 2$. We wish to show that the bundle of the theorem is trivial. Since the map $\alpha_{21} : S^1 \to \mathbb{F}_2(\mathbb{R}^2)$ is a homotopy equivalence, it is enough to show that the restriction $\mathbb{F}'(\mathbb{R}^2) \to S_{21}$ of the bundle

$$\mathrm{proj}_{k,2} : \mathbb{F}_k(\mathbb{R}^2) \to \mathbb{F}_2(\mathbb{R}^2)$$

to S_{21} is trivial. But the latter is the bundle associated with the tangent bundle of S^1, according to Corollary 3.2 of Chapter I. Hence, it is trivial. Explicitly, for $\xi \in S_{21}$, let $\rho(\xi)$ be the rotation defined by the condition that $\rho(\xi)(e_1) = \xi$. Then, a trivialization $\psi : S_{21} \times \mathbb{F}_{k-2,2}(\mathbb{R}^2) \to \mathbb{F}'_k(\mathbb{R}^2)$ is defined by the rule $\psi(\xi, x_3, \cdots, x_k) = (\xi, \rho(\xi)(x_3), \cdots, \rho(\xi)(x_k))$. This also proves the desired implication.

Proof of ($\Rightarrow$)

The proof is similar to that of Theorem 8.1 of Chapter III. First, denote by $proj'_{k,r} : \mathbb{F}'_{k,r}(\mathbb{R}^2) \to S_{21}$ the pull-back of the bundle $\mathrm{proj}_{k,r} : \mathbb{F}_k(\mathbb{R}^2) \to \mathbb{F}_r(\mathbb{R}^2)$ by the map $\alpha_{21} : S^1 \to \mathbb{F}_r(\mathbb{R}^2)$. Clearly, the fiber is $\mathbb{F}_{k-r,r}(\mathbb{R}^2)$. Suppose now that

$$\begin{array}{ccc} \mathbb{F}_r(\mathbb{R}^2) \times \mathbb{F}_{k-r,r}(\mathbb{R}^2) & \overset{\phi}{\to} & \mathbb{F}_k(\mathbb{R}^2) \\ \downarrow & & \downarrow \\ \mathbb{F}_r(\mathbb{R}^2) & \overset{id}{\to} & \mathbb{F}_r(\mathbb{R}^2) \end{array}$$

is a fiber homotopy equivalence over $\mathbb{F}_r(\mathbb{R}^2)$. Assume, without loss of generality, that $\phi|(\mathbb{R}^2_r)$ is the identity. Then ϕ induces the trivialization

$$\begin{array}{ccc} S_{21} \times \mathbb{F}_{k-r,r}(\mathbb{R}^2) & \overset{\phi}{\to} & \mathbb{F}'_k(\mathbb{R}^2) \\ \downarrow & & \downarrow \\ S_{21} & \overset{id}{\to} & S_{21}. \end{array}$$

Recall that $\pi_1(\mathbb{F}'_k(\mathbb{R}^2), q) \cong \pi_1(S_{21}) \rtimes \pi_1(\mathbb{F}_{k-r,r}(\mathbb{R}^2), q)$. Moreover, the representation of $\mathbb{F}_{k-r,r}(\mathbb{R}^2)$ as a twisted product leads to $\pi_1(\mathbb{F}_{k-r,r}(\mathbb{R}^2) \cong \rtimes_{s=r+1}^{k} \pi_1(\vee_{t=1}^{s-1} S_{st})$. For simplicity, put

$$\Gamma = \pi_1(\mathbb{F}_{k-r,r}(\mathbb{R}^2),\ \Gamma_1 = \Gamma, \text{ and } \Gamma_t = [\Gamma, \Gamma_{t-1}], \text{ for } t > 1.$$

Now the morphism $\pi_1(\phi) : \pi_1(S_{21}, 1) \times \Gamma \to \pi_1(\mathbb{F}'_k(\mathbb{R}^2), q)$ is such that

$$\text{(I)}\quad \alpha_{21} \mapsto \alpha_{21}\omega,$$

where ω is a word in the elements $\{\alpha_{js} | r+1 \leq j \leq k,\ s < j\}$. Rewriting the previous morphism modulo Γ_2, we obtain

$$\text{(I)}'\quad \alpha_{21} \mapsto \alpha_{21} \sum_{s=r+1}^{k} \omega_{s*},$$

where each β_{s*}, in $\pi_1(S_{s1} \vee \cdots \vee S_{ss-1}, q)$. As the commutators $[\alpha_{r+11}, \alpha_{21}] = 1$ in $\pi_1(S_{21} \times \mathbb{F}'_{k-r,r}(\mathbb{R}^2), (1, q))$, it follows that

$$\text{(II)}\quad [\alpha_{r+11}, \alpha_{21}] + \sum_{s=r+1}^{k} [\alpha_{r+11}, \beta_{s*}] = 1$$

modulo Γ_2. Now, the Y-B relations imply that

$$[\alpha_{r+11}, \alpha_{21}],\ [\alpha_{r+11}, \beta_{r+1*}] \in \pi_1(S_{r+11} \vee \cdots \vee S_{ss-1}, q),$$

while, for $s > (r+1)$,

$$[\alpha_{r+11}, \beta_{s*}] \in \pi_1(S_{s1} \vee \cdots \vee S_{ss-1}, q).$$

As these are two distinct factors, (II) implies that

$$\text{(III)} \quad [\alpha_{r+11}, \alpha_{21}] + \sum_{s+1}^{r} c_{r+1s}[\alpha_{r+1s}, \alpha_{21}] = 1$$

modulo Γ_2. But $[\alpha_{r+11}, \alpha_{21}]$ can be replaced by $[\alpha_{r+11}, \alpha_{r+12}]$, and, by virtue of the modulo Γ_2 Y-B relations

$$[\alpha_{r+11}, \alpha_{r+12} + \alpha_{21}] = 1,$$

we obtain the following version:

$$\text{(III)}^* \begin{cases} c_{r+11}[\alpha_{r+11}, \alpha_{r+11}] + & = 0 \\ (c_{r+12} - 1)[\alpha_{r+11}, \alpha_{r+12}] + & \\ \sum_{t>2}^{r} c_{r+1t}[\alpha_{r+11}, \alpha_{r+1t}] & = 0. \end{cases}$$

The first term is 0 because $S_{r+11} = S^1$ is a group. The other two terms are in distinct summands in Γ_1/Γ_2 because they are distinct basic commutators (cf. [53, Hall], Theorem 11.2.4, and [45, Falk-Randell]). So, we see that $c_{r+12} = 1$. Similarly, using α_{r+12} instead of α_{r+11}, one proves that $c_{r+12} = 1$.

We see that

$$\beta_{r+1*} = \alpha_{21} + \alpha_{r+11} + \alpha_{r+12}$$

modulo Γ_2. Now, assume to the contrary that $r > 2$. Note that the Y-B relations imply that

$$[\beta_{r+1}, \alpha_{r+1t}] = 1, \ [\alpha_{21}, \ [\alpha_{21}, \alpha_{r+1t} = 1$$

modulo Γ_2, for all $2 < t \leq r$. Hence,

$$\begin{aligned} 1 &= [\beta_{r+1*}, \alpha_{r+1t}] \\ &= [\alpha_{21} + \alpha_{r+11} + \alpha_{r+12}, \alpha_{r+1t} \\ &= [\alpha_{r+11}, \alpha_{r+1t}] + [\alpha_{r+12}, \alpha_{r+1t}]. \end{aligned}$$

However, the commutators

$$[\alpha_{r+11}, \alpha_{r+1t}], \ [\alpha_{r+12}, \alpha_{r+1t}]$$

belong to distinct summands in Γ_1/Γ_2, according to Theorem 11.2.2 of [53, Hall]. Hence, each vanishes; but that would be a contradiction, since the basic commutators are a basis for Γ_1/Γ_2 by Theorem 11.2.4 of [53, Hall]. Therefore, $r \leq 2$, as required,and the theorem is proved. □

The following result is related to that of [16, Chow] (also cf. Proposition 7.3 below).

Corollary 6.1 *The center of* $\pi_1(\mathbb{F}_k(\mathbb{R}^2), q)$ *is infinite cyclic. In fact, it is generated by* $\omega_2\omega_3\cdots\omega_k$*, where* $\omega_s = \alpha_{21}(\alpha_{31}\alpha_{32})\cdots(\alpha_{s1}\cdots\alpha_{ss-1})$.

Proof. Note that $\mathrm{proj}_{k,2} : \mathbb{F}_k(\mathbb{R}^2) \to \mathbb{F}_2(\mathbb{R}^2)$ is trivial with $\mathbb{F}_{k-2,2}(\mathbb{R}^2)$ as fiber. Therefore, $\mathbb{F}_k(\mathbb{R}^2) \simeq \mathbb{F}_2(\mathbb{R}^2) \times \mathbb{F}_{k-2,2}(\mathbb{R}^2)$, and

$$\mathrm{Cent}(\pi_1(\mathbb{F}_k(\mathbb{R}^2), \mathrm{q})) \cong \mathrm{Cent}(\pi_1(\mathbb{F}_2(\mathbb{R}^2), \mathrm{q})) \times \mathrm{Cent}(\pi_1(\mathbb{F}_{k-2,2}(\mathbb{R}^2), \mathrm{q})),$$

where Cent stands for the center. Now, a simple inductive argument based on the filtration Theorem 1.2 shows that $\mathrm{Cent}(\pi_1(\mathbb{F}_{k-2,2}(\mathbb{R}^2), \mathrm{q}))$ is trivial. Since $\pi_1\mathbb{F}_2(\mathbb{R}^2), q) \cong \pi_1(S^1, *) \cong \mathbb{Z}$, the center is infinite cyclic. Using the trivialization ψ introduced in the proof of Theorem 6.1, one easily sees that the generator α_{21} of $\pi_1(\mathbb{F}_2(\mathbb{R}^2), q)$ is carried to $\omega_2\omega_3\cdots\omega_k$. □

In the next section, we shall see that $\omega_2\omega_3\cdots\omega_k = \Delta_k(\mathbb{R}^2)$, where $\Delta_k(\mathbb{R}^2)$ is the Dirac braid on k strands.

7 The Group $\pi_1(\mathbb{F}_{k+1}(S^2), q^e)$

As $\pi_1(\mathbb{F}_1(S^2))\pi_1(S^2) = \pi_1(\mathbb{F}_2(S^2))$, and S^2 is simply connected, we need only consider the case when $k+1 \geq 3$. To begin, following the notation and conventions of Chapter III, regard $\mathbb{R}^3$ as $\mathbb{R}\times\mathbb{R}\times\mathbb{R}$, and denote the subspace $\{x \in \mathbb{R}\times\mathbb{R}\times\mathbb{R} | x_3 = 0\}$ by V^e. Let $\phi^e : S^2 - -e \to V^e = \mathbb{R}\times\mathbb{R} = \mathbb{R}^2$ be the stereographic projection from $-e$ to V^e, where $e = (0,0,1)$. Define q_i^e by the rule $q_1^e = e$, $q_2^e = e_1$, $\phi^e(q_i^e) = q_i = 4(i-1)e_1$, $3 \leq i \leq k$, where $e_1 = (1,0,0) \in \mathbb{R}\times\mathbb{R}$, and put $Q_0^e = \{-e\}$, $Q_i^e = \{q_1^e, \cdots, q_i^e, -e\}$, $1 \leq i < k$. We choose $(q_1^e, q_2^e, \cdots, q_k^e, -e)$ as a basepoint for $\mathbb{F}_{k+1}(S^2)$. Note that the points q_i^e lie on the great circle in S^2 cut by the plane spanned by the first and last basic vectors in $\mathbb{R}^2$ (see Fig.IV.7).

Consider the projection $\mathrm{proj}_{k+1,1} : \mathbb{F}_{k+1}(S^2) \to S^2$, where $(x_1, \cdots, x_{k+1}) \mapsto x_{k+1}$. Note that, on taking $-e$ as the basepoint of S^{n+1}, we see that the fiber of the $\mathrm{proj}_{k+1,1}$ can be identified with $\mathbb{F}_k(\mathbb{R}^2)$ by means of the map $\phi^e : S^2 - \{-e\} \to V^e = \mathbb{R}^2$.

Define $i_k : \mathbb{F}_k(\mathbb{R}^2) \to \mathbb{F}_{k+1}(S^2)$ to be the map where $(x_1, \cdots, x_k) \mapsto ((\phi^e)^{-1}(x_1), \cdots, (\phi^e)^{-1}(x_k), -e)$. As in §2 of Chapter III, and with $\mathbb{R}^2_r$ standing for $S^2 - Q_r^e$, the preceding fibration fits into the diagram

$$\mathcal{F}_{k+1}(S^2) : \begin{cases} \mathbb{F}_{k+1}(S^2) \leftarrow \mathbb{F}_{k,0} \leftarrow \cdots \ \mathbb{F}_{k-r,r} \leftarrow \\ \quad\downarrow \qquad\qquad \downarrow \qquad\quad \downarrow \quad \downarrow \\ \quad S^2 \qquad\qquad \mathbb{R}^2 \qquad \cdots \quad \mathbb{R}^2_r, \end{cases}$$

where $\mathbb{F}_{k-r,r} = \mathbb{F}_k(S^2 - Q_r^e)$ and the vertical maps are the projections $\mathrm{proj}_{k-r,r} : \mathbb{F}_{k-r,r}(S^2) \to \mathbb{R}_r^{n+1}$, with the property that

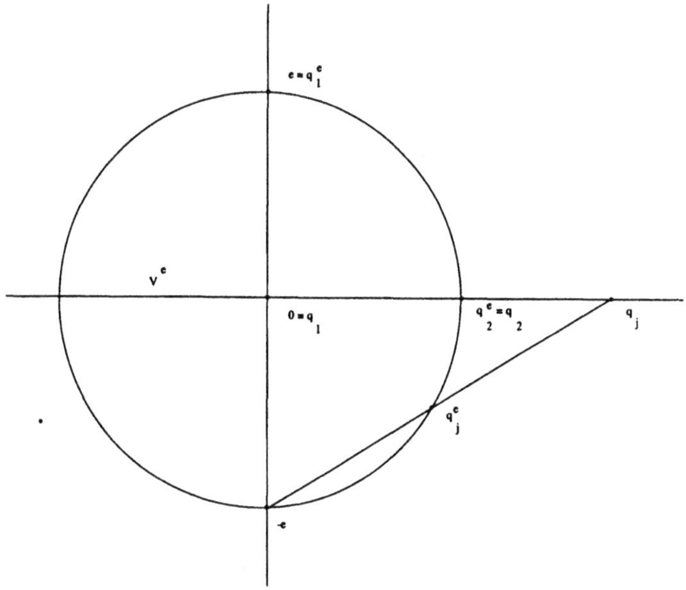

Fig. IV.7. Stereographic Projection

$$(q_1, \cdots, q_r, x_{r+1}, \cdots, x_k) \in \mathbb{F}_{k-r,r}(\mathbb{R}_2) \mapsto x_{r+1} \in \mathbb{R}^2.$$

The proofs of the following results are exactly like those of their analogues in §2 of Chapter III.

Proposition 7.1 *The map* $\psi_3 : O_{3,2} \to \mathbb{F}_3(S^2)$, $(x, y) \mapsto (x, y, -x)$ *is a homotopy equivalence.*

Corollary 7.1 *The map* $\psi_3 : O_{3,2} \to \mathbb{F}_3(S^2)$ *induces a fiber homotopy equivalence*

$$\begin{array}{ccc} O_{3,2} & \overset{\psi_3}{\to} & \mathbb{F}_3(S^2) \\ \downarrow & & \downarrow \\ S^2 & \overset{\text{id}}{\to} & S^2, \end{array}$$

where the vertical maps are projections on the first factor.

The fibrations $\mathrm{p}_r : \mathbb{F}_{k-r,r}(\mathbb{R}^{n+1}) \to \mathbb{R}^2_r$ in diagram $\mathcal{F}_{k+1}(S^2)$ admit sections, as we saw earlier in this chapter.

Proposition 7.2 *There is a map* $\psi_{k+1} : O_{3,2} \to \mathbb{F}_{k+1}(S^2)$ *such that* $\mathrm{proj}_{k+1,3} \circ \psi_{k+1} = \psi_3$, *where* $\mathrm{proj}_{k+1,3}$ *takes* $(x_1, \cdots, x_{k+1})$ *to* (x_1, x_2, x_{k+1}).

Going back to diagram $\mathcal{F}_{k+1}(S^2)$, we see that $\mathbb{F}_{k-2,2}$ is the fiber of the projection $\mathrm{proj}_{k+1,3} : \mathbb{F}_{k+1}(S^2) \to \mathbb{F}_3(S^2)$, and this fibration has a section by virtue of the previous propositions. So we get the representation

$$\mathbb{F}_{k+1}(S^2) \simeq O_{3,2} \rtimes \mathbb{F}_{k-2,2}(\mathbb{R}^2).$$

Theorem 7.1 *The pull-back* $\text{proj}'_{k+1,3} : \mathbb{F}'_{k+1}(S^2) \to O_{3,2}$ *of* $\text{proj}_{k+1,3} : \mathbb{F}_{k+1}(S^2) \to \mathbb{F}_3(S^2)$ *by* $\psi_3 : O_{3,2} \to \mathbb{F}_3(\mathbb{R}^2)$ *is a trivial bundle. In fact, there is a map* $\rho : O_{3,2} \times \mathbb{F}_{k-2,2}(\mathbb{R}^2) \to \mathbb{F}_{k+1}(S^2)$ *that carries the generator of* $\pi_1(O_{3,2})$ *to* $\pi(i_k)(\omega_2 \cdots \omega_k)$*, where* $i_k : \mathbb{R}^2 \to S^2$ *is the map defined above.*

Proof. Consider, first, the map $\lambda : O_{3,2} \to SO_3,\quad (x,y) \mapsto (x, y, x \times y)$, where $x \times y$ is the usual cross product in $\mathbb{R}^3$. One can easily verify that λ is a homeomorphism that takes the point $(e_2, e) \in O_{3,2}$ to the $1 \in SO_3$. Observe next that $\text{proj}_{k+1,3}^{-1}((e, e_2, -e)) = \{(e, e_2, x_3, \cdots, x_{k+1}, -e)\}$. Now, proceeding as in Chapter I (cf. Theorem 4.1 of Chapter I), consider the map $\rho : O_{3,2} \times \mathbb{F}_{k-2,2}(S^2) \to \mathbb{F}_{k+1}(S^2)$ defined so that $((x,y),(x_3, \cdots, x_{k+1}))$ maps to

$$(\lambda(x,y)(e), \lambda(x,y)(e_2), \lambda(x,y)(x_3), \cdots, (\lambda(x,y)(x_{k-1}), -\lambda(x,y)(e))),$$

where $e_2 = (0,1,0) \in \mathbb{R}^3$. It is easy to verify that ρ is a trivialization of the bundle under consideration. □

Consider again the fundamental fiber sequence introduced above. Identify the spaces $\mathbb{F}_{r,k-r}(\mathbb{R}^2)$ with their images in $\mathbb{F}_{k+1}(S^2)$. This being done, the filtration

$$\mathbb{F}_{1,k-1} \subset \cdots \mathbb{F}_{k-r,r} \subset \cdots \subset \mathbb{F}_{k-2,2} \subset \mathbb{F}_{k-1,1} \subset \mathbb{F}_k(\mathbb{R}^2) \subset \mathbb{F}_{k+1}(S^2)$$

leads naturally to a filtration of the group $\pi_1(\mathbb{F}_{k+1}(S^2), q^e)$ by subgroups as follows. First, put

$$\begin{aligned} L_{k+1} &= \pi_1(\mathbb{F}_{k+1}(S^2), q^e), \\ L_k &= \pi_1(\mathbb{F}_{k,0}, q), \\ L_{k-r} &= \pi_1(\mathbb{F}_{k-r,r}, q), \text{ and} \\ L_{rr-1} &= \pi_1(\mathbb{R}^2_r, q), \end{aligned}$$

with $\mathbb{R}^2_r = \mathbb{R}^2 - Q_r$. Each L_{rr-1} is the fundamental group of a bouquet of r circles. The space $\mathbb{F}_{k-2,2}$ can be identified with the fiber of the projection

$$\text{proj}_{k+1,3} : \mathbb{F}_{k+1}(S^2) \to \mathbb{F}_3(S^2)$$

that sends a tuple to its last three entries. Since this fibration is fiber homotopically trivial, the morphism $\pi_1(i_{k-2}) : L_{k-2} \to L_{k+1}$ is injective. Similarly, $\mathbb{F}_{k-r,r}$ can be identified with the fiber of $\text{proj}_{k,k-r} : \mathbb{F}_k(\mathbb{R}^2) \to \mathbb{F}_r(\mathbb{R}^2)$, and this fibration has a section. Hence, the induced morphism $\pi_1(i_{k-r}) : L_{k-r} \to \pi_1(\mathbb{F}_{k+1}(S^2), q^e)$ is injective for all $2 \le r \le (k-2)$. For simplicity, let us identify L_{k-r} with its image.

Theorem 7.2 *There is an ascending chain of normal subgroups*

$$L_1 \subset \cdots \subset L_{k-r} \subset \cdots \subset L_{k-2} \subset L_{k+1}$$

such that

(i) L_{rr-1} *is isomorphic to the free group on* r *letters,*
(ii) $L_{k-r} \cong L_{k-r-1} \rtimes L_{r+1r}$, *for* $r \leq (k-2)$, *and*
(iii) $L_k \cong L_{k-2} \times M_{3,2}$,

where $M_{3,2} = \pi_1(O_{3,2})$.

Proof. To see that L_{k-r}, for $r \geq 2$ is a normal subgroup in L_{k+1}, recall that

$$\mathbb{F}_{k-r,r}(S^2) \cong \mathbb{F}_{k-r}(S^2 - Q_r^e) \cong \mathbb{F}_{k-r}(\mathbb{R}^2 - Q_r)$$

can be identified with the fiber of $\mathrm{proj}_{k+1,r+1} : \mathbb{F}_{k+1}(S^2) \to \mathbb{F}_{r+1}(S^2)$ that takes a tuple to its first r entries. Hence, L_{k-r} is the kernel of the morphism $\pi_1(\mathrm{proj}_{k+1,r+1})$. Statement(i) follows from the fact that $\mathbb{R}^2_r = \mathbb{R}^2 - Q_r$ is homotopy equivalent to a bouquet of r circles; Statement (ii) follows easily from the fact the appropriate fibration admits a section; and Statement(iii) is an immediate consequence of Theorem 7.1. Moreover, the nature of the various split extensions is determined by the Y-B relations. □

We need the following special case of Proposition 6.3 in Chapter III. Let $\mu : S^1 \times S^1 \to S^1$ be the map where $(z_1, z_2) \mapsto z_2^2 z_1$.

Proposition 7.3 *There is a fiber homotopy equivalence*

$$S^1 \cup_\mu D^2 \times S^1 \overset{\simeq}{\to} O_{3,2},$$

where the space on the left is fibered by projection on the second factor, while that on the right fibered by the map $(x, y) \mapsto x$.

In particular, $\pi_1(O_{3,2}, *) \cong \mathbb{Z}/2\mathbb{Z}$. The following is an immediate consequence of Statement (iii) and the fact that $\pi_1(O_{3,2}, *) \cong \mathbb{Z}/2\mathbb{Z}$.

Corollary 7.2 *With the notation as above, we have*

$$\pi_1(\mathbb{F}_{k+1}(S^2), q^e) \cong L_{k-2} \times (\mathbb{Z}/2\mathbb{Z}).$$

Since $\pi_1(\mathbb{F}_{k-r,r}(\mathbb{R}^2), q)$ are centerless if $r > 1$, the following is an easy consequence (cf. [49, Gillette-Van Buskirk]).

Corollary 7.3 *For* $k+1 \geq 3$, *there is an isomorphism*

$$\mathbb{Z}/2\mathbb{Z} \overset{\cong}{\to} \mathrm{Cent}(\pi_1(\mathbb{F}_{k+1}(S^2), q^e))$$

that takes the generator to $\pi_1(i_k)(\omega_2 \cdots \omega_k)$.

Consider the map $\mathrm{proj}_{k+1,1} : \mathbb{F}_{k+1}(S^2) \to S^2$, where $(x_1, \cdots, x_{k+1}) \mapsto x_{k+1}$, and choose $-e = (0, 0, -1)$ as the basepoint of $S^2 \subset \mathbb{R}^3$. With the map $i_k : \mathbb{F}_k(\mathbb{R}^2) \to \mathbb{F}_{k+1}(S^2)$ being that defined above, put $\beta_{rs} = \pi_1(i_k)(\alpha_{rs})$, $1 \leq$

$s < r \leq k$, where $\alpha_{rs} \in \pi_1(\mathbb{F}_k(\mathbb{R}^2), q)$ are the generators introduced in Definition 2.2 above. Then the elements β_{rs} for all $1 \leq s < r \leq k$ are in $\pi_1(\mathbb{F}_{k+1}(S^2), q^e)$.

Again, with ψ_{k+1} being the map introduced earlier, put $\Delta'_{k+1}(S^2) = \psi_{k+1}|S^1$, where the circle S^1 is identified with the set $\{y|(e,y) \in O_{3,2}$. The *Dirac class on* $k+1$ *strands* $\Delta_{k+1}(S^2)$ is, by definition, the homotopy class $[\Delta'_{k+1}(S^{n+1})] \in \pi_1(\mathbb{F}_{k+1}(S^2), q^e)$.

There is an analogous class in $\pi_1(\mathbb{F}_k(\mathbb{R}^2), q)$: the Dirac class $\Delta_k(\mathbb{R}^2)$ on k strands ([29, Fadell]). To see this, note that the image of $\Delta'_{k+1}(S^2)$ consists of $k+1$ circles centered on the e-axis and with the first and last being degenerate. These degenerate circles correspond to the points e and $-e$, respectively. Define *the Dirac class on k strands* $\Delta(\mathbb{R}^2) \in \pi_1(\mathbb{F}_k(\mathbb{R}^2), q)$ to be the homotopy class of the map $\phi^e \circ \mathrm{proj}_{k+1,k} \circ \Delta'_{k+1}(S^2) : S^1 \to \mathbb{F}_k(\mathbb{R}^2)$, where $\mathrm{proj}_{k+1,k} : \mathbb{F}_{k+1}(S^2) \to \mathbb{F}_k(S^2)$ sends a $k+1$ tuple to its first k entries. Here we identify the space $V^e \subset \mathbb{R}^2$ orthogonal to e with $\mathbb{R}^2$ by means of the stereographic map ϕ^e. Note that the map $\phi^e \circ \mathrm{proj}_{k+1,k} \circ \Delta'_{k+1}(S^2)$ omits the degenerate point $-e$ and keeps k concentric circles in $\mathbb{R}^2$ centered at the origin, with the first being the degenerate circle at the origin.

The following is a simple consequence of this description of $\Delta_k(\mathbb{R}^2)$.

Proposition 7.4 *Let* $\omega_r = \alpha_{r1} \cdots \alpha_{rr-1}$. *Then, the relation* $\Delta_k(\mathbb{R}^2) = \omega_2\omega_3 \cdots \omega_k$ *holds in* $\pi_1(\mathbb{F}_{k+1}(\mathbb{R}^2), q)$.

Consider next the map $i_k : \mathbb{F}_k(\mathbb{R}^2) \to \mathbb{F}_k(S^2)$, where $(x_1, \cdots, x_k) \mapsto ((\phi^e)^{-1}(x_1), \cdots, (\phi^e)^{-1}(x_k), -e)$. The following is an immediate consequence of Proposition 7.1

Corollary 7.4 *The morphism* $\pi_n(i_k) : \pi_n(\mathbb{F}_k(\mathbb{R}^2, q)) \to \pi_n(\mathbb{F}_k(S^2), q^e)$ *takes* $\Delta_k(\mathbb{R}^2)$ *to* $\Delta_{k+1}(S^2)$. *In particular, it follows that*

$$\Delta_{k+1}(S^2) = \omega'_2\omega'_3 \cdots \omega'_k,$$

where $\omega'_r = \beta_{r1} \cdots \beta_{rr-1}$.

Summing up, we have the following result.

Theorem 7.3 *Let* $i_k : \mathbb{F}_k(\mathbb{R}^2) \to \mathbb{F}_{k+1}(S^2)$ *be the natural imbedding of* $\mathbb{F}_k(\mathbb{R}^2)$ *as the fiber at* $-e$. *Then,*

(*i*) *the set* $\{\beta_{rs} \mid 1 \leq s < r \leq k)\}$ *generates the group* $\pi_1(\mathbb{F}_{k+1}(S^2), q^e)$;
(*ii*) $\ker(\pi_1(\mathbb{F}_k(\mathbb{R}^2, q)) \overset{\pi_1(i_k)}{\to} \pi_1(\mathbb{F}_{k+1}(S^2)))$ *is infinite cyclic;*
(*iii*) $\pi_1(i_k)(\Delta_k(\mathbb{R}^2)) = \Delta_{k+1}(S^2)$ *and, moreover,* $\Delta_{k+1}(S^2)$ *has order two in* $\pi_1(\mathbb{F}_{k+1}(S^2), q^e)$; *and*
(*iv*) *the induced morphism* $\pi_r(\mathrm{proj}_{k+1,1}) : \pi_r(\mathbb{F}_{k+1}(S^2), q^e) \to \pi_r(S^2)$ *is an isomorphism for* $r > 2$.

Proof. Put $\bar{\psi}_{k+1} = \psi_{k+1}|S^1$, and observe that the homotopy class of $\bar{\psi}_{k+1}$ is $\Delta_k(\mathbb{R}^2)$. Since $\psi_{k+1}; O_{3,2} \to \mathbb{F}_{k+1}(S^2))$ is a map of fibrations, we obtain the commutative diagram

$$\begin{array}{ccccccc} \pi_2(S^2) & \xrightarrow{\partial} & \pi_1(S^1) & \to & \pi_1(O_{3,2}) & \to 0 \\ \downarrow = & & \downarrow \bar{\psi}_k & & \downarrow \psi_{k+1} & \\ \pi_2(S^2) & \xrightarrow{\partial} & \pi_1(\mathbb{F}_k(\mathbb{R}^2), q) & \to & \pi_1(\mathbb{F}_{k+1}(S^2), q^e) & \to 0 \end{array}$$

with exact rows. Here, the upper and lower horizontal sequences are the tails of the homotopy sequences of, respectively, the projections $O_{3,2} \to S^2$ and $\mathbb{F}_{k+1}(S^2) \to S^2$. Each of these projections sends a tuple to its first entry (see Propositions 7.1 and 7.2.)

According to Theorem 5.1, the set $\{\alpha_{rs} | 1 \le s < r \le k\}$ generates $\pi_1(\mathbb{F}_k(\mathbb{R}^2))$. Moreover, according to the remark after Corollary 7.3, we have $\pi_1(i_k)(\alpha_{rs}) = \beta_{rs}$. Thus, we see that the set $\{\beta_{rs} | 1 \le s < r \le k\}$ generates $\pi_1(\mathbb{F}_k(S^2), q^e)$. This proves Statement (i).

Statement (ii) is an immediate consequence of the fact that the group $\pi_1(\mathbb{F}_k(\mathbb{R}^2), q)$ can be represented as a multisplit product of free groups (cf. Theorem 1.2) and, hence, is torsion free.

The first assertion of Statement (iii) is obvious (see Corollary 7.4). To finish the proof of Statement (iii), note first that the third morphisms of both sequences are surjective, since $\pi_1(S^2, q^e)$ is trivial. Also, as we pointed out at the beginning of the proof, $\pi_1(\bar{\psi}_3)(\iota_1) = \Delta_2(\mathbb{R}^2)$. Observe also that the Dirac class $\Delta_2(\mathbb{R}^2)$ goes to the Dirac class $\Delta_k(\mathbb{R}^2)$. Hence, $\Delta_3(S^2)$ goes to $\Delta_{k+1}(S^2)$. Now, as a consequence of Proposition 7.1, we conclude that the order of $\Delta_3(S^2)$ in $\pi_1(O_{3,2}) \cong \pi_1(\mathbb{F}_3(S^2), q^e)$ is exactly two. Hence, the order of $\Delta_{k+1}(S^2)$ is also two. This proves Statement (iii).

To prove Statement (iv), consider the long exact homotopy sequence of the map $\mathrm{proj}_{k+1,1} : \mathbb{F}_{k+1}(S^2) \to S^2$, and note that $\mathbb{F}_k(\mathbb{R}^2)$ is aspherical according to Theorem 1.1. □

The preceding result and Theorem 5.1 together imply the following.

Theorem 7.4 *The pure braid group $\pi_1(\mathbb{F}_{k+1}(S^2), [q])$, $k+1 \ge 3$ is isomorphic to the free group generated by the symbols α_{rs} where $1 \le s < r \le k$ modulo the Y-B relations for the pure braid group of $\mathbb{F}_k(\mathbb{R}^2)$ and the additional relation*

$$(\omega_2 \cdots \omega_k)^2 = 1, \quad \omega_j = \alpha_{j1} \cdots \alpha_{jj-1}.$$

8 Historical Remarks

We have offered a fresh look at braid groups and their presentations. The method is more geometric but is equivalent to the mainly algebraic approach familiar to braid group theorists. We hope that this method is useful in the study of the fundamental groups of similar spaces, such as subspaces of configuration spaces of $\mathbb{R}^2$ (see for example [85, Moulton]) and also arrangements of the fiber type (see [62, Jambu-Papadima] and [86, Orlik-Terao]).

A presentation of $\pi_1(\mathbb{F}_{k+1}(\mathbb{R}^2), q)$ appears in [9, Birman] and is obtained algebraically by means of the Schreier-Reidemeister method, (see, [73, Magnus-Karrass-Solitar]). The generators are the same, but the relations take the form of the normality relations in Propositions 5.2 and 5.3 rather than the simpler Y-B relations. A presentation due to [7, Artin] appears in [73, Magnus-Karrass-Solitar]. Our presentation of $\pi_1(\mathbb{F}_{k+1}(S^2), q^e)$ is new and is distinguished by the fact that the set of generators $\{\beta_{rs} | 1 \leq s < r \leq k\}$ excludes the elements β_{k+1j}, $1 \leq j < k+1$. Another presentation is possible: if one considers S^2 as the one-point compactification $\mathbb{R}^2 \cup \infty$, then the natural imbedding $i : \mathbb{R}^2 \to S^2$ induces a surjective morphism $\pi_1(\mathbb{F}_{k+1}(\mathbb{R}^2)) \to \pi_1(\mathbb{F}_{k+1}(S^2)$, and, using the methods of this chapter, one can obtain a presentation of $\pi_1(\mathbb{F}_{k+1}(S^2)$, with all the elements $\{\beta_{rs} | 1 \leq s < r \leq k+1\}$ as generators and with the Y-B relations for $\pi_1(\mathbb{R}^2)$, among others, coming into play. In particular $\{\beta_{rs} | 1 \leq s < r \leq k+1\}$ is invariant under the symmetric group Σ_{k+1}.

The center of classical braid group $\pi_1(\mathcal{B}_k(\mathbb{R}^2), [q])$ is also generated by the Dirac braid $\Delta_k(\mathbb{R}^2)$ [16, Chow], and that of $\pi_1(\mathcal{B}_k(S^2), [q^e])$ by $\Delta_{k+1}(S^2)$ [49, Gillette-Van Buskirk]. Our result that $\Delta_k(\mathbb{R}^2)$ and $\Delta_{k+1}(S^2)$ generate the centers of $\pi_1(\mathbb{F}_{k+1}(\mathbb{R}^2), q)$ and $\pi_1(\mathbb{F}_{k+1}(S^2), q^e)$, respectively, implies both results for $\pi_1(\mathcal{B}_k(\mathbb{R}^2, [q])$ and $\pi_1(\mathcal{B}_k(S^2, [q^e])$.

See [44, Fadell-Van Buskirk] in relation to Theorem 7.2.

Part II

Homology and Cohomology of $\mathbb{F}_k(M)$

Introduction

In Part Two our aim is to define minimal CW-complexes X_k, Y_{k+1} homotopy equivalent to $\mathbb{F}_k(\mathbb{R}^{n+1})$ and $\mathbb{F}_{k+1}(S^{n+1})$, respectively. In Chapter V we determine the structure of $H^*(\mathbb{F}_k(M);\mathbb{Z})$, as an algebra, when M is $\mathbb{R}^{n+1}$ or S^{n+1}. We view the generators α_{rs} of the group $\pi_n(\mathbb{F}_k(\mathbb{R}^{n+1}),q)$, defined in Chapter II, §2, as spherical homology classes and introduce the elements $\{\alpha^*_{rs} \in H^n(\mathbb{F}_k(\mathbb{R}^{n+1};\mathbb{Z})|1 \leq s < r \leq r\}$ dual to the α_{rs}. These elements generate the group $H^n(\mathbb{F}_k(\mathbb{R}^{n+1});\mathbb{Z})$ and are invariant, set-wise, up to sign, under the action of the symmetric group Σ_k. Moreover, they satisfy the cohomological version of the Y-B relations of Chapter II, §3. We show that $H^*(\mathbb{F}_k(\mathbb{R}^{n+1});\mathbb{Z})$ is the universal, commutative, graded algebra generated by the set of all α^*_{rs} modulo the ideal generated by the Y-B relations. The proof is by induction on the natural filtration in diagram $\mathcal{F}_k(\mathbb{R}^{n+1})$ of Chapter II. The rest of Chapter V is devoted to determining the cohomology algebra of $\mathbb{F}_{k+1}(S^{n+1})$. These results lead to cohomology bases consisting of multifold products of the elements of $\{\alpha^*_{rs}|1 \leq s < r \leq k\}$.

In Chapter VI we describe the minimal cellular structures of $\mathbb{F}_k(M)$ naturally associated with the bases of $H^*(\mathbb{F}_k(M))$ for $M = \mathbb{R}^{n+1}$ or S^{n+1} of Chapter V. The basic idea is that the twisted product representation $\mathbb{F}_k(\mathbb{R}^{n+1}) \simeq \mathbb{R}^{n+1}_1 \rtimes \cdots \rtimes \mathbb{R}^{n+1}_{k-1}$ introduced in Chapter II, §5 leads to a twisted product

$$H_*(\mathbb{R}^{n+1}_1) \otimes \cdots \otimes H_*(\mathbb{R}^{n+1}_{k-1}) \cong H_*(\mathbb{F}_k(\mathbb{R}^{n+1});\mathbb{Z})$$

on homology, which we write as $\alpha_{r_1s_1} \otimes \cdots \otimes \alpha_{r_ps_p} \mapsto \omega = \alpha_{r_1s_1} \ltimes \cdots \ltimes \alpha_{r_ps_p}$, for all $1 \leq p \leq (k-1)$. For each p-fold twisted product ω there is an imbedding

$$\phi_\omega : S^n_1 \times \cdots \times S^n_p \to \mathbb{F}_k(\mathbb{R}^{n+1})$$

that agrees with $\alpha_{rs} : S^n \to \mathbb{F}_k(\mathbb{R}^{n+})$, when $p = 1$, and is $\phi_{\omega'}$, up to homotopy, when restricted to the last $(k-1)$ factors, with ω' being the $(p-1)$-fold twisted product obtained by deleting the first factor of ω. The idea underlying their definition is the same as in the definition of $\alpha_{r_ps_p}$ in §2 of Chapter II: the map

$$\phi_{\omega'} : S^n_1 \times \cdots \times S^n_{p-1} \to \mathbb{F}_{k-1}(\mathbb{R}^{n+1})$$

is modified by inserting in the r_p^{th}-position the unit sphere, suitably rescaled, centered at q'_{s_p}, where $\phi_{\omega'}(\xi_1, \cdots, \xi_{p-1}) \mapsto (q'_1, \cdots, q'_{s_p}, \cdots)$. We call these maps *affine*. Each such map ϕ_ω defines a table with rows that are the restrictions to the factors of ω. We then show that the jth row begins with the jth factor $\alpha_{r_j s_j}$. So the table is of the form

$$\mathcal{T}(\phi_\omega): \begin{cases} \alpha_{r_1 s_1} & + \sum_{m=2}^{p} \lambda_{1m} \\ \cdots & + \cdots \\ \alpha_{r_j s_j} & + \sum_{m=j+1}^{p} \lambda_{jm} \\ \cdots & + \cdots \\ \alpha_{r_p s_p}. \end{cases}$$

The remainder of the jth row is determined by the Y-B relations. This is described explicitly. Next, we prove, first, that $\omega = \alpha_{r_1 s_1} \ltimes \cdots \ltimes \alpha_{r_p s_p}$ is the dual of

$$\alpha^*_{\underline{r},\underline{s}} = \alpha^*_{r_1 s_1} \cdots \alpha^*_{r_p s_p},$$

and, then, that ω is carried by the fundamental class of the multifold products of the spheres imbedded by ϕ_ω. The structure of the factor table, $\mathcal{T}(\phi_\omega)$, plays a key role here.

We construct the cells of the CW-complex $X_k \simeq \mathbb{F}_k(\mathbb{R}^{n+1})$ that are in one-one correspondence with the basis of twisted products, and their characteristic maps come from the corresponding affine maps. Next, we carry out the analogous program for $\mathbb{F}_{k+1}(S^{n+1})$. As is to be expected, the Stiefel manifold $O_{n+2,2}$ plays a role here. The rest of the chapter is devoted to cellular structures of $\mathbb{F}_k(\mathbb{R}^2)$ and $\mathbb{F}_{k+1}(S^2)$. Again, the lack of simple connectivity in this case requires some adjustments.

Our aim in Chapter VII is to present cellular DG-models for the singular chain coalgebras $C_*S(M)$, where M is $\mathbb{R}^{n+1}$ or S^{n+1}. The models are based on the cellular models $X_k \simeq \mathbb{F}_k(\mathbb{R}^{n+1})$ and $Y_{k+1} \simeq \mathbb{F}_{k+1}(S^{n+1})$ of Chapter VI, with $n > 1$. These results play a crucial role in Part Three, where cellular models for the loop spaces of the configuration spaces are required. The central fact here is that the coproduct is given in terms of the rows of the corresponding factor table.

V. The Algebra $H^*(\mathbb{F}_k(M);\mathbb{Z})$

Our aim in this chapter is to determine the structure of $H^*(\mathbb{F}_k(M);\mathbb{Z})$, as an algebra, when M is $\mathbb{R}^{n+1}$ or S^{n+1} (cf. [17, Cohen], [19, Cohen-Taylor]). Note that we are including the case $n = 1$. In this case some notational change is necessary.

In §2 we interpret the generators α_{rs} of the group $\pi_n(\mathbb{F}_k(\mathbb{R}^{n+1}), q)$, defined in Chapter II, §2, as homology classes, and we introduce the elements

$$\{\alpha^*_{rs} \in H^n(\mathbb{F}_k(\mathbb{R}^{n+1};\mathbb{Z}) \mid 1 \le s < r \le r\}$$

dual to the α_{rs}. These generate the group $H^n(\mathbb{F}_k(\mathbb{R}^{n+1});\mathbb{Z})$. Then we show that the multifold twisted product representation

$$\mathbb{F}_k(\mathbb{R}^{n+1}) \simeq \rtimes_{r=2}^{k}(S_{r1} \vee \cdots \vee S_{rr-})$$

leads to an isomorphism $H^*(\mathbb{F}_k(\mathbb{R}^{n+1});\mathbb{Z}) \cong \bigotimes_{r=2}^{k} H^*(S_{r1} \vee \cdots \vee S_{rr-1};\mathbb{Z})$ of abelian groups.

In §3 we show that $H^n(\sigma)(\alpha^*_{rs}) = \alpha^*_{\sigma^{-1}r\sigma^{-1}s}$ for all $\sigma \in \Sigma_k$ and then give the cohomological version of the homotopy Y-B relation of Chapter II, §3.

In §4 we turn to the structure of $H^*(\mathbb{F}_k(\mathbb{R}^{n+1});\mathbb{Z})$. We show that it is the universal, associative, commutative, graded algebra generated by the set of α^*_{rs} modulo the ideal generated by the Y-B relations. The proof is by induction on the natural filtration of Chapter II.

In §6, using the results of Chapter III, we show that there is an isomorphism

$$\Phi^*_{k-2,3} : H^*(O_{n+2,2};\mathbb{Z}) \otimes H^*(\mathbb{F}_{k-2}(\mathbb{R}^{n+1}_2);\mathbb{Z}) \cong H^*(\mathbb{F}_{k+1}(S^{n+1});\mathbb{Z})$$

of $H^*(O_{n+2,2};\mathbb{Z})$ modules, where $O_{n+2,2}$ is the Stiefel manifold of orthonormal 2-frames in $\mathbb{R}^{n+2}$.

In §§6, 7, and 8 we investigate when the isomorphism $\Phi^*_{k-2,3}$ is one of algebras. The answer depends on the parity of $(n+1)$. We show that, if $(n+1)$ is even, then $\Phi^*_{k-2,3}$ is, in fact, an isomorphism of algebras; but this is not so if $(n+1)$ is odd. However, in the latter case we show that there is another isomorphism

$$H^*(S^{n+1};\mathbb{Z}) \otimes H^*(\mathbb{F}_k(\mathbb{R}^{n+1});\mathbb{Z}) \cong H^*(\mathbb{F}_{k+1}(S^{n+1});\mathbb{Z})$$

of algebras.

1 The Group $H^*(\mathbb{F}_k(\mathbb{R}^{n+1});\mathbb{Z})$

In this section we aim to describe the module structure of the cohomology of $\mathbb{F}_k(\mathbb{R}^{n+1})$ in terms of the twisted-product structure

$$\mathbb{F}_k(\mathbb{R}^{n+1}) \simeq \rtimes_{r=2}^{k}(S_{r1} \vee \cdots \vee S_{rr-1})$$

given in Chapter II. So, consider the map of Chapter II, §1,

$$alpha'_{rs} : S^n \to \mathbb{F}_k(\mathbb{R}^{n+1}), \ \ \xi \mapsto (q_1, \cdots, q_{r-1}, q_s + \xi, q_r, \cdots, q_{k-1}),$$

where $1 \le s \ne r \le k$, and put $\alpha_{rs} = H_n(\alpha'_{rs})(\iota_n)$, where ι_n is the generator of $H_n(S^n;\mathbb{Z})$. In effect, we identify α_{rs}, the element of $\pi_n(\mathbb{F}_k(\mathbb{R}^{n+1}, q))$, with its image in $H_n(\mathbb{F}_k(\mathbb{R}^{n+1});\mathbb{Z})$ under the Hurewicz map. When $n = 1$, the morphism sends an element $\alpha_{rs} \in \pi_1(\mathbb{F}_k(\mathbb{R}^2, q))$ to its image in the abelianization $H_1(\mathbb{F}_k(\mathbb{R}^2);\mathbb{Z})$ of $\pi_1(\mathbb{F}_k(\mathbb{R}^2, q))$. We adopt this convention, except where there is cause for confusion.

Next, consider the map

$$g_{rs} : \mathbb{F}_k(\mathbb{R}^{n+1}) \to S^n, \ (x_1, \cdots, x_k) \mapsto (x_r - x_s)/\|(x_r - x_s)\|.$$

Note that the map $g_{21} : \mathbb{F}_2(\mathbb{R}^{n+1}) \to S^n$ is a homotopy equivalence. By definition, let

$$\alpha^*_{rs} = H^n(g_{rs})(\iota^*_n), \quad 1 \le s < r \le k,$$

where $\iota^*_n \in H^n(S^n;\mathbb{Z})$ is the dual cohomology class with respect to the Kronecker product

$$H^*(\mathbb{F}_k(\mathbb{R}^{n+1});\mathbb{Z}) \otimes H_*(\mathbb{F}_k(\mathbb{R}^{n+1});\mathbb{Z}) \to \mathbb{Z}.$$

Thus, $< \iota^*_n, \iota_n >= 1$, where $< .,. >$ denotes the Kronecker product.

Observe that $H^n(\mathbb{F}_k(\mathbb{R}^{n+1});\mathbb{Z}) \cong \mathrm{Hom}(H_n(\mathbb{F}_k(\mathbb{R}^{n+1});\mathbb{Z}),\mathbb{Z})$.

Proposition 1.1 *The class α^*_{rs} is the Kronecker dual of α_{rs}. That is to say, $< \alpha^*_{rs}, \alpha_{uv} >= 1$, if $(u,v) = (r,s)$, and $< \alpha^*_{rs}, \alpha_{uv} >= 0$, otherwise.*

Proof. Observe that

$$\begin{aligned} < \alpha^*_{rs}, \alpha_{uv} > &= < H^n(g_{rs})(\iota^*_n), \alpha_{uv} > \\ &= < \iota_n, H_n(g_{uv})(\alpha_{uv}) > \\ &= < \iota_n, H_n(g_{uv} \circ \alpha'_{rs} > \\ &= 1, \end{aligned}$$

since, as it is easy to see, the composite $S^n \overset{\alpha'}{\to} \mathbb{F}_k(\mathbb{R}^{n+1}) \overset{g_{uv}}{\to} S^n$ is homotopic to the identity when $(r,s) = (u,v)$ and is null homotopic otherwise. □

Proposition 1.2 of Chapter II immediately implies the following.

Corollary 1.1 *For all* $1 \le s < r \le k$, *we have* $\alpha^*_{rs} = (-1)^{n+1}\alpha^*_{sr}$.

Proof. Note that $\alpha_{sr} = \tau_{rs} \circ \alpha_{rs}$, where $\tau_{rs} : \mathbb{F}_k(\mathbb{R}^{n+1}) \to \mathbb{F}_k(\mathbb{R}^{n+1})$ is the map that interchanges the rth and s^{th} factors. On projecting to $\mathbb{F}_2(\mathbb{R}^{n+1})$, one can easily see that τ_{rs} has the degree of the antipodal map of S^n. This implies the proposition. □

As in Chapter II, let $\mathbb{F}_{k-r,r}(\mathbb{R}^{n+1}) = \mathbb{F}_{k-r}(\mathbb{R}^{n+1}_r)$, and $\mathbb{R}^{n+1}_r = \mathbb{R}^{n+1} - Q_r$. For each $0 \le r < k$, regard $\mathbb{F}_{k-r,r}(\mathbb{R}^{n+1})$ as a subspace of $\mathbb{F}_k(\mathbb{R}^{n+1})$ by means of the imbeddings

$$i_{k-r} : \mathbb{F}_{k-r,r}(\mathbb{R}^{n+1}) \to \mathbb{F}_k(\mathbb{R}^{n+1}),$$

where $(x_{r+1}, \cdots, x_k) \mapsto (q_1, \cdots, q_r, x_{r+1}, \cdots, x_k)$. Note that $\mathbb{F}_{k-r,r}(\mathbb{R}^{n+1})$ is the fiber, at the point $(q_1, \cdots, q_r)$, of the projection

$$\text{proj}_{k,r} : \mathbb{F}_k(\mathbb{R}^{n+1}) \to \mathbb{F}_r(\mathbb{R}^{n+1}),\ (x_1, \cdots, x_k) \mapsto (x_1, \cdots, x_r).$$

Thus, the natural sequence

$$\mathbb{F}_k(\mathbb{R}^{n+1}) \to \mathbb{F}_{k-1}(\mathbb{R}^{n+1}) \to \cdots \to \mathbb{F}_1(\mathbb{R}^{n+1})$$

leads to the fundamental fiber sequence

$$\mathcal{F}_k(\mathbb{R}^{n+1}) : \left\{ \begin{array}{ccccccc} \mathbb{F}_k & \leftarrow \mathbb{F}_{k-1,1} & \leftarrow \cdots & \leftarrow \mathbb{F}_{k-r,r} & \leftarrow \\ \downarrow & \downarrow & \downarrow & \downarrow & \\ \mathbb{R}^{n+1} & \mathbb{R}^{n+1}_1 & \cdots & \mathbb{R}^{n+1}_r & \end{array} \right.$$

introduced in §2 of Chapter II. The vertical maps are the projections on the first variable entry.

As the map $\mathbb{F}_{k-1,1}(\mathbb{R}^{n+1}) \to \mathbb{F}_k(\mathbb{R}^{n+1})$ is a homotopy equivalence, we shall assume that $r \ge 1$. Define $\mathcal{A}^*_{k-r,r}$ to be $\{\alpha^*_{ts} | 1 \le s < t \le k,\ t > r\}$, and put

$$\mathcal{A}^{*\wedge p}_{k-r,r} = \{\alpha^*_{r_1 s_1}\alpha^*_{r_2 s_2} \cdots \alpha^*_{r_p s_p} | 1 \le s_1 < r_1, \cdots, 1 \le s_p < r_p\},$$

where $\alpha^*_{r_j s_j} \in \mathcal{A}^*_{k-r,r}$, $r+1 \le r_1 < r_2 < \cdots < r_p \le k$ and $1 \le p \le k-r$. Observe that $\mathcal{A}^*_{k-0,0} = \mathcal{A}^*_{k-1,1}$, and $\mathcal{A}^*_{k-r,r} = \mathcal{A}^{*\wedge 1}_{k-r,r}$.

Theorem 1.1 *For* $1 \le p < k$, *the module* $H^{pn}(\mathbb{F}_{k,k-r}(\mathbb{R}^{n+1});\mathbb{Z})$ *is a free module, with* $\mathcal{A}^{*\wedge p}_{k-r,r}$ *as a basis.*

Proof. The proof is by decreasing induction on r. The theorem is certainly true for $r = k-1$, since $\mathbb{F}_{1,k-1}(\mathbb{R}^{n+1}) = \mathbb{R}^{n+1}_{k-1} \simeq (S_{k1} \vee \cdots \vee S_{kk-1})$.

Assume that the theorem is true for t such that $r+1 \le t \le k$, and consider the fibration

$$\mathrm{p}_{k-t,t} : \mathbb{F}_{k-t,t}(\mathbb{R}^{n+1}) \to \mathbb{R}_t^{n+1}, \ (q_1, \cdots, q_t, x_{t+1}, \cdots, x_k) \mapsto x_{t+1}.$$

Note that $\mathbb{F}_{k-t+1,t-1}(\mathbb{R}^{n+1})$ is the fiber, and denote by

$$i_{k+t-1} : \mathbb{F}_{k-t+1,t-1}(\mathbb{R}^{n+1}) \to \mathbb{F}_{k-t,t}(\mathbb{R}^{n+1})$$

the natural imbedding defined above. The module $H^{pn}(\mathbb{F}_{k-t+1,t-1}(\mathbb{R}^{n+1});\mathbb{Z})$ is a free module, by induction, with $\mathcal{A}^{*\wedge p}_{k-t+1,t-1}$ as a basis, where $1 \leq p \leq k-t+1$. Recall that we are using the same symbols for the elements α^*_{t+rs}, no matter where they lie. Therefore, regarding $\mathcal{A}^{*\wedge p}_{k-t+1,t-1}$ as a subset of $H^{pn}(\mathbb{F}_{k-t,t}(\mathbb{R}^{n+1});\mathbb{Z})$, we see that the morphism

$$H^*(i_{k-t+1}) : H^*(\mathbb{F}_{k-t,t}(\mathbb{R}^{n+1});\mathbb{Z}) \to H^*(\mathbb{F}_{k-t+1,t-1}(\mathbb{R}^{n+1});\mathbb{Z})$$

is surjective. Hence, the fiber $\mathbb{F}_{k-t+1,t-1}(\mathbb{R}^{n+1})$ is totally nonhomologous to zero in $\mathbb{F}_{k-t,t}(\mathbb{R}^{n+1})$ and, on applying the Leray-Hirsch Theorem, we obtain an isomorphism

$$H^*(\mathbb{R}_t^{n+1});\mathbb{Z}) \otimes H^*(\mathbb{F}_{k-t+1,t-1}(\mathbb{R}^{n+1});\mathbb{Z}) \to H^*(\mathbb{F}_r(\mathbb{R}_t^{n+1});\mathbb{Z})$$

of $H^*(\mathbb{R}_t^{n+1};\mathbb{Z})$-modules. The module $H^{pn}(\mathbb{F}_r(\mathbb{R}_t^{n+1});\mathbb{Z})$ is certainly is free. Since $H^{pn}(\mathbb{F}_{k-t+1,t-1}(\mathbb{R}^{n+1});\mathbb{Z})$ is free by assumption, it follows easily that $H^{pn}(\mathbb{F}_{k-t}(\mathbb{R}^{n+1});\mathbb{Z})$ is also free. The assertion that $\mathcal{A}^{*\wedge p}_{k-t,t}$ is a basis follows easily, and, thus, the induction step is established. □

One may regard the subconfiguration space $\mathbb{F}_{k+t,t}(\mathbb{R}^{n+1})$ as the configuration space $\mathbb{F}_k(\mathbb{R}^{n+1} - Q_t)$. The preceding theorem provides us with tools to study $H^*(\mathbb{F}_k(\mathbb{R}^{n+1} - Q_t);\mathbb{Z})$ and the natural projections

$$\mathrm{proj}_{k,r} : \mathbb{F}_k(\mathbb{R}^{n+1} - Q_t) \to \mathbb{F}_r(\mathbb{R}_t^{n+1}).$$

To see this, for $1+t \leq r \leq k$ and $1 \leq s < s$, define $g_{rs} : \mathbb{F}_k(\mathbb{R}^{n+1} - Q_t) \to S^n$ to be the map given by the rule

$$(x_{t+1}, \cdots, x_{t+k}) \to \begin{cases} (x_r - q_s)/|x_r - q_s|, & \text{if } s \leq t \\ |x_r - x_s|/|x_r - q_s|, & \text{if } s \geq t+1, \end{cases}$$

where $(x_{t+1}, \cdots, x_{t+k}) \in \mathbb{F}_k(\mathbb{R}^{n+1} - Q_t)$. Moreover, for $1+t \leq r \leq k$ and $1 \leq s < r$, set $\alpha^*_{rs} = H^n(g_{rs})(\iota^*_n)$. The map

$$i_{kt} : \mathbb{F}_k(\mathbb{R}^{n+1} - Q_t) \to \mathbb{F}_{k+t,t}(\mathbb{R}^{n+1}),$$

where $(x_{t+1}, \cdots, x_{t+k})$ goes to $(q_1, \cdots, q_t, x_{t+1}, \cdots, x_{t+k})$, identifies the relative configuration space $\mathbb{F}_k(\mathbb{R}^{n+1} - Q_t)$ with $\mathbb{F}_{k+t,t}(\mathbb{R}^{n+1})$ in the canonical manner. One sees that the relative case maps g_{rs} are just the restrictions of the analogous maps $g_{rs} : \mathbb{F}_{k+t}(\mathbb{R}^{n+1}) \to S^n$ to $\mathbb{F}_{k+t,t}(\mathbb{R}^{n+1})$. Therefore, the elements $\alpha^*_{rs} = H^n(g_{rs})(\iota^*_n)$ may be naturally identified with their counterparts in $H^n(\mathbb{F}_{k+t,t}(\mathbb{R}^{n+1});\mathbb{Z})$. For $t \geq 0$, by definition, let

$$\mathcal{A}^*_{k+t,t}(\mathbb{R}^{n+1}-Q_t)=\{\alpha^*_{rs}|1+t\le r\le k, 1\le s<r\}.$$

Note that $Q_0=\emptyset$, when $t=0$. So, in this case, $\mathcal{A}^*_{k+t,t}(\mathbb{R}^{n+1}-Q_t)$ coincides with $\mathcal{A}^*_k$ of $H^*(\mathbb{F}_k(\mathbb{R}^{n+1});\mathbb{Z})$. Put

$$\mathcal{A}^{*\wedge p}_{k+t,t}(\mathbb{R}^{n+1}-Q_t)=\{\alpha^*_{r_1s_1}\alpha^*_{r_2s_2}\cdots\alpha^*_{r_ps_p}|1\le s_1<r_1,\cdots,1\le s_p<r_p\},$$

where $t+1\le r_1<r_2<\cdots<r_p\le k$ and $1\le p\le k-t$. Thus we see that, for $1\le p<k$, the module $H^{pn}(\mathbb{F}_k(\mathbb{R}^{n+1}_t);\mathbb{Z})$ is a free module with $\mathcal{A}^{*\wedge p}_{k+t,t}(\mathbb{R}^{n+1}-Q_t)$ as a basis.

The corollary follows immediately. Observe that $\mathbb{F}_{k-r}(\mathbb{R}^{n+1}-Q_{t+r})$ is the fiber of the fibration $\mathrm{proj}_{k,r}:\mathbb{F}_k(\mathbb{R}^{n+1}_t)\to\mathbb{F}_r(\mathbb{R}^{n+1}_t)$.

Corollary 1.2 *The fiber $\mathbb{F}_{k-r}(\mathbb{R}^{n+1}-Q_{t+r})$ is totally nonhomologous to zero. Moreover, the Leray-Hirsch morphism,*

$$\Phi^*_{k-r,r}:H^*(\mathbb{F}_r(\mathbb{R}^{n+1}-Q_t)\otimes H^*(\mathbb{F}_{k-r}(\mathbb{R}^{n+1}_{t+r}))\to H^*(\mathbb{F}_k(\mathbb{R}^{n+1}-Q_{t+r})),$$

is an isomorphism of $\mathbb{F}_r(\mathbb{R}^{n+1}-Q_t)$-modules.

Since the study of the cohomology of the configuration spaces $\mathbb{F}_k(\mathbb{R}^{n+1}-Q_t)$ is a special case of the analogous study of the subspaces $\mathbb{F}_{k-r,r}(\mathbb{R}^{n+1})$ of the configuration space $\mathbb{F}_k(\mathbb{R}^{n+1})$, we shall hereon restrict our attention to the latter and its subspaces.

The following corollary is an immediate consequence of the preceding theorem.

Corollary 1.3 *There are isomorphisms of modules*

$$H^*(\mathbb{F}_{k-r,r}(\mathbb{R}^{n+1});\mathbb{Z})=\bigotimes_{j=r}^{k-1}H^*(\mathbb{R}^{n+1}_j;\mathbb{Z}).$$

Proof. Consider the fundamental fiber sequence

$$\begin{array}{ccccc}\mathbb{F}_k & \leftarrow\mathbb{F}_{k-1,1}\leftarrow & \cdots & \leftarrow\mathbb{F}_{k-r,r}\leftarrow & \cdots\\ \downarrow & \downarrow & \downarrow & \downarrow & \downarrow\\ \mathbb{R}^{n+1} & \mathbb{R}^{n+1}_1 & \cdots & \mathbb{R}^{n+1}_r & \cdots,\end{array}$$

where $\mathbb{F}_{k-r,r}=\mathbb{F}_{k-r,r}(\mathbb{R}^{n+1})$. An inductive argument starting from the left proves the assertion. □

Corollary 1.4 *The Poincaré polynomial of $\mathbb{F}_{k-r,r}(\mathbb{R}^{n+1})$ is given by the relation* $\mathrm{P}(\mathbb{F}_{k-r,r},T)=\bigotimes_{t=r+1}^{k}(1+(t-1)T^n)$.

Proof. The Poincaré polynomial of $(S_{t1}\vee\cdots\vee S_{tt-1})$ is $1+(t-1)T^n$. The rest follows by virtue of Corollary 1.4. □

An immediate corollary is the following.

Corollary 1.5 *The Euler characteristic,* $\chi(\mathbb{F}_{k-r,r}(\mathbb{R}^{n+1}))$ *is* $\neq 0$ *for* $r \geq 2$.

We shall see in the next chapter that $\mathbb{F}_{k-r,r}(\mathbb{R}^2)$ is of the homotopy type of a finite complex. Moreover, according to the proof of Theorem 1.1 of Chapter IV, the space $\mathbb{F}_{k-r,r}(\mathbb{R}^2)$ is aspherical. Hence, by [50, Gottlieb] (see also [61, Husseini], [102, Stallings]), the center of $\pi_1(\mathbb{F}_{k-r,r}, q)$ for $t \geq 3$ is trivial. This is another proof of the result proved in Chapter IV.

Corollary 1.6 *Let* $\text{proj}_{k,r} : \mathbb{F}_k(\mathbb{R}^{n+1}) \to \mathbb{F}_r(\mathbb{R}^{n+1})$ *be the projection that sends* $(x_1, \cdots, x_r, x_{r+1}, \cdots, x_k)$ *to* $(x_1, \cdots, x_r)$. *Then, the local coefficient system over* $\mathbb{F}_r(\mathbb{R}^{n+1})$, *defined by the integral cohomology* $H^*(\mathbb{F}_{k-r}(\mathbb{R}^{n+1});\mathbb{Z})$ *of the fibers* $\text{proj}_{k,r}$, *is simple.*

Proof. If $n > 1$, then the space $\mathbb{F}_r(\mathbb{R}^{n+1})$ is simply connected, and the assertion of the lemma is obvious. So, assume that $n = 1$. Now, by [98, Serre] and Corollary 2.2, the local coefficient system is trivial. □

It is worthwhile to note that, alternatively, one can deduce the preceding corollary from the Y-B relations. For, according to Proposition 6.1 of Chapter IV, the action of $\pi_1(\mathbb{R}^2_t, q)$ on $\pi_1(\mathbb{F}_{k-1}(\mathbb{R}^2_{t+1}, q)$ is by conjugation in $\pi_1(\mathbb{F}_k(\mathbb{R}^2_t, q)$. Now, by Propositions 5.2 and 5.3 of Chapter IV, this conjugation is the same as conjugation in $\pi_1(\mathbb{F}_{k-1}(\mathbb{R}^2_{t+1}), q)$. Therefore, on passing to the homology groups $H_1(\mathbb{F}_{k-1}(\mathbb{R}^2_{t+1});\mathbb{Z})$, the action becomes trivial. Hence, the action on $H^1(\mathbb{F}_{k-t-1,t+1};\mathbb{Z})$ is trivial, and this implies that the action on the algebra $H^*(\mathbb{F}_{k-t-1,t+1};\mathbb{Z})$ is also trivial.

2 Invariance Under Σ_k

In Chapter II we discussed the effect of the natural action of the symmetric Σ_k on the group $\pi_n(\mathbb{F}_k(\mathbb{R}^{n+1}))$. Here, we shall describe the induced action on cohomology. As the cohomology is generated by the elements α^*_{rs}, it is enough to describe the action on $H^n(\mathbb{F}_k(\mathbb{R}^{n+1});\mathbb{Z})$. This is quite easy to do, because of the results of §2 of Chapter II.

Theorem 2.1 *For all* $\sigma \in \Sigma_k$ *and* $\alpha^*_{rs} \in \mathcal{A}_k$, *the relation* $H^n(\sigma)(\alpha^*_{rs}) = \alpha^*_{\sigma^{-1}r\sigma^{-1}s}$ *holds in* $H^n(\mathbb{F}_k(\mathbb{R}^{n+1});\mathbb{Z})$.

Proof. Observe that $H_n(\sigma)(\alpha^*_{rs}) = H_n(\sigma \circ \alpha'_{rs})$, and the right hand side equals $H_n(\alpha'_{\sigma r\sigma s}) = \alpha_{\sigma r\sigma s}$ by Theorem 2.1 of Chapter II. Therefore,

$$\begin{aligned} < H^n(\sigma)(\alpha^*_{rs}), \alpha_{\sigma^{-1}r\sigma^{-1}s} > &= < \alpha^*_{rs}, H_n(\sigma)(\alpha_{\sigma^{-1}r\sigma^{-1}s}>) \\ &= < \alpha^*_{rs}, \alpha_{rs} > \\ &= 1. \end{aligned}$$

Now, suppose that $(u, v) \neq (\sigma^{-1}r, \sigma^{-1}s)$. Then,

$$\begin{aligned} < H^n(\sigma)(\alpha^*_{rs}), \alpha_{uv} > &= < \alpha^*_{rs}, H_n(\sigma)(\alpha_{uv}) > \\ &= < \alpha^*_{rs}, \alpha_{\sigma u \sigma v} > \\ &= 0, \end{aligned}$$

since, by assumption, $(r, s) \neq (\sigma u, \sigma v)$. Thus the theorem is established. □

3 The Cohomological Y-B Relations

The Y-B relations in $\pi_*(\mathbb{F}_k(\mathbb{R}^{n+1}))$ imply similar relations among the corresponding cohomology classes. Our object here is to establish these relations, which we call *cohomological Y-B relations.*

Theorem 3.1 *For all $\sigma \in \Sigma_k$, the relations*

$$\begin{aligned} &(i)\ (\alpha^*_{\sigma 2 \sigma 1})^2 = 0 \ \ and \\ &(ii)\ \alpha^*_{\sigma 3 \sigma 1}\alpha^*_{\sigma 3 \sigma 2} = \alpha^*_{\sigma 2 \sigma 1 s}(\alpha^*_{\sigma 3 \sigma 2} - \alpha^*_{\sigma 3 \sigma 1}) \end{aligned}$$

hold in the algebra $H^*(\mathbb{F}_k(\mathbb{R}^{n+1}); \mathbb{Z})$.

Proof. To prove (i), recall that $\alpha^*_{rs} = H^n(g_{rs})(\iota^*_n)$, where ι^*_n is the fundamental cohomology class of S^n. Note that $(\iota^*_n)^2 = 0$ and, hence, $(\alpha^*_{21})^2 = 0$. To prove (ii), we first prove the special case

$$(*) \quad \alpha^*_{31}\alpha^*_{32} = \alpha^*_{21}(\alpha^*_{32} - \alpha^*_{31}).$$

Proof of the Special Case (*)

Consider the fibration $\text{proj}_{2,1} : \mathbb{F}_3(\mathbb{R}^{n+1}) \to \mathbb{F}_2(\mathbb{R}^{n+1})$, where $x_1, x_2, x_3) \mapsto (x_1, x_2)$, the fiber of which is $\mathbb{R}^{n+1}_2 \simeq (S_{31} \vee S_{32})$. By Corollary 1.2, we have the Leray-Hirsch isomorphism, $\Phi^*_{2,1} : H^*(\mathbb{F}_2(\mathbb{R}^{n+1}); \mathbb{Z}) \otimes H^*(\mathbb{R}^{n+1}_2; \mathbb{Z}) \cong H^*(\mathbb{F}_3(\mathbb{R}^{n+1}); \mathbb{Z})$ of $H^*(\mathbb{F}_2(\mathbb{R}^{n+1}); \mathbb{Z})$-modules. Therefore,

$$\text{(I)} \quad \alpha^*_{31}\alpha^*_{32} = c_{31}\alpha^*_{21}\alpha^*_{31} + c_{32}\alpha^*_{21}\alpha^*_{32},$$

where the coefficients, c_{31} c_{32}, are integers. To calculate the coefficients, consider the map $\phi_1 : S^n_1 \times S^n_2 \to \mathbb{F}_3(\mathbb{R}^{n+1})$ that takes (ξ_1, ξ_2) to $(q_1, q_1 + \xi_1, q_1 + 1/2\xi_2)$, where S^n_1 and S^n_2 are unit spheres. One verifies that, as in the proof of Proposition 1.2 of Chapter II,

$$[\phi_1 | S^n_1] = \alpha_{21} + (-1)^{n+1}\alpha_{32}, \quad [\phi_1 | S^n_1] = \alpha_{31}.$$

The map ϕ_1 yields the Y-B relation

$$[\alpha_{21} + (-1)^{n+1}\alpha_{32}, \alpha_{31}] = 0$$

in $\pi_n(\mathbb{F}_3(\mathbb{R}^{n+1}), q)$, for $n > 1$, and the analogous relation when $n = 1$ (see Chapter II, §3 and Chapter IV, §4). Passing to cohomology, we obtain the relations,

$$\begin{cases} \alpha_{21}^* \mapsto & \iota_1^* \otimes 1 \\ \alpha_{31}^* \mapsto & 1 \otimes \iota_2^* \\ \alpha_{32}^* \mapsto & (-1)^{n+1}\iota_1^* \otimes 1, \end{cases}$$

where ι_1^* and ι_2^* are, respectively, the duals of the classes $\iota_1 \in H_n(S_1;\mathbb{Z})$ and $\iota_2 \in H_n(S_2;\mathbb{Z})$ represented by the identity $S^n \overset{\text{id}}{\to} S^n$. Applying $H^n(\phi_1)$ to (I), we see that

$$(1 \otimes \iota_2^*)((-1)^{n+1}\iota_1^* \otimes 1) = (\iota_1^* \otimes 1)(c_{31}(1 \otimes \iota_2^*) + c_{32}(-1)^{n+1}(\iota_1^* \otimes 1)).$$

The left hand side of the latter expression is equal to $(-1)^{n^2+n+1}\iota_1^* \otimes \iota_2^*)$, while the right hand side is $c_{31}\iota_1^* \otimes \iota_2^*$. Hence, $c_{31} = -1$.

Next consider the map $\phi_2 : S_1^n \times S_2^n \to \mathbb{F}_3(\mathbb{R}^{n+1})$, where $(\xi_1, \xi_2) \mapsto (q_1, q_1 + \xi_1, q_1 + \xi_1 + 1/2\xi_2)$. Again, as in the case of ϕ_1, we see that $[\phi_2|S_1^n] = \alpha_{21} + \alpha_{31}$ and $[\phi_2|S_2^n] = \alpha_{32}$. The map ϕ_2 yields the Y-B relation

$$\text{(II)} \quad [\alpha_{21} + \alpha_{31}, \alpha_{32}] = 0$$

in $\pi_n(\mathbb{F}_3(\mathbb{R}^{n+1}), q)$ for $n > 1$, and the corresponding relation for $n = 1$ (see Chapter II, §3 and Chapter IV, §4). Passing to cohomology, we obtain the morphisms

$$\begin{cases} \alpha_{21}^* \mapsto \iota_1^* \otimes 1 \\ \alpha_{31}^* \mapsto \iota_1^* \otimes 1 \\ \alpha_{32}^* \mapsto 1 \otimes \iota_2^*. \end{cases}$$

Again, on applying $H^n(\phi_2)$ to (II), we find, after calculating the products, that $c_{32} = 1$. This proves that

$$\begin{aligned} \alpha_{31}^*\alpha_{32}^* &= -\alpha_{21}^*(\alpha_{31}^* - \alpha_{32}^*) \\ &= \alpha_{21}^*(\alpha_{32}^* - \alpha_{31}^*), \end{aligned}$$

as required. □

Proof of the general case

The projection $\text{proj}_{k,3} : \mathbb{F}_k(\mathbb{R}^{n+1}) \to \mathbb{F}_3(\mathbb{R}^{n+1})$, where $(x_1, \cdots, x_k)$ goes to (x_1, x_2, x_3), induces a monomorphism on cohomology that takes $\alpha_{21}^*, \alpha_{31}^*$ and α_{32}^* in $H^n(\mathbb{F}_3(\mathbb{R}^{n+1});\mathbb{Z})$ to their counterparts in $H^*(\mathbb{F}_k(\mathbb{R}^{n+1});\mathbb{Z})$, and, therefore, the same relation $\alpha_{31}^*\alpha_{32}^* = \alpha_{21}^*(\alpha_{32}^* - \alpha_{31}^*)$ holds in $H^*(\mathbb{F}_k(\mathbb{R}^{n+1});\mathbb{Z})$. Now, on applying the permutation σ^{-1} to the preceding equality, we see that Theorem 2.1 implies the assertion of the theorem. □

For an another proof of Theorem 3.1 see [17, Cohen]. Similar relations hold in $H^*(\mathbb{F}_{k-r,r}(\mathbb{R}^{n+1});\mathbb{Z})$.

Theorem 3.2 *The following relations,*

$$(i)\ (\alpha^*_{ts})^2 = 0,\ \text{for}\ \alpha^*_{ts} \in \mathcal{A}^*_{k-r,r},$$
$$(ii)\ \alpha^*_{ts}\alpha^*_{tu} = \alpha^*_{us}(\alpha^*_{tu} - \alpha^*_{ts}),\ \text{for}\ s < u,\ u \geq r+1,$$
$$(iii)\ \alpha^*_{ts}\alpha^*_{tu} = 0,\ \text{if}\ s < u,\ u < r+1,$$

hold in $H^*(\mathbb{F}_{k-r,r}(\mathbb{R}^{n+1}); \mathbb{Z}$, *for* $1 \leq r \leq k$.

Proof. Consider the fibration $\text{proj}_{k,r} : \mathbb{F}_k(\mathbb{R}^{n+1}) \to \mathbb{F}_r(\mathbb{R}^{n+1})$ the fiber of which is $\mathbb{F}_{k-r,r}(\mathbb{R}^{n+1})$. Then $H^*(i_{k-r})(\alpha^*_{ts_1}\alpha^*_{ts_2}) = \alpha^*_{s_2s_1}(\alpha^*_{ts_2} - \alpha^*_{ts_1})$, for $t \geq r+1$, if $s_2 \geq r+1$, and is zero for $s_2 < r+1$. □

4 The Structure of $H^*(\mathbb{F}_k(\mathbb{R}^{n+1}))$

All homology and cohomology groups in this section are, unless it is mentioned otherwise, with integral coefficients. We wish to show that the algebra $H^*(\mathbb{F}_k(\mathbb{R}^{n+1}))$ is the universal commutative algebra generated by $\{\alpha_{rs} | 1 \leq s < r \leq k\}$ modulo the cohomological Y-B relations (cf. [17, Cohen]).

First, recall that a sequence

$$0 \to A'^* \to A^* \to A''^* \to 0$$

of graded, augmented algebras is said to be *exact* if, and only if,

$$A^*//A'^* = A^*/A^*\bar{A'}^*A^* \cong A''^*$$

is an isomorphism of algebras, where $\bar{A'}^*$ is the ideal $\bar{A'}^* \subset A'^*$ of augmentation 0 ([83, Milnor-Moore]).

Recall also that $\mathcal{A}^*_r = \{\alpha^*_{ts} \mid 1 \leq s < t \leq r \leq k\}$, $\mathcal{A}^*_{k-r,r} = \mathcal{A}^*_k - \mathcal{A}^*_r$ (see §1 above). Denote by $P(\mathcal{A}^*_{k-r,r})$ the universal, commutative and graded algebra generated by the set $\mathcal{A}^*_{k-r,r}$. Let $I_{k-r,r}$ be the ideal in $P(\mathcal{A}^*_{k-r,r})$ generated by the following relations:

(i) for all $\alpha^*_{ts} \in \mathcal{A}^*_{k-r,r}$, $(\alpha^*_{ts})^2 = 0$;
(ii) for all α^*_{ts} and α^*_{tu} in $\mathcal{A}^*_{k-r,r}$ with $s < u$, $\alpha^*_{ts}\alpha^*_{tu} = 0$, if $u \leq r$, and $\alpha^*_{ts}\alpha^*_{tu} = \alpha^*_{us}(\alpha^*_{tu} - \alpha^*_{ts})$ otherwise.

Observe that the relations above are just the cohomological Y-B relations given in Theorem 3.2, where the right hand side of expressions of the type $\alpha^*_{ts}\alpha^*_{tu} = \alpha^*_{us}(\alpha^*_{tu} - \alpha^*_{ts})$ are set equal to 0 whenever α^*_{us} is not in $\mathcal{A}^*_{k-r,r}$. Put $A_{k-r,r} = P(\mathcal{A}^*_{k-r,r})/I_{k-r,r}$. Observe that $A_{1,r-1} = \bigoplus_{s=1}^{r-1} E(\alpha^*_{rs})$, where $E(\mathbb{Z}\alpha^*_{rs})$ is the exterior algebra of the $\mathbb{Z}$-module generated by α^*_{rs}, with the direct sum being one of algebras.

By definition, let $\phi_{k-r,r} : A_{k-r,r} \to H^*(\mathbb{F}_{k-r,r}(\mathbb{R}^{n+1}))$ be the morphism that takes an element $\alpha^*_{rs} \in A_{k-r,r}$ to the element α^*_{rs} in $H^*(\mathbb{F}_{k-r,r}(\mathbb{R}^{n+1}))$.

Theorem 4.1 *For all* $1 \le r < k$*, the morphism* $\phi_{k-r,r}$ *is an isomorphism of algebras.*

Proof. If $r = k-1$, then $\mathbb{F}_{1,k-1}(\mathbb{R}^{n+1}) \simeq (S_{k1} \vee \cdots \vee S_{kk-1})$ and, therefore, $H^*(\mathbb{F}_{1,k-1}(\mathbb{R}^{n+1})) \cong \bigoplus_{s=1}^{k-1} E(\alpha^*_{ks})$. The right hand side of the expression is isomorphic to the algebra $A_{1,k-1}$. This proves the case $r = k-1$.

Proceeding inductively, assume that

$$\phi_{k-r-1,r+1} : A_{k-r-1,r+1} \to H^*(\mathbb{F}_{k-r-1,r+1})$$

is an isomorphism of algebras. We wish to show that the morphism $\phi_{k-r,r} : A_{k-r,r} \to H^*(\mathbb{F}_{k-r,r})$ is also an isomorphism of algebras. So consider the commutative diagram $\mathcal{D}_r$,

$$\begin{array}{ccccccccc} 0 \to & A_{r,r-1} & \to & A_{k-r,r} & \to & A_{k-r-1,r+1} & \to 0 \\ & \downarrow \phi_{r,r-1} & & \downarrow \phi_{k-r,r} & & \downarrow \phi_{k-r-1,r+1} & \\ 0 \to & H^*(\mathbb{R}^{n+1}_r) & \to & H^*(\mathbb{F}_{k-r,r}) & \to & H^*(\mathbb{F}_{k-r-1,r+1}) & \to 0, \end{array}$$

where $A_{r,r-1}$ is the subalgebra of $A_{k-r,r}$ generated by $\{\alpha^* r + 1s | 1 \le s \le r\}$. The lower horizontal maps are those induced by the map $\mathrm{p}_{k-r} : \mathbb{F}_{k-r,r} \to \mathbb{R}^{n+1}_r$ with fiber $\mathbb{F}_{k-r-1,r+1}(\mathbb{R}^{n+1})$. It is easy to verify that

$$\phi_{r,r-1} : A_{r,r-1} \to H^*(\mathbb{R}^{n+1}_r) \cong H^*(S_{r+11} \vee \cdots \vee S_{r+1r})$$

is an isomorphism. Now, one deduces immediately that the left horizontal morphism of the upper row of $\mathcal{D}_r$ is injective. The second horizontal morphism is clearly surjective, because it is the morphism that takes the elements of $\mathcal{A}^*_{r,r-1} = \mathcal{A}^*_r - \mathcal{A}^*_{r-1} = \{\alpha_{r+11}, \cdots, \alpha_{r+1r}\}$ to zero.

Let us prove now that the first row is exact. For convenience, put $\underline{r} = \{r_1, \cdots, r_p\}$ and $\underline{s} = \{s_1, \cdots, s_p | s_1 < r_1, \cdots, s_p < r_p\}$. Then, given $\omega \in A^*_{k-r,r}$, we can write $\omega = \sum_{\underline{r},\underline{s}} \nu_{\underline{r},\underline{s}} \alpha_{\underline{r},\underline{s}}$, where $\nu_{\underline{r},\underline{s}} \in \mathbb{Z}$ and $\alpha_{\underline{r},\underline{s}} = \alpha_{r_1 s_1} \cdots \alpha_{r_1 s_1}$ with each $r_i > r$. Moreover, using the Y-B relations of Theorem 3.2 we can ensure that $r_1 < r_2 < \cdots < r_p \le k$. Now write $\omega = \omega_1 + \omega_2$, where ω_1 does not involve any of the elements $\mathcal{A}^*_{r,r-1}$ and ω_2 is in the ideal generated by $\mathcal{A}^*_{r,r-1}$. We wish to show that $\omega_1 = 0$. So, if the morphism $A^*_{k-r,r} \to A^*_{k-r-1,r+1}$ takes ω to zero, it must take ω_1 to zero. Assume, by way of contradiction, that $\omega_1 \ne 0$, and consider the expression $\omega_1 = \sum_{\underline{r},\underline{s}} \nu_{\underline{r},\underline{s}} \alpha_{\underline{r},\underline{s}}$, where $r+1 \le r_1 < \cdots < r_p$. However, according to Theorem 1.1, the terms $\alpha_{\underline{r},\underline{s}}$ are linearly independent in $H^*(\mathbb{F}_{k-r-1,r+1})$. Hence, they are linearly independent in $A_{k-r-1,r+1}$ and, consequently, $\omega_1 \ne 0$ in $A^*_{k-r-1,r+1}$. This implies that ω_1 does not lie in $\ker(A_{k-r,r} \to A_{k-r-1,r+1})$. But this is a contradiction. Therefore, $\omega_1 = 0$, and, hence, ω lies in the ideal in $\sum_{s=1}^{r-1} \alpha^*_{rs} A_{k-r-1,r+1}$. The latter is clearly the ideal in $A_{k-r,r}$ generated by the elements $\alpha^*_{r+1s_1}, \cdots, \alpha_{r+1s_p}$. All of the this implies that the upper sequence in $\mathcal{D}_r$ is exact.

Finally, the lower row is exact because the spectral sequence of the fibration $\mathrm{p}_r : \mathbb{F}_{k-r,r} \to \mathbb{F}_{1,r} = \mathbb{R}^{n+1}_r$ collapses, and the Leray-Hirsch Theorem implies the rest. This establishes the induction step, proving the theorem. □

An immediate consequence is the following.

Theorem 4.2 *The algebra $H^*(\mathbb{F}_{k-r,r}(\mathbb{R}^{n+1}))$ is the universal, commutative, and graded algebra generated by the set $\mathcal{A}^*{}_{k-r,r} = \{\alpha^*_{ts} \mid 1 \leq s < t \leq k, t > r\}$ subject to the cohomological Y-B relations*

$$\begin{aligned} (i)\ (\alpha^*_{ts})^2 &= 0, \\ (ii)\ \alpha^*_{ts}\alpha^*_{tu} &= \alpha^*_{us}(\alpha^*_{tu} - \alpha^*_{ts}),\ \textit{for } u > r, \textit{ and} \\ (iii)\ \alpha^*_{ts}\alpha^*_{tu} &= 0,\ \textit{for } u \leq r. \end{aligned}$$

Recall that $\mathcal{A}^{*\wedge p}_{k-r,r} = \{\alpha^*_{r_1s_1}\alpha^*_{r_2s_2}\cdots\alpha^*_{r_ps_p}\}$, where $r < r_1 < r_2 < \cdots < r_p \leq k$ and $1 \leq p \leq (k-1)$. Put $\mathcal{A}^{*\wedge 0}_{k-r,r} = \{1\}$.

Applying Theorem 1.1, we see that the set $\mathcal{A}^{*\wedge}_{k-r,r} = \cup_{p=0}^{k-1}\mathcal{A}^{*\wedge p}$ is a basis for $H^*(\mathbb{F}_{k-r,r}(\mathbb{R}^{n+1});\mathbb{Z})$, where $r < r_1 < r_2 < \cdots < r_p \leq k\}$.

The following notation is suggested by the preceding result.

Definition 4.1 *Let*

$$\begin{aligned} \mathcal{R}_p &= \{\underline{r} = (r_1, \cdots, r_p) \mid 1 < r_1 < \cdots < r_p \leq k\} \text{ and} \\ \mathcal{S}_p &= \{\underline{s} = (s_1, \cdots, s_p) \mid 1 \leq s_1, s_2, \cdots, s_p < k\}. \end{aligned}$$

Put $|\underline{r}| = p = |\underline{s}|$. We shall say that $\underline{s} < \underline{r}$ if, and only if, $s_i < r_i$ for all $i = 1, \ldots, p$. Also, for $\underline{s} < \underline{r}$, put $\alpha^*_{\underline{r}\,\underline{s}} = \alpha^*_{r_1s_1}\cdots\alpha^*_{r_ps_p}$.

Using the indexing sets $\mathcal{R}$ and $\mathcal{S}$, one can restate Theorem 1.1 as follows.

Theorem 4.3 *The set of elements $\mathcal{A}^{*\wedge p}_{k-r,r} = \{\alpha^*_{\underline{r}\,\underline{s}} \mid \underline{r} \in \mathcal{R}_p, \underline{s} \in \mathcal{S}_p, \underline{s} < \underline{r}\}$ is a basis for $H^{pn}(\mathbb{F}_{k-r,r}(\mathbb{R}^{n+1}))$.*

5 The group $H^n(\mathbb{F}_{k+1}(S^{n+1}))$

In this section we seek to describe the group $H^n(\mathbb{F}_{k+1}(S^{n+1}))$ and the action of Σ_{k+1} thereon, induced by the natural action of Σ_{k+1} on $\mathbb{F}_{k+1}(S^{n+1})$. We shall do this by means of a suitable basis for $H^n(\mathbb{F}_{k+1}(S^{n+1});\mathbb{Z})$. The answer depends on the parity of $n+1$. Before discussing the separate cases, we first study $H_n(\mathbb{F}_{k+1}(S^{n+1}))$ in general, and, then, establish facts valid for $H^n(\mathbb{F}_{k+1}(S^{n+1}))$ in both cases.

Recall that the definitions and results of §2 of Chapter III are subject to the assumption that $n > 1$. One can easily verifies that the same results hold for integral homology when $n = 1$. In what follows we shall assume throughout that $n > 1$, leaving the case $n = 1$ to the reader.

Consider the elements $\{\beta_{rs} | 1 \leq s < r \leq (k+1)\} \subset \pi_n(\mathbb{F}_{k+1}(S^{n+1}))$, defined in §2 of Chapter III, and, as in that section, let $\phi^e : S^{n+1} - \{-e\} \to V^e$, be the stereographic projection, with $e = (0, \cdots, 0, 1) \in \mathbb{R}^{n+2}$ and $V^e = \mathbb{R} \times \mathbb{R}^n$. The composite map of $(\phi^e)^{-1}$ with the natural inclusion $S^{n+1} - \{-e\} \to S^{n+1}$ induces, in turn, a map $j^e_{k+1} : \mathbb{F}_{k+1}(\mathbb{R}^{n+1}) \to \mathbb{F}_{k+1}(S^{n+1})$. By (i) of Theorem 2.2 of Chapter III, the morphism

$$\pi_n(j^e_{k+1}) : \pi_n(\mathbb{F}_{k+1}(\mathbb{R}^{n+1})) \to \pi_n(\mathbb{F}_{k+1}(S^{n+1}))$$

is such that $\pi_n(j^e_{k+1})(\alpha_{rs}) = \beta_{rs}$, where $1 \leq s < r \leq k+1$. The elements of $\{\beta_{rs} \mid 1 \leq s < r \leq k+1$ of $\pi_n(\mathbb{F}_{k+1}(S^{n+1}, q^e)$ define naturally homology classes in $H_n(\mathbb{F}_{k+1}(S^{n+1}))$, which we denote by the same symbols. Theorem 2.1 of Chapter III immediately implies the following.

Theorem 5.1 *The relations*

$$\begin{aligned} \beta_{rs} &= (-1)^{n+1}\beta_{sr} \ \ \textit{and} \\ H_n(\sigma)(\beta_{rs}) &= \beta_{\sigma r \sigma s} \end{aligned}$$

hold in $H_n(\mathbb{F}_{k+1}(S^{n+1}))$, *for all* $1 \leq s < r \leq k+1$, *and* $\sigma \in \Sigma_{k+1}$.

Similarly, one deduces the following result.

Theorem 5.2 *The elements* $\{\beta_{rs} \mid 1 \leq s < r \leq (k+1)\}$ *satisfy the following relations:*

$$\begin{aligned} &(i)\ H_n(i_k)(\alpha_{rs}) = \beta_{rs}, \quad \textit{for } 1 \leq s < r \leq k, \\ &(ii)\ \textstyle\sum_{s=1}^{k} \beta_{k+1s} = 0, \\ &(iii) \qquad \beta_{k+1k} = (-1)^n \textstyle\sum_{s=1}^{k-1} \beta_{ks}, \ \ \textit{and} \\ &(iv) \qquad \beta_{k+1t} = (-1)^n \textstyle\sum_{s=1}^{k-1} \beta_{\sigma k \sigma s} \end{aligned}$$

in $H_n(\mathbb{F}_{k+1}(S^{n+1}))$.

The following is an immediate corollary.

Corollary 5.1 *The elements* $\{\beta_{rs}|1 \leq s < r \leq k\}$ *generate* $H_n(\mathbb{F}_{k+1}(S^{n+1}))$.

Proof. Now, by Theorem 2.2 of Chapter III, the morphism $\pi_n(j^e_{k+1})$: $\pi_n(\mathbb{F}_{k+1}(\mathbb{R}^{n+1})) \to \pi_n(\mathbb{F}_{k+1}(S^{n+1}))$ is surjective. Hence, the induced morphism $H_n(j^e_{k+1}) : H_n(\mathbb{F}_{k+1}(\mathbb{R}^{n+1})) \to H_n(\mathbb{F}_{k+1}(S^{n+1}))$ of homology groups is also surjective, and, therefore, the elements $\{\beta_{rs}|1 \leq s < r \leq k+1\}$ generate $H_n(\mathbb{F}_{k+1}(S^{n+1}))$. But each β_{k+1s} is expressible as a linear combination of the elements β_{rs} with $r \leq k$ and, consequently, the corollary follows. □

Proposition 5.1 *Let $i_k : \mathbb{F}_k(\mathbb{R}^{n+1}) \to \mathbb{F}_{k+1}(S^{n+1})$ be the imbedding defined in §2 of Chapter III. The following statements are valid.*

(*i*) *The induced morphism* $H_n(i_k) : H_n(\mathbb{F}_k(\mathbb{R}^{n+1})) \to H_n(\mathbb{F}_{k+1}(S^{n+1}))$ *is surjective;*

(*ii*) *if $n+1$ is odd, then $H_n(i_k)$ is an isomorphism; and*

(*iii*) *if $n+1$ is even, then* $\ker(H_n(i_k))$ *is the infinite cyclic group generated by $2\Delta_k(\mathbb{R}^{n+1})$, where $\Delta_k(\mathbb{R}^{n+1})$ is the homology Dirac class.*

Proof. Observe that $H_n(i_k)$ takes α_{rs} to β_{rs}, for $1 \leq s < r \leq k$. Also, note that Corollary 3.1 and Theorem 6.1 of Chapter III imply that $H_n(i_k)$ is surjective. This proves (i).

To prove (ii), note that the fibration $\text{proj}_{k+1,1} : \mathbb{F}_{k+1}(S^{n+1}) \to S^{n+1}$ admits a section, and, hence, the homotopy exact sequence of the fibration yields the short exact sequence

$$0 \to \pi_n(\mathbb{F}_k(\mathbb{R}^{n+1})) \to \pi_n(\mathbb{F}_{k+1}(S^{n+1})) \to \pi_n(S^{n+1}) \to 0.$$

Hence, $H_n(i_k)$ is injective.

To prove (iii), note that, according to Theorem 6.1 of Chapter III, $\ker(\pi_n(i_k))$ is generated by $2\Delta_k(\mathbb{R}^{n+1})$, where $\Delta_k(\mathbb{R}^{n+1}) \in \pi_n(\mathbb{R}^{n+1})$ is the homotopy Dirac class. Passing to homology, one obtains the required assertion. □

Let us turn now to the description of $H^n(\mathbb{F}_{k+1}(S^{n+1}))$. Consider, therefore, the induced morphism $H^n(i_k) : H^n(\mathbb{F}_{k+1}(S^{n+1})) \to H^n(\mathbb{F}_k(\mathbb{R}^{n+1}))$. As the induced morphism of homology groups is surjective, it follows immediately that the morphism $H^n(i_k)$ is injective. Using this fact, we define appropriate generators $\beta^*_{rs} \in H^n(\mathbb{F}_{k+1}(S^{n+1}))$. There are two cases, depending on the parity of $n+1$.

Case $(n+1)$ odd.

To start, observe that (ii) of Proposition 5.1 implies immediately the following.

Corollary 5.2 *The morphism $H^n(i_k) : H^n(\mathbb{F}_{k+1}(S^{n+1})) \to H^n(\mathbb{F}_k(\mathbb{R}^{n+1}))$ is an isomorphism.*

Let $\beta^*_{rs} \in H^n(\mathbb{F}_{k+1}(S^{n+1}))$ be the elements defined by the equations $\alpha^*_{rs} = H^n(i_k)(\beta^*_{rs})$, for $1 \leq s < r \leq k$. It is easy to verify that the elements $\{\beta^*_{rs} | 1 \leq s < r \leq k\}$ are the Kronecker duals of the elements $\{\beta_{rs} | 1 \leq s < r \leq k\}$ of $H_n(\mathbb{F}_{k+1}(S^{n+1}))$.

To study the effect of Σ_{k+1} on $H^n(\mathbb{F}_{k+1}(S^{n+1}))$, we turn to the morphism $H^n(j^e_{k+1}) : H^n(\mathbb{F}_{k+1}(S^{n+1})) \to H^n(\mathbb{F}_{k+1}(\mathbb{R}^{n+1}))$ induced by the map $j^e_{k+1} : \mathbb{F}_{k+1}(\mathbb{R}^{n+1}) \to \mathbb{F}_{k+1}(S^{n+1})$.

Theorem 5.3 *The morphism $H^n(j^e_{k+1})$ is injective, and*

$$H^n(j^e_{k+1})(\beta^*_{rs}) = \alpha^*_{k+1r} - \alpha^*_{k+1s} + \alpha^*_{rs} \text{ and}$$
$$H^n(\sigma)(\beta^*_{rs}) = \alpha^*_{\sigma^{-1}(k+1)\sigma^{-1}r} - \alpha^*_{\sigma^{-1}(k+1)\sigma^{-1}s} + \alpha^*_{\sigma^{-1}r\sigma^{-1}s},$$

for all $1 \leq s < r \leq k$ and $\sigma \in \Sigma_{k+1}$.

Proof. The map $j^e_{k+1} : \mathbb{F}_{k+1}(\mathbb{R}^{n+1}) \to \mathbb{F}_{k+1}(S^{n+1})$ induces a surjective morphism on homology and, hence,

$$H^n(j^e_{k+1}) : H^n(\mathbb{F}_{k+1}(S^{n+1})) \to H^n(\mathbb{F}_{k+1}(\mathbb{R}^{n+1}))$$

is injective.

According to Theorem 5.2, we have $\beta_{k+1s} = \sum_{t=1}^{k-1} \beta_{\sigma k \sigma t}$, where $\sigma = (k, s) \in \Sigma_{k+1}$, since $(-1)^n = 1$, when $n+1$ is odd. Rewriting the expression, we obtain $\beta_{k+1s} = \sum_{t=1}^{s-1} \beta_{s\sigma t} - \beta_{ks} - \sum_{t=s+1}^{s} \beta_{ts}$. Hence,

$$\alpha_{k+1s} = \sum_{t=1}^{s-1} \alpha_{st} - \alpha_{ks} - \sum_{t=s+1}^{k-1} \alpha_{ts}$$

modulo $\ker(H_n(j^e_{k+1}) : H_n(\mathbb{F}_{k+1}(\mathbb{R}^{n+1})) \to H_n(\mathbb{F}_{k+1}(S^{n+1}))$. Now one can easily see that $H^n(j^e_{k+1})(\beta^*_{rs}) = \alpha^*_{k+1r} - \alpha^*_{k+1s} + \alpha^*_{rs}$, as required.

Finally, note that j^e_{k+1} is Σ_{k+1}-equivariant. Consequently, the induced morphism

$$H^n(j^e_{k+1}) : H^n(\mathbb{F}_{k+1}(S^{n+1})) \to H^n(\mathbb{F}_{k+1}(\mathbb{R}^{n+1}))$$

is also Σ_{k+1}-equivariant. Thus the second formula above follows directly from the first and Theorem 2.1. □

Case 2: $n+1$ even

First, note that $H^n(\mathbb{F}_k(\mathbb{R}^{n+1});\mathbb{Z}) \cong \mathrm{Hom}(H_n(\mathbb{F}_k(\mathbb{R}^{n+1}), \mathbb{Z}))$.

Proposition 5.2 *The morphism $H^n(i_k)$ induces an isomorphism*

$$H^n(i_k) : H^n(\mathbb{F}_{k+1}(S^{n+1})) \to \mathfrak{A}(\Delta_k),$$

where $\mathfrak{A}(\Delta_k) = \{\omega \in H^n(\mathbb{F}_k(\mathbb{R}^{n+1}))| < \omega, \Delta_k >= 0\}$.

Proof. The proof is easy. By (iii) of Proposition 5.1, $H^n(\mathbb{F}_{k+1}(S^{n+1}))$ is isomorphic to the annihilator of $2\Delta_k(\mathbb{R}^{n+1})$, which is clearly the same as the annihilator of $\Delta_k(\mathbb{R}^{n+1})$. □

Define $\beta^*_{rs} \in H^n(\mathbb{F}_{k+1}(S^{n+1}))$ to be the elements given by the rule

$$H^n(i_k)(\beta^*_{rs}) = \alpha^*_{rs} - \alpha^*_{21},\ 1 \le s < r \le k, \text{ and } r > 2.$$

Observe that each $\alpha^*_{rs} - \alpha^*_{21} \in \mathfrak{A}(\Delta_k)$ and, therefore, the elements β^*_{rs} are well-defined. The torsion subgroup of $H_n(\mathbb{F}_{k+1}(S^{n+1}))$ is generated by the Dirac class $\Delta_{k+1}(S^{n+1})$. Denote by $\bar{\beta}_{rs}$ the image of β_{rs} under the quotient morphism $H_n(\mathbb{F}_{k+1}(S^{n+1})) \to H_n(\mathbb{F}_{k+1}(S^{n+1}))/(\Delta_{k+1}(S^{n+1}))$. Note that the Kronecker product induces a pairing $H^n(\mathbb{F}_{k+1}(S^{n+1})) \otimes H_n(\mathbb{F}_{k+1}(S^{n+1}))/(\Delta_{k+1}(S^{n+1})) \to \mathbb{Z}$, which we denote by the same symbol $< .,. >$.

Assertion 5.1 *For all* $1 \le s < r \le k; r > 2$ *and* $\le v < u \le k$,

$$< \beta^*_{rs}, \bar{\beta}_{uv} >= \begin{cases} 1, & \textit{if } (r,s) = (u,v) \\ 0, & \textit{if } (r,s) \ne (u,v), \end{cases}$$

if $(u,v) \ne (2,1)$, *and* $< \beta^*_{rs}, \bar{\beta}_{21} >= -1$.

Proof. Observe that

$$\begin{aligned} < \beta^*_{rs}, \bar{\beta}_{uv} > &= < \beta^*_{rs}, \beta_{uv} > \\ &= < \beta^*_{rs}, H_n(i_k)(\alpha_{uv}) > \\ &= < H^n(i_k)(\beta^*_{rs}), \alpha_{uv} > \\ &= < \alpha^*_{rs} - \alpha^*_{21}, \alpha_{uv} > . \end{aligned}$$

The assertion is now evidently true. □

Proposition 5.3 *The group* $H^n(\mathbb{F}_{k+1}(S^{n+1}))$ *is generated by* $\{\beta^*_{rs} | 1 \le s < r \le k,\ r > 2\}$ *and is isomorphic to the subgroup of* $H^n(\mathbb{F}_k(\mathbb{R}^{n+1}))$ *generated by the elements of* $\{\alpha^*_{rs} - \alpha^*_{21} | 1 \le s < r \le k,\ r > 2\}$.

Proof. The group $H_n(\mathbb{F}_{k+1}(S^{n+1}))/(\Delta_{k+1}(S^{n+1}))$ is clearly generated by $\{\bar{\beta}_{uv} | 1 \le v < u \le k\}$. But

$$\Delta_{k+1}(S^{n+1}) = \beta_{21} + (\beta_{31} + \beta_{32}) + \cdots + (\beta_{k1} + \cdots + \beta_{kk-1}),$$

which implies that $\bar{\beta}_{21}$ is expressible in terms of the elements of $\{\bar{\beta}_{uv} | 1 \le v < u \le k\}$ with $u > 2$. Therefore, $\{\bar{\beta}_{uv} | 1 \le v < u \le k,\ u > 2\}$ generates $H_n(\mathbb{F}_{k+1}(S^{n+1}))$. Since the elements $\{\beta^*_{rs} | 1 \le s < r \le k,\ r > 2\}$ are the Kronecker duals of $\{\bar{\beta}_{uv} | 1 \le v < u \le k,\ u > 2\}$, one concludes immediately that the former set generates $H^n(\mathbb{F}_{k+1}(S^{n+1}))$.
To finish the proof, just note that $H^n(i_k)(\beta^*_{rs}) = \alpha^*_{rs} - \alpha^*_{21}$. □

To describe the effect of Σ_{k+1} on $\{\beta^*_{rs} | 1 \le s < r \le k+1\}$ when $n+1$ is even, we first identify $H^n(\mathbb{F}_{k+1}(S^{n+1}))$ with its image under $H^n(j^e_{k+1})$ in $H^n(\mathbb{F}_{k+1}(\mathbb{R}^{n+1}))$. Put

$$N_{k+1} = \ker(H_n(j^e_{k+1}) : H_n(\mathbb{F}_{k+1}(\mathbb{R}^{n+1})) \to H_n(\mathbb{F}_{k+1}(S^{n+1}))).$$

Recall that $H_n(j^e_{k+1})$ is surjective. As $j^e_{k+1} : \mathbb{F}_{k+1}(\mathbb{R}^{n+1}) \to \mathbb{F}_{k+1}(S^{n+1})$ is Σ_{k+1}-equivariant, it follows easily that N_{k+1} is Σ_{k+1}-invariant. In fact, N_{k+1} is generated by the relations expressed in (ii), (iii), and (iv) of Theorem 3.2 and Corollary 2.1 of Chapter III. Put

$$\mathcal{A}(N_{k+1}) = \{\phi \in \mathrm{Hom}(H_n(\mathbb{F}_{k+1}(\mathbb{R}^{n+1})),\mathbb{Z}) | <\phi,\omega>= 0, \omega \in N_{k+1}\}.$$

The following result is an easy consequence of the preceding discussion.

Proposition 5.4 *The morphism of modules*

$$H^n(j^e_{k+1}) : H^n(\mathbb{F}_{k+1}(S^{n+1});\mathbb{Z}) \to H^n(\mathbb{F}_{k+1}(\mathbb{R}^{n+1}))$$

is an isomorphism onto $\mathcal{A}(N_{k+1}) \subset H^n(\mathbb{F}_{k+1}(\mathbb{R}^{n+1}))$.

Proof. The map $j^e_{k+1} : \mathbb{F}_{k+1}(\mathbb{R}^{n+1}) \to \mathbb{F}_{k+1}(S^{n+1})$ induces a surjective morphism $H_n(j^e_{k+1}) : H_n(\mathbb{F}_{k+1}(\mathbb{R}^{n+1})) \to H_n(\mathbb{F}_{k+1}(S^{n+1})/(\Delta_{k+1}(S^{n+1}))$, and, hence, the cohomology morphism

$$H^n(j^e_{k+1}) : H^n(\mathbb{F}_{k+1}(S^{n+1})) \to H^n(\mathbb{F}_{k+1}(\mathbb{R}^{n+1}))$$

is injective. □

Theorem 5.4 *The morphism $H^n(j^e_{k+1})$ satisfies the following relations:*

$$\begin{aligned} H^n(j^e_{k+1})(\beta^*_{rs}) &= \alpha^*_{rs} - (\alpha^*_{k+1r} + \alpha^*_{k+1s} + \alpha^*_{21}) \\ H^n(\sigma)(\beta^*_{rs}) &= \alpha^*_{\sigma^{-1}r\sigma^{-1}s} - (\alpha^*_{\sigma^{-1}(k+1)\sigma^{-1}r} + \alpha^*_{\sigma^{-1}(k+1)\sigma^{-1}s} + \\ &\quad \alpha^*_{\sigma^{-1}2\sigma^{-1}1}), \end{aligned}$$

for all $1 \le s < r \le k$, $r > 2$ and $\sigma \in \Sigma_{k+1}$.

Proof. The proof is analogous to that of Theorem 5.3. According to Theorem 5.2, we have $\bar{\beta}_{k+1s} = -\sum_{t=1}^{k-1} \bar{\beta}_{\sigma k \sigma t}$, where $\sigma = (k,s) \in \Sigma_{k+1}$, since $(-1)^n = -1$ when $n+1$ is even. Rewriting the expression, we obtain $\bar{\beta}_{k+1s} = \sum_{t=1}^{s-1} \bar{\beta}_{s\sigma t} - \bar{\beta}_{ks} - \sum_{t=s+1}^{s} \bar{\beta}_{ts}$. Hence,

$$\alpha_{k+1s} = \sum_{t=1}^{s-1} \alpha_{st} - \alpha_{ks} - \sum_{t=s+1}^{k-1} \alpha_{ts}$$

modulo $\ker(H_n(j^e_{k+1}) : H_n(\mathbb{F}_{k+1}(\mathbb{R}^{n+1})) \to H_n(\mathbb{F}_{k+1}(S^{n+1})/(\Delta_{k+1}(S^{n+1}))$. Now one can easily see that $H^n(j^e_{k+1})(\beta^*_{rs}) = \alpha^*_{rs} - (\alpha^*_{k+1r} + \alpha^*_{k+1s} + \alpha^*_{21})$, as required.

Finally, note that j^e_{k+1} is Σ_{k+1}-equivariant. Consequently, the morphism $H^n(j^e_{k+1}) : H^n(\mathbb{F}_{k+1}(S^{n+1})) \to H^n(\mathbb{F}_{k+1}(\mathbb{R}^{n+1}))$ is also Σ_{k+1}-equivariant. Thus, the second formula above follows directly from the first and Theorem 2.1, and the proof is finished. □

Let $\mathbb{F}_{k-2,3}(S^{n+1}) = \mathbb{F}_{k-2}(S^{n+1} - Q_2^e) \cong \mathbb{F}_{k-2,2}(\mathbb{R}^{n+1})$ be the natural homeomorphisms given in §1 of Chapter III, and put

$$i_k|\mathbb{F}_{k-2,2}(\mathbb{R}^{n+1}) = i_{k-2} : \mathbb{F}_{k-2,2}(\mathbb{R}^{n+1}) \to \mathbb{F}_{k+1}(S^{n+1}).$$

Consider the morphism $H^n(i_{k-2}) : H^n(\mathbb{F}_{k+1}(S^{n+1})) \to H^n(\mathbb{F}_{k-2,2}(\mathbb{R}^{n+1}))$, with $n > 1$.

Corollary 5.3 *The morphism $H^n(i_{k-2})$ takes β^*_{rs} to α^*_{rs}, for $1 \le s < r \le k$ and $r > 2$.*

Proof. Note that the natural imbedding $\mathbb{F}_{k-2,2}(\mathbb{R}^{n+1}) \to \mathbb{F}_{k+1}(\mathbb{R}^{n+1})$ induces a morphism $H^n(\mathbb{F}_{k+1}(\mathbb{R}^{n+1})) \to H^n(\mathbb{F}_{k-2,2}(\mathbb{R}^{n+1}))$ that takes $\alpha^*_{rs} \in H^n(\mathbb{F}_{k+1}(\mathbb{R}^{n+1}))$ to its counterpart in $H^n(\mathbb{F}_{k-2,2}(\mathbb{R}^{n+1}))$. □

6 $H^*(\mathbb{F}_{k+1}(S^{n+1}))$ as an $H^*(\mathbb{F}_3(S^{n+1}))$-Module

The twisted product representation $\mathbb{F}_{k+1}(S^{n+1}) \simeq O_{n+2,2} \rtimes \mathbb{F}_{k-2,2}(S^{n+1})$, with $k+1 > 3$ (see §1 of Chapter III for $n > 1$ and §7 of Chapter IV for $n = 1$) has some interesting consequences. Assume throughout this section that the homology and cohomology groups are with integral coefficients.

Theorem 6.1 *For $n \ge 1$, there is an isomorphism*

$$\Phi^*_{k-2,3} : H^*(\mathbb{F}_3(S^{n+1})) \otimes H^*(\mathbb{F}_{k-2}(\mathbb{R}_2^{n+1})) \cong H^*(\mathbb{F}_{k+1}(S^{n+1}))$$

of $H^*(\mathbb{F}_3(S^{n+1}))$-modules.

Proof. Consider the projection $\mathrm{proj}_{k+1,3} : \mathbb{F}_{k+1}(S^{n+1}) \to \mathbb{F}_3(S^{n+1})$. Its fiber at $(q_1^e, q_2^e, -q_1^e)$ is clearly $\mathbb{F}_{k-2,3}(S^{n+1})$, and the latter is naturally homeomorphic to $\mathbb{F}_{k-2,2}(\mathbb{R}^{n+1})$, according to the conventions of §1 of Chapter III.

For the case $n = 1$, Theorem 7.1 of Chapter IV implies that $\mathbb{F}_{k+1}(S^2) \simeq O_{3,2} \times \mathbb{F}_{k-2,2}(\mathbb{R}^2)$, and the theorem is therefore an immediate consequence.

Assume that $n > 1$. According to Corollary 5.3, the morphism

$$H^n(i_{k-2}) : H^n(\mathbb{F}_{k+1}(S^{n+1})) \to H^n(\mathbb{F}_{n-2,2}(\mathbb{R}^{n+1}))$$

induced by the restriction $i_{k-2} = i_k|\mathbb{F}_{k-2,2}$ of the imbedding $i_k : \mathbb{F}_k(\mathbb{R}^{n+1}) \to \mathbb{F}_{k+1}(S^{n+1})$ is surjective. But, according to Theorem 5.2, the elements

$$\mathcal{A}^*_{k-2,2} = \{\alpha^*_{rs} \mid 1 \le s < r \le k,\ 2 < r\}$$

generate the algebra $H^*(\mathbb{F}_{k-2,2}(\mathbb{R}^{n+1}))$. Hence, the morphism $H^*(i_{k-2}) : H^*(\mathbb{F}_{k+1}(S^{n+1})) \to H^*(\mathbb{F}_{n-2,2}(\mathbb{R}^{n+1}))$ is surjective. Therefore, the Leray-Hirsch Theorem applies, and we obtain an isomorphism

$$\Phi^*_{k-2,2} : H^*(\mathbb{F}_3(S^{n+1})) \otimes H^*(\mathbb{F}_{k-2,2}) \cong H^*(\mathbb{F}_{k+1}(S^{n+1}))$$

of $H^*(\mathbb{F}_3(S^{n+1}); \mathbb{Z})$-modules, where $\Phi^*_{k-2,3}(\omega \otimes \gamma) = H^*(\mathrm{proj}_{k+1,3})(\omega)\theta(\gamma)$, with θ being a cohomology extension of the fiber. □

The following is an easy consequence.

Corollary 6.1 *There is an isomorphism*

$$\Phi^*_{k-2,3} : H^*(O_{n+2,2}) \otimes H^*(\mathbb{F}_{k-2,2}(S^{n+1})) \to H^*(\mathbb{F}_{k+1}(S^{n+1}))$$

of $H^*(O_{n+2,2})$-modules, where $O_{n+2,2}$ is the Stiefel manifold of orthonormal 2-frames in $\mathbb{R}^{n+2}$.

Proof. Let $\mathrm{proj}'_{k+1,3} : \mathbb{F}'_{k+1}(S^{n+1}) \to O_{n+2,2}$ be the pull-back of the fibration in the preceding theorem. By Proposition 1.1 of Chapter III, $\psi_3 : O_{n+2,2} \to \mathbb{F}_3(S^{n+1})$ is a homotopy equivalence. Hence, the pull-back $\mathbb{F}'_{k+1}(S^{n+1}) \to \mathbb{F}_{k+1}(S^{n+1})$ is homotopy equivalent to the fibration of the preceding theorem. The morphism $\Phi^*_{k-2,3}$ leads readily to the isomorphism, denoted by the same symbol, $\Phi^*_{k-2,3} : H^*(O_{n+2,2})) \otimes H^*(\mathbb{F}_{k-2,2}(S^{n+1})) \to H^*(\mathbb{F}_{k+1}(S^{n+1}))$ of $H^*(O_{n+2,2})$-modules. □

Before proceeding any further, we need the following notation. Put

$$\mathcal{B}^{\wedge *}_{k-2,2} = \cup_{p=1}^{(k-2)} \{\beta^*_{\underline{r}\,\underline{s}} \mid \underline{r} \in \mathcal{R}_p, \underline{s} \in \mathcal{S}_p, \underline{s} < \underline{r}\},$$

where $3 \le r_1$, with $\beta^*_{\underline{r}\,\underline{s}} = \beta^*_{r_1 s_1} \cdots \beta^*_{r_p s_p}$ and $\underline{r}, \underline{s}$ ranging over the sets

$$\mathcal{R}_p = \{\underline{r} = (r_1, \cdots, r_p) | 1 < r_1 < \cdots < r_p \le k\} \text{ and}$$
$$\mathcal{S}_p = \{\underline{s} = (s_1, \cdots, s_p) | 1 \le s_1, s_2, \cdots, s_p < k\},$$

respectively.

Theorem 6.2 *There is an isomorphism,*

$$\Phi^*_{k-2,3} : H^*(O_{n+2,2}) \otimes H^*(\mathbb{F}_{k-2,2}) \to H^*(\mathbb{F}_{k+1}(S^{n+1})),$$

of $H^*(O_{n+2,2})$-modules. In particular, the elements of the set $\mathcal{B}^{*\wedge}_{k-2,2}$ is a basis for the group $H^*(\mathbb{F}_{k+1}(S^{n+1}))$ as a module over $H^*(O_{n+2,2})$.

Proof. It is easy to verify that the restriction of $\beta^*_{r_i s_i}$ to $\mathbb{F}_{k-2,2}$ is just $\alpha^*_{r_i s_i}$. As the set $\mathcal{B}^\wedge$ is a basis for $H^*(\mathbb{F}_{k-2,2}(\mathbb{R}^{n+1}))$, it follows that the set of elements $\mathcal{A}^{\wedge *}_{k-2,2}$ of $H^*(\mathbb{F}_{k+1}(S^{n+1}))$ constitutes a cohomology extension of the fiber. Now the Leray-Hirsch Theorem implies the rest. □

7 The Algebra $H^*(\mathbb{F}_{k+1}(S^{n+1}))$, $n+1$ Even

A natural question to ask is whether the morphism $\Phi^*_{k-2,3}$ of Corollary 6.1 is, in fact, an isomorphism of algebras. The answer depends on the parity of $(n+1)$.

Theorem 7.1 *If $(n+1)$ is even, then the isomorphism*

$$\Phi^*_{k-2,3} : H^*(O_{n+2,2};\mathbb{Z}) \otimes H^*(\mathbb{F}_{k-2,2}(\mathbb{R}^{n+1});\mathbb{Z}) \to H^*(\mathbb{F}_{k+1}(S^{n+1});\mathbb{Z})$$

of Corollary 6.2 is an isomorphism of algebras.

Proof. Assume, first, that $(n+1) \neq 2$. According to Proposition 6.3 of Chapter III, we have

$$H^j(O_{n+2,2};\mathbb{Z}) \cong \begin{cases} \mathbb{Z}/2\mathbb{Z}, \text{ for} & (n+1) \\ \mathbb{Z}, \quad \text{for} & j = 0, 2n+1 \\ 0, \quad \text{otherwise.} & \end{cases}$$

The morphism $\theta : H^*(\mathbb{F}_{k-2,2}(\mathbb{R}^{n+1});\mathbb{Z}) \to H^*(\mathbb{F}_{k+1}(S^{n+1});\mathbb{Z})$ induced by sending α^*_{rs} to β^*_{rs} yields an isomorphism of modules on $P[B^*_{k-2,2}]$, where $P[B^*_{k-2,2}]$ is the universal graded algebra generated by $\mathcal{B}^*_{k,2,2} = \{\beta^*_{rs} | 1 \leq s < r \leq k, 2 < r\}$ modulo the Y-B relations in $H^*(\mathbb{F}_{k-2,2}(\mathbb{R}^{n+1})$ (see Theorem 4.1). The morphism θ is a cohomology extension of the fiber and, hence, induces the isomorphism

$$\Phi'_{k-2,3} : H^*(O_{n+2,2};\mathbb{Z}) \otimes H^*(\mathbb{F}_{k-2,2}(\mathbb{R}^{n+1});\mathbb{Z}) \to H^*(\mathbb{F}_{k+1}(S^{n+1});\mathbb{Z}),$$

where Φ_{k+1} is the morphism such that $(\omega, \alpha^*_{rs}) \mapsto H^*(\mathrm{proj}_{k+1,3})(\omega)\theta(\alpha^*_{rs})$. As the restriction of $\beta^*_{rs} = \theta(\alpha^*_{rs}$ to the fiber of the fibration $\mathrm{proj}'_{k+1,3} : \mathbb{F}'_{k+1}(S^{n+1}) \to O_{n+2,2}$ is just $\alpha^*_{rs} \in H^n(\mathbb{F}_{k-2,2})$, the Y-B cohomological relations (see Theorem 4.1) imply that the restriction of $\gamma = \beta^*_{rs}\beta^*_{rt} - \beta^*_{ts}(\beta^*_{rt} - \beta^*_{rs})$ to the fiber is trivial. As the element γ is of dimension $2n$, it follows readily, for dimensional reasons, that $\gamma = 0$. This implies that the Y-B relations in the algebra $H^*(\mathbb{F}_{k-2,2}(\mathbb{R}^{n+1}))$ hold in the algebra $H^*(\mathbb{F}_{k+1}(S^{n+1}))$, as well. Therefore, $\theta : H^*(\mathbb{F}_{k-2,2}(\mathbb{R}^{n+1})) \to P[B^*_{k-2,2}] \subset H^*(\mathbb{F}_{k+1}(S^{n+1}))$ is an isomorphism of algebras. Hence, the Leray-Hirsch morphism $\Phi'_{k-2,2}$ is actually one of algebras.

To finish the proof, we need to consider the case when $(n+1) = 2$. But, by Theorem 7.2 of Chapter IV, the bundle $\mathrm{proj}_{k-2,3} : \mathbb{F}'_{k+1}(S^2) \to O_{3,2}$ is trivial. Hence, $\mathbb{F}'_{k+1}(S^2) \cong O_{3,2} \times \mathbb{F}'_{k+1}$, and, therefore, $\Phi^*_{k-2,3}$ is an isomorphism of algebras. This completes the proof. □

8 The Algebra $H^*(\mathbb{F}_{k+1}(S^{n+1}))$, $n+1$ Odd

Homology and cohomology groups are assumed throughout this section to have integral coefficients. Consider the fibration $\mathrm{proj}_{k+1,1} : \mathbb{F}_{k+1}(S^{n+1}) \to S^{n+1}$ that sends a tuple to its last entry (see Chapter III, §3), and let $\delta'_{k+1} : S^{n+1} \to \mathbb{F}_{k+1}(S^{n+1})$ be the section defined in Chapter III, §3. The existence of the section δ'_{k+1} implies readily that the induced morphism $H^*(\mathrm{proj}_{k+1,1}) : H^*(S^{n+1};\mathbb{Z}) \to H^n(\mathbb{F}_{k+1}(S^{n+1});\mathbb{Z})$ is injective. Put $\delta^*_{k+1} = H^{n+1}(\mathrm{proj}_{k+1,1})(\iota^*_{n+1})$, where ι^*_{n+1}, the generator of $H^{n+1}(S^{n+1};\mathbb{Z})$.

Theorem 8.1 *The elements* $\{\beta^*_{rs}, \delta^*_{k+1}\}$ *satisfy the following relations in* $H^*(\mathbb{F}_{k+1}(S^{n+1});\mathbb{Z})$*:*

$$\begin{aligned} &(i)\ {\delta^*_{k+1}}^2 = 0,\\ &(ii)\ {\beta^*_{rs}}^2 = 0, \text{ and}\\ &(iii)\ \beta^*_{rs}\beta^*_{rt} = \beta^*_{ts}(\beta^*_{rt} - \beta^*_{rs}), \end{aligned}$$

for all r, s, t *such that* $1 \le s < t < r \le k$.

Proof. Recall that the fiber of $\mathrm{proj}_{k+1,1} : \mathbb{F}_{k+1}(S^{n+1}) \to S^{n+1}$ is $\mathbb{F}_k(\mathbb{R}^{n+1})$, and that the restriction of each of the elements β^*_{rs} to the space $\mathbb{F}_k(\mathbb{R}^{n+1})$ is just the corresponding α^*_{rs}. The latter elements generate the algebra $H^*(\mathbb{F}_k(\mathbb{R}^{n+1});\mathbb{Z})$. Therefore, the Leray-Hirsch Theorem applies, and we have the isomorphism

$$\Phi^*_{k,1} : H^*(S^{n+1};\mathbb{Z}) \otimes H^*(\mathbb{F}_k(\mathbb{R}^{n+1});\mathbb{Z}) \to H^*(\mathbb{F}_{k+1}(S^{n+1});\mathbb{Z})$$

of $H^*(S^{n+1};\mathbb{Z})$-modules (see §6). Now, as $(\iota^*_{n+1})^2 = 0$, it follows that $(\delta^*_{k+1})^2 = 0$. This proves (i).

To prove (ii), note that ${\beta^*_{rs}}^2 = \Phi^*_{k,1}(a \otimes 1 + b \otimes u + 1 \otimes {\alpha^*_{rs}}^2)$, where $a \in H^{2n}(S^{n+1};\mathbb{Z})$, $b \in H^n(S^{n+1};\mathbb{Z})$ and $u \in H^n(\mathbb{F}_k(\mathbb{R}^{n+1});\mathbb{Z})$, since the restriction of β^*_{rs} to the fiber is α^*_{rs}. For dimensional reasons, $a = 0 = b$. Now the assertion follows because, according to the Y-B cohomological relations, $(\alpha^*_{rs})^2 = 0$ (see Theorem 3.1). Hence, ${\beta^*_{rs}}^2 = 0$, as required.

The proof of (iii) is similar. In fact, as in the previous case, since $H^n(\beta^*_{rs} = \alpha^*_{rs})$, we have $\beta^*_{rs}\beta^*_{rt} = \Phi^*_{k,1}(1 \otimes \alpha^*_{rs}\alpha^*_{rt} + a \otimes 1 + b \otimes u)$, $s < t$, where $a \in H^{2n}(S^{n+1};\mathbb{Z})$, $b \in H^n(S^{n+1};\mathbb{Z})$ and $u \in H^n(\mathbb{F}_{k,1}(S^{n+1});\mathbb{Z})$. Again, for dimensional reasons, $a = 0 = b$. But, by virtue of the Y-B relations, $\alpha^*_{rs}\alpha^*_{rt} = -\alpha^*_{ts}(\alpha^*_{rs} - \alpha^*_{rt})$ (see Theorem 3.1). Thus (iii) is proved and, hence, the theorem also. □

Let $E(\delta^*_{k+1})$ be the exterior algebra of the module $\mathbb{Z}\delta^*_{k+1}$. Also, let $\mathcal{B}^*_{k-1,1}$ be the algebra generated by the set $\{\beta^*_{rs} \mid 1 \le s < r \le k\}$ modulo the Y-B relations given in (ii) and (iii) (see also Theorem 3.1).

Theorem 8.2 *The morphism*

$$\Phi^*_{k,1} : E(\delta^*_{k+1}) \otimes \mathcal{B}^*_{k-1,1} \to H^*(\mathbb{F}_{k+1}(S^{n+1})),$$

where $\delta^*_{k+1} \otimes \gamma \mapsto \delta^*_{k+1}\gamma$, *is an isomorphism of algebras.*

Proof. The morphism whereby $\alpha^*_{rs} \mapsto \beta^*_{rs}$ defines a cohomology extension of the fiber $\theta : H^*(\mathbb{F}_k(\mathbb{R}^{n+1})) \to H^*(\mathbb{F}_{k+1}(S^{n+1}))$. The Leray-Hirsch Theorem implies that the morphism $\Phi^*_{k,1} : H^*(S^{n+1}) \otimes H^*(\mathbb{F}_k(\mathbb{R}^{n+1})) \to H^*(\mathbb{F}_{k+1}(S^{n+1}))$, where $(\delta^*_{k+1}, \alpha^*_{rs}) \mapsto \delta^*_{k+1}\theta(\alpha^*_{rs})$, is an isomorphism of $H^*(\mathbb{F}_{k+1}(S^{n+1}))$-modules. The Y-B relations imply that $\Phi^*_{k,1}$ is actually an isomorphism of algebras. □

On the other hand, we have the following result.

Corollary 8.1 *The morphism*

$$\Phi^*_{k-2,3} : H^*(O_{n+2,2};\mathbb{Z}) \otimes H^*(\mathbb{F}_{k-2,2}(\mathbb{R}^{n+1});\mathbb{Z}) \to H^*(\mathbb{F}_{k+1}(S^{n+1});\mathbb{Z})$$

is not one of algebras.

Proof. Note that $\alpha^*_{31}\alpha^*_{32} = 0$ in the algebra $H^*(\mathbb{F}_{k-2,2}(S^{n+1});\mathbb{Z})$, while it is $\neq 0$ in $H^*(\mathbb{F}_{k+1}(S^{n+1});\mathbb{Z})$. □

9 Historical Remarks

Most of the results here on the cohomology of $\mathbb{F}_k(\mathbb{R}^{n+1})$ are due to [17, Cohen]. The proofs are different in places. Results concerning the cohomology $H^*(\mathbb{F}_{k+1}(S^{n+1}))$, with coefficients in $\mathbb{Z}/2\mathbb{Z}$ when $n+1$ is even, are announced in [19, Cohen-Taylor], with the details given in [20, Cohen-Taylor]. An alternative approach, based on the theory of arrangements, for studying the cohomology algebra $H^*(\mathbb{F}_{k+1}(S^{n+1});\mathbb{Z})$ appeared recently in [46, Feichtner-Ziegler].

An ad hoc method for determining the Betti numbers in the special case $\mathbb{F}_3(M)$, where M is a compact orientable manifold, is in [15, Brown-White].

VI. Cellular Models

Our objective here is to describe cellular structures of $\mathbb{F}_k(M)$ naturally associated with the bases of $H^*(\mathbb{F}_k(M))$ for $M = \mathbb{R}^{n+1},\ S^{n+1}$ of Theorem 4.2 and §6 of Chapter V. The basic ideas are the following: first, that the twisted product representation $\mathbb{F}_k(\mathbb{R}^{n+1}) \simeq \mathbb{R}_1^{n+1} \ltimes \cdots \ltimes \mathbb{R}_{k-1}^{n+1}$ introduced in Chapter II, §4 leads to a twisted product

$$H_*(\mathbb{R}_1^{n+1}) \otimes \cdots \otimes H_*(\mathbb{R}_{k-1}^{n+1}) \cong H_*(\mathbb{F}_k(\mathbb{R}^{n+1}); \mathbb{Z})$$

on homology, which we write as

$$\alpha_{r_1 s_1} \otimes \cdots \otimes \alpha_{r_p s_p} \mapsto \omega = \alpha_{r_1 s_1} \ltimes \cdots \ltimes \alpha_{r_p s_p}, r_i < r_{i+1}$$

for all $1 \leq p \leq (k-1)$; and, second, that each p-fold twisted product ω leads to an imbedding

$$\phi_\omega : S_1^n \times \cdots \times S_p^n \to \mathbb{F}_k(\mathbb{R}^{n+1})$$

of a certain kind. These maps provide us with the cells and attaching maps of the desired complex.

In §1, to illustrate and motivate the general procedure, we take up, first, the case of $\mathbb{F}_3(\mathbb{R}^{n+1})$, with $n > 1$. We consider the maps

$$\phi, \psi : S_1^n \times S_2^n \to \mathbb{F}_3(\mathbb{R}^{n+1}),$$

where $S_1^n, S_2^n \subset \mathbb{R}^{n+1}$ are n-dimensional spheres used in the proof of the Y-B relations

$$[\alpha_{21} + \alpha_{31}, \alpha_{32}] = 0 \text{ and} [\alpha_{21} + (-1)^{n+1}\alpha_{32}, \alpha_{31}] = 0.$$

We show that the morphisms

$$H_{2n}(\phi), H_{2n}(\psi) : H_{2n}(S_1^n \times S_2^n; \mathbb{Z}) \to H_{2n}(\mathbb{F}_3(\mathbb{R}^{n+1}); \mathbb{Z})$$

take the fundamental class $\iota_1 \times \iota_2$ to the duals, with respect to the Kronecker product, of $\alpha_{21}^*\alpha_{32}^*$ and $\alpha_{21}^*\alpha_{31}^*$, respectively. The maps α_{rs} such that $1 \leq s < r \leq 3$ provide us with the n-skeleton, while ϕ, ψ give the $(2n)$-dimensional cells.

In §2 we consider the fundamental fiber sequence

$$\mathcal{F}_k(\mathbb{R}^{n+1}) : \begin{cases} \mathbb{F}_k & \leftarrow \mathbb{F}_{k-1,1} \leftarrow \cdots \leftarrow \mathbb{F}_{k-r,r} \leftarrow \\ \downarrow \qquad\quad \downarrow \qquad\quad \downarrow \qquad\quad \downarrow \\ \mathbb{R}^{n+1} \qquad \mathbb{R}^{n+1}_1 \qquad \cdots \qquad \mathbb{R}^{n+1}_r \end{cases}$$

introduced in Chapter II, §1. Recall that $\mathbb{F}_{k-r,r} = \mathbb{F}_{k-r}(\mathbb{R}^{n+1}_r)$ and $\mathbb{R}^{n+1}_r = \mathbb{R}^{n+1} - Q_r$ and that the vertical maps $\mathrm{p}_r : \mathbb{F}_{k-r,r} \to \mathbb{R}^{n+1}_r$ are the projections on the first nonconstant entry. Then, noting that the fibers are totally nonhomologous to zero, we define the twisted product inductively by means of the corresponding Leray-Hirsch isomorphism

$$\Phi_{k-r,r} : H_*(\mathbb{F}_{k-r,r};\mathbb{Z}) \to H_*(\mathbb{R}^{n+1}_r;\mathbb{Z}) \otimes H_*(\mathbb{F}_{k-r-1,r+1};\mathbb{Z}).$$

In §3, for a given twisted p-fold product, $\omega = \alpha_{r_1 s_1} \ltimes \cdots \ltimes \alpha_{r_p s_p}$, and $1 \leq p \leq (k-1)$, we define a map

$$\phi_\omega : S^n_1 \times \cdots \times S^n_p \to \mathbb{F}_k(\mathbb{R}^{n+1})$$

that agrees with $\alpha_{rs} : S^n \to \mathbb{F}_k(\mathbb{R}^{n+})$, when $p = 1$, and with $\phi_{\omega'}$, when restricted to the last $(k-1)$ factors, where ω' is the $(p-1)$-fold twisted product obtained by deleting the first factor of ω. The basic idea is the same as in the definition of α_{rs} in Chapter II, §1: the map

$$\phi_{\omega'} : S^n_1 \times \cdots \times S^n_{p-1} \to \mathbb{F}_{k-1}(\mathbb{R}^{n+1})$$

is modified by inserting in the r_p^{th}-position the unit sphere, suitably rescaled, centered at q'_s, where $\phi_{\omega'}(\xi_1, \cdots, \xi_{p-1}) = (q'_1, \cdots, q'_s, \cdots)$. The map ϕ_ω defines a table with rows that are the restrictions to the factors of ω. We then show that the jth row begins with the jth factor $\alpha_{r_j s_j}$. Hence the table is of the form

$$\mathcal{T}(\phi_\omega) : \begin{cases} \alpha_{r_1 s_1} & + \sum_{m=2}^{p} \lambda_{1m}(\omega_m) \\ \cdots & + \cdots \\ \alpha_{r_j s_j} & + \sum_{m=j+1}^{p} \lambda_{jm}(\omega_m) \\ \cdots & + \cdots \\ \alpha_{r_p s_p}, & \end{cases}$$

where $\omega_m = \alpha_{r_1 s_1} \ltimes \cdots \ltimes \alpha_{r_m s_m}$. To describe the remainder of the jth row, we look at the Y-B relations differently. Recall that those of the first kind come from three particles, where the second moves around the first acting as a center, while the third moves around the second. These measure the *perturbation* of the second by the third. For example, the perturbation of α_{21} by α_{32} is α_{31}. Those of the second kind are interpreted similarly. The idea is formulated precisely in §3. Next, we prove that the rows of the factor table $\mathcal{T}(\phi_{\underline{r}\,\underline{s}})$ are of a special form: the term in the jth-row, for $1 \leq j < p$, succeeding $\alpha_{r_j s_j}$ breaks up as a sum $\sum_{m=j+1}^{p} \lambda_{jm}(\omega_m)$, with $\lambda_{jm}(\omega_m)$ being the perturbation of $\alpha_{r_j s_j} + \sum_{u=j+1}^{m-1} \lambda_{ju}(\omega_u)$ by the factor $\alpha_{r_{m+1} s_{m+1}}$. The structure of the factor table $\mathcal{T}(\phi_{\underline{r}\,\underline{s}})$ plays a key role in Part Three .

In §4 we give an illustrative example. In §5 we first prove that $\omega = \alpha_{\underline{r},\,\underline{s}}$ is the Kronecker dual of $\alpha^*_{\underline{r},\,\underline{s}}$ and then show that ω is carried by the fundamental class of the multifold products of the spheres imbedded by ϕ_ω. We conclude that they are the Kronecker duals of the basis $\{\alpha^*_{\underline{r},\,\underline{s}}\}$ of the cohomology groups.

Next, in §§6 and 7 we turn to the space, $\mathbb{F}_{k+1}(S^{n+1})$. We show how the twisted product representation $O_{n+2,2} \ltimes \mathbb{F}_{k-2,3}(S^{n+1}) \simeq \mathbb{F}_{k+1}(S^{n+1})$ of Chapter III §1, where $O_{n+2,2}$ is the Stiefel manifold of orthonormal 2-frames in $\mathbb{R}^{n+2}$, leads to a twisted product

$$H_*(O_{n+2,2};\mathbb{Z}) \otimes H_*(\mathbb{F}_{k-2,3}(S^{n+1});\mathbb{Z}) \cong H_*(\mathbb{F}_{k+1}(S^{n+1});\mathbb{Z}),$$

and $(\gamma, \beta_{\underline{r},\,\underline{s}})$ is mapped to $\gamma \ltimes \beta_{\underline{r},\,\underline{s}}$, with $\gamma \in H_*(O_{n+2,2};\mathbb{Z})$ and $\beta_{\underline{r},\,\underline{s}}$ in $H_*(\mathbb{F}_{k-2,3}(S^{n+1});\mathbb{Z})$. For each $(p-1)$-fold twisted product

$$\begin{aligned} \omega &= \beta_{\underline{r},\,\underline{s}} \\ &= \beta_{r_2 s_2} \ltimes \cdots \beta_{r_p s_p} \end{aligned}$$

in the integral homology of $\mathbb{F}_{k-2,3}(S^{n+1}) \cong \mathbb{F}_{k-2,2}(\mathbb{R}^{n+1})$), we define an affine map

$$\psi_\omega : O_{n+2,2} \times (S^n_2 \times \cdots \times S^n_p) \to \mathbb{F}_{k+1}(S^{n+1}),\ 1 \leq p \leq k$$

such that the induced map on homology takes the cross product ([101, Spanier]) $\gamma \times (\iota_2 \times, \cdots, \iota_p)$ to the twisted product $\gamma \ltimes \beta_{\underline{r},\,\underline{s}}$, where ι_j stands for the fundamental class of the jth-factor. The bundle

$$\text{proj}_{k+1,1} : \mathbb{F}_{k+1}(S^{n+1}) \to S^{n+1},$$

with its the fiber at $-e$ identified with $\mathbb{F}_k(\mathbb{R}^{n+1})$ in the manner of Chapter III §2, is homeomorphic to

$$\text{proj} : D^{n+1} \times \mathbb{F}_k(\mathbb{R}^{n+1}) \cup_\mu \mathbb{F}_k(\mathbb{R}^{n+1}) \to S^{n+1},$$

as a bundle, where $\mu : \partial D^{n+1} \times \mathbb{F}_k(\mathbb{R}^{n+1}) \to \mathbb{F}_k(\mathbb{R}^{n+1})$ is the characteristic map and proj is induced by projection on the first factor.

In §8, with $n > 1$, we describe an inductive procedure to construct a cell complex $X_k \simeq \mathbb{F}_k(\mathbb{R}^{n+1})$ by use of the affine maps ϕ_ω, as ω ranges over the set of all twisted products of the basis. We describe its cells and characteristic maps. The procedure is based on the diagram $\mathcal{F}_k(\mathbb{R}^{n+1})$. We construct a filtered complex

$$X_k = X_{k,0} \supset X_{k-1,1} \supset \cdots \supset X_{1,k-1}$$

homotopy equivalent to the filtration in $\mathcal{F}_k(\mathbb{R}^{n+1})$. In fact, for each X_{k-rr}, with $S_{r,*}$ being the bouquet $S_{r+11} \vee \cdots \vee S_{r+1r}$, we construct a sequence

$$X_{k-r-1,r+1} \to X_{k-r,r} \to S_{r,*},$$

homotopy equivalent to the fibration sequence

$$\mathbb{F}_{k-r-1,r+1} \to \mathbb{F}_{k-r,r} \stackrel{\text{proj}_{k-r,1}}{\to} \mathbb{R}^{n+1}_r.$$

The procedure is by induction: we take $X_{1,k-1}$ to be $S_{k,*}$, and the inductive step is to pull back the fibration, $\text{proj}_{k-r,1} : \mathbb{F}_{k-r,r} \to \mathbb{R}^{n+1}_r$, to the bouquet $S_{r,*}$ by the map $\rho_{r,*}S_{r,*} \to \mathbb{R}^{n+1}_r$, defined by the maps $\{\alpha_{r+11} \vee \cdots \vee \alpha_{r+1r}\}$. The pull-back, denoted by $\text{p}'_r : E'_r \to S_{r,*}$, is homeomorphic as a bundle to the fibration

$$\text{p}_{k-r} : S_{r,*} \times \mathbb{F}_{k-r-1,r+1} \cup_\mu \mathbb{F}_{k-r-1,r+1} \to S_{r,*}$$

induced by the projection on the first factor, where

$$\mu : (\bigvee_{s=1}^{r} \partial D^n_{r+1s}) \times \mathbb{F}_{k-r-1,r+1} \to \mathbb{F}_{k-r-1,r+1}$$

is the characteristic map of the fibration $\text{proj}_{k-r+1,1} : \mathbb{F}_{k-r+1,r+1} \to \mathbb{R}^{n+1}_{r+1}$. Assuming that $X_{k-r-1,r+1} \simeq \mathbb{F}_{k-r-1,r+1}$ has been constructed, we show how to obtain $X_{k-r,r} \simeq \mathbb{F}_{k-r,r}$ by adding cells e_ω in one-one correspondence with the twisted products that begin with α_{r+1s}. The attaching maps come from the corresponding affine maps ϕ_ω.

With X_k being the complex homotopy equivalent to $\mathbb{F}_k(\mathbb{R}^{n+1})$, as indicated, we construct in §9 a complex Y_{k+1} containing X_k as a subcomplex and homotopy equivalent to $\mathbb{F}_{k+1}(S^{n+1})$. The cells come from the cells of $O_{n+2,2}$, as described in Chapter III, §§5 and 6, and the cells of X_k. Also, the complex Y_{k+1} depends on the parity of $(n+1)$.

In §§10 and 11 we deal with the cases $\mathbb{F}_k(\mathbb{R}^2)$ and $\mathbb{F}_{k+1}(S^2)$. The procedure of the previous sections is modified to take care of lack of simple connectivity. First, as in the case of $n > 1$, we have a homeomorphism,

$$\text{p} : D^1 \times \mathbb{F}_{k-1}(\mathbb{R}^2) \cup_\mu \mathbb{F}_{k-1}(\mathbb{R}^2) \to \mathbb{F}_k(\mathbb{R}^2).$$

The attaching map $\mu : \partial D^1 \times \mathbb{F}_{k-1}(\mathbb{R}^2) \to \mathbb{F}_{k-1}(\mathbb{R}^2)$ is given by the formulas of Propositions 5.2 and 5.3 of Chapter IV (and see also Theorem 5.1 of Chapter IV). Now, the cellular complex X_k homotopy equivalent to $\mathbb{F}_k(\mathbb{R}^2)$ is built up inductively, starting with $X_{1k-1} \simeq (S_{k1} \vee \cdots \vee S_{kk-1})$. The case of $\mathbb{F}_{k+1}(S^2) \to S^2$ is handled in an analogous manner to that of $n > 1$ by means of the representation

$$\mathrm{p} : D^2 \times \mathbb{F}_k(\mathbb{R}^2) \cup_\mu \mathbb{F}_k(\mathbb{R}^2) \to S^2,$$

where $\mu : \partial D^2 \times \mathbb{F}_k(\mathbb{R}^2) \to \mathbb{F}_k(\mathbb{R}^2)$ is the characteristic map and p is induced by the projection on the first factor. With X_k being the cell complex for $\mathbb{F}_k(\mathbb{R}^2)$, one proceeds to construct the complex Y_{k+1} that contains X_k and quasi fibers over S^2. Moreover, Y_{k+1} fits in a fiber sequence homotopy equivalent to the fibration of $\mathbb{F}_{k+1}(S^2)$ over S^2.

1 A Model for $\mathbb{F}_3(\mathbb{R}^{n+1})$

To motivate the procedure, let us consider the special case of $\mathbb{F}_3(\mathbb{R}^{n+1})$. According to §1 of Chapter V, the group $H^n(\mathbb{F}_3(\mathbb{R}^{n+1}); \mathbb{Z})$ has the set of three elements, $\mathcal{A}_3^* = \{\alpha_{21}^*, \alpha_{31}^*, \alpha_{32}^*\}$, as a basis, while the set $\mathcal{A}_3^{*\wedge 2} = \{\alpha_{21}^*\alpha_{31}^*, \alpha_{21}^*\alpha_{32}^*\}$ is a basis for $H^{2n}(\mathbb{F}_3(\mathbb{R}^{n+1}); \mathbb{Z})$. Dually, the basis for $H_n(\mathbb{F}_3(\mathbb{R}^{n+1}); \mathbb{Z})$ is clearly $\mathcal{A}_3 = \{\alpha_{21}, \alpha_{31}, \alpha_{32}\}$.

Consider next the fibration

$$\mathrm{proj}_{3,2} : \mathbb{F}_3(\mathbb{R}^{n+1}) \to \mathbb{F}_2(\mathbb{R}^{n+1}), \ (x_1, x_2, x_3) \mapsto (x_1, x_2),$$

with $\mathbb{R}_2^{n+1} = \mathbb{R}^{n+1} - Q_2$ as fiber. Observe that the restriction of the set $\mathcal{A}_3^* = \mathcal{A}_3^{*\wedge 1} \cup \mathcal{A}_3^{*\wedge 2}$ to $\mathbb{R}_2^{n+1}$ generates $H^*(\mathbb{R}_2^{n+1}; \mathbb{Z})$; and, therefore, it defines naturally a cohomology extension

$$\theta : H^*(\mathbb{R}_2^{n+1}; \mathbb{Z}) \to H^*(\mathbb{F}_3(\mathbb{R}^{n+1}); \mathbb{Z})$$

of the fiber in the sense of [101, Spanier]. Let

$$\Phi_3 : H_*(\mathbb{F}_3(\mathbb{R}^{n+1}); \mathbb{Z}) \to H_*(\mathbb{F}_2(\mathbb{R}^{n+1}); \mathbb{Z}) \otimes H_*(\mathbb{R}_2^{n+1}; \mathbb{Z})$$

be the isomorphism $z \mapsto \sum_{u^*} H_*(\mathrm{proj}_{3,2})(\theta(u^*) \cap z) \otimes u$, where u stands for the dual of u^*, and u^* ranges over the set $\mathcal{A}_3^*$. Observe that

$$\Phi_3(\alpha_{21}) = \alpha_{21} \otimes 1, \ \Phi_3(\alpha_{31}) = 1 \otimes \alpha_{31}, \ \Phi_3(\alpha_{32}) = 1 \otimes \alpha_{32}.$$

Define the elements $\alpha_{21} \ltimes \alpha_{31}$, $\alpha_{21} \ltimes \alpha_{32}$ in $H_{2n}(\mathbb{F}_3(\mathbb{R}^{n+1}); \mathbb{Z})$ by the formulas

$$\Phi_3(\alpha_{21} \ltimes \alpha_{31}) = \alpha_{21} \otimes \alpha_{31}, \ \Phi_3(\alpha_{21} \ltimes \alpha_{32}) = \alpha_{21} \otimes \alpha_{32}.$$

Let

$$\Phi_3^* : H^*(\mathbb{F}_2(\mathbb{R}^{n+1}); \mathbb{Z}) \otimes H^*(\mathbb{R}_2^{n+1}; \mathbb{Z}) \to H^*(\mathbb{F}_3(\mathbb{R}^{n+1}); \mathbb{Z})$$

be the Leray-Hirsch morphism $\omega^* \otimes u^* \to H^*(\mathrm{proj}_{3,2})(\omega^*)\theta(u^*))$.

Assertion 1.1 *For all* $\omega^* \in H^*(\mathbb{F}_2(\mathbb{R}^{n+1}))$, $u^* \in H^*(\mathbb{R}_2^{n+1})$, *and* $z \in H_*(\mathbb{F}_3(\mathbb{R}^{n+1}))$, *the relation*

$$< \omega^* \otimes u^*, \Phi_3(z) > = < \Phi_3^*(\omega^* \otimes u^*), z >$$

holds.

Proof. The proof is easy. In fact, the equations

$$\begin{aligned} < \omega^* \otimes u^*, \Phi_3(z) > &= < \omega^* \otimes u^*, \textstyle\sum_u H_*(\mathrm{proj}_{3,2})(u^* \cap z) \otimes u > \\ &= < \omega^*, H_*(\mathrm{proj}_{3,2})(u^* \cap z) > \\ &= < H^*(\mathrm{proj}_{3,2})(\omega^*)u^*, z >, \\ &= < \Phi_3^*(\omega^* \otimes u^*), z > \end{aligned}$$

are valid by virtue of the general properties of the products in the expressions. □

Consider the map $\phi : S_1^n \times S_2^n \to \mathbb{F}_3(\mathbb{R}^{n+1})$ such that

$$(\xi_1, \xi_2) \mapsto (q_1, q_1 + \xi_1, q_1 + \xi_1 + 1/2\xi_2), \quad |\xi_1| = 1 = |\xi_2|,$$

as given in Chapter II, §3. Using Proposition 2.2 of Chapter II, one can easily compute the homotopy classes, when $n > 1$, and the homology classes, when $n = 1$, of the restrictions of ϕ to the first and second factors. The results are summerized in the table

$$\mathcal{T}(\phi) : \begin{cases} \alpha_{21} + \alpha_{31}\cdot \\ \alpha_{32}, \end{cases}$$

where the first and second rows correspond, respectively, to the first and second factors. We shall call it *the factor table of* ϕ. With $\iota_1 \times \iota_2$ being the fundamental class of $S^n \times S^n$, we need to verify that $H_{2n}(\phi)(\iota_1 \times \iota_2) = \alpha_{21} \ltimes \alpha_{32}$, We proceed as follows.

Assertion 1.2 *The class* $\alpha_{21}^* \alpha_{32}^* \in H^{2n}(\mathbb{F}_3(\mathbb{R}^{n+1}))$ *is the Kronecker dual of* $\alpha_{21} \ltimes \alpha_{32} \in H_{2n}(\mathbb{F}_3(\mathbb{R}^{n+1}))$.

Proof. Consider the expressions

$$\begin{aligned} < \alpha_{rs}^* \alpha_{tu}^*, \alpha_{vw} \ltimes \alpha_{xy} > &= < H^n(\mathrm{proj}_{3,2}(\alpha_{rs}^*)\theta(\alpha_{tu}^*), \alpha_{vw} \ltimes \alpha_{xy} > \\ &= < \Phi_3^*(\alpha_{rs}^* \otimes \alpha_{tu}^*), \alpha_{vw} \ltimes \alpha_{xy} > \\ &= < \alpha_{rs}^* \otimes \alpha_{tu}^*, \Phi_3(\alpha_{vw} \otimes \alpha_{xy}) > \\ &= < \alpha_{rs}^* \otimes \alpha_{tu}^*, \alpha_{vw} \otimes \alpha_{xy} > \\ &= \delta_{rs,vw}\delta_{tu,xy}. \end{aligned}$$

This proves the Assertion □

Now, put $S_\phi = H_{2n}(\phi)(\iota_1 \times \iota_2))$.

Assertion 1.3 *The class $\alpha_{21}^*\alpha_{32}^* \in H^{2n}(\mathbb{F}_3(\mathbb{R}^{n+1}))$ is the Kronecker dual of S_ϕ.*

Proof. On inspecting the factor table of ϕ, one sees that

$$\begin{aligned} H^n(\phi)(\alpha_{21}^*) &= \iota_1^* \times 1 = H^n(\phi)(\alpha_{31}^*) \\ H^n(\phi)(\alpha_{32}^*) &= 1 \times \iota_2^*. \end{aligned}$$

Hence,

$$\begin{aligned} < \alpha_{21}^*\alpha_{32}^*, H_{2n}(\phi)(\iota_1 \times \iota_2) > &= < H^{2n}(\phi)(\alpha_{21}^*\alpha_{32}^*), \iota_1 \times \iota_2 > \\ &= < H^n(\phi)(\alpha_{21}^*)H^n(\phi)(\alpha_{32}^*), \iota_1 \times \iota_2 > \\ &= < H^n(\phi)(\alpha_{21}^*), \iota_1 >< H^n(\phi)(\alpha_{32}^*), \iota_2 > \\ &= 1. \end{aligned}$$

Similarly, one shows that $< \alpha_{21}^*\alpha_{31}^*, H_{2n}(\phi)(\iota_1 \times \iota_2) >= 0$, since the restriction of α_{31}^* to the second factor is trivial. This proves the Assertion. □

Again similarly, the map $\psi : S_1^n \times S_2^n \to \mathbb{F}_3(\mathbb{R}^{n+1})$, where

$$(\xi_1, \xi_2) \mapsto (q_1, q_1 + \xi_1, q_1 + 1/2\xi_2), \quad |\xi_1| = 1 = |\xi_2|,$$

induces a morphism on homology represented by the table

$$\mathcal{T}(\psi) : \begin{cases} \alpha_{21} + (-1)^{n+1}\alpha_{32} \\ \alpha_{31}, \end{cases}$$

as one can verify by appealing to Proposition 1.2 of Chapter II. Again, it follows that $H_{2n}(\psi)(\iota_1 \times \iota_2) = \alpha_{21} \ltimes \alpha_{31}$.

Next, we turn to the construction of a CW-complex X that is homotopy equivalent to $\mathbb{F}_3(\mathbb{R}^{n+1})$. We offer two constructions, dealing first with the case $n > 1$. The procedure for the case $n = 1$ is somewhat different.

The First Construction

Recall that the elements α_{31}, α_{32} induce a homotopy equivalence

$$S_{31} \vee S_{32} \to \mathbb{R}_2^{n+1},$$

as in Chapter II, §1. Define the n-skeleton $X^{(n)}$ of the complex X to be the bouquet

$$X^{(n)} = S_{21} \vee (S_{31} \vee S_{32})$$

of n-spheres.

Assertion 1.4 *The map $\alpha : X^{(n)} = S_{21} \vee (S_{31} \vee S_{32}) \to \mathbb{F}_3(\mathbb{R}^{n+1})$ induced by $\{\alpha_{21}, \alpha_{31}, \alpha_{32}\}$ is $(2n-1)$-connected.*

Proof. Observe that α is homologically $(2n-1)$-connected. As the spaces in question are simply connected, it follows that α is homotopically $(2n-1)$-connected. □

Recall that

$$S^n \times S^n = (S^n \times S^n) \cup_{[\iota_n, \iota_n]} D^{(2n)},$$

where, by abuse of notation, $[\iota_n, \iota_n]$ stands for a map $S^{2n-1} \to S^n \vee S^n$ representing the Whitehead product of the fundamental class ι_n by itself. Note that Assertion 1.4 implies that the maps

$$\phi|(S_1 \vee S_2) : S_1 \vee S_2 \to \mathbb{F}_3(\mathbb{R}^{n+1}),\ \psi|(S_1 \vee S_2) : S_1 \vee S_2 \to \mathbb{F}_3(\mathbb{R}^{n+1})$$

can be deformed so as to make their images lie in $X^{(n)}$. As $\mathbb{F}_k(\mathbb{R}^{n+1})$ is simply connected, the rows of the factor tables of $\alpha_{21} \ltimes \alpha_{32}$ and $\alpha_{21} \ltimes \alpha_{31}$ can be regarded as elements in $\pi_n(\mathbb{F}_3(\mathbb{R}^{n+1}))$. This being so, note that

$$\begin{aligned} \phi \circ [\iota_n, \iota_n] &= [\lambda_1(\phi), \lambda_2(\phi)] \\ \psi \circ [\iota_n, \iota_n] &= [\lambda_1(\psi), \lambda_2(\psi)], \end{aligned}$$

where λ_1, λ_2 are the first and second rows, respectively, of the factor tables of ϕ and ψ. Put

$$X = X^{(n)} \cup_{[\lambda_1(\phi), \lambda_2(\phi)]} D^{2n} \cup_{[\lambda_1(\psi), \lambda_2(\psi)]} D^{2n}.$$

The space X can be expressed as the disjoint union

$$X = e_0 \cup (e_{21} \cup (e_{31} \cup e_{32})) \cup e_{21} \ltimes e_{31} \cup e_{21} \ltimes e_{32}$$

of cells, where e_{21}, e_{31}, e_{32} are the n-cells corresponding to the spheres S_{21}, S_{31}, S_{32} and $e_{21} \ltimes e_{31}$, $e_{21} \ltimes e_{32}$ are the $2n$-cells corresponding to the two-fold spherical cycles $\phi(\iota_1 \times \iota_2)$ and $\psi(\iota_1 \times \iota_2)$ defined above. The space X, with the decomposition defined above and the characteristic maps ϕ and ψ, is certainly a CW complex.

Let $h_3 : X \to \mathbb{F}_3(\mathbb{R}^{n+1})$ be the map induced by α, ϕ, and ψ. It is clear that h_3 induces an isomorphism on homology, and, as the spaces are simply connected, h_3 is, in fact, a homotopy equivalence.

The Second Construction

Recall that $\mathbb{F}_3(\mathbb{R}^{n+1}) \to \mathbb{F}_2(\mathbb{R}^{n+1})$ is a fiber bundle with $\mathbb{R}^{n+1}_2$ as fiber (see Theorem 1.1 of Chapter I). The characteristic map of the fiber bundle restricted to $S_{21} \subset \mathbb{F}_2(\mathbb{R}^{n+1})$ is a map $\hat{\mu} : S^{n-1} \to \mathrm{Top}_{s2}(\mathbb{R}^{n+1})$. Also, recall that the elements α_{31}, α_{32} induce a homotopy equivalence $h : S_{31} \vee S_{32} \to \mathbb{R}^{n+1}_2$. Denote by $G(\mathbb{R}^{n+1}_2)$ the space of homotopy equivalences of $\mathbb{R}^{n+1}_2$ and, by $G(S_{31} \vee S_{32})$ those of $S_{31} \vee S_{32}$. Choose a homotopy equivalence $\gamma : G(\mathbb{R}^{n+1}_2) \to G(S_{31} \vee S_{32})$ of monoids, and consider the composite

$$\rho : \mathrm{Top}_{s2}(\mathbb{R}^{n+1}) \xrightarrow{j} G(\mathbb{R}_2^{n+1}) \xrightarrow{\gamma} G(S_{31} \vee S_{32}),$$

where $j : \mathrm{Top}_{s2}(\mathbb{R}^{n+1}) \to G(\mathbb{R}_2^{n+1})$ is the natural imbedding. Observe that the map $\rho \circ \hat{\mu} : S^{n-1} \to G(S_{31} \vee S_{32})$, in turn, induces an action

$$\mu : \partial D^n \times (S_{31} \vee S_{32}) \to (S_{31} \vee S_{32}).$$

The following is easily proved (see for example [26, Dold-Lashof]).

Assertion 1.5 *There is a fiber-preserving homotopy equivalence*

$$\begin{array}{ccc} D^n \times (S_{31} \vee S_{32}) \cup_\mu (S_{31} \vee S_{32}) & \xrightarrow{h_3} & \mathbb{F}_3(\mathbb{R}^{n+1}) \\ \downarrow & & \downarrow \\ S_{21} & \xrightarrow{\alpha_{21}} & \mathbb{F}_2(\mathbb{R}^{n+1}), \end{array}$$

where the first vertical map $\mathrm{proj} : X \to S_{21}$ *is the map induced by the projection on* S_{21}*. Moreover,* $\mathrm{proj} : X \to S_{21}$ *is a quasi fibration.*

Put $X = D^n \times (S_{31} \vee S_{32}) \cup_\mu (S_{31} \vee S_{32})$. The quasi fibration $\mathrm{proj} : X \to S_{21}$ has a section and, hence, a natural twisted-product structure $X \simeq S_{21} \ltimes (S_{31} \vee S_{32})$. Regard $\alpha_{21}^*, \alpha_{31}^*, \alpha_{32}^*$ as elements in $H^n(X)$. Then, as in the case of $\mathbb{F}_3(\mathbb{R}^{n+1})$, there are twisted products $\alpha_{21} \ltimes \alpha_{32}$ and $\alpha_{21} \ltimes \alpha_{31}$ in $H_*(X)$. It can easily be seen that these products correspond to their counterparts in $H_*(\mathbb{F}_3(\mathbb{R}^{n+1}))$. Moreover, the maps $\phi, \psi : S_1 \times S_2 \to \mathbb{F}_3(\mathbb{R}^{n+1})$ can be factored, up to homotopy, through X.

Thus the space X has a natural cellular structure

$$X = e_0 \cup (e_{21} \cup (e_{31} \cup e_{32})) \cup e_{21} \ltimes e_{31} \cup e_{21} \ltimes e_{32}$$

of cells, where e_{21}, e_{31}, e_{32} are the n-cells corresponding to the spheres S_{21}, S_{31}, S_{32} and $e_{21} \ltimes e_{31}$, $e_{21} \ltimes e_{32}$ are the $2n$-cells that correspond under h_3 to the two-fold spherical cycles $\phi(\iota_1 \times \iota_2)$ and $\psi(\iota_1 \times \iota_2)$ defined above.

It is worthwhile noting here that one can compute the attaching maps for $e_{21} \ltimes e_{31}$, $e_{21} \ltimes e_{32}$ along the lines of Proposition 5.1 of Chapter III. Doing this, one sees that the term $[\alpha_{31}, \alpha_{32}]$ is the analogue of $J(\xi)$ in that proposition.

If $n = 1$, we put $X = S_{21} \cup_{\phi_{rs}} (D_{21} \times (S_{31} \vee S_{32})$, as in Theorem 5.1 of Chapter IV. It is a complex of dimension 2, the attaching maps of which are given by conjuguating by α_{21}. Recall that the projection on S_{21} is a quasi fibration with $S_{31} \vee S_{32}$ as fiber. Hence, it is aspherical. Then one constructs a fiber-preserving map $h_3 : X \to \mathbb{F}_3(\mathbb{R}^2)$ that induces an isomorphism on the homotopy groups. This proves that h_3 is a homotopy equivalence.

2 The Twisted-Product Structure on $H_*(\mathbb{F}_{k-r,r})$

All through this section, unless indicated otherwise, homology and cohomology have integral coefficients. We shall give here two constructions for, respectively, a left and right twisted-product structure on the group $H_*(\mathbb{F}_{k-r,r}(\mathbb{R}^{n+1}))$.

The left twisted product is based on the fundamental fiber sequence

$$\mathcal{F}_k(\mathbb{R}^{n+1}) : \begin{cases} \mathbb{F}_k & \leftarrow \mathbb{F}_{k-1,1} \leftarrow \cdots \leftarrow \mathbb{F}_{k-r,r} \leftarrow \\ \downarrow & \quad\downarrow \qquad\quad \downarrow \qquad\quad \downarrow \\ \mathbb{R}^{n+1} & \quad \mathbb{R}^{n+1}_1 \qquad \cdots \qquad \mathbb{R}^{n+1}_r \end{cases}$$

introduced in §1 of Chapter II. According to our notation,

$$\mathbb{F}_{k-r,r} = \{(x_1, \cdots, x_k) \in \mathbb{F}_k(\mathbb{R}^{n+1}) | x_1 = q_1, \cdots, x_r = q_r\},$$

and the vertical projections $p_r : \mathbb{F}_{k-r,r} \to \mathbb{R}^{n+1}_r$ are the maps where $(q_1, \cdots, q_r, x_{r+1}, \cdots, x_k) \mapsto x_{r+1}$. The fiber of the projection p_r at the point $q_{r+1} \in \mathbb{R}^{n+1}_r$ is just the space $\mathbb{F}_{k-r-1,r+1}(\mathbb{R}^{n+1})$.
Theorem 4.2 of Chapter V asserts that $H^*(\mathbb{F}_{k-r-1,r+1})$ is generated by $\mathcal{A}^{*\wedge m} = \{\alpha^*_{\underline{r}\,\underline{s}}\}$, where $\underline{r} = (r_1, \cdots, r_m)$, with $m \leq (k-r-1)$, $r_1 > r+1$, and $\alpha^*_{\underline{r}\,\underline{s}} = \alpha^*_{r_1 s_1} \cdots \alpha^*_{r_m s_m}$. The set $\mathcal{A}^{*\wedge m}$ defines, therefore, a cohomology extension of the fiber $\theta : H^*(\mathbb{F}_{k-r-1,r+1}(\mathbb{R}^{n+1})) \to H_*(\mathbb{F}_{k-r,r})$ (see [101, Spanier], p.258). Denote the corresponding Leray-Hirsch isomorphism by

$$\Phi_{k-r,r} : H_*(\mathbb{F}_{k-r,r}) \to H_*(\mathbb{R}^{n+1}_r) \otimes H_*(\mathbb{F}_{k-r-1,r+1}).$$

Proceeding naturally, by descending induction on r, the formula

$$\Phi_{k-r,r}(\alpha_{r_1 s_1} \ltimes \cdots \ltimes \alpha_{r_m s_m}) = \alpha_{r_1 s_1} \otimes (\alpha_{r_2 s_2} \ltimes \cdots \ltimes \alpha_{r_m s_m})$$

leads to the left twisted product in $H_*(\mathbb{F}_{k-r,r})$, with $r_1 = r+1$. The set of all such elements where $r_1 \geq r+1$ is a basis for the group $H_*(\mathbb{F}_{k-r,r})$.

The construction of the right twisted product is based on the fibration

$$\text{proj}_{k,k-1} : \mathbb{F}_k(\mathbb{R}^{n+1}) \to \mathbb{F}_{k-1}(\mathbb{R}^{n+1}),\ (x_1, \cdots, x_k) \mapsto (x_1, \cdots, x_{k-1}),$$

the fiber of which at $(q_1, \cdots, q_{k-1})$ is the space $\mathbb{R}^{n+1}_{k-1} = \mathbb{R}^{n+1} - Q_{k-1}$. The set $\mathcal{A}^* = \{\alpha^*_{rs} | 1 \leq s < r \leq k\}$ gives a cohomology extension of the fiber. Using the corresponding Leray-Hirsch isomorphism

$$\Phi_k : H_*(\mathbb{F}_k(\mathbb{R}^{n+1})) \to H_*(\mathbb{F}_{k-1}(\mathbb{R}^{n+1})) \otimes H_*(\mathbb{R}^{n+1}_{k-1}),$$

one defines the right twisted product in $H_*(\mathbb{F}_k(\mathbb{R}^{n+1}))$, proceeding inductively. Using a section $\sigma : \mathbb{F}_{k-1}(\mathbb{R}^{n+1}) \to \mathbb{F}_k(\mathbb{R}^{n+1})$, one regards the coalgebra $H_*(\mathbb{F}_{k-1}(\mathbb{R}^{n+1}))$ as a sub-coalgebra of $H_*(\mathbb{F}_k(\mathbb{R}^{n+1}))$ and, then, defines the right twisted product $\alpha_{r_1 s_1} \rtimes \alpha_{r_2 s_2} \rtimes \cdots \rtimes \alpha_{r_p s_p}$, $p \leq (k-1)$, which agrees with the right twisted product inductively defined on $H_*(\mathbb{F}_{k-1}(\mathbb{R}^{n+1}))$ and is such that

$$\Phi_k(\alpha_{r_1 s_1} \rtimes \cdots \rtimes \alpha_{r_{p-1} s_{p-1}} \rtimes \alpha_{ks}) = (\alpha_{r_1 s_1} \rtimes \cdots \rtimes \alpha_{r_{p-1} s_{p-1}}) \otimes \alpha_{ks}.$$

We shall see in §5 below that these two twisted products agree, and that the bases they give are the dual of the basis $\mathcal{A}^{*\wedge}$ of the group $H^*(\mathbb{F}_k(\mathbb{R}^{n+1}))$ described above. For the sake of brevity, put $\mathcal{A}^{p\wedge} = \{\alpha_{\underline{r}\,\underline{s}}\}$ where $\underline{r} \in \mathcal{R}_p, \underline{s} \in \mathcal{S}_p, \underline{s} < \underline{r}$.

3 Perturbation and Affine Maps

Again, unless indicated otherwise, homology and cohomology are understood throughout this section to have integral coefficients.

Our aim in this section is to define, for each p-fold twisted product $\alpha_{\underline{r}\,\underline{s}}$, a map

$$\phi_{\alpha_{\underline{r}\,\underline{s}}} : S_1^n \times \cdots \times S_p^n \to \mathbb{F}_k(\mathbb{R}^{n+1})$$

of a certain kind. When $p = 1$, these maps are just the maps α_{rs} of Chapter II, §2. In the next section, we shall see that the map $\phi_{\alpha_{\underline{r}\,\underline{s}}}$ takes the fundamental class of $S_1^n \times \cdots \times S_p^n$ to the homology class $\alpha_{\underline{r}\,\underline{s}}$. In order to show this, we need first to develop suitable concepts.

Let us consider the case of $\mathbb{F}_3(\mathbb{R}^{n+1})$ first. There are two twisted products, $\alpha_{21} \rtimes \alpha_{32}$ and $\alpha_{21} \rtimes \alpha_{31}$. The former is represented by the map

$$\phi : S_1 \times S_2 \to \mathbb{F}_3(\mathbb{R}^{n+1})$$

and the latter, by

$$\psi : S_1 \times S_2 \to \mathbb{F}_3(\mathbb{R}^{n+1})$$

(see §2). At the homotopy level, the map ϕ transcribes the commutativity, under Whitehead product, of $\alpha_{21} + \alpha_{31}$ with α_{32}, while ψ does the same for α_{31} with $\alpha_{21} + (-1)^{n+1}\alpha_{32}$. This information is summerized by the factor tables $\mathcal{T}(\phi)$ and $\mathcal{T}(\psi)$, respectively.

The idea underlying the factor tables is to look at the Y-B relations of Chapter II, §3 in a different manner. While α_{32} does not commute with α_{21}, it will do so if α_{21} is *perturbed* by α_{31}, and one may view $(-1)^{n+1}\alpha_{32}$ as the *perturbation* of α_{21} by α_{31}. The concept can be precisely expressed as follows.

To start, we define a parameterized form of α_{rs} in analogy with the definition given in Chapter II, §1. Let

$$\beta : S_1^n \times \cdots \times S_m^n \to \mathbb{F}_r(\mathbb{R}^{n+1}) \subset \mathbb{F}_k(\mathbb{R}^{n+1})$$

be an imbedding, where each S_i^n, $i = 1, \cdots, m$ is the unit sphere in $\mathbb{R}^{n+1}$. Thus,

$$\beta(\xi_1, \cdots, \xi_m) = (q_1(\xi), q_2(\xi), \cdots, q_r(\xi), q_{r+1}(\xi), \cdots, q_k(\xi)),$$

where the elements $q_i(\xi)$ for $1 \leq i \leq k$ are functions of $\xi = (\xi_1, \cdots, \xi_m)$. For $1 \leq s < r \leq k$, define

$$\beta \bowtie \alpha_{rs} : S_1^n \times \cdots S_m^n \times S_{m+1}^n \to \mathbb{F}_k(\mathbb{R}^{n+1})$$

to be the map where

$$(\xi_1, \cdots, \xi_{m+1}) \mapsto (q_1(\xi), q_2(\xi), \cdots, q_{r-1}(\xi), q_s(\xi) + 2^{-\nu}\xi_{m+1}, q_r(\xi), \cdots),$$

with $2^{-\nu} < \min_{\xi, s \neq r}(|q_r(\xi) - q_s(\xi)|)$. For example, $\alpha_{21} \bowtie \alpha_{32} = \phi$ and $\alpha_{21} \bowtie \alpha_{31} = \psi$, where ϕ and ψ are the maps defined in §1.

Consider now the map

$$\beta : S^n \to \mathbb{F}_r(\mathbb{R}^{n+1}) \subset \mathbb{F}_k(\mathbb{R}^{n+1}),$$

where $\xi \mapsto (q_1, q_2', \cdots, q_{r-1}', q_r, \cdots, q_k)$ and, q_1 and q_t, for $t \geq r$, are independent of ξ.

Definition 3.1 *The* perturbation *of β by α_{tu}, where $u < t$, is the class*

$$[(\beta \bowtie \alpha_{tu})|S_1] - [\beta],$$

where $[\gamma]$ denotes the homology class of γ.

The following is an immediate consequence of the relevant Y-B relations.

Proposition 3.1 *The perturbation of α_{rs} by α_{tu} is equal to*

$$\delta_{ur}\alpha_{ts} + \delta_{us}(-1)^{n+1}\alpha_{tr},$$

where $s < r < t$ and $u < t$, and $\delta_{ij} = 1$, if $i = j$, and $= 0$ otherwise.

At the geometric level, the perturbation of α_{21} by α_{32} is caused by the third particle q_3' moving on a sphere of radius $2^{-1}\xi_2$ centered at the second particle q_2', which is moving on a sphere of radius 1 centered at q_1 (see Fig. VI. 1). On the other hand, the perturbation of α_{21} by α_{32} is caused by constraining the third particle q_3' to a sphere of radius 2^{-1} centered at q_1 inside the concentric sphere of unit radius described by α_{21} (see Fig. VI. 2). Note that the concept of perturbation, as defined here, is not symmetric. In fact, α_{rs} is perturbed by another α_{tu} if, and only if, $t > r > s$ and $u = r$ or $u = s$. Thus, the perturbation of α_{rs} by α_{tu} is 0, if $\{r, s\} \cap \{t, u\} = \emptyset$. It is worthwhile to

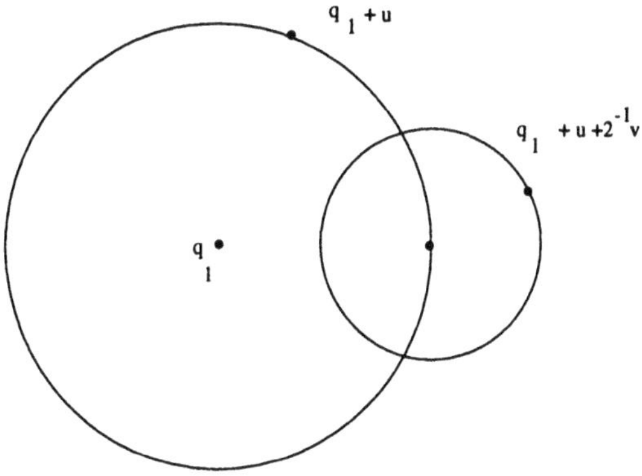

Fig. VI.1. The perturbation of α_{21} by α_{32}

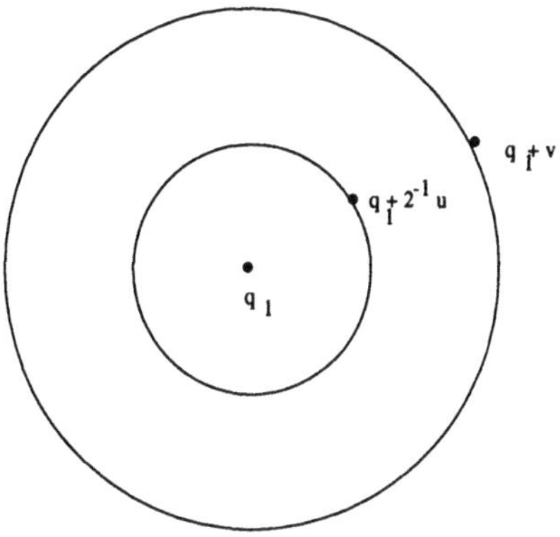

Fig. VI.2. The perturbation of α_{21} by α_{31}

note that the perturbation of α_{rs} by α_{tu} is just the term in the relevant Y-B relation that is added to the perturbed term α_{rs}. Therefore, perturbation is invariant under the action of certain elements of the symmetric group, in the following sense.

If σ is a permutation that preserves the order $t > r > s$ and $t > u$, then the perturbation of $\alpha_{\sigma r \sigma s}$ by $\alpha_{\sigma t \sigma u}$, is obtained by applying σ to the perturbation in the definition above. Thus, if one takes $t = 3$, $r = 2$, $s = 1$, and σ to be a permutation that preserves order, then one sees that the perturbation of $\alpha_{\sigma 2 \sigma 1}$ by $\alpha_{\sigma 3 \sigma 2}$ is given by the composite of

$$\xi \mapsto (q_1, q_1 + \xi, q_1 + \xi + 1/2, q_3, \cdots)$$

with σ, where $\xi \in S^n$.

Proposition 3.2 *Let $\beta : S_1^n \to \mathbb{F}_{k-1}(\mathbb{R}^{n+1})$. Then the perturbation of β by α_{ks} is a linear expression in the elements of the set $\{\alpha_{kt} | 1 \leq t < k\}$.*

Proof. Consider the map $\beta \bowtie \alpha_{ks} : S_1^n \times S_2^n \to \mathbb{F}_k(\mathbb{R}^{n+1})$. Note that

$$H_n(\mathrm{proj}_{k,k-1})(\beta \bowtie \alpha_{ks}) = \beta,$$

where $\mathrm{proj}_{k,k-1} : \mathbb{F}_k(\mathbb{R}^{n+1}) \to \mathbb{F}_{k-1}(\mathbb{R}^{n+1})$ is the projection that deletes the last entry. Recall now that the sequence

$$1 \to H_n(\mathbb{R}_{k-1}^{n+1}) \to H_n(\mathbb{F}_k(\mathbb{R}^{n+1})) \to H_n(\mathbb{F}_{k-1}(\mathbb{R}^{n+1})) \to 1,$$

where the morphisms are those induced by the natural imbedding $i : \mathbb{R}_{k-1}^{n+1} \to \mathbb{F}_k(\mathbb{R}^{n+1}))$ and projection $\mathrm{proj}_{k,k-1} : \mathbb{F}_k(\mathbb{R}^{n+1}) \to \mathbb{F}_{k-1}(\mathbb{R}^{n+1})$, respectively, is a split exact sequence. Hence,

$$H_n(\mathbb{F}_k(\mathbb{R}^{n+1})) \cong H_n(\mathbb{F}_{k-1}(\mathbb{R}^{n+1}) \oplus H_n(\mathbb{R}_{k-1}^{n+1}),$$

and $[\beta \bowtie \alpha_{ks} \mid S_1^n] = \beta + \gamma$, where γ in $H_n(\mathbb{R}_{k-1}^{n+1})$. But the latter module is generated by the elements of the set $\{\alpha_{kt} | 1 \leq t < k\}$, as Theorem 1.5 of Chapter V easily implies. This proves the proposition. □

Let us consider now the p-fold twisted product,

$$\omega = \alpha_{\underline{r}\,\underline{s}} = \alpha_{r_1 s_1} \rtimes \cdots \rtimes \alpha_{r_p s_p}.$$

The elements $\alpha_{r_i s_i}$ of which the p-fold twisted product ω is constructed are called the *factors* of ω. Using the bowtie construction described above, we shall define a class of maps corresponding to the elements of the basis

$$\alpha_{\underline{r}\,\underline{s}} = \alpha_{r_1 s_1} \rtimes \cdots \rtimes \alpha_{r_p s_p} \in H_{pn}(\mathbb{F}_k(\mathbb{R}^{n+1}); \mathbb{Z}),$$

where, as in Definition 4.1 of Chapter V,

$$\begin{aligned} \underline{r} \in \mathcal{R}_p &= \{\underline{r} = (r_1, \cdots, r_p) | 1 < r_1 < \cdots < r_p \leq k\} \\ \underline{s} \in \mathcal{S}_p &= \{\underline{s} = (s_1, \cdots, s_p) | 1 \leq s_1, s_2, \cdots, s_p < k\}. \end{aligned}$$

Theorem 3.1 *For each $1 \leq m \leq p$, there is a map*

$$\phi_{\omega_m} : S_1^n \times \cdots \times S_m^n \to \mathbb{F}_k(\mathbb{R}^{n+1})$$

such that $\phi_{\omega_m} = \phi_{\omega_{m-1}} \bowtie \alpha_{r_m s_m}$, with $\omega_{m-1} = \alpha_{r_1 s_1} \rtimes \cdots \rtimes \alpha_{r_{m-1} s_{m-1}}$, and $\omega_m = \omega_{m-1} \rtimes \alpha_{r_m s_m}$.

Proof. To prove the theorem, proceed by induction on m. When $\omega = \alpha_{r_1 s_1}$, put $\phi_\omega = \alpha_{r_1 s_1}$. Assume that $\phi_{\omega_{m-1}} : S_1^n \times \cdots \times S_{m-1}^n \to \mathbb{F}_k(\mathbb{R}^{n+1})$ is a map subject to the condition in the theorem, where $\omega_{m-1} = \alpha_{r_1 s_1} \rtimes \cdots \rtimes \alpha_{r_{m-1} s_{m-1}}$. Put $\phi_{\omega_m} = \phi_{\omega_{m-1}} \bowtie \alpha_{r_m s_m}$. One can easily verify that ϕ_{ω_m} satisfies the condition of the theorem. □

A map satisfying the preceding condition will be said to be *affine*. Now, let $\phi_\omega : S_1^n \times \cdots \times S_m^n \to \mathbb{F}_k(\mathbb{R}^{n+1})$ be the affine map corresponding to the m-fold $\omega_m = \alpha_{\underline{r}\,\underline{s}} = \alpha_{r_1 s_1} \rtimes \cdots \rtimes \alpha_{r_m s_m}$, and put $\lambda_j(\omega_m) = \phi_{\omega_m}|S_j^n$. For simplicity of notation, we shall denote the homology class of $\lambda_j(\omega_m) \in H_n(\mathbb{F}_k(\mathbb{R}^{n+1}))$ by the same symbol.

Theorem 3.2 *Suppose that $\omega_{m-1} = \alpha_{\underline{r}\,\underline{s}} = \alpha_{r_1 s_1} \rtimes \cdots \rtimes \alpha_{r_{m-1} s_{m-1}}$ is an $(m-1)$-fold twisted basic product, and consider $\omega_m = \omega_{m-1} \rtimes \alpha_{r_m s_m}$, where $r_m > r_{m-1}$. Then*

$$\begin{aligned} \lambda_j(\omega_m) &= \lambda_j(\omega_{m-1}) + \lambda_{mj}(\omega_m), \quad 1 \le j \le (m-1) \\ \lambda_m(\omega_m) &= \alpha_{r_m s_m}, \end{aligned}$$

where $\lambda_{mj}(\omega_m)$ is the perturbation of $\lambda_j(\omega_{m-1})$ by $\alpha_{r_m s_m}$.

Proof. Recall that $\lambda_j(\omega_1) \bowtie \alpha_{r_m, s_m} : S_j^n \times S_m^n \to \mathbb{F}_k(\mathbb{R}^{n+1})$, and observe that

$$\phi_{\omega_m}|S_j^n = \lambda_j(\omega_{m-1}) \bowtie \alpha_{r_m, s_m}|S_j^n$$

for $1 \le j \le (m-1)$, where S_j^n stands for the j factor in $S_1^n \times \cdots \times S_m^n$. Therefore, $\lambda_j(\omega_m)$ is the sum of $\lambda_j(\omega_{m-1})$, with the perturbation of the latter by $\alpha_{r_m s_m}$. This proves the first assertion.

To prove the second, just note that $\phi_\omega|S_m = \alpha_{r_m s_m}$. □

Let us note here that if $n > 1$, then $\mathbb{F}_k(\mathbb{R}^{n+1})$ is simply connected, and the previous theorem remains valid if the elements in question are regarded as elements in $\pi_n(\mathbb{F}_k(\mathbb{R}^{n+1}))$. But, for $n = 1$ the theorem is valid only in the abelianization $H_1(\mathbb{F}_k(\mathbb{R}^2))$ of $\pi_1(\mathbb{F}_k(\mathbb{R}^2), q)$.

Theorem 3.3 *Let $\omega_m = \alpha_{r_1 s_1} \rtimes \cdots \rtimes \alpha_{r_m s_m}$ be an m-fold twisted basic product. Then, for each j such that $1 \le j \le m$,*

$$\lambda_j(\omega_m) = \alpha_{r_j s_j} + \sum_{u=j+1}^{m} \lambda_{ju}(\omega_u), \quad 1 \le j \le m,$$

where $\omega_u = \alpha_{r_1 s_1} \rtimes \cdots \rtimes \alpha_{r_u s_u}$, with $\lambda_{ju}(\omega_u)$ being the perturbation of $\alpha_{r_j s_j} + \sum_{v=j+1}^{u-1} \lambda_{jv}(\omega_v)$ by $\alpha_{r_u s_u}$.

Proof. Proceed by induction on m. For $m = 1$ the lemma is obvious. Assume that the lemma has been proved for $m-1$, and let $\omega = \alpha_{r_1 s_1} \rtimes \cdots \rtimes \alpha_{r_m s_m}$ be an m-fold basic twisted product. Define $\omega_{m-1} = \alpha_{r_1 s_1} \rtimes \cdots \rtimes \alpha_{r_{m-1}, s_{m-1}}$ to be the $(m-1)$-fold twisted product obtained from ω by deleting the last factor $\alpha_{r_m s_m}$. According to Theorem 3.2,

$$\begin{aligned} \lambda_j(\omega_m) &= \lambda_j(\omega_{m-1}) + \lambda_{jm}(\omega_m), \quad 1 \le j \le (m-1) \\ \lambda_m(\omega_m) &= \alpha_{r_m s_m}, \end{aligned}$$

where $\lambda_{jm}(\omega_m)$ is the perturbation of $\lambda_j(\omega_{m-1})$ by $\alpha_{r_m s_m}$. By induction,

$$\lambda_j(\omega_{m-1}) = \alpha_{r_js_j} + \sum_{u=j+1}^{m-1} \lambda_{ju}(\omega_u),$$

with $\lambda_{ju}(\omega_u)$ being the perturbation of $\alpha_{r_js_j} + \sum_{v=j+1}^{u-1} \lambda_{jv}(\omega_v)$ by α_{r_u,s_u}. The preceding two facts imply the assertion in the theorem. □

The m-fold twisted product $\omega = \alpha_{r_1s_1} \rtimes \cdots \rtimes \alpha_{r_ms_m}$ leads to the table

$$\mathcal{T}(\phi_\omega) : \begin{cases} \alpha_{r_1s_1} & +\sum_{u=2}^{m} \lambda_{1u}(\omega_u) \\ \cdots & +\cdots, \\ \alpha_{r_js_j} & +\sum_{u=j+1}^{m} \lambda_{ju}(\omega_u) \\ \cdots & +\cdots \\ \alpha_{r_{m-1}s_{m-1}} + \lambda_{m-1m}(\omega_m) & \\ \alpha_{r_ms_m}, & \end{cases}$$

where $\omega_u = \alpha_{r_1s_1} \rtimes \cdots \rtimes \alpha_{r_us_u}$ for each $1 \leq u \leq m$, and the jth-row is the homotopy class of $\phi_\omega \mid S_j^n$ for $n > 1$, and the homology class for $n = 1$. We shall call the table above the *factor table* of the m-fold twisted product $\omega = \alpha_{\underline{r}\,\underline{s}} = \alpha_{r_1s_1} \rtimes \cdots \rtimes \alpha_{r_ms_m}$ (see §4 for an illustrated example).

The terms $\lambda_{ju}(\omega_u)$ appearing in $\mathcal{T}(\phi_\omega)$ can be decomposed further in terms of the generators α_{rs}.

Corollary 3.1 *Let $\lambda_{ju}(\omega_u)$ be as in factor table $\mathcal{T}(\phi_\omega)$. Then, for all j,u such that $j < u \leq m$, we have $\lambda_{ju}(\omega_u) = \sum_{s=1}^{r_u-1} \nu_s\alpha_{r_us}$, for some $\nu_s \in \mathbb{Z}$.*

Proof. To start, let us express the factor table in the form

$$\mathcal{T}(\phi_\omega) : \begin{cases} \alpha_{r_1s_1} & +\sum_{u=2}^{m-1} \lambda_{1u}(\omega_u) + \lambda_{1m}(\omega_m) \\ \cdots & +\cdots, \\ \alpha_{r_js_j} & +\sum_{u=j+1}^{m-1} \lambda_{ju}(\omega_u) + \lambda_{jm}(\omega_m) \\ \cdots & +\cdots \\ \alpha_{r_{m-1}s_{m-1}} & +\lambda_{m-1m}(\omega_m) \\ \alpha_{r_ms_m}, & \end{cases}$$

where $\omega_m = \omega$. Proceed by induction on m. Assume that the corollary is true for $u < m$. Then, there are integers ν_s such that $\lambda_{ju}(\omega_u) = \sum_{s=1}^{r_u-1} \nu_s\alpha_{r_us}$ for all $u \leq (m-1)$. Now, according to Theorem 3.2, the element $\lambda_{jm}(\omega_m)$ is the perturbation of $\alpha_{r_js_j} + \sum_{u=j+1}^{m-1} \lambda_{ju}(\omega_u)$ by $\alpha_{r_ms_m}$. Applying Proposition 3.1, one sees that the said perturbation is a linear combination in the elements $\{\alpha_{r_ms} \mid 1 \leq s < r_m\}$. This proves the induction assumption and, hence, the corollary. □

4 An Illustrated Example

By way of illustration, let us consider the element

$$\omega = \alpha_{21} \rtimes \alpha_{32} \rtimes \alpha_{41} \rtimes \alpha_{53}.$$

To compute its factor table and the corresponding affine map, proceed as in the previous section. The table is of the form

$$\mathcal{T}(\phi_\omega) : \begin{cases} \alpha_{21} & +\lambda_{12}(\omega_2) + \lambda_{12}(\omega_2) + \lambda_{13}(\omega_3) + \lambda_{14}(\omega_4) \\ \alpha_{32} & +\lambda_{23}(\omega_3) + \lambda_{24}(\omega_4) \\ \alpha_{41} & +\lambda_{34}(\omega_4) \\ \alpha_{53}, \end{cases}$$

where, following the notation of the previous section,

$$\begin{aligned} \omega_1 &= \alpha_{21} \\ \omega_2 &= \alpha_{21} \rtimes \alpha_{32} \\ \omega_3 &= \alpha_{21} \rtimes \alpha_{32} \rtimes \alpha_{41} \\ \omega_4 &= \alpha_{21} \rtimes \alpha_{32} \rtimes \alpha_{41} \rtimes \alpha_{53}. \end{aligned}$$

Now, clearly, the factor table of ω_1 is

$$\mathcal{T}(\phi_{\omega_1}) : \{ \alpha_{21}.$$

Since $\lambda_{12}(\omega_2) = \alpha_{31}$, the perturbation of α_{21} by α_{32}, we have

$$\mathcal{T}(\phi_{\omega_2}) : \begin{cases} \alpha_{21} + \alpha_{31} \\ \alpha_{32}, \end{cases}$$

and the corresponding affine map $\phi_{\omega_2} : S_1^n \times S_2^n \to \mathbb{F}_k(\mathbb{R}^{n+1})$ is that under which

$$(\xi_1, \xi_2) \mapsto (q_1, q_1 + \xi_1, q_1 + \xi_1 + 2^{-1}\xi_2, q_3, \cdots).$$

Note that the perturbation $\lambda_{13}(\omega_3)$ of α_{32} by α_{41} is $\epsilon\alpha_{42} + \epsilon\alpha_{43}$, where $\epsilon = (-1)^{n+1}$, while $\lambda_{23}(\omega_3) = 0$ is that of α_{32} by α_{41}. Thus we obtain

$$\mathcal{T}(\phi_{\omega_3}) : \begin{cases} \alpha_{21} + \alpha_{31} + (\epsilon\alpha_{42} + \epsilon\alpha_{43}) \\ \alpha_{32} + 0 \\ \alpha_{41} \end{cases}$$

as the factor table of ω_3. The corresponding affine map is

$$\phi_{\omega_3} : S_1^n \times S_2^n \times S_3^n \to \mathbb{F}_k(\mathbb{R}^{n+1}),$$

where $(\xi_1, \xi_2, \xi_3, \xi_4)$ is mapped to

$$(q_1, q_1 + \xi_1, q_1 + \xi_1 + 2^{-1}\xi_2, q_1 + 2^{-2}\xi_3, q_4, \cdots).$$

Finally, one sees that $(\alpha_{51}+\epsilon^2\alpha_{54}), (\alpha_{51}+\epsilon^2\alpha_{54}), \alpha_{52}$, and 0 are the perturbations of $\lambda_1(\omega_3), \lambda_2(\omega_3), \lambda_3(\omega_3)$, and $\lambda_4(\omega_3)$, respectively, caused by α_{53}. Thus the factor table of ω_4 is

$$\mathcal{T}(\phi_{\omega_4}): \begin{cases} \alpha_{21}+\alpha_{31}+(\epsilon\alpha_{42}+\epsilon\alpha_{43})+(\alpha_{51}+\epsilon^2\alpha_{54}) \\ \alpha_{32}+0+\alpha_{52} \\ \alpha_{41}+0 \\ \alpha_{53}, \end{cases}$$

and its corresponding affine map

$$\phi_{\omega_4}: S_1^n \times S_2^n \times S_3^n \times S_4^n \to \mathbb{F}_k(\mathbb{R}^{n+1})$$

is that where $(\xi_1, \xi_2, \xi_3, \xi_4)$ is mapped to

$$(q_1, q_1+\xi_1, q_1+\xi_1+2^{-1}\xi_2,$$
$$q_1+2^{-2}\xi_3, q_1+\xi_1+2^{-1}\xi_2+2^{-3}\xi_4, q_5, \cdots).$$

5 Multispherical Cycles

Our aim in this section is to show that the p-fold twisted homology products in $H_*(\mathbb{F}_{k-r,r}(\mathbb{R}^{n+1}))$ are carried by the fundamental classes of suitable p-products of spheres and that they are the Kronecker dual to the cohomology base, $\alpha^*_{\underline{r},\underline{s}}$. Throughout this section homology is understood to have integral coefficients.

As in §§1 and 4 above, let, $\iota_1 \in H_n(S_1^n), \cdots, \iota_p \in H_n(S_p^n)$ be the fundamental classes of the factors of $S_1^n \times \cdots \times S_p^n$. Then $\iota_1 \times \cdots \times \iota_p$ is the fundamental class of $S_1^n \times \cdots \times S_p^n$. Put

$$S_\omega = H_{pn}(\phi_\omega)(\iota_1 \times \cdots \times \iota_p),$$

where $\phi_\omega : S_1^n \times \cdots \times S_p^n \to \mathbb{F}_{k-r,r}(\mathbb{R}^{n+1})$ is the affine map of ω, as in Theorem 3.1, and

$$\omega = \alpha_{r_1 s_1} \ltimes \cdots \ltimes \alpha_{r_p s_p}, \; r < r_1 < \cdots < r_p \leq k, \; 1 \leq p \leq (k-r)$$

is a p-fold twisted product in $H_*(\mathbb{F}_{k-r,r}(\mathbb{R}^{n+1}))$.

Theorem 5.1 *The elements $\omega^* = \alpha^*_{u_1 v_1} \cdots \alpha^*_{u_p v_p} \in H^*(\mathbb{F}_{k-r,r}(\mathbb{R}^{n+1}))$ and ω are Kronecker duals. In particular, if γ^* is a basic product, where $r < u_1 < \cdots < u_p$. Then $< S_\omega, \gamma^* > \neq 0$ if, and only if, $\gamma^* = \pm\omega^*$, where $< .,. >$ is the Kronecker product, and $\omega^* = \alpha^*_{r_1 s_1} \cdots \alpha^*_{r_p s_p}$.*

The argument is similar to that given in §1 to establish special cases. Denote by $\iota_j^* \in H^n(S_1^n \times \cdots \times S_p^n)$ the Kronecker dual of ι_j. Let

$$\mathrm{p}_r : \mathbb{F}_{k-r,r}(\mathbb{R}^{n+1}) \to \mathbb{R}^{n+1}_r$$

be the natural projection occurring in $\mathcal{F}(\mathbb{R}^{n+1})$. Denote by θ the cohomology extension of the fiber defined by the $(p-1)$-fold basic products and, by $\Phi^*_{k-r,r}, \Phi_{k-r,r}$ the corresponding Leray-Hirsch morphisms in cohomology and homology, respectively (see §2).

Assertion 5.1 *For all* $\omega^* \in H^*(\mathbb{R}^{n+1}_r)$, $u^* \in H^*(\mathbb{F}_{k-r-1,r+1}(\mathbb{R}^{n+1}))$ *and* $z \in H_*(\mathbb{F}_{k-r,r}(\mathbb{R}^{n+1}))$, *the relation*

$$< \omega^* \otimes u^*, \Phi_{k-r,r}(z) >=< \Phi^*_{k-r,r}(\omega^* \otimes u^*), z >$$

holds.

Proof. The proof is exactly the same as that of Assertion 2.1. In fact, the equations

$$\begin{aligned} < \omega^* \otimes u^*, \Phi_{k-r,r}(z) > &= < \omega^* \otimes u^*, \textstyle\sum_v H_*(\mathrm{p}_r)(v^* \cap z) \otimes v > \\ &= < \omega^*, H_*(\mathrm{p}_r)(u^* \cap z) > \\ &= < H^*(\mathrm{p}_r)(\omega^*)u^*, z > \\ &= < \Phi^*_{k-r,r}(\omega^* \otimes u^*), z > \end{aligned}$$

are valid by virtue of the general properties of the products in the expressions. □

Assertion 5.2 *Let* ω, ω^* *be as in the theorem. Then the relation* $< \omega, \omega^* >= 1$ *holds.*

Proof. The proof is by induction on p. It is certainly true when $p = 1$. Assume it is true for $(p-1)$-fold left twisted products, and write ω as $\alpha_{r_1 s_1} \ltimes \omega_1$, where $\omega_1 = \alpha_{r_2 s_2} \rtimes \cdots \rtimes \alpha_{r_p s_p}$, $r < r_1 < \cdots < r_p \leq k$, $1 \leq p \leq (k-r)$. Then $\omega_1^* = \alpha^*_{r_2 s_2} \cdots \alpha^*_{r_p s_p}$. As in Assertion 1.2, consider the expressions

$$\begin{aligned} < \alpha^*_{r_1 s_1}\omega_1^*, \alpha_{r_1 s_1} \ltimes \omega_1 > &= < H^n(\mathrm{p}_r(\alpha^*_{r_1 s_1})\theta(\omega_1^*), \alpha_{r_1 s_1} \ltimes \omega_1 > \\ &= < \Phi^*_{k-r,r}(\alpha^*_{r_1 s_1} \otimes \omega_1^*), \alpha_{r_1 s_1} \ltimes \omega_1 > \\ &= < \alpha^*_{r_1 s_1} \otimes \omega_1^*, \Phi_{k-r,r}(\alpha_{r_1 s_1} \ltimes \omega_1) > \\ &= < \alpha^*_{r_1 s_1} \otimes \omega_1^*, \alpha_{r_1 s_1} \otimes \omega_1 > \\ &= 1. \end{aligned}$$

This proves the induction step, and, thus, the Assertion is established. □

Assertion 5.3 *The relation* $< \omega^*, S_\omega >= 1$ *holds.*

Proof. On inspecting the factor table

$$\mathcal{T}(\phi_\omega) : \begin{cases} \alpha_{r_1 s_1} & + \sum_{m=2}^{p} \lambda_{1m}(\omega_m) \\ \cdots & + \cdots \\ \alpha_{r_j s_j} & + \sum_{m=j+1}^{p} \lambda_{jm}(\omega_m) \\ \cdots & + \cdots \\ \alpha_{r_p s_p} & \end{cases}$$

of ϕ_ω, one sees that $H_n(\phi_\omega)(\iota_j) = \alpha_{r_j s_j} + \sum_{m=j+1}^{p} \lambda_{jm}(\omega_m)$, where $\iota_j \in H_n(S_1^n \times \cdots \times S_p^n)$ stands for the class of the jth factor, with $1 \leq j \leq p$. Hence,

$$\begin{aligned} H^n(\phi_\omega)(\alpha^*_{r_j s_j}) &= 1 \times \cdots \times \iota_j^* \times 1 \times \cdots \times 1 \\ &= \pm H^n(\phi_\omega))(\alpha^*_{uv}) \end{aligned}$$

for all α_{uv} appearing as terms in $\lambda_j(\omega)$. The expression

$$< \alpha^*_{r_1 s_1} \cdots \alpha^*_{r_p s_p}, H_{pn}(\phi_\omega)(\iota_1 \times \cdots \times \iota_p) >$$

is equal to

$$\begin{aligned} & < H^{pn}(\phi_\omega)(\alpha^*_{r_1 s_1} \cdots \alpha^*_{r_p s_p}, \iota_1 \times \cdots \times \iota_p > \\ &= < H^n(\phi_\omega)(\alpha^*_{r_1 s_1}) \cdots H^n(\phi_\omega)(\alpha^*_{r_p s_p}), \iota_1 \times \cdots \times \iota_p > \\ &= < H^n(\phi_\omega)(\alpha^*_{r_1 s_1}), \iota_1 > \cdots << H^n(\phi_\omega)(\alpha^*_{r_p s_p}), \iota_p > \\ &= 1. \end{aligned}$$

Hence, $< \omega^*, S_\omega >= 1$. □

Next, let

$$\Gamma = \{\alpha_{u_1 v_1}, \cdots, \alpha_{u_p v_p} \mid r < u_1 < \cdots < u_p \leq k\},$$

denote the set of elements that are the Kronecker duals of the factors of γ^*.

Assertion 5.4 *Assume that $< \gamma^*, \omega > \neq 0$, and let $\alpha_{u_i v_i} \in \Gamma$. Then $\alpha_{u_i v_i}$ appears in one of the rows of the factor table $\mathcal{T}(\phi_\omega)$ of ω.*

Proof. Assume that $\alpha_{u_i v_i}$ does not appear in any of the rows of the factor table $\mathcal{T}(\phi_\omega)$ of ω. Then, arguing as above, we see that $H^(\phi_\omega)(\alpha^*_{u_i v_i}) = 0$. This clearly implies that

$$H^{pn}(\phi_\omega)(\gamma^*) = H^n(\phi_\omega)(\alpha^*_{u_1 v_1}) \cdots H^n(\phi_\omega)(\alpha^*_{u_p v_p}) = 0,$$

which contradicts the assumption that $< \gamma^*, \omega > \neq 0$. □

Assertion 5.5 *Suppose that $\alpha_{u_i v_i}, \alpha_{u_j v_j} \in \Gamma$ are distinct. Then $\alpha_{u_i v_i}, \alpha_{u_j v_j}$ cannot belong to same row of $\mathcal{T}(\phi_\omega)$.*

Proof. Assume, by way of contradiction, that $\alpha_{u_i v_i}, \alpha_{u_j v_j}$ belong to, say, the rth row. Then

$$w_r^* = H^n(\phi_\omega)(\alpha^*_{u_i v_i}) = H^n(\phi_\omega)(\alpha^*_{u_j v_j}).$$

Hence, $\gamma^* = H^n(\alpha^*_{r_1 s_1} \cdots H^*(\alpha^*_{r_p s_p}) = 0$. This is a contradiction. □

Assertion 5.6 *Suppose that $\alpha_{u_1 v_1}, \cdots, \alpha_{u_p v_p}$ belong to distinct rows of the factor table $\mathcal{T}(\phi_\omega)$ of ω, where $u_1 < u_2 < \cdots < u_p$. Then $\gamma^* = \pm\omega^*$.*

Proof. To see this, note that, as the last row of $\mathcal{T}(\phi_\omega)$ is just $\alpha_{r_p s_p}$, it must be that $\alpha_{r_p s_p} = \alpha_{u_p v_p}$. Moving to the $(p-1)^{\text{st}}$, we see, by virtue of Theorem 3.2, that this row is the sum of $\alpha_{r_{p-1} s_{p-1}}$ together with its perturbation by $\alpha_{r_p s_p}$. Thus, according to Proposition 3.1, we conclude that $\alpha_{u_{p-1} v_{p-1}}$ is either $\alpha_{r_{p-1} s_{p-1}}$ or $\pm\alpha_{r_p, s}$ for some s such that $1 \leq s < r_p$. But, $u_{p-1} < u_p = r_p$. Hence, $\alpha_{u_{p-1} v_{p-1}} = \pm\alpha_{r_{p-1} s_{p-1}}$. Proceeding in this vein, we conclude that

$$\alpha_{u_p v_p} = \pm\alpha_{r_p s_p}, \cdots, \alpha_{u_1 v_1} = \pm\alpha_{r_1 s_1}.$$

Clearly, this implies the assertion. □

Now, note that Theorem 5.1 is a consequence of the preceding Assertions. Thus the proof of the theorem is completed. □

Following the notation of Chapter V, §4, let $\alpha_{\underline{r}, \underline{s}}$ denote $\alpha_{r_1 s_1} \ltimes \cdots \ltimes \alpha_{r_p s_p}$, where $(\underline{r}, \underline{s}) \in (\mathcal{R}_p, \mathcal{S}_p)$. Observe here that the set $\mathcal{A}^{\wedge p} = \{\alpha_{\underline{r}, \underline{s}} \mid (\underline{r}, \underline{s}) \in (\mathcal{R}_p, \mathcal{S}_p)\}$ is a basis for $H_{pn}(\mathbb{F}_k(\mathbb{R}^{n+1}))$ for all $p > 0$.

Let us consider next the right twisted products on homology. Denote by

$$\omega' = \alpha_{r_1 s_1} \rtimes \cdots \rtimes \alpha_{r_p s_p}, \; 1 < r_1 < \cdots < r_p \leq k$$

the right twisted product corresponding to the right twisted product ω'. The proof of the following theorem is similar to that of Theorem 5.1 and is left to the reader.

Theorem 5.2 *Let $\omega = \alpha_{r_1 s_1} \rtimes \cdots \rtimes \alpha_{r_p s_p}$. Then the relation $S_\omega = \omega$ holds in $H_*(\mathbb{F}_k(\mathbb{R}^{n+1}))$.*

Corollary 5.1 *In the group $H_*(\mathbb{F}_k(\mathbb{R}^{n+1}))$, the left and right twisted products agree up to sign.*

Proof. Note that the left, ω, and right, ω', twisted products are carried by the multispherical class $\pm S_\omega$ and that S_ω is the Kronecker dual to the same cohomology class $\alpha^*_{r_1 s_1} \cdots \alpha^*_{r_p s_p}$. □

6 Twisted Products in $H_*(\mathbb{F}_{k+1}(S^{n+1}))$, $n+1$ Odd

The cohomology and homology groups in this section, unless stated otherwise, have integral coefficients.

In this section we define the homology twisted products in $H_*(\mathbb{F}_{k+1}(S^{n+1}))$. Consider the projection

$$\text{proj}_{k+1,1} : \mathbb{F}_{k+1}(S^{n+1}) \to S^{n+1},$$

that takes $(x_1, \cdots, x_{k+1})$ to x_{k+1}. As in Chapter III, §1, identify the fiber at $-e$ by $\mathbb{F}_k(\mathbb{R}^{n+1})$. Note that the fiber is totally nonhomologous to zero. In fact, Theorem 8.2 of Chapter V implies that $\mathcal{B}_k^{\wedge *} = \{1, \beta^*_{\underline{r},\underline{s}}\}$, where $(\underline{r},\underline{s}) \in (\mathcal{R},\mathcal{S})\}$ defines a cohomology extension of the fiber, since their restrictions to the fiber is a basis for the cohomology $H^*(\mathbb{F}_k(\mathbb{R}^{n+1}))$. Let

$$\Phi_{k+1} : H_*(\mathbb{F}_{k+1}(S^{n+1})) \to H_*(S^{n+1}) \otimes H_*(\mathbb{F}_k(\mathbb{R}^{n+1}))$$

be the corresponding Leray-Hirsch isomorphism.

Following Chapter V, §8, define the class $\delta^*_{k+1} \in H^{n+1}(\mathbb{F}_{k+1}(S^{n+1}))$ by the equation $H^{n+1}(\mathrm{proj}_{k+1,1})(\iota^*_{n+1}) = \delta^*_{k+1}$, where ι^*_{n+1} is the canonical generator of $H^{n+1}(S^{n+1})$. Observe now that the set

$$\{\delta^*_{k+1}, \beta^*_{rs}, \delta^*_{k+1}\beta^*_{rs} | 1 \leq s < r \leq k\}$$

is a basis for $H^*(\mathbb{F}_{k+1}(S^{n+1}))$, as Theorem 8.2 of Chapter V implies. We wish to define the twisted products among the duals of these generators.

We shall do so by cases. First, by induction on r and using the fundamental fiber sequence

$$\mathcal{F}_{k+1}(S^{n+1}) : \left\{ \begin{array}{cccc} \mathbb{F}_{k+1}(S^{n+1}) \leftarrow & \mathbb{F}_k & \leftarrow \cdots \leftarrow & \mathbb{F}_{k-r,r} \leftarrow \\ \downarrow & \downarrow & \downarrow \ \downarrow & \\ S^{n+1} & \mathbb{R}^{n+1} & \cdots & \mathbb{R}^{n+1}_r , \end{array} \right.$$

define the left twisted products

$$\begin{aligned} \omega &= \beta_{r_1 s_1} \ltimes \cdots \ltimes \beta_{r_p s_p} \\ &= \beta_{\underline{r},\underline{s}} \end{aligned}$$

for $(\underline{r}, \underline{s}) \in (\mathcal{R},\mathcal{S})$ in the manner of the twisted products in the module $H_*(\mathbb{F}_k(\mathbb{R}^{n+1}))$ (see §2). Thus, we have

$$\Phi_{k+1}(\beta_{r_1 s_1} \ltimes \cdots \ltimes \beta_{r_p s_p}) = 1 \otimes (\alpha_{r_2 s_2} \ltimes \cdots \ltimes \alpha_{r_p s_p}).$$

Next, define $\delta_{k+1} \ltimes \omega$ to be the element in $H_*(\mathbb{F}_{k+1}(S^{n+1}))$ such that

$$\Phi_{k+1}(\delta_{k+1} \ltimes \omega) = \delta_{k+1} \otimes \alpha_{r_1 s_1} \ltimes \cdots \ltimes \alpha_{r_p s_p},$$

where $\omega = \beta_{r_1 s_1} \ltimes \cdots \ltimes \beta_{r_p s_p}$ is a p-fold twisted product.

Consider next the set

$$\{1, \delta_{k+1}, \beta_{\underline{r},\underline{s}}, \delta_{k+1} \ltimes \beta_{\underline{r},\underline{s}}\},$$

with $(\underline{r}, \underline{s}) \in (\mathcal{R},\mathcal{S})$. Observe that it constitutes an additive basis for $H_*(\mathbb{F}_{k+1}(S^{n+1}))$. We wish to define their affine maps and multispherical maps. These depend on whether the twisted product has δ_{k+1} as a factor.

Consider the former case first. Let

$$\begin{aligned}\omega &= \beta_{\underline{r},\underline{s}}\\ &= \beta_{r_1 s_1} \ltimes \cdots \ltimes \beta_{r_p s_p}\end{aligned}$$

be a p-fold twisted product. With $i_k : \mathbb{F}_k(\mathbb{R}^{n+1}) \to \mathbb{F}_{k+1}(S^{n+1})$ being the imbedding defined in Chapter III, §1, note that there is a p-fold twisted product

$$\begin{aligned}\varpi &= \alpha_{\underline{r},\underline{s}}\\ &= \alpha_{r_1 s_1} \ltimes \cdots \ltimes \alpha_{r_p s_p}\end{aligned}$$

in $H_*(\mathbb{F}_k(\mathbb{R}^{n+1}))$ such that $H_*(i_k)(\varpi) = \omega$. The corresponding spherical map is, by definition, $\phi_\omega = (\phi^e)^{-1} \circ \phi_\varpi$, where $\phi_\varpi : S_1^n \times \cdots \times S_p^n \to \mathbb{R}^{n+1}$ is the affine map of ϖ and

$$\phi^e : S^{n+1} - \{-e\} \to V^e = \mathbb{R} \times \mathbb{R}^n = \mathbb{R}^{n+1}$$

is the stereographic projection, as in Chapter III, §1. By definition, let

$$S_\omega = H_*(\phi_\omega)(\iota_1 \times \times \cdots \times \iota_p),$$

where ι_j is the fundamental class of the jth factor.

Theorem 6.1 *The element* $\omega^* = \beta^*_{r_1 s_1} \cdots \beta^*_{r_p s_p}$ *in* $H^*(\mathbb{F}_{k+1}(S^{n+1}))$ *is the Kronecker dual to* S_ω. *Moreover,* $S_\omega = \omega$.

The proof is entirely similar to those of Theorems 5.1 and 5.2. We leave it to the reader.

Consider next the general case. As when $(n+1)$ is odd, using the notation of Chapter III, §3, it follows that the projection $p_{21} : O_{n+2,2} \to S^{n+1}$, where $(x, y) \mapsto x$ and with fiber at e being S^n, has a section $\delta_3' : S^{n+1} \to O_{n+2,2}$ (see Chapter III, §3). The fiber is totally nonhomologous to zero, and, hence, we a have a Leray-Hirsch isomorphism

$$\Phi_3 : H_*(O_{n+2,2}) \to H_*(S^{n+1}) \otimes H_*(S^n).$$

Denote by the same symbol, β_n, the homology class in $H_n(O_{n+2,2})$ corresponding to the homotopy class $\beta_{21} \in \pi_n(O_{n+2,2}) \cong \pi_n(\mathbb{F}_3(S^{n+1}))$ (see Chapter III, §2). Also, let $\delta_3 \in H_{n+1}(O_{n+2,2})$ be the class of the section $\delta_3' : S^{n+1} \to O_{n+2,2}$. Define $\delta_3 \ltimes \beta_n$ to be the element such that

$$\Phi_3(\delta_3 \ltimes \beta_n) = \delta_3 \otimes \beta_n.$$

It is easy to verify that the element $\mu = \delta_3 \ltimes \beta_n$ is a fundamental class of $O_{n+2,2}$.

Consider now the twisted product is of the form

$$\begin{aligned}\omega &= \delta_{k+1} \ltimes \omega'\\ &= \delta_{k+1} \ltimes \beta_{21} \ltimes \varpi.\end{aligned}$$

We need certain maps and constructions introduced in Chapter III. Following Proposition 1.2 of that chapter, let $\psi_{k+1} : O_{n+2,2} \to \mathbb{F}_{k+1}(S^{n+1})$ be the map defined by the rule where

$$(x, y) \mapsto (x, q_1^{xy}, q_2^{xy}, \cdots, q_k^{xy}, -x).$$

Also, recall that the map $\phi^x : S^{n+1} - \{-x\} \to V^x$ is the stereographic projection onto the plane orthogonal to x. It takes the great circle cut by (x, y) onto a ray $\mathbb{R}^{xy} \subset V^x$. The set $\{x, q_1^{xy}, \cdots, q_k^{xy}\}$ on this ray corresponds to the set $\{q_0, q_1, \cdots, q_{k+1}\}$ on the line $\mathbb{R} \subset \mathbb{R}^\times \mathbb{R}^n = \mathbb{R}^{n+1}$. Now, proceed by induction on the length of the partial products of ω, using the bowtie construction in the manner of the definition of the affine map ϕ_ω (see Theorem 3.1) to define the map $\psi_{\omega,xy} : S_2^n \times \cdots \times S_p^n \to \mathbb{F}_k(V^x)$, where $\mathbb{F}_k(V^x)$ stands for the configuration space of k particles in V^x. Now, put

$$\psi_\omega : O_{n+2,2} \times (S_2^n \times \cdots \times S_p^n) \to \mathbb{F}_{k+1}(S^{n+1}),$$

where $(x, y, \xi_1, \cdots, \xi_p) \mapsto (x, (\phi^x)^{-1}(\psi_{\omega,xy}(\xi_1, \cdots, \xi_p,)) - x)$. The map ψ_ω is called the *spherical map* of ω.

By definition, let

$$S_\omega = H_*(\psi_\omega)(\mu \times \iota_2 \times \cdots \times \iota_p),$$

where $\mu \times (\iota_2 \times \cdots \times \iota_p)$ is the cross product of the fundamental class $\mu = \delta_3 \rtimes \beta_n$ of $O_{n+2,2}$, with the fundamental classes of the other factors S_i, with $2 \leq i \leq p$, of $S_{2,\cdots p}$.

Theorem 6.2 *Suppose that*

$$\begin{aligned} \omega &= \delta_{k+1} \ltimes \omega' \\ &= \delta_{k+1} \ltimes \beta_{21} \ltimes \varpi. \end{aligned}$$

Then, the relation $S_\omega = \omega$ holds in $H_*(\mathbb{F}_{k+1}(S^{n+1}))$.

Proof. The proof is similar to that of Theorem 5.1. Consider the factor table

$$T(\phi_\omega) : \begin{cases} \delta_{k+1} & \\ \beta_n & + \sum_{m=3}^{k} \lambda_m \\ \beta_{r_2 s_2} & + \sum_{m=3}^{p} \lambda_{2m}(\omega_m) \\ \cdots & + \cdots \\ \beta_{r_j s_j} & + \sum_{m=j+1}^{p} \lambda_{jm}(\omega_m) \\ \cdots & + \cdots \\ \beta_{r_p s_p} & \end{cases}$$

of ψ_ω, where $\lambda_m(\omega_m) = \sum_{s=1}^{s=m-1} \beta_{ms}$, as the restriction of ψ_ω to the first factor is just the Dirac class $\Delta_{k+1}(S^{n+1})$, while its restriction to S^{n+1} is δ_{k+1} (see Corollary 6.1 of Chapter III). Now, proceed along the lines of the proof of Theorem 5.2, using the factor table above to show that the left and right hand sides of the statement in the theorem have the same image under the Leray-Hirsch isomorphism. □

Finally, consider the case when

$$\begin{aligned}\omega &= \delta_{k+1} \ltimes \varpi \\ &= \delta_{k+1} \ltimes \beta_{r_2 s_2} \ltimes \cdots \ltimes \beta_{r_p s_p},\end{aligned}$$

where $r_2 > 2$. Put

$$S_\omega = H_*(\phi_\omega)(\iota \times \iota_2 \times \cdots \times \iota_p),$$

where ι stands for $\beta_n \in H_n(O_{n+2,2}))$, and ι_j, with $2 \le j \le p$, is the fundamental class of the factor, S_j, of $S_2^n \times \cdots S_p^n$.

Theorem 6.3 *Suppose that* $\omega = \delta_{k+1} \ltimes \beta_{r_2 s_2} \ltimes \cdots \ltimes \beta_{r_p s_p}$, *with* $r_2 > 2$. *Then the relation* $S_\omega = \omega$ *holds in* $H_*(\mathbb{F}_{k+1}(S^{n+1}))$.

The proof is similar to that of the preceding theorem and is left to the reader.

7 Twisted Products in $H_*(\mathbb{F}_{k+1}(S^{n+1}))$, $n+1$ Even

Assume that $(n+1)$ is even. Our aim in this section is to define the homology twisted on $H_*(\mathbb{F}_{k+1}(S^{n+1}))$, using the twisted-product structure

$$O_{n+2,2} \ltimes \mathbb{F}_{k-2,2}(\mathbb{R}^{n+1}) \simeq \mathbb{F}_{k+1}(S^{n+1})$$

of Chapter III, §1. It is an extension of that defined in §2 on $H_*(\mathbb{F}_{k-2,2}(\mathbb{R}^{n+1}))$. Consequently, we obtain a set of generators for the homology when $(n+1)$ is even, and, for each such element, we define a multispherical carrier in the manner of §§3 and 6.

The fiber of the projection

$$\mathrm{proj}_{k+1,3} : \mathbb{F}_{k+1}(S^{n+1}) \to \mathbb{F}_3(S^{n+1}),$$

where $(x_1, x_2 \cdot \cdots, x_{k+1}) \mapsto (x_1, x_2, -x_1)$ is homeomorphic to $\mathbb{F}_{k-2,2}(\mathbb{R}^{n+1})$ (see Chapter III, §1). Now, according to Theorem 6.1 of Chapter V, the basis

$$\mathcal{B}^{\wedge *}_{k-2,2} = \{1, \beta^*_{\underline{r},\underline{s}} \mid \in (\mathcal{R}, \mathcal{S}), r_1 > 3\},$$

where r_1 is the first entry of $\underline{r}$, defines a cohomology extension of the fiber and, in turn, leads to the Leray-Hirsch isomorphism

$$\Phi_{k+1,3} : H_*(\mathbb{F}_{k+1}(S^{n+1}); \mathbb{Z}) \to H_*(\mathbb{F}_3(S^{n+1}); \mathbb{Z}) \otimes H_*(\mathbb{F}_{k-2}(\mathbb{R}_2^{n+1}); \mathbb{Z}).$$

Following Proposition 1.2 of Chapter III , let $\psi_{k+1} : O_{n+2,2} \to \mathbb{F}_{k+1}(S^{n+1})$ be the map defined by the rule, where

$$(x, y) \mapsto (x, q_1^{xy}, q_2^{xy}, \cdots, q_k^{xy}, -x).$$

For $k = 2$, the map $\psi_3 : O_{n+2,2} \to \mathbb{F}_3(S^{n+1})$ is a homotopy equivalence (see Proposition 1.1 of Chapter III). Replacing the module $H_*(\mathbb{F}_3(S^{n+1}); \mathbb{Z})$ by $H_*(O_{n+2,2}; \mathbb{Z})$ in the expression given above, we obtain the Leray-Hirsch isomorphism

$$\Phi_{k+1,3} : H_*(\mathbb{F}_{k+1}(S^{n+1}); \mathbb{Z}) \to H_*(O_{n+2,2}; \mathbb{Z}) \otimes H_*(\mathbb{F}_{k-2}(\mathbb{R}^{n+1}_2); \mathbb{Z}).$$

Here, for simplicity, we indicate the latter by the same symbol.

Now, let

$$\begin{aligned} \omega &= \beta_{\underline{r}, \underline{s}} \\ &= \beta_{r_2 s_2} \ltimes \ltimes \cdots \ltimes \beta_{r_p s_p} \end{aligned}$$

be a $(p-1)$-fold twisted product in $H_*(\mathbb{F}_{k-2,2}(\mathbb{R}^{n+1}); \mathbb{Z})$.

Definition 7.1 *Let $\xi \ltimes \omega$ be the element in $H_*(\mathbb{F}_{k+1}(S^{n+1}); \mathbb{Z})$ such that*

$$\Phi_{k+1,3}(\xi \ltimes \omega) = \xi \otimes \omega$$

for $\xi \in H_*(O_{n+2,2})$.

Denote by the same symbol, β_n, the homology class in $H_n(O_{n+2,2}; \mathbb{Z})$ corresponding to the homotopy class $\beta_{21} \in \pi_n(O_{n+2,2})$ (see Chapter III, §2).

Consider again the map

$$\psi_{k+1} : O_{n+2,2} \to \mathbb{F}_{k+1}(S^{n+1})$$

defined in the proof of Proposition 1.2 of Chapter III. Now, put

$$\psi_\omega : O_{n+2,2} \times (S^n_2 \times \cdots \times S^n_p) \to \mathbb{F}_{k+1}(S^{n+1}),$$

where $(x, y, \xi_1, \cdots, \xi_p) \mapsto (x, (\phi^x)^{-1}(\psi_{\omega,xy}(\xi_1, \cdots, \xi_p,)) - x)$. We shall call the map ψ_ω the *multispherical map* corresponding to $\omega \in H_*(\mathbb{F}_{k-2,2})$.

Assume that homology and cohomology have coefficients in $\mathbb{Z}/2\mathbb{Z}$. Then, β_n generates $H_n(O_{n+2,2})$, and $\delta_{k+1} = \Delta_{k+1}(S^{n+1})$, the Dirac class mod 2, generates $H_n(\mathbb{F}_{k+1}(S^{n+1}))$ (see Theorem 6.1 of Chapter III). Moreover, the fundamental class of $O_{n+2,2}$ is $\delta_{k+1} \ltimes \beta_n$. Denote by ξ, the class δ_{k+1}, β_n, or $\delta_{k+1} \ltimes \beta_n$. By definition, let

$$H_*(\psi_\omega)(\xi \times \iota_2 \times \cdots \times \iota_p) = S_{\xi \ltimes \omega}.$$

Theorem 7.1 *Suppose that $(n+1)$ is even. Then the relation $S_{\xi \ltimes \omega} = \xi \ltimes \omega$ holds in $H_*(\mathbb{F}_{k+1}(S^{n+1}); \mathbb{Z}/2\mathbb{Z})$, if $\xi = \Delta_{k+1}(S^{n+1}), \beta_n$, or $\delta_{k+1} \ltimes \beta_n$, where $\omega = \beta_{r_2 s_2} \ltimes \ltimes \cdots \ltimes \beta_{r_p s_p}$.*

Proof. Assume that $\xi = \beta_n$. The proof is similar to that of Theorem 6.1. One shows that the left and right hands side of the equality have the same image under the Leray-Hirsch isomorphism, using the factor table

$$\mathcal{T}(\phi_\omega) : \begin{cases} \beta_n & + \sum_{m=3}^{k} \lambda_m(\omega_m) \\ \beta_{r_2 s_2} & + \sum_{m=3}^{p} \lambda_{2m}(\omega_m) \\ \cdots & + \cdots \\ \beta_{r_j s_j} & + \sum_{m=j+1}^{p} \lambda_{jm}(\omega_m) \\ \cdots & + \cdots \\ \beta_{r_p s_p} \end{cases}$$

of ψ_ω, where $\lambda_m(\omega_m) = \sum_{s=1}^{s=m-1} \beta_{ms}$, since the restriction of ψ_ω to the first factor is just the Dirac class $\Delta_{k+1}(S^{n+1})$ (see Corollary 6.1 of Chapter III). Now proceed along the lines of the proof of Theorem 6.1. □

By definition, let

$$H_*(\psi_\omega)(\mu \times \iota_2 \times \cdots \times \iota_p) = S_{\mu \ltimes \omega},$$

where homology is assumed to be with rational coefficients.

Theorem 7.2 *Suppose that* $(n+1)$ *is even. Then the relation*

$$S_{\mu \ltimes \omega} = \mu \ltimes \omega$$

holds in $H_*(\mathbb{F}_{k+1}(S^{n+1}); \mathbb{Q})$, *where* $\omega = \beta_{r_2 s_2} \ltimes \cdots \ltimes \beta_{r_p s_p}$.

The proof is similar to that of the preceding theorem. It is left to the reader.

8 The Cellular Structure of $\mathbb{F}_k(\mathbb{R}^{n+1}), n > 1$

Following the notation of Chapter V, §4, let $\alpha_{\underline{r}, \underline{s}}$ denote $\alpha_{r_1 s_1} \ltimes \cdots \ltimes \alpha_{r_m s_m}$, where $(\underline{r}, \underline{s}) \in (\mathcal{R}_m, \mathcal{S}_m)$. Recall that

$$\mathcal{A}^{\wedge m} = \{\alpha_{\underline{r}\,\underline{s}} | (\underline{r}, \underline{s}) \in (\mathcal{R}, \mathcal{S}),\ |\underline{r}| = m\}$$

is a basis for $H_{mn}(\mathbb{F}_k(\mathbb{R}^{n+1}); \mathbb{Z})$. Our aim in this section is to describe a cellular complex X_k of the homotopy type of $\mathbb{F}_k(\mathbb{R}^{n+1})$. The mn-cells of X_k will be in one-one correspondence with the elements of $\mathcal{A}^{\wedge m}$. For brevity, put

$$e_\omega = e_{r_1 s_1} \times \cdots \times e_{r_m s_m}, \ |\omega| = m,$$

where $\omega = \alpha_{\underline{r}\,\underline{s}}$. We shall follow the pattern of the constructions in §1.

Theorem 8.1 *There is a* CW*-complex* X_k *and a homotopy equivalence* $h_k : X_k \to \mathbb{F}_k(\mathbb{R}^{n+1})$ *with the following properties:*

(*i*) *for all* $m \geq 0$, $X^{(mn)} - X^{(m-1)n} = \coprod_\omega = e_{r_1s_1} \times \cdots \times e_{r_ms_m}$, *where* $\omega \in \mathcal{A}^{\wedge m}$ *and* $|\omega| = m$, *and*

(*ii*) *for each* $\omega \in \mathcal{A}^{\wedge m}$ *there is a map* $\chi_\omega : S_1^n \times \cdots S_m^n \to X_k$ *such that* $h_{k,m} \circ \chi_\omega \simeq \phi_\omega$, *where* $h_{k,m} = h_k|X^{(mn)}$, *and* ϕ_ω *is the affine map corresponding to* ω.

Proof. Write the Poincaré polynomial of $\mathbb{F}_k(\mathbb{R}^{n+1})$ in the form

$$\begin{aligned} \mathrm{P}(\mathbb{F}_k(\mathbb{R}^{n+1}), T) &= \bigotimes_{t=2}^{k}(1 + (t-1)T^n) \\ &= \sum_m \nu_m T^{mn)}, \end{aligned}$$

where $\nu_m = \operatorname{rank} H_{mn}(\mathbb{F}_k(\mathbb{R}^{n+1}); \mathbb{Z})$ (see Corollary 1.4 of Chapter V). Now, we define the complex X inductively. The number of mn-cells will be ν_m.

To start, denote the bouquet of n-spheres $\bigvee_{s=1}^{r-1} S_{rs}$ by S_{r*}, where S_{rs} is the n-sphere corresponding to α_{rs} (see Proposition 1.1 of Chapter II). Define the n-skeleton $X^{(n)}$ of X to be

$$\begin{aligned} X^{(n)} &= \bigvee_{2=1}^{k} S_{r*}, \\ &= S_{21} \vee (S_{31} \vee S_{32}) \vee \cdots \vee (S_{k1} \vee \cdots \vee S_{kk-1}). \end{aligned}$$

Clearly, $X^{(n)} = e_0 \cup (\bigcup_{r,s} e_{rs})$, where each e_{rs} is an n-cell and (r, s) ranges over all pairs $r, s)$ such that $\alpha_{rs} \in \mathcal{A}^{\wedge 1}$. The elements α_{rs}, $1 \leq s < r \leq k$ together induce a map

$$h_{k,n} : X^{(n)} \to \mathbb{F}_k(\mathbb{R}^{n+1}),$$

where $h_{k,n}|S_{rs}$ is in the homotopy class of the α_{rs}. The map $h_{k,n}$ induces an isomorphism on homology through dimension $2n-1$ and, hence, is $2n-1$-connected. For each $\omega = \alpha_{rs} \in \mathcal{A}^{\wedge 1}$, define χ_ω to be $h_{k,n}|S_{rs}$.

Proceeding inductively, suppose that, for $1 \leq j < m$, the $(j-1)n$-skeleton $X^{(j-1)n}$ and the maps

$$h_{k,(j-1)n} : X^{(j-1)n} \to \mathbb{F}_k(\mathbb{R}^{n+1}),\ \chi_\omega : S_1 \times \cdots \times S_j \to \mathbb{F}_k(\mathbb{R}^{n+1})$$

are defined so that

$$\begin{aligned} &(i)\ h_{k,(j-1)n} \text{ is } jn-1\text{-connected, and} \\ &(ii)\ \chi_\omega \simeq \phi_\omega, \end{aligned}$$

where $\omega \in \mathcal{A}^{\wedge j}$, and ϕ_ω is the corresponding affine map. To define X^{mn}, regard $S_1 \times \cdots \times S_m$ as the m-fold product

$$(e^0 \cup e^n) \times \cdots \times (e^0 \cup e^n)$$

of the CW-complex $S = e^0 \cup e^n$. Let $\omega \in \mathcal{A}^{\wedge m}$, and consider the affine map

$$\phi_\omega : S_1^n \times \cdots \times S_m^n \to \mathbb{F}_k(\mathbb{R}^{n+1}).$$

Put $W_m = \{(x_1, \cdots, x_m) | x_i = e^0, \text{ for some } i\}$, and note that, as $h_{k,(m-1)n}$ is $mn-1$-connected, it follows that the map ϕ_ω is homotopic to a map ϕ'_ω, where $\phi'_\omega | W_m$ can be factored through $X^{(m-1)n}$. Define

$$X^{mn} = X^{(m-1)n} \cup_\omega \cup_{f_\omega} D^{(mn)}_\omega, \; \omega \in \mathcal{A}^{\wedge m},$$

where $f_\omega = (\phi'_\omega) \circ \eta_m$, with η_m being the attaching map of the mn-dimensional cell of $S_1 \times \cdots \times S_m$. Put

$$h_{k,mn} : X^{mn} \to \mathbb{F}_k(\mathbb{R}^{n+1})$$

equal to the natural map defined by $h_{k,(m-1)}n$ and the maps ϕ'_ω, and let

$$\chi_\omega : S_1 \times \cdots S_m \to X^{(mn)}_k$$

be that defined by ϕ'_ω. It is easy to verify that $h_{k,mn} \circ \chi_\omega \simeq \phi_\omega$. This finishes the inductive step and, hence, completes the proof. □

Consider the fundamental fiber sequence

$$\mathcal{F}_k(\mathbb{R}^{n+1}) : \left\{ \begin{array}{cccc} \mathbb{F}_k & \leftarrow \mathbb{F}_{k-1,1} \leftarrow & \cdots \leftarrow & \mathbb{F}_{k-r,r} \leftarrow \\ \downarrow & \downarrow & \downarrow & \downarrow \\ \mathbb{R}^{n+1} & \mathbb{R}^{n+1}_1 & \cdots & \mathbb{R}^{n+1}_r \end{array} \right.$$

of $\mathbb{R}^{n+1}$, introduced in Chapter II, §1. We wish to find an equivalent filtration where the spaces are replaced by homotopy equivalent CW-complexes and the fibrations, by fiber sequences. We shall proceed along the lines of the second construction in §1.

Theorem 8.2 *There is a filtered* CW*-complex* E_k*,*

$$\mathrm{E}_k = \mathrm{E}_{k,0} \supset \mathrm{E}_{k-1,1} \supset \cdots \supset \mathrm{E}_{1,k-1},$$

where the cells of $\mathrm{E}_{k-r,r}$ *are enumerated by the basic products* $\mathcal{A}^{\wedge k-r,r}$*. The complexes fit together in the commutative diagram*

$$\mathcal{E}_k(\mathbb{R}^{n+1}) : \left\{ \begin{array}{cccc} \cdots \leftarrow & \mathrm{E}_{k-rr} \leftarrow & \mathrm{E}_{k-r-1,r+1} \leftarrow & \cdots \\ \downarrow & \downarrow & \downarrow & \downarrow \\ \cdots & S_{r+1,*} & S_{r+2,*} & \cdots, \end{array} \right.$$

where, as above, $S_{r*} =$ *stands for the bouquet of* n*-spheres* $S_{r1} \vee \cdots \vee S_{rr-1}$*, the sequences* $\mathrm{E}_{k-r-1,r+1} \to \mathrm{E}_{k-r,r} \to S_{r+1,*}$ *are fiber sequences, and the diagram* $\mathcal{E}_k(\mathbb{R}^{n+1})$ *is homotopy equivalent to the diagram* $\mathcal{F}_k(\mathbb{R}^{n+1})$*.*

Proof. To start, put $\mathrm{E}_{1,k-1} = S_{k*}$, and let

$$g_{1,k-1} : \mathrm{E}_{1,k-1} \to \mathbb{F}_{1,k-1} = \mathbb{R}^{n+1}_{k-1}$$

be the map defined by the condition that $g_{1,k-1} \mid S_{ks} = \alpha_{ks}$. The natural cell structure $S^n = e^0 \cup e^n$ endows $\mathrm{E}_{1,k-1}$ with the structure of a CW-complex with cells enumerated by $\{\alpha_{k1}, \cdots, \alpha_{kk-1}\}$. Clearly, $g_{1,k-1}$ is a homotopy equivalence.

Suppose that we have defined a CW-complex $\mathrm{E}_{k-r-1,r+1}$ and a homotopy equivalence $g_{k-r-1,r+1} : \mathrm{E}_{k-r-1,r+1} \to \mathbb{F}_{k-r-1,r+1}$ so that

$$\begin{array}{ccccccc} & & \mathrm{E}_{k-r-1,r+1} & \leftarrow & \cdots & \leftarrow & \mathrm{E}_{1,k-1} \\ & & \downarrow & & \downarrow & & \downarrow \\ \mathbb{F}_{k-r,r} & \leftarrow & \mathbb{F}_{k-r-1,r+1} & \leftarrow & \cdots & \leftarrow & \mathbb{F}_{1k-1} \\ \downarrow & & \downarrow & & \downarrow & & \downarrow \\ \mathbb{R}^{n+1}_r & & \mathbb{R}^{n+1}_{r+1} & & \cdots & & \mathbb{R}^{n+1}_{k-1} \end{array}$$

is homotopy commutative and

$$\mathrm{E}_{k-r-1,r+1} - \mathrm{E}_{k-r-2,r+2} = \coprod e_\omega$$

for all $\omega \in \{\alpha_{\underline{r}\,\underline{s}} | r_1 = r+2\}$, where the indices $(\underline{r}\,\underline{s})$ range over all admissible pairs $\underline{r}\,\underline{s}$ in $(\mathcal{R}, \mathcal{S})$ such that $m \leq (k-r-1)$, $r_1 = (r+2)$, with r_1 being the first entry in $\underline{r} = (r_1, \cdots, r_m)$.

To establish the induction step, define $\mathrm{E}'_{k-r,r}$ to be the pull-back

$$\mathrm{p}_r : \mathrm{E}'_{k-r,r} \to S_{r+1,*}$$

of $\mathrm{p}_r : \mathbb{F}_{k-r,r} \to \mathbb{R}^{n+1}_r$ by the map $S_{r+1,*} \to \mathbb{R}^{n+1}_r$. Then, on replacing the fiber $\mathbb{F}_{k-r-1,r+1}$ by the homotopy equivalent complex $\mathrm{E}_{k-r-1,r+1}$, we obtain a homotopy equivalent quasi fibration

$$\mathrm{p}_r : \mathrm{E}_{k-r,r} \to S_{r+1,*}$$

(cf. Assertion 1.5) and a fiber-preserving homotopy equivalence $g_{k-r,r} : \mathrm{E}_{k-r,r} \to \mathbb{F}_{k-r,r}(\mathbb{R}^{n+1})$. The fibrations admit sections, and, on identifying the image in $\mathrm{E}_{k-r,r}$ with S_{r*}, we can arrange it so that $g_{k-r,r}$ is also section preserving. The space $\mathrm{E}_{k-r,r}$ admits a natural cell structure that contains S_{r*} as a subcomplex. The cells of $\mathrm{E}_{k-r,r}$ consist of the cells of $\mathrm{E}_{k-r-1,r+1} \vee S_{r*}$, as well as the product cells of the base S_{r*} by those of the fiber $\mathrm{E}_{k-r-1,r+1}$. The cells are clearly enumerated by $\mathcal{A}^{\wedge k-r,r} = \{\alpha_{\underline{r}\,\underline{s}} | r_1 \geq r+1\}$, since this set constitutes a basis for $H_*(\mathbb{F}_{k-r,r}; \mathbb{Z})$.

Let $\omega \in \mathcal{A}^{\wedge k-r,r}$. Note that there is a map

$$\chi_\omega : S_1 \times \cdots \times S_m \to \mathrm{E}_{k-r,r}$$

such that $g_{k-r,r} \circ \chi_\omega \simeq \phi_\omega$. Proceeding by induction on m, in the manner of the proof of Theorem 8.1, one can replace $\mathrm{E}_{k-r,r}$ by a homotopy equivalent CW-complex the cells of which are the images of the mn-dimensional cell in $S_1 \times \cdots \times S_m$, with $\chi_\omega \circ \eta$ as characteristic maps. The proof of the theorem is thereby complete. □

9 The Cellular Structure of $\mathbb{F}_{k+1}(S^{n+1})$

Assume, first, that $(n+1)$ is odd. Let $X_k = \coprod_{\underline{r},\underline{s}} e_{\underline{r},\underline{s}}$ be the CW-complex $\simeq \mathbb{F}_k(\mathbb{R}^{n+1})$ of §8. Recall that its cells are indexed by the set $\cup_m \mathcal{A}^{\wedge m}$, while their attaching maps are the corresponding spherical maps $\phi_{\beta_{\underline{r},\underline{s}}}$. Identify S^{n+1} with its image in $\mathbb{F}_{k+1}(S^{n+1})$ under the section $\delta' : S^{n+1} \to \mathbb{F}_{k+1}(S^{n+1})$ (see Chapter III, §4), and put $\Gamma = \cup_m \{1, \delta_{k+1} \times \mathcal{A}^{\wedge m}\}$.

Theorem 9.1 *There is a CW-complex $Y_{k+1} = \coprod_\omega e_\omega$, with $\omega \in \Gamma$, that contains $S^{n+1} \vee X_k$ as a subcomplex. Moreover, there is a map $\mathrm{p}_{k+1} : Y_{k+1} \to S^{n+1}$ that fits in the commutative diagram*

$$\begin{array}{ccccc} X_k & \to & Y_{k+1} & \to & S^{n+1} \\ \downarrow & & \downarrow & & \downarrow \\ \mathbb{F}_k(\mathbb{R}^{n+1}) & \to & \mathbb{F}_{k+1}(S^{n+1}) & \to & S^{n+1}, \end{array}$$

where the vertical maps are homotopy equivalences.

Proof. As in §8, we proceed by induction on the length of the basic elements in Γ. The construction of Y is similar: the only novelty is the presence of basis elements of the form

$$\omega = \delta_{k+1} \rtimes \beta_{21} \rtimes \beta_{r_2,s_2} \cdots \rtimes \beta_{r_p s_p},$$

with spherical maps

$$\psi_\omega : O_{n+2,2} \times S_2^n \times \cdots \times S_p^n \to \mathbb{F}_{k+1}(S^{n+1}).$$

In these cases, one uses the cellular decomposition $O_{n+2,2} = (S^n \cup S^{n+1}) \cup e^{2n+1}$, where the characteristic map of e^{2n+1} is like that given in Proposition 5.1 of Chapter III. Then, the multispherical class that realizes ω is obtained from $O_{n+2,2} \times e_{\underline{r}',\underline{s}'}$, where $\underline{r}' = (r_2, \cdots, r_p)$, $\underline{s}' = (s_2, \cdots, s_p)$. The details are left for the reader. □

Assume next that $(n+1)$ is even. The cellular structure for $\mathbb{F}_{k+1}(S^{n+1})$ for this case is constructed in the manner of that for the previous case. First, note that $H_*(\mathbb{F}_{k+1}(S^{n+1}))$ is generated by

$$\Gamma = \{1, \tilde{\Delta}_{k+1}, \mu\} \ltimes \{1, \mathcal{B}_{k-2,2}\},$$

where $\mu \in H_{2n+1}(O_{n+2,2}; \mathbb{Z})$ is the fundamental class (see the remark before Theorem 7.1).

Theorem 9.2 *There is a CW-complex $Y_{k+1} = \coprod_\omega e_\omega$, with $\omega \in \Gamma$, that contains $O_{n+2,2} \vee X_{k-2,2}$ as a subcomplex. Moreover, there is a map $\mathrm{p}_{k+1} : Y_{k+1} \to S^{n+1}$ that fits in the commutative diagram*

$$\begin{array}{ccccc} X_{k-2,2} & \to & Y_{k+1} & \to & S^{n+1} \\ \downarrow & & \downarrow & & \downarrow \\ \mathbb{F}_{k-2,2}(\mathbb{R}^{n+1}) & \to & \mathbb{F}'_{k+1}(S^{n+1}) & \to & O_{n+2,2}, \end{array}$$

where the vertical maps are homotopy equivalences.

The details are left for the reader.

There is another method that proceeds in the manner of Theorem 8.2 to deal with the case when $n+1$ is even. Recall that there is bundle equivalence

$$\begin{array}{ccc} O_{n+2,2} \times_{O_n} \mathbb{F}_k(\mathbb{R}^{n+1}) & \longrightarrow & \mathbb{F}_{k+1}(\mathbb{R}^{n+1}) \\ \downarrow & & \downarrow \\ S^{n+1} & \xrightarrow{\text{id}} & S^{n+1} \end{array}$$

(see Corollary 3.1 of Chapter I). Now, as in Theorem 8.2 (cf. Assertion 1.5), we can replace the fiber $\mathbb{F}_k(\mathbb{R}^{n+1})$ by the homotopy equivalent CW-complex X_k of Theorem 8.1 to obtain the commutative diagram

$$\begin{array}{ccc} O_{n+2,2} \times_{O_n} X_k & \longrightarrow & \mathbb{F}_{k+1}(\mathbb{R}^{n+1}) \\ \downarrow & & \downarrow \\ S^{n+1} & \xrightarrow{\text{id}} & S^{n+1}, \end{array}$$

where the vertical map on the right is a quasi fibration. Thus we obtain the cellular decomposition

$$\begin{aligned} O_{n+2,2} \times_{O_n} X_k &= X_k \cup_{\tilde{\mu}} D^{n+1} \times X_k \\ &= X_k \cup e^{n+1} \cup \coprod e_\omega, \end{aligned}$$

where $\tilde{\mu} : S^n \times X_k \to X_k$ is the map induced by the map $\mu : S^n \times S^n \to S^n$ of Proposition 6.3 of Chapter III. The cells correspond to

$$\{1, \Delta(S^{n+1})\} \times \{\omega | \omega \in \mathcal{B}^{\wedge}_{k-2,2}\},$$

and the corresponding multispherical maps are

$$\psi_\omega : O_{n+2,2} \times S^n_2 \times \cdots \times S^n_m \to \mathbb{F}_{k+1}(S^{n+1})$$

(see Theorem 7.1). Now one modifies the attaching maps, turning

$$O_{n+2,2} \times_{O_n} X_k$$

into a CW-complex. Thus we obtain the following theorem.

Theorem 9.3 *There is a* CW*-complex* $\mathrm{E}_{k+1} = \coprod_\omega e_\omega$*, with* $\omega \in \Gamma$*, that contains* X_k *as a subcomplex. Moreover, there is a map* $\mathrm{p}_{k+1} : \mathrm{E}_{k+1} \to S^{n+1}$ *that fits in the commutative diagram*

$$\begin{array}{ccccc} X_k & \to & \mathrm{E}_{k+1} & \to & S^{n+1} \\ \downarrow & & \downarrow & & \downarrow \\ \mathbb{F}_k(\mathbb{R}^{n+1}) & \to & \mathbb{F}_{k+1}(S^{n+1}) & \to & S^{n+1}, \end{array}$$

where the vertical maps are homotopy equivalences.

It is easy to see that Y_{k+1} and E_{k+1} are homotopy equivalent.

10 The Cellular Structure of $\mathbb{F}_k(\mathbb{R}^2)$

The space $\mathbb{F}_k(\mathbb{R}^2)$ is not simply connected, and the methods above need be modified accordingly. The complex X_k is defined in the manner of the complex constructed in Chapter IV for the presentation of the pure braid group.

Consider the diagram

$$\mathcal{F}_k(\mathbb{R}^2) : \left\{ \begin{array}{lllllll} \mathbb{F}_k & \leftarrow \mathbb{F}_{k-1,1} & \leftarrow \cdots & \leftarrow \mathbb{F}_{k-r,r} & \leftarrow \\ \downarrow & \quad\downarrow & \quad\downarrow & \quad\downarrow & \\ \mathbb{R}^2 & \quad\mathbb{R}^2_1 & \quad\cdots & \quad\mathbb{R}^2_r & \end{array} \right.$$

introduced in Chapter II, §1. The vertical maps are the projections on the last entry.

Theorem 10.1 *There is a filtered CW-complex* X_k*,*

$$X_k = X_{k0} \supset X_{k-11} \supset \cdots \supset X_{1k-1},$$

and homotopy equivalences $\psi_{k-rr} : X_{k-rr} \to \mathbb{F}_{k-rr}$ *such that the diagram*

$$\begin{array}{lllll} X_{k0} & \leftarrow X_{k-11} & \leftarrow \cdots & X_{k-rr} & \cdots \\ \downarrow & \quad\downarrow & & \quad\downarrow & \\ \mathbb{F}_{k0} & \leftarrow \mathbb{F}_{k-11} & \leftarrow \cdots & \mathbb{F}_{k-rr} & \cdots \\ \downarrow & \quad\downarrow & & \quad\downarrow & \\ \mathbb{R}^2_0 & \quad\mathbb{R}^2_1 & \quad\cdots & \mathbb{R}^2_r & \end{array}$$

is homotopy commutative, where the maps from the second row to the third are the usual projections. Moreover, $X_{k-rr} - X_{k-r-1r+1} = \coprod e_{\underline{r}\,\underline{s}}$*, where* $(\underline{r}\,\underline{s})$ *ranges over all admissible pairs such that* $r_1 \geq (r+1)$*, where* r_1 *is the first entry in* $\underline{r} = (r_1, \cdots, r_m)$.

Proof. We start by defining $X_{1k-1} = S_{k1} \vee \cdots \vee S_{kk-1}$, where each S_{kj} is a sphere of dimension 1, and by letting $\psi_{1k-1} : X_{1k-1} \to \mathbb{F}_1(\mathbb{R}^{n+1}_{k-1})$ be the map defined by the condition that $\psi_{1k-1} \mid S_{ks} = \alpha_{ks}$. It is clear that ψ_{1k-1} is a homotopy equivalence.

Suppose that we have defined $X_{k-r-1r+1}$ and $\psi_{k-r-1r+1}$ so that

$$\begin{array}{ccccc} & X_{k-r-1r+1} & \leftarrow \cdots \leftarrow & X_{1k-1} \\ & \downarrow & & \downarrow \\ \mathbb{F}_{k-rr} \leftarrow & \mathbb{F}_{k-r-1r+1} & \leftarrow \cdots & \mathbb{F}_{1k-1} \\ \downarrow & \downarrow & & \downarrow \\ \mathbb{R}^2_r & \mathbb{R}^2_{r+1} & & \mathbb{R}^2_{k-1} \end{array}$$

is homotopy commutative and the conditions of the theorem are satisfied. The fibration $p_r : \mathbb{F}_{k-rr} \to \mathbb{R}^2_r$ leads to the split-exact sequence

$$1 \to\to \pi_1(\mathbb{F}_{k-r-1r+1}, q) \to \pi_1(\mathbb{F}_{k-rr}, q) \to \pi_1(\mathbb{R}^2_r, q) \to 1$$

(cf. Theorem 1.2 of Chapter IV). Thus we have the split extension

$$\pi_1(\mathbb{R}^2_r, q) \rtimes \pi_1(\mathbb{F}_{k-r-1r+1}, q) \cong \pi_1(\mathbb{F}_{k-rr}, q),$$

where $\pi_1(\mathbb{R}^2_r, q)$ acts on $\pi_1\mathbb{F}_{k-r-1r+1}, q)$ by automorphisms. This action can be described explicitly, as in the second proof of Theorem 5.1 of Chapter IV. To do so, note that $\pi_1(\mathbb{R}^2_r, q)$ is the free group generated by $\{\alpha_{r_1 1}, \cdots, \alpha_{r+1r}\}$ and $\pi_1(\mathbb{F}_{k-r-1r+1}, q)$ is that generated by $\{\alpha_{ts} | 1 \leq s < t \leq k,\ r+1 < t\}$. Next, arguing as in the proof of Propositions 5.2 and 5.3 of Chapter IV, we obtain the relations

$$\begin{aligned} \alpha^{-1}_{r+1s}\alpha_{ts}\alpha_{r+1s} &= \alpha_{tr+1}\alpha_{ts}\alpha^{-1}_{tr+1} \\ \alpha^{-1}_{r+1s}\alpha_{tr+1}\alpha_{r+1s} &= \alpha_{ts}\alpha_{tr}\alpha^{-1}_{ts} \end{aligned}$$

for $1 \leq s < r+1$, $1 \leq s < t \leq k, r+1 < t$, and

$$\alpha^{-1}_{r+1s}\alpha_{tu}\alpha_{r+1s} = \omega_{tr+1su}\alpha_{tu}\omega^{-1}_{tr+1su},$$

where

$$\begin{aligned} \omega_{tr+1su} &= [\alpha_{ts}, \alpha_{tr+1}], \text{ if } 1 \leq s < u < r+1 < t \text{ and} \\ \omega_{tr+1su} &= 1, \text{ if } 1 \leq s < r+1 < u < t \text{ or } 1 \leq u < s < r+1 < t. \end{aligned}$$

Now, for every $(r+1, s)$ such that $1 \leq s < r+1 \leq k$, denote by ϕ_{r+1s} the automorphism

$$\phi_{r+1s} : \pi_1(X_{k-r-1r+1}, q) \to \pi_1(X_{k-r-1r+1}, q), \alpha_{tu} \mapsto \gamma_{tr+1su}\alpha_{tu}\gamma^{-1}_{tr+1su},$$

where γ_{tr+1su} is the appropriate term in the expressions above describing the action of $\pi_1(\mathbb{R}^{n+1}_r, q)$ on $\pi_1\mathbb{F}_{k-r-1r+1}, q) \cong \pi_1(X_{k-r-1r+1}, q)$. Now, define X_{k-rr} to be

$$X_{k-rr} = X_{k-r-1r+1} \cup_{\mu_{r*}} D_{rs} \times X_{k-r-1r+1},$$

where D_{rs} is a 1-disk and

$$\mu_{r*} : \partial D_{rs} \times X_{k-r-1r+1}$$

is the map equal to the identity at $X_{k-r+1r+1} \times 0$ and μ_{rs} at $X_{k-r-1r+1} \times \{1\}$. The projection

$$\mathrm{proj}_{k-r} : X_{k-rr} \to (S_{r1} \vee \cdots \vee S_{rr-1})$$

is a quasi fibration. The resulting diagram

$$\begin{array}{ccc} X^{(m-1)}_{k-rr} & \leftarrow & X^{(m-1)}_{k-r-1r+1} \\ \downarrow & & \downarrow \\ \mathbb{F}_{k-rr} & \leftarrow & \mathbb{F}_{k-r-1r+1} \end{array}$$

homotopy commutes, and the vertical maps induce isomorphisms on the fundamental groups. The spaces are aspherical. Hence the vertical maps are homotopy equivalences. □

11 The Cellular Structure for $\mathbb{F}_{k+1}(S^2)$

The construction here follows the pattern of the construction for the space $\mathbb{F}_k(\mathbb{R}^2)$ given above. There are two natural cellular structures here.

To define the first, note that, according to Theorem 7.2 of Chapter IV, the pull-back

$$\mathrm{proj}_{k+1,3} : \mathbb{F}'(S^2) \to O_{3,2} \cong SO_3$$

of $\mathrm{proj}_{k+1,3} : \mathbb{F}_{k+1}(S^2) \to \mathbb{F}_3(S^2)$ is homeomorphic to the trivial bundles $O_{3,2} \times \mathbb{F}_{k-2,2}(\mathbb{R}^2)$. The cellular structure is that of the product.

The second comes from the equivalence

$$D^2 \times \mathbb{F}_k(\mathbb{R}^2) \cup_\mu \mathbb{F}_k(\mathbb{R}^2) \to \mathbb{F}_{k+1}(S^2),$$

where the attaching map is that induced by Proposition 6.3 of Chapter III.

12 Historical Remarks

Techniques used in the computations of the homology of $\Omega(\mathbb{F}_k(\mathbb{R}^{n+1}))$ in [18, Cohen-Gitler] and [113, Xicoténcatl Merino] depend on constructions similar to our multispherical cycles $S_1 \times \cdots \times S_m$ of §§3, and 6. Ours is the first attempt to develop the idea systematically.

VII. Cellular Chain Models

Our aim in this chapter is to present cellular chain models for the singular chain coalgebras $C_*S(\mathbb{F}_k(M))$, where M is $\mathbb{R}^{n+1}$ or S^{n+1}. The models are based on the cellular models $X \simeq \mathbb{F}_k(\mathbb{R}^{n+1})$ and $Y \simeq \mathbb{F}_{k+1}(S^{n+1})$ of Chapter VI, with $n > 1$. The results here are to be used in Part Three, where cellular models for the loop spaces of the configuration spaces are studied.

In §1 we start with the complex X. First we show that a cellular approximation $\nabla^x : X \to X \times X$ makes the cellular chain complex $C_*(X)$ a coassociative and cocommutative coalgebra with a two-sided counit and coaugmentation. Next we express $\nabla^x(\omega)$ in terms of the rows of the factor table of the multifold twisted product ω.

In §2 we consider the case of $Y \simeq \mathbb{F}_{k+1}(S^{n+1})$, with $n > 1$, and odd. Again we show that the cellular chains $C_*(Y)$ form a coassociative, cocommutative coalgebra with a two-sided counit and coaugmentation. Then we compute $C_*(\nabla^y)(\omega)$, where ω ranges over the set of multifold twisted products.

Finally, in §3 we turn to the case of $Y \simeq \mathbb{F}_{k+1}(S^{n+1})$ where $(n+1)$ is even and greater than two. The results here, like those of §2, are in terms of the Stiefel manifold $O_{n+2,2}$ and $\mathbb{F}_{k-2,2}(\mathbb{R}^{n+1})$.

1 Cellular Chain Coalgebras

Following [63, James], we shall call a CW-complex *special*, if it is connected, simply connected, countable, and has a single vertex. Let X be a special CW-complex. Recall that the cellular chain complex $\{C_*(X), \partial^x\}$ with integral coefficients or, equivalently, the DG-module of the cell complex X is defined by setting $C_m(X) = H_m(X^{(m)}, X^{(m-1)}; \mathbb{Z})$ and letting $\partial^x : C_m(X) \to C_{m-1}(X)$ be the boundary homomorphism of the triple (X^m, X^{m-1}, X^{m-2}). Let $\rho : C_*(X) \to C_*S(X)$ be a chain morphism, with $C_*S(X)$ the singular chain complex with integral coefficients of X that induces an isomorphism on homology. As the chain modules $C_*(X)$ and $C_*S(X)$ are free, the chain morphism ρ is actually a chain homotopy equivalence. Recall that the module $[C_*(X), C_*S(X)]$ of chain homotopy classes of chain morphisms of degree 0 is isomorphic to the module of morphisms

$\mathrm{Hom}(H_*C(X), H_*CS(X))$ of degree 0, provided that $H_*C(X)$ is also free. Thus, in this case, the morphism ρ has a chain homotopy inverse. One can use such an inverse to study the possible natural coalgebra structures on $C_*(X)$. However, we shall proceed in an alternative and, we believe, more direct manner.

Let X be a special complex, and denote its single 0-vertex by e_0. Assume that $H_*(X;\mathbb{Z})$ is free. Then, the diagonal map $X \to X \times X$ induces a coproduct

$$H_*(X;\mathbb{Z}) \to H_*(X;\mathbb{Z}) \otimes H_*(X;\mathbb{Z})$$

that turns $H_*(X;\mathbb{Z})$ into a coassociative, cocommutative coalgebra. Moreover, the constant map $X \to \{e_0\}$, gives rise to a two-sided counit, while the imbedding $\{e_0\} \to X$ leads to a coaugmentation.

Theorem 1.1 *Assume that the boundary homomorphism ∂^x of the cellular chain complex $C_*(X)$ is trivial. Let $\nabla^x : X \to X \times X$ be a cellular approximation to the diagonal map. Then, the induced chain morphism $C_*(\nabla^x) : C_*(X) \to C_*(X) \otimes C_*(X)$ turns $C_*(X)$ into a coassciative, cocommutative coalgebra. Moreover, the constant map, $X \to \{e_0\}$, induces a two-sided counit $\epsilon : C_*(X) \to \mathbb{Z}$, while the canonical imbedding $\{e_0\} \to X$ gives a coaugmentation $\eta : \mathbb{Z} \to C_*(X)$.*

Proof. The diagonal approximation $\nabla^x : X \to X \times X$ leads to the chain map $C_*(\nabla^x) : C_*(X) \to C_*(X) \otimes C_*(X)$, and this, in turn, induces the commutative diagram

$$\begin{array}{ccc} C_*(X) & \xrightarrow{C_*(\nabla^x)} & C_*(X) \otimes C_*(X) \\ \downarrow & & \downarrow \\ H_*(X;\mathbb{Z}) & \xrightarrow{H_*(\nabla^x)} & H_*(X;\mathbb{Z}) \otimes H_*(X;\mathbb{Z}). \end{array}$$

That is to say, the natural isomorphism $C_*(X) \to H_*(X;\mathbb{Z})$ takes the coproduct $C_*(\nabla^x)$ to the coproduct $H_*(\nabla^x)$. Consider next the diagram

$$\begin{array}{ccc} C_*(X) & \xrightarrow{C_*(\nabla^x)} & C_*(X) \otimes C_*(X) \\ \downarrow & & \downarrow \\ CS_*(X) & \xrightarrow{CS_*(\nabla^x)} & CS_*(X) \otimes CS_*(X). \end{array}$$

Note that the diagram obtained on passing to homology commutes, and, as the DG-modules $C_*(X)$ and $CS_*(X)$ are free, it follows that the latter diagram commutes up to chain homotopy. Finally, consider the diagram

$$\begin{array}{ccc} CS_*(X) & \xrightarrow{CS_*(\nabla^x)} & CS_*(X) \otimes CS_*(X) \\ \downarrow & & \downarrow \\ CS_*(X) & \xrightarrow{CS_*(\nabla)} & CS_*(X) \otimes CS_*(X), \end{array}$$

where $\nabla : X \to X \times X$ is the diagonal map. The diagram commutes up to DG-homotopy, since ∇^x and ∇ are homotopic. Thus, we conclude that the

$$C_*(\nabla^x) = H_*(\nabla^x) = H_*(\nabla)$$

are equivalent coproducts. The coalgebra $\{H_*(X;\mathbb{Z}), H_*(\nabla)\}$ certainly satisfies the conditions stated in the theorem, and this implies that the coalgebra $\{C_*(X), C_*(\nabla^x)\}$ satisfies the conditions of the theorem. □

The argument above shows, provided that the conditions of the theorem are satisfied, that the coproduct $C_*(\nabla^x)$ is unique and, in particular, is independent of the cellular approximation ∇^x chosen.

2 The Coalgebra of $\mathbb{F}_k(\mathbb{R}^{n+1})$

In §8 of Chapter VI we constructed a minimal and finite CW-complex X homotopy equivalent to the configuration spaces $\mathbb{F}_k(\mathbb{R}^{n+1})$. Now, by definition,

$$C_p(X) = H_p(X^{(p)}, X^{(p-1)}; \mathbb{Z}),$$

and, according to the results of §1 of Chapter V, $C_p(X) \neq 0$ if, and only if, $p = 0 \mod n$ and $p \leq \dim X$. Hence, the set $\mathcal{A}^\wedge = \{1\} \bigcup_m \mathcal{A}^{\wedge m}$, where

$$\mathcal{A}^{\wedge m} = \{\alpha_{r_1 s_1} \ltimes \cdots \ltimes \alpha_{r_m s_m} | p = mn\}$$

for all $\underline{s} \in \mathcal{R}$ and $\underline{s} \in \mathcal{S}$ such that $\underline{s} < \underline{r}$ and $|\underline{r}| = p = |\underline{s}|$, is a basis for $C_*(X) = \sum_p C_p(X)$.

Let $\epsilon : C_*(X) \to \mathbb{Z}$ be the morphism induced by the constant map $X \to e_0$, and let $\eta : \mathbb{Z} \to C_*(X)$ be that induced by the imbedding $e_0 \to X$. Choose a cellular approximation $\nabla^x : X \to X \times X$ to the diagonal map.

Theorem 2.1 *The induced coproduct $C_*(\nabla^x) : C_*(X) \to C_*(X) \otimes C_*(X)$ turns $\{C_*(X), C_*(\nabla^x), \epsilon, \eta\}$ into a cocommutative, coassociative coalgebra with a two-sided counit ϵ and coaugmentation η. Moreover, this structure is unique.*

Proof. The differential ∂^x is trivial for dimensional reasons, if $n \geq 2$. On the other hand, for $n = 1$, one notes that ∂^x is the morphism induced by the abelianization of the Y-B relations. Hence, the chain module $\{C_*(X), \partial^x\}$ satisfies the conditions of Theorem 1.1, and this is enough to imply the theorem. □

Next we wish to obtain the expression for the coproduct

$$C_*(\nabla) : C_*(X) \to C_*(X) \otimes C_*(X)$$

in terms of the basis $\mathcal{A}^\wedge$. Denote by $\lambda_j(\omega)$ the jth row of the factor table

$$\mathcal{T}(\phi_\omega): \begin{cases} \alpha_{r_1 s_1} & + \sum_{t=2}^{m} \lambda_{1t}(\omega_t) \\ \cdots & + \cdots \\ \alpha_{r_j s_j} & + \sum_{t=j+1}^{m} \lambda_{1t}(\omega_t) \\ \cdots & + \cdots \\ \alpha_{r_m s_m} & \end{cases}$$

of the m-fold twisted product $\omega = \alpha_{\underline{r}\,\underline{s}} = \alpha_{r_1 s_1} \ltimes \cdots \ltimes \alpha_{r_m s_m}$, where $\omega_t = \alpha_{r_1 s_1} \ltimes \cdots \ltimes \alpha_{r_t s_t}$.

Denote by $\iota_1, \cdots, \iota_m$ the elements in $C_*(X) = H_*(S_1^n \times \cdots \times S_m^n; \mathbb{Z})$ corresponding to the fundamental classes of the various factors. Then, $\iota_1 \times \cdots \times \iota_m$ is the fundamental class of $S_1^n \times \cdots \times S_m^n$. Put

$$H_n(\phi_\omega)(\iota_{i_1} \times \cdots \times \iota_{i_t}) = \begin{cases} \lambda_{i_1,\cdots,i_t}, \text{ for } 1 \leq i_1 < i_2 < \cdots < i_t \leq m \\ 1, \text{ for } t = 0 \end{cases}$$

in $H_{tn}(\mathbb{F}_k(\mathbb{R}^{n+1}); \mathbb{Z}) \cong C_{tn}(X)$, where

$$H_*(\phi_\omega) : H_*(S_1^n \times \cdots \times S_m^n; \mathbb{Z}) \to H_*(\mathbb{F}_k(\mathbb{R}^{n+1}); \mathbb{Z})$$

is the morphism induced by ϕ_ω.

Recall that a (p,q)-shuffle of the set of integers $\{1, \cdots, m\}$ is a partition into two disjoint subsets, $\mu_1 < \cdots < \mu_p$ and $\nu_1 < \cdots < \nu_q$, of p and q integers, respectively (see [27, Eilenberg-Mac Lane], [72, Mac Lane]). A (p,q)-shuffle defines a permutation σ_μ of $\{1, \cdots, m\}$ with the property that $\sigma_\mu(i) < \sigma_\mu(j)$, if $i < j \leq p$ or $p < i < j$. By definition, let $\epsilon(\mu) = \sum_{i=1}^{p} \mu_i - i$. Then $\text{sign}(\sigma_\mu) = (-1)^{\epsilon(\mu)}$. Put

$$\eta(\mu) = \text{sign}(\sigma_\mu)^{n^2} = (-1)^{n^2\epsilon(\mu)}.$$

Observe that $\eta(\mu)$ is 1 if n is even or the associated permutation σ_μ is even; otherwise, $\eta(\mu) = -1$.

Theorem 2.2 *The coproduct $C_*(\nabla^x) : C_*(X) \to C_*(X) \otimes C_*(X)$ is such that*

$$\omega \mapsto \sum_{(p,q)} \sum_{(\mu,\nu)} \eta(\mu) \lambda_{\mu(1)\cdots\mu(p)} \otimes \lambda_{\nu(1)\cdots\nu(q)},$$

where (μ, ν) ranges over the set of all (p, q)-shuffles of the set $\{1, \cdots, m\}$, for all $p, q \geq 0$ such that $p + q = m$.

Proof. Consider the product cell complex $S_1^n \times \cdots \times S_m^n$, where each factor $S_i^n = e^0 \cup e^n$. Then the coproduct

$$C_*(S_1^n \times \cdots \times S_m^n) \to C_*(S_1^n \times \cdots \times S_m^n) \otimes C_*(S_1^n \times \cdots \times S_m^n)$$

is such that

$$\iota_1 \times \cdots \times \iota_m \mapsto \sum_{(p,q)} \sum_{(\mu,\nu)} \eta(\mu)\iota_{\mu(1)} \times \cdots \times \iota_{\mu(p)} \otimes \iota_{\nu(1)} \times \cdots \times \iota_{\nu(q)},$$

where (μ, ν) ranges over all shuffles (p, q) of $\{1, \cdots, m\}$ for all $p, q \geq 0$ such that $p + q = m$. Note that

$$H_{pn}(\phi_\omega)(\iota_{\mu(1)} \times \cdots \times \iota_{\mu(p)}) = \lambda_{\mu(1)} \cdots \iota_{\mu(p)}$$

for all p. By naturality and uniqueness, the coproduct in $C_*(S_1^n \times \cdots \times S_m^n)$ goes to the coproduct $C_*(\nabla^x)$. This implies the theorem. □

Note that each of the terms $\lambda_{\mu(1)\cdots\mu(p)}$ and $\lambda_{\nu(t+1)\cdots\nu(q)}$ is a sum of products of terms of the form $\pm\alpha_{\underline{r}'\,\underline{s}'}$.

As an example, we see that for $\alpha_{\underline{r}\,\underline{s}}$, with $|\,\underline{r}\,| = 2$, the coproduct is determined by the Y-B relations: the element $\alpha_{r_1s_1} \ltimes \alpha_{r_2s_2}$ maps to

$$(\alpha_{r_1s_1} \ltimes \alpha_{r_2s_2}) \otimes 1 + \lambda_1 \otimes \lambda_2 + (-1)^{n^2}\lambda_2 \otimes \lambda_1 + 1 \otimes \alpha_{r_1s_1} \ltimes \alpha_{r_2s_2},$$

where λ_1 and λ_2 are the two rows of the factor table

$$\mathcal{T}_2 : \begin{cases} \alpha_{r_1s_1} + \lambda_{12}(\alpha_{r_1s_1}) = \lambda_1 \\ \alpha_{r_2s_2} \qquad\qquad\quad\; = \lambda_2 \end{cases}$$

and $\lambda_{12}(\alpha_{r_1s_1}) = \delta_{s_2r_1}\alpha_{r_2s_1} + \delta_{s_1s_2}(-1)^{(n+1)}\alpha_{r_2r_1}$.

3 The Coalgebra of $\mathbb{F}_{k+1}(S^{n+1})$, (n+1) Odd

As $(n+1)$ is odd, there is a section $\delta_3' : S^{n+1} \to O_{n+2,2}$ that leads to a cellular decomposition $O_{n+2,2} \simeq (S^n \vee S^{n+1}) \cup_f D^{2n+1}$, as is shown in Chapter III, §5.

Recall that $H_*(\mathbb{F}_{k-2}(\mathbb{R}_2^{n+1}); \mathbb{Z})$ has the set $\mathcal{B}^{\wedge k-2,2} = \{1\} \cup_m \mathcal{B}^{\wedge m}$ as a basis, where, in the notation of Chapter VI, §6, $\mathcal{B}^{\wedge m} = \{\beta_{r_2s_2} \ltimes \cdots \ltimes \beta_{r_ms_m} | r_2 > 2\}$. Also, recall that $H_*(O_{n+2,2}; \mathbb{Z})$ has the set $\mathcal{S} = \{1, \beta_n, \delta_3, \delta_3 \ltimes \beta_n\}$ as a basis. Hence,

$$H_*(\mathbb{F}_{k+1}(S^{n+1}); \mathbb{Z}) \cong H_*(O_{n+2,2}; \mathbb{Z}) \otimes H_*(\mathbb{F}_{k-1}(\mathbb{R}_1^{n+1}); \mathbb{Z})$$

has the set $\mathcal{S} \times \mathcal{B}^{\wedge k-2,2}$ as a basis.

In §8 of Chapter VI we defined a CW-complex $X_{k-2,2} \simeq \mathbb{F}_{k-2,2}(\mathbb{R}^{n+1})$ the cells of which are in one-one correspondence with the elements of the set $\mathcal{B}^{\wedge k-2,2}$. Also, in §9 of that Chapter, we defined a finite CW-complex Y_{k+1} that contains the complex $O_{n+2,2} \vee X_{k-2,2}$ as a subcomplex and is of the homotopy type of $\mathbb{F}_{k+1}(S^{n+1})$. Moreover, Y_{k+1} can be represented, symbolically, as a twisted product $Y_{k+1} = O_{n+2,2} \ltimes X_{k-2,2}$, and its cells are in one-one correspondence with the elements of the set $\mathcal{S} \times \mathcal{B}^{\wedge k-2,2}$. Note that there is an isomorphism $C_*(Y_{k+1}) \cong C_*(O_{n+2,2}) \otimes C_*(X_{k-2,2})$ of DG-modules, where $C_*(X_{k-2,2})$ is the cellular chain complex of the CW-complex $X_{k-2,2}$.

Next, let

$$\nabla^y : Y_{k+1} \to Y_{k+1} \times Y_{k+1}$$

be a cellular approximation to the diagonal map. We can assume, without loss of generality, that ∇^y is an extension of the already chosen cellular approximation $\nabla^x : X_{k-2,2} \to X_{k-2,2} \times X_{k-2,2}$. Then, let

$$\nabla^s : O_{n+2,2} \to O_{n+2,2} \times O_{n+2,2}$$

be a cellular approximation to the diagonal map. We can also assume, without loss of generality, that ∇^y is an extension of ∇^s as well. Observe that

$$\begin{aligned} H_*(Y_{k+1};\mathbb{Z}) &\cong H_*(O_{n+2,2};\mathbb{Z}) \otimes H_*(X_{k-2,2};\mathbb{Z}), \\ H_*(X_{k-2,2};\mathbb{Z}) &\cong C_*(X_{k-2,2}), \\ H_*(O_{n+2,2};\mathbb{Z}) &\cong C_*(O_{n+2,2}), \end{aligned}$$

since $(n+1)$ is assumed to be odd. Therefore,

$$C_*(Y_{k+1}) = H_*(C(Y_{k+1})) \cong H_*(Y_{k+1};\mathbb{Z}).$$

Thus, the differential ∂^y of $C_*(Y_{k+1})$ is trivial, Theorem 1.1 above applies, and we obtain the following result.

Theorem 3.1 *The morphism $C_*(\nabla^y) : C_*(Y_{k+1}) \to C_*(Y_{k+1}) \otimes C_*(Y_{k+1})$ turns $C_*(Y_{k+1})$ into a coassociative, cocommutative, coaugmented coalgebra with a two-sided counit.*

To express the coproduct $C_*(\nabla^y)$ in terms of the elements of the set $\mathcal{S} \times \mathcal{B}^{\wedge k-2,2}$, we proceed as follows. Now, each such $\rho \rtimes \omega$ in $\mathcal{S} \times \mathcal{B}^{\wedge k-2,2}$ determines a cell $e_{\rho,\omega}$ of Y_{k+1}. Let

$$\psi_\omega : O_{n+2,2} \times S^n_2 \times \cdots \times S^n_m \to \mathbb{F}_{k+1}(S^{n+1})$$

be the multispherical map corresponding to the $(m-1)$-fold twisted product $\omega = \beta_{r_2,s_2} \ltimes \cdots \ltimes \beta_{r_m s_m}$ (see Chapter VI, §6). Put

$$\begin{aligned} \lambda_j(\omega) &= (\psi_\omega | S_j) : S^n_j \to \mathbb{F}_{k+1}(S^{n+1}) \text{ for } j \geq 2 \\ \lambda(O_{n+2,2}) &= (\psi_\omega | O_{n+2,2}) : O_{n+2,2} \to \mathbb{F}_{k+1}(S^{n+1}) \\ \lambda(O_{n+2,2})|S^{n+1} &= \lambda_0 \\ \lambda(O_{n+2,2})|S^n &= \lambda_1, \end{aligned}$$

where, by abuse of notation, we make no distinction between the map and its homotopy class and where $S^{n+1} \subset O_{n+2,2}$ is the image of the section δ'_3 :

$S^{n+1} \to O_{n+2,2}$, while $S^n \subset O_{n+2,2}$ is fiber of the projection, p : $O_{n+2,2} \to S^{n+1}$ that takes (x, y) to x. Hence, the factor table of ψ_ω is

$$\mathcal{T}(\phi_\omega) : \begin{cases} \lambda_0 & \\ \lambda_1 & \\ \beta_{r_2 s_2} & + \sum_{t=3}^m \lambda_{2t}(\omega_t) \\ \cdots & + \cdots \\ \beta_{r_j s_j} & + \sum_{t=j+1}^m \lambda_{jt}(\omega_t) \\ \cdots & + \cdots \\ \alpha_{r_m s_m}, & \end{cases}$$

where $\omega_t = \beta_{r_2,s_2} \ltimes \cdots \ltimes \beta_{r_t s_t}$. Observe that $\lambda_0 = \delta_{k+1}$ and that

$$\lambda_1 = \beta_n + \sum_{r=3}^{k} (\sum_{s=1}^{r-1} \beta_{r,s}),$$

according to Proposition 6.1 of Chapter III.

The following result describes the coproduct in $C_*(O_{n+2,2})$.

Proposition 3.1 *For β_n, δ_{k+1} and $\delta_{k+1} \ltimes \beta_n$ in $C_*(O_{n+2,2}; \mathbb{Z})$, we have*

$$\begin{aligned} C_*(\nabla^y)(\beta_n) &= \beta_n \otimes 1 + 1 \otimes \beta_n \\ C_*(\nabla^y)(\delta_{k+1}) &= \delta_{k+1} \otimes 1 + 1 \otimes \delta_{k+1} \\ C_*(\nabla^y)(\delta_{k+1} \ltimes \beta_n) &= \delta_{k+1} \ltimes \beta_n \otimes 1 + 1 \otimes \delta_{k+1} \ltimes \beta_n \\ &\quad + \delta_{k+1} \otimes \beta_n + \beta_n \otimes \delta_{k+1}. \end{aligned}$$

Proof. First, we note that the coproduct in $C_*(O_{n+2,2})$ coincides with that of $H_*(O_{n+2,2}; \mathbb{Z})$, since $C_*(O_{n+2,2}) = H_*(O_{n+2,2}; \mathbb{Z})$ when $(n+1)$ is odd and $O_{n+2,2}$ is given the cellular structure described above. The first two formulas are valid because both β_n and δ_{n+1} are primitive in $H_*(O_{n+2,2}; \mathbb{Z})$. For the third case, note that $\delta_{k+1} \ltimes \beta_n$ is the fundamental class of the manifold $O_{n+2,2}$, and that $H^*(O_{n+2,2}; \mathbb{Z}) \cong H^*(S^{n+1}; \mathbb{Z}) \otimes H^*(S^n; \mathbb{Z})$ is an isomorphism of algebras. □

Denote $O_{n+2,2}$ by $S^{n+1} \ltimes S^n$, and rewrite the multispherical map ψ_ω in the form

$$\psi_\omega : (S^{n+1} \ltimes S^n) \times (S_2^n \times \cdots \times S_m^n) \to \mathbb{F}_{k+1}(S^{n+1}).$$

Denote by ι_0 and ι_1, respectively, the elements δ_3 and β_n of $H_*(O_{n+2,2}; \mathbb{Z})$. Let $j_1, \cdots, j_p \in \{\iota_0, \iota_1, \cdots, \iota_m\}$, where $j_1 < j_2 < \cdots < j_p$, and put

$$\lambda_{j_1,\cdots,j_p} = \begin{cases} H_*(\psi_\omega)(\iota_{j_1} \times \cdots \times \iota_{j_p}) \text{ if } \iota_{j_1} = \beta_n \text{ or } \delta_3 \\ H_*(\psi_\omega)((\delta_3 \ltimes \beta_n) \times \iota_{j_3} \times \cdots \times \iota_{j_p} \text{ if } \iota_{j_1} = \delta_3 \text{ and } \iota_{j_2} = \beta_n. \end{cases}$$

The following results describe the coproduct $C_*(\nabla^y)$. Here $\eta(\mu) = (-1)^{n^2 \epsilon(\mu)}$ (see Theorem 2.1 above).

Theorem 3.2 *Let* $\omega = \beta_{r_2 s_2} \ltimes \cdots \ltimes \beta_{r_m s_m}$, *with* $r_2 > 2$ *and* $\rho = \beta_n$. *Then the coproduct of the basic element* $\rho \rtimes \omega$ *is given by the expression*

$$C_*(\nabla^y)(\rho \rtimes \omega) = \sum_{(p,q)} \sum_{(\mu,\nu)} \eta(\mu)\lambda_{\mu(1),\cdots,\mu(p)} \otimes \lambda_{\nu(1)\cdots\nu(q)},$$

and (μ, ν) *ranges over all* (p, q)*-shuffles of* $\{1, \cdots, m\}$, *for all* $p, q \geq 0$ *such that* $p + q = m$.

The proof is entirely similar to that of Theorem 2.1 and is left to the reader.

Let (μ, ν) be a (p,q)-shuffle of $\{0, 2, \cdots, m\}$. Thus, $\mu, \nu \subset \{0, 2, \cdots, m\}$ such that $\mu_1 < \mu_2 < \cdots < \mu_p$ and $\nu_1 < \cdots < \nu_q$. By definition, let

$$\epsilon(\mu) = \begin{cases} \sum_{i=2}^{p} \mu_i - i, \text{ if } \mu_1 = 0 \text{ and} \\ \sum_{i=2}^{p} \mu_i - i - 1, \text{ if } \mu_1 > 0. \end{cases}$$

Set $\eta(\mu) = (-1)^{n^2 \epsilon(\mu)}$.

Theorem 3.3 *Let* $\omega = \beta_{r_2 s_2} \ltimes \cdots \ltimes \beta_{r_m s_m}$, *with* $r_2 > 2$ *and* $\rho = \delta_k$. *Then the coproduct of the basic element* $\rho \rtimes \omega$ *is given by the expression*

$$C_*(\nabla^y)(\rho \rtimes \omega) = \sum_{(p,q)} \sum_{(\mu,\nu)} \eta(\mu)\lambda_{\mu(1),\cdots,\mu(p)} \otimes \lambda_{\nu(1)\cdots\nu(q)},$$

and (μ, ν) *ranges over all* (p, q)*-shuffles of* $\{0, 2, \cdots, m\}$, *for all* $p, q \geq 0$ *such that* $p + q = m$.

Proof. Consider the map

$$\psi_\omega : (S^{n+1} \ltimes S^n) \times (S_2^n \times \cdots \times S_m^n) \to \mathbb{F}_{k+1}(S^{n+1}).$$

Assume that $S_2^n \times \cdots \times S_m^n$ and $O_{n+2,2} = S^{n+1} \ltimes S^n$ have the cellular structures defined above, and give $(S^{n+1} \ltimes S^n) \times (S_2^n \times \cdots \times S_m^n)$ the inherited the product cellular structure. Now, choose a cellular approximation

$$\begin{aligned} &\nabla^m : (S^{n+1} \ltimes S^n) \times (S_2^n \times \cdots \times S_m^n) \to \\ &(S^{n+1} \ltimes S^n) \times (S_2^n \times \cdots \times S_m^n) \times (S^{n+1} \ltimes S^n) \times (S_2^n \times \cdots \times S_m^n). \end{aligned}$$

Note that

$$C_*((S^{n+1} \ltimes S^n) \times (S_2^n \times \cdots \times S_m^n) \cong H_*(S^{n+1} \ltimes S^n) \otimes H_*(S_2^n \times \cdots \times S_m^n)$$

is an isomorphism of DG-modules. Moreover, it satisfies the conditions of Theorem 1.1. Therefore, the preceding morphism is a morphism of coalgebras. Assume now that ψ_ω has been deformed into a cellular map $\psi_\omega : (S^{n+1} \ltimes S^n) \times (S_2^n \times \cdots \times S_m^n) \to Y_{k+1}$. Observe that the induced morphism

$$C_*(\psi_\omega) : C_*((S^{n+1} \ltimes S^n) \times (S_2^n \times \cdots \times S_m^n)) \to C_*(Y_{k+1})$$

is a morphism of coalgebras. Theorem 6.2 of Chapter VI implies that

$$C_*(\psi_\omega)(\iota_1 \times \iota_2 \times \cdots \times \iota_m) = \rho \ltimes \omega.$$

We also have $C_*(\nabla^m)(\iota_1 \times \iota_2 \times \cdots \times \iota_m)$ equal to

$$\sum_{\mu,\nu} \eta(\mu)\iota_{\mu(1)} \times \cdots \times \iota_{\mu(p)} \otimes \iota_{\nu(1)} \times \cdots \times \iota_{\nu(q)},$$

where the sum is over all (p,q)-shuffles (μ,ν) of $\{0,2,\cdots,m\}$. Applying $C_*(\psi_\omega)$, and noting that $\iota_{\mu(1)} \times \cdots \times \iota_{\mu(p)} \mapsto \lambda_{\mu(1)\cdots\mu(p)}$, as well as that $\iota_{\nu(p+1)} \times \cdots \times \iota_{\nu(m)} \mapsto \lambda_{\nu(1)\cdots\nu(q)}$, for all (p,q)-shuffles (μ,ν) such that $p+q=m+1$, one obtains the desired result. □

The next result deals with the remaining case.

Theorem 3.4 *Let $\omega = \beta_{r_2s_2} \ltimes \cdots \ltimes \beta_{r_ms_m}$, with $r_2 > 3$ and $\rho = \delta_3 \ltimes \beta_n$. Then, the coproduct of the basic element $\rho \rtimes\!\!\!\ltimes \omega$ is given by the expression*

$$C_*(\nabla^y)(\rho \rtimes\!\!\!\ltimes \omega) = \sum_{(p,q)} \sum_{(\mu,\nu)} \eta(\mu))\lambda_{\mu(1),\cdots,\mu(p)} \otimes \lambda_{\nu(1)\cdots\nu(q)},$$

where $\eta(\mu)$ is an expression analogous to that in Theorem 3.3 and (μ,ν) ranges over all (p,q)-shuffles of $\{0,\cdots,m\}$.

The proof is similar to that of Theorem 3.3 and is left to the reader.

4 The Coalgebra $C_*(Y)$, $Y \simeq \mathbb{F}_{k+1}(S^{n+1})$, $(n+1)$ Even

The situation here is somewhat different from the case when $n+1$ is odd. For one thing, the Stiefel manifold $O_{n+2,2}$ has the following minimal cellular decomposition: $O_{n+2,2} = e^0 \cup e^n \cup e^{n+1} \cup e^{2n+1} = S^n \cup e^{n+1} \cup e^{2n+1}$, with the characteristic map of e^{n+1} being $f : (D^{n+1}, \partial D^{n+1}) \to (S^n \cup e^{n+1}, S^n)$, where $f : S^n = \partial D^{n+1} = S^n \to S^n$ is a map of degree 2.

Consider the fibration $\mathbb{F}_{k+1}(S^{n+1}) \to \mathbb{F}_3(\mathbb{R}^{n+1})$, with $\mathbb{F}_{k-2,2}(\mathbb{R}^{n+1})$ as fiber. Replacing $\mathbb{F}_3(\mathbb{R}^{n+1})$ by the homotopy equivalent $O_{n+2,2}$, and noting that $\mathbb{F}_{k-2,2}(\mathbb{R}^{n+1})$ is totally nonhomologous to 0, one sees that the Leray-Hirsch homomorphism

$$\Phi_{k+1,3} : H_*(\mathbb{F}_{k+1}(S^{n+1}); \mathbb{Z}) \cong H_*(O_{n+2,2}; \mathbb{Z}) \otimes H_*(\mathbb{F}_{k-2,2}(\mathbb{R}^{n+1}); \mathbb{Z})$$

is an isomorphism (see Chapter VI, §6). The following is easy to prove.

Proposition 4.1 *The module $H_*(O_{n+2,2}; \mathbb{Z})$ is generated by the set $\mathcal{S} = \{1,\ \beta_n,\ \mu_{2n+1}\}$, where $\beta_n \in H_n(O_{n+2,2}; \mathbb{Z})$ and $\mu_{2n+1} \in H_{2n+1}(O_{n+2,2}; \mathbb{Z})$.*

The preceding discussion leads directly to the following.

Proposition 4.2 *The set* $\mathcal{B}^\wedge = \{1, \rho \ltimes \beta_{\underline{r}\,\underline{s}}\}$*, where* ρ *is in* $\mathcal{S}$ *and*

$$\beta_{\underline{r}\,\underline{s}} = \beta_{r_2 s_2} \rtimes \cdots \ltimes \beta_{r_m s_m},\ (k-1) > m \geq 3,$$

is a set of generators for $H_*(\mathbb{F}_{k+1}(S^{n+1}); \mathbb{Z})$.

In Chapter VI, §8 we constructed a finite CW-complex $X_{k-2,2}$ of the homotopy type of $\mathbb{F}_{k-2,2}(\mathbb{R}^{n+1})$ and in §9, a CW-complex Y_{k+1} that contains $O_{n+2,2} \vee X_{k-2,2}$ as a subcomplex and has cells e_ω in one-one correspondence with the elements $\rho \ltimes \omega$ of the set $\mathcal{S} \times \mathcal{B}^\wedge$.

Recall that the characteristic of the cell e_ω, $\omega \in \mathcal{B}^\wedge$ is the composite of the map

$$\psi_\omega^{x,y} : M = S_2 \times \cdots \times S_m \to \mathbb{F}_{k-2,2}(V^x),$$

with the natural homeomorphism

$$\phi^x : \mathbb{F}_{k-2,2}(V^x) \to \mathbb{F}_{k-2,3}(S^{n+1}) \cong \mathbb{F}_{k-2,2}(\mathbb{R}^{n+1})$$

induced by the stereographic projection ϕ^x, which, by abuse of notation, is denoted by the same symbol.

Also recall that the characteristic map of a given element $\beta_n \ltimes \omega$ is defined using the map

$$\psi_\omega : O_{n+2,2} \times (S_2 \times \cdots \times S_m) \to \mathbb{F}_{k+1}(S^{n+1}),$$

where $(x, y, \xi_2, \cdots, \xi_m)$ goes to $(-x, x, \psi_\omega^{x,y}(\xi_2, \cdots, \xi_m))$. Give $(S_2 \times \cdots \times S_m)$ and $O_{n+2,2}$ the cellular structures as above, and $O_{n+2,2} \times (S_2 \times \cdots \times S_m)$ the product cellular structure. Then, the characteristic maps of the cells of Y_{k+1} are obtained by composing ψ_ω with the characteristic map of the cells $e^t \times e_\omega \subset O_{n+2,2} \times (S_2 \times \cdots \times S_m)$, where e^t ranges over the cells of $O_{n+2,2}$ and e_ω is the top cell of $(S_2 \times \cdots \times S_m)$. As $H_*(O_{n+2,2}; \mathbb{Z})$ has 2-torsion, the situation when $p \neq 2$ differs from that when $p = 2$.

So, let us consider the case when $p = 2$. Here we have an isomorphism $H_*(O_{n+2,2}; \mathbb{Z}_2) \cong E(\beta_n) \otimes E(\delta_{k+1})$ of coalgebras, with $E(\beta_n)$, $E(\delta_{k+1})$ being the exterior coalgebras generated by the elements β_n, δ_{k+1} of dimension n and $(n+1)$, respectively. (The exterior coalgebras, are by definition, the duals of exterior algebras.) Deform ψ_ω into a cellular map, if necessary, and let

$$\begin{aligned} \lambda_0 &= C_*(\psi_\omega)(\delta_{k+1}) \\ \lambda_1 &= C_*(\psi_\omega)(\beta_n) \\ \lambda_i &= C_*(\psi_\omega)(\iota_i), \end{aligned}$$

for $j = 2, \cdots, m$. Here, ι_j is the jth-factor of $(S_2 \times \cdots \times S_m))$. By definition,

$$\lambda_{i_1 \cdots i_t} = C_*(\psi_\omega)(\iota_{i_1} \times \cdots \times \iota_{i_t}).$$

Clearly, $\lambda_{i_1 \cdots i_t} \in C_*(O_{n+2,2} \times (S_2 \times \cdots \times S_m))$.

Theorem 4.1 *The induced coproduct $C_*(\nabla^y)$ takes the fundamental class $\delta_{k+1} \ltimes \beta_n$ of $O_{n+2,2} \times (S_2 \times \cdots \times S_m)$ to the sum*

$$\sum_{(t,u)} \sum_{(\mu,\nu)} \lambda_{\mu(1)\cdots\mu(t)} \otimes \lambda_{\nu(1)\cdots\nu(u)},$$

where (μ,ν) ranges over all (t,u)-shuffles of $\{0,1,2,\cdots,m\}$, with $t+u = m+1$.

The proof follows exactly as in the previous section.

The case $p \neq 2$ is even easier. Here $H_*(O_{n+2,2}, \mathbb{Z}_p) \cong E(\gamma_{2n+1})$, where $E(\mu_{2n+1})$ is the exterior coalgebra generated by the element μ_{2n+1} of dimension $(2n+1)$. Define $\lambda_1 = C_*(\psi_\omega)(\mu_{2n+1})$ and λ_i for $i > 1$ as in the case $p = 2$.

The following theorem is proved similarly. Here

$$\epsilon(\mu) = \begin{cases} \sum_{i=2}^{p} \mu_i - i, \text{ if } \mu_1 = 0 \\ \sum_{i=2}^{p} \mu_i - i - 1, \text{ if } \mu_1 > 0, \end{cases}$$

and set $\eta(\mu) = (-1)^{n^2\epsilon(\mu)}$.

Theorem 4.2 *The coproduct $C_*(\nabla^y)$ takes the fundamental class $\mu_{2n+1} \times \iota_2 \times \cdots \times \iota_m$ of $O_{n+2,2} \times (S_2 \times \cdots \times S_m)$ to the sum*

$$\sum_{(t,u)} \sum_{(\mu,\nu)} \eta(\mu)\lambda_{\mu(1)\cdots\mu(t)} \otimes \lambda_{\nu(1)\cdots\nu(u)},$$

where (μ,ν) ranges over all (t,u)-shuffles of $\{0,2,\cdots,m\}$, with $t+u = m$.

Part III

Homology and Cohomology of Loop Spaces

Introduction

In Part Three we turn to the loop spaces, first based and then free, of the configuration spaces.

In Chapter VIII we determine the structure of integral homology groups $H_*(\Omega F_k(M))$, regarded as a Pontryagin algebra, when M is $\mathbb{R}^{n+1}$, or S^{n+1}. We consider the case of $\mathbb{R}^{n+1}$ first. Proceeding by induction on k, we prove that the algebra $H_*(\Omega \mathbb{F}_k(\mathbb{R}^{n+1}))$ is the universal Hopf algebra generated by the set $\Xi_k = \{\bar{\alpha}_{rs} | 1 \leq s < r \leq k\}$ modulo a set of relations that are homology analogues of the Y-B relations in the loop space. Moreover, the space of primitive elements in the Hopf algebra $H_*(\Omega \mathbb{F}_k(\mathbb{R}^{n+1}); \mathbb{Q})$ is obtained from the graded Lie algebra $\pi_*(\mathbb{F}_k(\mathbb{R}^{n+1})) \otimes \mathbb{Q}$ by desuspension. In the remainder of Chapter VIII we consider the analogous questions for S^{n+1}.

The rest of the monograph is devoted to topology of the free loop spaces of $\mathbb{F}_k(\mathbb{R}^{n+1})$ and $\mathbb{F}_{k+1}(S^{n+1})$.

In Chapter IX we construct the cellular models for the based loop spaces, the free loop spaces, and the space of paths Λ_σ, which go from a tuple to its image under a homoeomorphism σ of the configuration space.

In Chapter X we study the cellular models of $\Omega(M)$, where M stands for $\mathbb{F}_k(\mathbb{R}^{n+1})$ and $\mathbb{F}_{k+1}(S^{n+1})$, respectively, and in Chapter X, §6 we prove that the Eilenberg-Moore spectral sequence of the evaluation map

$$\mathrm{eval}_k : \Lambda \mathbb{F}_k(\mathbb{R}^{n+1}) \to \mathbb{F}_k(\mathbb{R}^{n+1}),\ \omega \mapsto \omega(1)$$

collapses.

In Chapter XI we consider the mod 2 Serre spectral sequence $\{\mathrm{E}^r_{*,*}, \mathrm{d}^r\}$ of the fibration

$$\mathrm{eval}_k : \Lambda \mathbb{F}_k(\mathbb{R}^{n+1}) \to \mathbb{F}_k(\mathbb{R}^{n+1}),\ \omega \mapsto \omega(1).$$

The spectral sequence is easily seen to have the following properties:

$$(i)\ \mathrm{E}^2_{*,*} \cong H_*(\mathbb{F}_k(\mathbb{R}^{n+1})) \otimes H_*(\Omega \mathbb{F}_k(\mathbb{R}^{n+1})) \ \text{ and}$$
$$(ii)\ \mathrm{E}^2_{*,*} \cong \mathrm{E}^n_{*,*}$$

To determine how the differential operator $d^n : E^n_{*,*} \to E^n_{*,*}$ and the higher differential operators behave is essential for a successful attempt to compute the homology and cohomology of $\Lambda\mathbb{F}_k(\mathbb{R}^{n+1})$. Using Chapter X, §6, we show in Chapter XI, §1 that $d^r = 0$ for all $r > n$. Then we show that the differential

$$d^n : E^n_{*,*} \to E^n_{*,*}$$

can be described explicitly in terms of the factor tables of Chapter VI.

In Chapter XII we show that

$$H_*(\Lambda(\mathbb{F}_3(\mathbb{R}^{n+1})); \mathbb{Z}_2) \cong H_*(\Lambda(S_{21}); \mathbb{Z}_2) \otimes H_*(\Lambda(S_{31} \vee S_{32}); \mathbb{Z}_2)$$

as modules. The splitting reflects the twisted product splitting

$$\mathbb{F}_3(\mathbb{R}^{n+1}) \simeq S_{21} \ltimes (S_{31} \vee S_{32})$$

of $\mathbb{F}_3(\mathbb{R}^{n+1})$. For the case where $k > 3$, we show that the morphism $H_*(\Lambda(\mathbb{R}^{n+1}_{k-1})) \to H_*(\Lambda\mathbb{F}_k(\mathbb{R}^{n+1}))$ of the homology groups mod 2 is not injective for $k > 3$. Therefore, the splitting for the case $k = 3$ described above, does not hold for $k > 3$, and the situation is considerably more complicated. Nevertheless, we are able to estimate the rate of growth of the Poincaré series.

In Chapters XIII and XIV we consider some applications to certain variational problems of the k-body type. Such problems concern suitable action integrals f, defined on a Hilbert manifold homotopy equivalent to $\Lambda\mathbb{F}_k(\mathbb{R}^{n+1})$. Typically, one is interested in the existence of compact subspaces with arbitrarily high LS(Lusternik-Schnirelman)-category and in the growth of the Poincaré series of the homology of $\Lambda\mathbb{F}_k(M)$.

We present here two points of view. One is based on [8, Bahri-Rabinowitz], while the other is a modified version of [75, Majer-Terracini]. A basic inequality asserts that for $u = (u_1, \cdots, u_k)$ in $\Lambda\mathbb{F}_k(\mathbb{R}^{n+1})$ each loop u_i lies within a spherical disk in $\mathbb{R}^{k(n+1)}$, centered at $[u_i]$, the average value of u_i, and with radius depending on $f(u)$ and the period T. So configurations where the sum of the distances separating the k particles is above a certain bound must have subconfigurations that separate from the others. Using this idea, an interesting index, called a *collision* index, was introduced in [75, Majer-Terracini] and employed to prove the existence of critical points of f. However, that index, which involves a generalized notion of collision, is somewhat cumbersome and hard to relate to the topology of $\Lambda\mathbb{F}_k(\mathbb{R}^{n+1})$. We offer here an adaptation based on what we call the Γ-category and give an alternative proof of the main result in [75, Majer-Terracini]. On the other hand, the point of view in [8, Bahri-Rabinowitz] is that Palais-Smale sequences, which fail to be precompact, end up in the region where the configurations of k particles break up into subgroups of fewer particles. This region may be thought of as an *end* in $\Lambda\mathbb{F}_k(\mathbb{R}^{n+1})$. Exploiting the difference in the rates of growth of the Poincaré series of $\Lambda\mathbb{F}_3(\mathbb{R}^{n+1})$ and that of the subspace consisting of the configurations that break up into subsets of fewer particles, they prove that existence of an infinite set of critical points of f. In [8, Bahri-Rabinowitz] this program is carried out for $k = 3$, while the general case is treated in [91, Riahi]. We offer here a proof, in the case of $k = 3$, based on relative category and a precisely defined theory of *ends*.

VIII. The Algebra $H_*(\Omega\mathbb{F}_k(M))$

The object of this chapter is to determine the structure of the Pontryagin algebra $H_*(\Omega\mathrm{F}_k(\mathrm{M});\mathbb{K})$ when M is $\mathbb{R}^{n+1}$, or S^{n+1}, and $\mathbb{K} = \mathbb{Z}$ or $\mathbb{Z}_2$. As we have seen in the previous chapters, the two cases of $\mathbb{R}^{n+1}$, or S^{n+1} are best treated separately.

To begin, let us consider the case of $\mathbb{R}^{n+1}$. Take $\mathbb{K}$ to be $\mathbb{Z}$. For simplicity, let us denote $\mathbb{F}_{k-r,r}(\mathbb{R}^{n+1})$ by $\mathbb{F}_{k-r,r}$. As in Chapter II, §2, note that the twisted product representation

$$\mathbb{F}_{k-r,r} \simeq \rtimes_{t=r+1}^{k}(S_{t1} \vee \cdots \vee S_{tt-1})$$

of Chapter II,§4 leads to a homotopy equivalence

$$\Omega\mathbb{F}_{k-r,r} \simeq \prod_{t=r+1}^{k} \Omega(S_{t1} \vee \cdots \vee S_{tt-1}).$$

We note that $H_*(\Omega(S_{r1} \vee \cdots \vee S_{rr-1});\mathbb{Z})$ is torsion free, and, therefore, the induced morphism

$$H_*(\Omega\mathbb{F}_{k-r,r};\mathbb{Z}) \cong \bigotimes_{t=r+1}^{k} H_*(\Omega(S_{t1} \vee \cdots \vee S_{tt-1});\mathbb{Z})$$

is an isomorphism of homology groups. Moreover, $H_*(\Omega\mathbb{F}_{k-r,r})$ is a coalgebra isomorphic to the product of the coalgebras of bouquets of spheres, as indicated above. This isomorphism allows one to determine the module of primitive elements in terms of those of the homology of the loop spaces of bouquets of spheres. Our results in this regard imply that the module of the primitive elements $\mathfrak{P}(H_*(\Omega\mathbb{F}_{k-r,r};\mathbb{Q}))$ is isomorphic to $\mathcal{L}_k$, where $\mathcal{L}_k$ is the graded Lie algebra obtained by lowering degrees by one from the algebra $\pi_*(\mathbb{F}_k(\mathbb{R}^{n+1})) \otimes \mathbb{Q}$ of Chapter II, §4.

In §3 we study $H_*(\Omega\mathbb{F}_{k-r,r};\mathbb{Z})$ as a Hopf algebra. Denote by $\bar{\alpha}_{rs} : S^{n-1} \to \Omega\mathbb{F}_{k-r,r}$ the adjoint of α_{rs}. By virtue of [96, Samelson], the Y-B relations of Chapter II, §3 lead to similar relations among the elements $\mathfrak{X}_k = \{\bar{\alpha}_{rs} | 1 \leq s < r \leq k\}$. We show that the graded associative algebra over $\mathbb{Z}$, generated by $\mathfrak{X}_k$ modulo the Y-B relations, is a Hopf algebra isomorphic to the Pontryagin algebra $H_*(\Omega\mathbb{F}_{k-r,r});\mathbb{Z})$.

In §4, using the results of Chapter III and assuming that $(n+1)$ is odd, we study the algebra $H_*(\Omega\mathbb{F}_{k+1}(S^{n+1});\mathbb{Z})$. First, we note that morphism

$$\Omega(S^{n+1}) \times \Omega\mathbb{F}_k(\mathbb{R}^{n+1}) \simeq \Omega\mathbb{F}_{k+1}(S^{n+1})$$

induced by the twisted product representation $S^{n+1} \ltimes \mathbb{F}_k(\mathbb{R}^{n+1})$ of Chapter III, §4 is a homotopy equivalence. Then, using the Y-B relations of Chapter III, §5, we show that

$$H_*(\Omega(S^{n+1});\mathbb{Z}) \otimes H_*(\Omega\mathbb{F}_k(\mathbb{R}^{n+1});\mathbb{Z}) \to H_*(\Omega\mathbb{F}_{k+1}(S^{n+1});\mathbb{Z})$$

is an isomorphism of Hopf algebras.

In §5 we consider $H_*(\Omega\mathbb{F}_{k+1}(S^{n+1});\mathbb{Z})$, with $(n+1)$ being even. Using the fibration $\mathrm{proj}_{k+1,1} : \mathbb{F}_{k+1}(S^{n+1}) \to S^{n+1}$ with fiber $\mathbb{F}_k(\mathbb{R}^{n+1})$, we show that the natural imbedding $i_k : \mathbb{F}_k(\mathbb{R}^{n+1}) \to \mathbb{F}_{k+1}(S^{n+1})$ leads to an isomorphism

$$H_*(\Omega\mathbb{F}_k(\mathbb{R}^{n+1});\mathbb{Z})/(2\bar{\Delta}_k) \otimes H_*(\Omega(S^{2n+1});\mathbb{Z}) \to H_*(\Omega\mathbb{F}_{k+1}(S^{n+1});\mathbb{Z})$$

of $H_*(\Omega\mathbb{F}_{k+1}(\mathbb{R}^{n+1});\mathbb{Z})/(2\bar{\Delta}_k)$-modules, where $(2\bar{\Delta}_k)$ is the ideal generated by twice the desuspension $\bar{\Delta}_k$ of the Dirac class of Chapter III, §6, and the homology algebras have integral coefficients. Next we show that the representation

$$\Omega(O_{n+2,2}) \times \Omega\mathbb{F}_{k-2,2}(\mathbb{R}^{n+1}) \simeq \Omega\mathbb{F}_{k+1}(S^{n+1})$$

induced by the twisted product representation of Chapter III, §2, induces an isomorphism

$$H_*(\Omega(O_{n+2,2});\mathbb{Z}) \otimes H_*(\Omega\mathbb{F}_{k-2,2}(\mathbb{R}^{n+1});\mathbb{Z}) \to H_*(\Omega\mathbb{F}_{k+1}(S^{n+1});\mathbb{Z})$$

of algebras.

1 The Coalgebra $H_*(\Omega\mathbb{F}_{k-r,r})$

Throughout this section, unless indicated otherwise, homology is assumed to have integral coefficients. For the sake of simplicity, denote the space $\mathbb{F}_{k-r,r}(\mathbb{R}^{n+1})$ by $\mathbb{F}_{k-r,r}$.

Recall that the product in the Pontryagin algebra of $\Omega\mathbb{F}_{k-r,r}$ is that induced by the loop product $\Omega\mathbb{F}_{k-r,r} \times \Omega\mathbb{F}_{k-r,r} \to \Omega\mathbb{F}_{k-r,r}$. The induced product

$$H_*(\Omega\mathbb{F}_{k-r,r}) \otimes H_*(\Omega\mathbb{F}_{k-r,r}) \to H_*(\Omega\mathbb{F}_{k-r,r})$$

turns $H_*(\Omega\mathbb{F}_{k-r,r})$ into a graded associative algebra.

When $n=1$, note that the set of path components of $\Omega\mathbb{F}_k(\mathbb{R}^2)$ is isomorphic as a group to the fundamental group $\pi_1(\mathbb{F}_k(\mathbb{R}^2))$. But the latter is exactly the pure braid group (see [43, Fadell-Neuwirth] or [9, Birman]). This proves the following proposition.

Proposition 1.1 *The algebra $H_*(\Omega\mathbb{F}_k(\mathbb{R}^2);\mathbb{Z})$ is isomorphic to the group algebra of the pure braid group on k strings.*

Assume hereafter that $n > 1$. This implies that the space $\mathbb{F}_{k-r,r}$ is simply connected. Consider the fundamental fiber space diagram (Chapter II, §2),

$$\mathcal{F}_{k-r}(\mathbb{R}^{n+1}) : \begin{cases} \mathbb{F}_{k-r,r} \leftarrow \cdots \leftarrow \mathbb{F}_{2,k-2} \leftarrow \mathbb{F}_{1,k-1} \\ \quad\downarrow \qquad\qquad \downarrow \qquad\qquad \downarrow \\ \mathbb{R}^{n+1}_r \qquad \cdots \qquad \mathbb{R}^{n+1}_{k-2}, \end{cases}$$

where the projections are those that take a tuple to its first entry, and $1 \le r \le (k-1)$. As in Chapter II, §1, the set of elements $\{\alpha_{ts} | \alpha_{ts} : S^n \to \mathbb{F}_k(\mathbb{R}^{n+1}), r < t \le k, 1 \le s < t\}$ together define a section $\zeta_r : \mathbb{R}^{n+1}_r \to \mathbb{F}_{k-r,r}$, and the sections $\{\zeta_r\}$ induce a homotopy equivalence $\mathbb{F}_{k-r,r} \simeq \rtimes_{t=r+1}^{k-1} \mathbb{R}^{n+1}_t$. Passing to based loops, we obtain immediately a homotopy equivalence, $\Omega\mathbb{F}_{k-r,r} \simeq \prod_{t=r+1}^{k-1} \Omega(\mathbb{R}^{n+1}_t)$. But, for each t we have the homotopy equivalence $\mathbb{R}^{n+1}_t \simeq S_{t+11} \vee \cdots \vee S_{tt+1}$, and, hence, the induced map $\Omega(\mathbb{R}^{n+1}_t) \to \Omega(S_{t+11} \vee \cdots \vee S_{tt+1})$, is also a homotopy equivalence.
Summing up, we obtain the following proposition.

Proposition 1.2 *The sections $\{\zeta_t \mid (r+1) \le t \le k\}$ define a splitting $\Omega\mathbb{F}_k(\mathbb{R}^{n+1}) \simeq \prod_{t=r+1}^{k} \Omega(S_{t1} \vee \cdots \vee S_{tt-1})$ of topological spaces.*

In fact, we shall see later that this splitting is not that of H-spaces. But we proceed first to determine the additive structure of $H_*(\Omega\mathbb{F}_{k-r,r})$. It is completely described by the following proposition.

Proposition 1.3 *The homology groups $H_*(\Omega\mathbb{F}_{k-r,r})$ have the following properties:*

(i) $H_m(\Omega\mathbb{F}_{k-r,r}) \ne 0$ if, and only if, $m = 0 \bmod (n-1)$, and
(ii) $H_*(\Omega\mathbb{F}_{k-r,r})$ is torsion-free.

Proof. According to [12, Bott-Samelson], $H_*(\Omega(S_{t1} \vee \cdots \vee S_{tt-1}))$ is just the free associative algebra on the set $\{\bar{\alpha}_{t1}, \cdots, \bar{\alpha}_{tt-1}\}$, where $\bar{\alpha}_{ts}$ is the desuspension in $H_{n-1}(\Omega(S_{t1} \vee \cdots \vee S_{tt-1}))$ of α_{ts}. Clearly, this algebra is nontrivial exactly in the dimensions m where $m = 0 \bmod (n-1)$. As $n > 1$, the Universal Coefficient Theorem implies that $H_*(\Omega(S_{t1} \vee \cdots \vee S_{tt-1}))$ is torsion-free, and then (i) follows directly from the Künneth formula. The latter immediately implies (ii). □

Theorem 1.1

(i) The diagonal map $\Omega\mathbb{F}_{k-r,r} \to \Omega\mathbb{F}_{k-r,r} \times \Omega\mathbb{F}_{k-r,r}$ turns $H_*(\Omega\mathbb{F}_{k-r,r})$ into a coalgebra.
(ii) The twisted product representation induced by $\mathcal{F}_{k-r}$ induces an isomorphism $H_*(\Omega\mathbb{F}_{k-r,r}) \cong \bigotimes_{t=r+1}^{k} H_*(\Omega(S_{t1} \vee \cdots \vee S_{tt-1}))$ of coalgebras.

Proof. Note that the diagonal map $\Omega\mathbb{F}_{k-r,r} \to \Omega\mathbb{F}_{k-r,r} \times \Omega\mathbb{F}_{k-r,r}$ induces a morphism of modules $H_*(\Omega\mathbb{F}_{k-r,r}) \to H_*(\Omega\mathbb{F}_{k-r,r} \times \Omega\mathbb{F}_{k-r,r})$. Combining this morphism with the isomorphism of the Künneth formula, one verifies the first assertion.

To prove the second assertion, observe fact that the cross product in homology,

$$\bigotimes_{t=r+1}^{k} H_*(\Omega(S_{t1} \vee \cdots \vee S_{tt-1})) \to H_*(\prod_{t=r+1}^{k} \Omega(S_{t1} \vee \cdots \vee S_{tt-1})),$$

is one of coalgebras (see [101, Spanier], Chapter 5). Hence the homotopy equivalence $\Omega\mathbb{F}_{k-r,r}(\mathbb{R}^{n+1}) \simeq \prod_{t=r+1}^{k} \Omega(S_{t1} \vee \cdots \vee S_{tt-1})$ induces an isomorphism of coalgebras, and this establishes the assertion of (ii). □

2 The Primitives in $H_*(\Omega\mathbb{F}_{k-r,r})$

In this section, unless it is mentioned otherwise, homology is with integral coefficients. As we have seen in the previous section, the diagonal map $\Omega\mathbb{F}_{k-r,r} \to \Omega\mathbb{F}_{k-r,r} \times \Omega(\mathbb{F}_{k-r,r})$ induces a coproduct

$$H_*(\Omega\mathbb{F}_{k-r,r}) \to H_*(\Omega\mathbb{F}_{k-r,r}) \otimes H_*(\Omega\mathbb{F}_{k-r,r})$$

that turns the module $H_*(\Omega\mathbb{F}_{k-r,r})$ into a coalgebra.

Consider the pairing

$$\pi_*(\Omega\mathbb{F}_{k-r,r}) \times \pi_*(\Omega\mathbb{F}_{k-r,r}) \to \pi_*(\Omega\mathbb{F}_{k-r,r}),\ (\xi_1, \xi_2) \to < \xi_1, \xi_2 > .$$

Now, see §5 of Chapter X of [110, Whitehead], where it is proved (Theorem 5.1) that the pairing above turns $\pi_*(\Omega\mathbb{F}_{k-r,r})$ into a graded Lie algebra. Recall that the Hurewicz morphism $\phi : \pi_*(\Omega\mathbb{F}_{k-r,r}) \to H_*(\Omega\mathbb{F}_{k-r,r})$ is such that

$$\phi < \xi_1, \xi_2 >= (-1)^{pq+1}(\phi(\xi_1)\phi(\xi_2) - \phi(\xi_2)\phi(\xi_1)$$

for all $\xi_1 \in \pi_p(\Omega\mathbb{F}_{k-r,r})$ and $\xi_2 \in \pi_q(\Omega\mathbb{F}_{k-r,r})$, where ϕ is the Hurewicz morphism (see [96, Samelson]; also [110, Whitehead], §5 of Chapter X). The analogous pairing on $H_*(\Omega\mathbb{F}_{k-r,r})$ endows the latter with the structure of a graded Lie algebra, and the morphism *phi* is a morphism of graded Lie algebras. Denote by $\mathfrak{P}(H_*(\Omega\mathbb{F}_{k-r,r}))$ the module of primitive elements of this coalgebra.

We wish to describe $\mathfrak{P}(H_*(\Omega\mathbb{F}_{k-r,r}))$ in terms of the Lie algebra $L_{k-r} = \pi_*(\mathbb{F}_{k-r,r})$. Let $\tau : \pi_{p+1}(\mathbb{F}_{k-r,r}) \to \pi_p(\Omega\mathbb{F}_{k-r,r})$ be the adjoint isomorphism. Then, $\tau[\alpha, \beta] = (-1)^p < \tau\alpha, \tau\beta >$ for all $\alpha \in \pi_{p+1}(\mathbb{F}_{k-r,r})$ and

$\beta \in \pi_{q+1}(\mathbb{F}_{k-r,r})$ (see Theorem 7.10, Chapter X in [110, Whitehead]). Hence, τ induces an isomorphism $L_{k-r,r}/T \to \pi_*(\Omega\mathbb{F}_{k-r,r})/\bar{T}$ of Lie algebras, where T is the torsion submodule. By definition, let $\mathfrak{L}_{k-r} = \pi_*(\Omega\mathbb{F}_{k-r,r})/\bar{T}$. Denote by

$$\phi_{k-r,r} : \mathfrak{L}_{k-r} \to H_*(\Omega\mathbb{F}_{k-r,r}).$$

the morphism induced by ϕ. It is quite easy to verify that the image of $\phi_{k-r,r}$ consists of primitive elements.

Theorem 2.1 *The morphism $\phi_{k-r,r}$ satisfies the following properties:*

(i) $\phi_{k-r,r}$ is a monomorphism, and
(ii) $\operatorname{coker}(\phi_{k-r,r})$ is a 2-torsion group.

Proof. The proof is by descending induction on r. For $r = k-1$, we need to prove that $\phi_{1,k-1} : \mathfrak{L}_{1,k-1} \to \mathfrak{P}(H_*(\Omega\mathbb{R}^{n+1}_{k-1}))$ is injective and that $\operatorname{coker}\phi_{1,k-1}$ is a torsion group the exponent of which is two. First, recall that

$$\mathbb{R}^{n+1}_{k-1} \simeq S_{k1} \vee \cdots \vee S_{kk-1}$$

(see Chapter II, §1.) Next, recall that we have a decomposition

$$\prod_w \Omega(S^{|w|}) \simeq \Omega(S_{k1} \vee \cdots \vee S_{kk-1})$$

[55, Hilton], where w ranges over the set of basic Whitehead products $\mathcal{W}_{k-1}$ on set of elements $\{\bar{\alpha}_{rs} | 1 \leq s < r \leq (k-1)\}$ (see Chapter II, §4). The decomposition above is one of spaces and, hence, induces an isomorphism

$$\bigotimes_w H_*(\Omega(S^{|w|}) \cong H_*(\Omega(S_{k1} \vee \cdots \vee S_{kk-1}))$$

of coalgebras. Therefore,

$$\mathfrak{P}(H_*(\Omega(S_{k1} \vee \cdots \vee S_{kk-1}))) \cong \sum_w \mathfrak{P}(H_*(\Omega(S^{|w|})),$$

where w ranges over $\mathcal{W}$, the set of basic Whitehead products on $\{\bar{\alpha}_{rs} | 1 \leq s < r \leq k\}$.

Assertion 2.1 *The morphism*

$$\phi : \pi_*(\Omega(S^n))/T \to \mathfrak{P}(H_*(\Omega(S^n)))$$

is injective, and its cokernel is of order two. Here T is the torsion subgroup of $\pi_*(\Omega(S^n))$.

Recall that $H_*(\Omega(S^n)) \cong \mathcal{A}(\xi)$, where $\mathcal{A}(\xi)$ is the associative algebra in one variable ξ that generates $H_{n-1}(\Omega(S^{n-1}))$. If n is odd, $\mathcal{A}(\xi)$ is commutative, and $H_{n-1}(\Omega(S^{n-1}))$ is the module of primitives. Consequently, in this case ϕ : $(\pi_*(\Omega(S^n)/T) \to \mathfrak{P}(H_*(\Omega(S^n)))$ is an isomorphism. If n is even, then ξ and ξ^2 generate the module of primitives, and $2\xi^2$ is spherical ([110, Whitehead], Chapter XII). This proves the assertion, and the initial step of the induction follows immediately.

For simplicity, put $\Omega_{k-t} = \Omega\mathbb{F}_{k-t,t}$, and suppose that we have shown inductively that $\phi_{k-t,t} : \mathfrak{L}_{k-t,t} \to \mathfrak{P}(H_*(\Omega_{k-t}))$ for $r < t \leq (k-1)$ is injective and that every element of the torsion subgroup of coker$\phi_{k-t,t})$ is of order two. We need to prove the analogous assertions for

$$\phi_{k-t+1,t-1} : \mathfrak{L}_{k-t+1,t-1} \to \mathfrak{P}(H_*(\Omega_{k-t+1})).$$

The fibration $\mathbb{F}_{k-t+1} \to \mathbb{R}^{n+1}_{t-1}$, where a tuple is sent to the first entry, leads to the fibration $\Omega_{k-t} \to \Omega_{k-t+1}, \to \Omega(\mathbb{R}^{n+1}_{t-1})$. As $\mathbb{F}_{k-t+1} \to \mathbb{R}^{n+1}_{t-1}$ admits a section, we obtain the equivalence $\Omega_{k-t+1} \simeq \Omega_{k-t,t} \times \Omega(\mathbb{R}^{n+1}_{t-1})$. This in turn leads to the commutative diagram,

$$\begin{array}{ccccccccc} 0 \to & \mathfrak{L}_{k-t} & \to & \mathfrak{L}_{k-t+1} & \to & \mathfrak{L}_{t,t-1} & \to 0 \\ & \downarrow & & \downarrow & & \downarrow & \\ 0 \to & \mathfrak{P}(H_*(\Omega_{k-t})) & \to & \mathfrak{P}(H_*(\Omega_{k-t+1})) & \to & \mathfrak{P}(H_*(\Omega(\mathbb{R}^{n+1}_{t-1}))) & \to 0, \end{array}$$

where $\mathfrak{L}_{t,t-1}$ is $\pi_*(\Omega(\mathbb{R}^{n+1}_{t-1}))$ modulo torsion. The rows of the preceding diagram are easily seen to be split-exact, establishing the induction step. □

Next, let us consider the problem of determining the primitive submodule of the mod 2 homology $H_*(\Omega_{k-r,r};\mathbb{Z}_2)$. The procedure is similar to that of Theorem 2.1: observe that the splitting in Proposition 2.1 leads to an isomorphism of coalgebras

$$\otimes^k_{t=r+1} H_*(\Omega(\mathbb{R}^{n+1}_t;\mathbb{Z}_2) \to H_*(\Omega_{k-r,r};\mathbb{Z}_2).$$

This implies readily that

$$\mathfrak{P}(H_*(\Omega\mathbb{F}_{k-r,r});\mathbb{Z}_2) \cong \sum_{t=r+1}^{k-1} \mathfrak{P}(H_*(\Omega(\mathbb{R}^{n+1}_t);\mathbb{Z}_2)),$$

where $\mathfrak{P}$ is the module of primitive elements of the homology coalgebras. Thus the problem is reduced to finding the primitives of $H_*(\Omega(\mathbb{R}^{n+1}_t));\mathbb{Z}_2)$. This is done similarly by appealing to Hilton's decomposition ([55, Hilton]).

Theorem 2.2 *The set $\{\xi_w^{2^i} \mid i \geq 0, w \in \mathcal{W}_t\}$ is a basis for the module $\mathfrak{P}(H_*(\Omega(\mathbb{R}^{n+1}_t);\mathbb{Z}_2))$, where $\xi_w \in H_{n-1}(\Omega(S^{|w|},\mathbb{Z}_2))$ is a generator and w ranges over the set $\mathcal{W}_t$ of basic Whitehead products in $\{\bar{\alpha}_{ts} | 1 \leq s < t \leq k\}$.*

Proof. Consider the isomorphism

$$\bigotimes_w H_*(\Omega(S^{|w|})) \cong H_*(\Omega(S_{k1} \vee \cdots \vee S_{kk-1}))$$

of coalgebras induced by Hilton's decomposition,

$$\prod_w \Omega(S^{|w|}) \simeq \Omega(S_{k1} \vee \cdots \vee S_{kk-1}).$$

Recall that $H_*(\Omega(S^{|w|}); \mathbb{Z}_2) \cong \mathcal{A}(\xi_w) \otimes \mathbb{Z}_2$, where $\mathcal{A}(\xi_w)$ is the associative algebra generated by ξ_w. One easily verifies that the primitives of the latter algebra are exactly those listed in the theorem. □

3 The Hopf Algebra $H_*(\Omega\mathbb{F}_{k-r,r})$

Next, we wish to show that the algebra $H_*(\Omega\mathbb{F}_{k-r,r})$ is generated by a set of elements in one-one correspondence with the set $\mathcal{A}_{k-r,r} = \mathcal{A}_k - \mathcal{A}_r$, where $\mathcal{A}_r = \{\alpha_{ts} | 1 \leq s < t \leq r \leq k\}$, modulo a set of relations defined by the Y-B relations of Chapter II, §3. In order to do so, we need to introduce suitable notation (see Chapter V, §1).

In analogy with Chapter II, §4 and Chapter VI, §2, let

$$\Xi_m = \{\bar{\alpha}_{21}; \bar{\alpha}_{31}, \bar{\alpha}_{32}; \cdots; \bar{\alpha}_{m1}, \cdots, \bar{\alpha}_{mm-1}\},\ m \leq k$$

be a set of elements in one-one correspondence with the set of elements $\mathcal{A}_k$. Set

$$\Xi_{k-r,r} = \Xi_k - \Xi_r = \{\bar{\alpha}_{ts},\ (r+1) \leq t \leq k\}.$$

Regarding each $\bar{\alpha}_{ts}$ as the adjoint of the corresponding α_{ts}, we get maps $\bar{\alpha}_{ts} : S^{n-1} \to \Omega\mathbb{F}_k(\mathbb{R}^{n+1})$. The set $\Xi_{k-r,r}$ defines naturally elements in $H_{n-1}(\Omega\mathbb{F}_{k-r,r}(\mathbb{R}^{n+1}))$, which we denote by the same symbols.

Definition 3.1 *For $\xi_1 \in H_p(\Omega\mathbb{F}_k(\mathbb{R}^{n+1}))$ and $\xi_2 \in H_q(\Omega\mathbb{F}_k(\mathbb{R}^{n+1}))$, put $[\xi_1, \xi_2] = \xi_1\xi_2 - (-1)^{pq}\xi_2\xi_1$ The adjoint Y-B relations are the set*

$$\{\rho_\sigma \mid \rho_\sigma = (-1)^n[\bar{\alpha}_{\sigma 2\sigma 1}, \bar{\alpha}_{\sigma 3\sigma 1} + \bar{\alpha}_{\sigma 3\sigma 2}]\},$$

where σ ranges over all permutations on k letters.

Note that these relations are exactly the transcriptions, according to [96, Samelson], of the Y-B relations in the Lie algebra $\pi_*(\mathbb{F}_k(\mathbb{R}^{n+1})$ given in Chapter II, §3.

Denote by $\mathbb{Z}[\Xi_{k-r,r}]$ the graded, associative algebra generated by $\Xi_{k-r,r}$ over $\mathbb{Z}$. Also, let $\mathfrak{I}_k \subset \mathbb{Z}[\Xi_k]$ be the ideal generated by the Y-B relations, and put

$$\mathfrak{A}_{k-r,r} = \mathbb{Z}[\Xi_{k-r,r}]/\mathfrak{I}_{k-r,r}, \ \mathfrak{A}_{rr-1} = \mathbb{Z}[\Xi_{rr-1}],$$

where $\mathfrak{I}_{k-r,r} = \mathfrak{I}_k \bigcap \mathbb{Z}[\Xi_{k-r,r}]$. Before proceeding any further, we need the following concept.

Definition 3.2 *If $\mathfrak{A}$ is a an augmented, graded algebra and $\mathfrak{B} \subset \mathfrak{A}$ is a subgraded algebra, then we put*

$$\mathfrak{A}//\mathfrak{B} = \mathfrak{A}/\mathfrak{A}\bar{\mathfrak{B}}\mathfrak{A},$$

where $\bar{\mathfrak{B}}$ is the subalgebra of $\mathfrak{B}$ of augmentation 0. A sequence

$$0 \to \mathfrak{A}' \to \mathfrak{A} \to \mathfrak{A}'' \to 0$$

is said to be exact if, and only if, $\mathfrak{A}'$ is a subalgebra of $\mathfrak{A}$, and $\mathfrak{A}//\mathfrak{A}' \cong \mathfrak{A}''$.

The algebras $\mathfrak{A}_{t,t-1}$, $(r+1) \leq t \leq k$, defined above, are the factors that constitute $\mathfrak{A}_{k-r,r}$. But the product in $\mathfrak{A}_{t,t-1}$ is twisted in a fashion reflecting the Y-B relations.

Proposition 3.1 *The condition that $\bar{\alpha}_{ts} \mapsto \bar{\alpha}_{ts} \otimes 1 + 1 \otimes \bar{\alpha}_{tr}$ defines uniquely a coproduct $\mathfrak{A}_{k-r,r} \to \mathfrak{A}_{k-r,r} \otimes \mathfrak{A}_{k-r,r}$ that turns $\mathfrak{A}_{k-r,r}$ into a Hopf algebra. Moreover, $\mathfrak{A}_{k-r,r}//\mathfrak{A}_{k-r-1,r+1} \cong \mathfrak{A}_{rr-1}$, as Hopf algebras.*

Proof. First, one proves that the coproduct described in the proposition, turns the free associative algebra $\mathbb{Z}[\Xi_{k-r,r}]$ into a Hopf algebra. Next, one shows that this coproduct takes the ideal $\mathfrak{I}_{k-r,r}$ to itself. Then the proposition follows immediately. □

Now, we have the main theorem.

Theorem 3.1 *The morphism $\psi_{k-r,r} : \mathfrak{A}_{k-r,r} \to H_*(\Omega\mathbb{F}_{k-r,r})$, defined by sending each $\bar{\alpha}_{rs}$ to its counterpart in $H_*(\Omega\mathbb{F}_{k-r,r})$, is an isomorphism of Hopf algebras.*

Proof. First, note that $\psi_{k-r,r}$ is a morphism of Hopf algebras because the algebra $\mathfrak{A}_{k-r,r}$ is primitively generated by $\{\bar{\alpha}_{ts}|(r+1) \leq t \leq k\}$ and the elements $\bar{\alpha_{ts}}$ are also primitive in $H_*(\Omega\mathbb{F}_{k-r,r})$. Now, Diagram $\mathcal{F}_{k-r,r}(\mathbb{R}^{n+1})$ above has a corresponding diagram relating the algebras $\mathfrak{A}_{k-r,r}$ and $\mathfrak{A}_{rr-1}$ to the fibrations of Diagram $\mathcal{F}_{k-r,r}(\mathbb{R}^{n+1})$. To describe the companion diagram, let us put, for the sake of brevity,

$$\Omega\mathbb{F}_{k-r,r} = \Omega_{k-r,r} \ \ \Omega_r = \Omega(\mathbb{R}^{n+1}_r).$$

Then, the projection, $\mathbb{F}_{k-r,r} \to \mathbb{R}^{n+1}_r$, the fiber of which is $\mathbb{F}_{k-r-1,r+1}$, gives, in turn, the projection $\Omega_{k-r,r} \to \Omega S_r$, with fiber $\Omega_{k-r-1,r+1}$. On passing to homology, one obtains the second row of the commutative diagram

$$\begin{array}{ccccccccc} 0 \to & \mathfrak{A}_{k-r-1,r+1} & \to & \mathfrak{A}_{k-r,r} & \to & \mathfrak{A}_{rr-1} & \to 0 \\ & \downarrow & & \downarrow & & \downarrow & \\ 0 \to & H_*(\Omega_{k-r-1,r+1}) & \to & H_*(\Omega_{k-r,r}) & \to & H_*(\Omega S_r) & \to 0, \end{array}$$

where homology is with integral coefficients.

We shall prove the theorem by descending induction on r. To begin with, observe that $\Omega_{1,k-1} \simeq \Omega(S_{k1} \vee \cdots \vee S_{kk-1})$. Therefore, the morphism $\psi_{1,k-1} : \mathfrak{A}_{1,k-1} \to H_*(\Omega_{1,k-1})$ is clearly an isomorphism of Hopf algebras, as $H_*(\Omega_{1,k-1})$ is the free associative algebra on the set $\{\Xi_{kk-1}\}$. Now, assume that we have shown inductively that

$$\psi_{k-t-1,t+1} : \mathfrak{A}_{k-t-1,t+1} \to H_*(\Omega_{k-t-1,t+1}), \ (r+1) \le t \le k$$

is an isomorphism of Hopf algebras. Next, note that the principal fibration $\Omega(\mathrm{p}_t) : \Omega_{k-t,t} \to \Omega S_t$, $(r+1) \le t \le k$ is trivial, because it admits a section. Also, the fiber $\Omega_{k-t-1,t+1}$ is a sub-H-space of $\Omega_{k-t,t}$. Moreover, the projection $\Omega(\mathrm{p}_t) : \Omega_{k-t,t} \to \Omega(S_t)$ is a map of H-spaces. Therefore, the lower sequence in the diagram is an exact sequence of Hopf algebras, and

$$H_*(\Omega_{k-t-1,t+1}) \otimes H_*(\Omega(S_{t+1})) \cong H_*(\Omega_{k-t,t})$$

is an isomorphism of $H_*(\Omega_{k-t-1,t+1})$- modules. The induction assumption, together with the exactness of the lower row, implies that

$$\mathfrak{A}_{k-t-1,t+1} \to \mathfrak{A}_{k-t,t} \to 0$$

is exact. To verify that $\mathfrak{A}_{k-t,t}//\mathfrak{A}_{k-t-1,t+1} \cong \mathfrak{A}_{tt-1}$, note that, by virtue of the Y-B relations, any element $x \in \mathfrak{A}_{k-t,t}$ can be written in the form

$$x = x' + y',$$

where x' is an expression in the elements Ξ_{tt-1} alone, while y' is an expression that lies in the two-sided ideal $\mathfrak{A}_{k-t,t}(\bar{\mathfrak{A}}_{r-1})\mathfrak{A}_{k-t,t} \subset \mathfrak{A}_{k-t,t}$. Consequently, the upper sequence of the diagram is also exact. Also, the map

$$\psi_{1,t} : \mathfrak{A}_{tt-1} \to H_*(\Omega S_t)$$

is an isomorphism of Hopf algebras. Now, the five-lemma implies that the morphism $\phi_{k-t,t} : \mathfrak{A}_{k-t,t} \to H_*(\Omega_{k-t,t})$ is an isomorphism. Thus the induction step is proved. This finishes the proof of the theorem. □

Denote by $\mathfrak{P}_{k-r,r}$ the module of primitive elements in $H_*(\Omega_{k-r,r};\mathbb{Z})$. Recall that $\mathfrak{P}_{k-r,r}$ is a graded Lie algebra. Let $U(\mathfrak{P}_{k-r,r})$ be the universal enveloping algebra of $\mathfrak{P}_{k-r,r}$.

Theorem 3.2 *The natural morphism $U(\mathfrak{P}_{k-r,r}) \to H_*(\Omega_{k-r,r};\mathbb{Z})$ is an isomorphism of algebras. In particular,*

$$\begin{aligned} U(\mathfrak{P}_{k-r,r}\otimes\mathbb{Q}) &\cong U(\mathcal{L}_{k-r,r}\otimes\mathbb{Q}) \\ &\cong H_*(\Omega_{k-r,r};\mathbb{Q}), \end{aligned}$$

as Hopf algebras, where $\mathcal{L}_{k-r,r}$ is the Lie algebra $\pi_*(\mathbb{F}_{k-r,r})$ of Chapter II, §4.

Proof. Observe that it suffices to prove that $U(\mathfrak{L}_{k-r,r}) \cong \mathfrak{A}_{k-r,r}$, since $\psi_{k-r,r} : \mathfrak{A}_{k-r,r} \to H_*(\Omega_{k-r,r};\mathbb{Z})$ is an isomorphism of Hopf algebras. Arguing as in proof of the preceding theorem, we proceed inductively on t. The initial step is that $U(\mathfrak{L}_{1,k-1}) \to \mathfrak{A}_{1,k-1}$ is an isomorphism of algebras. This is so because both algebras are the free associative algebra on k letters. Suppose that we have proved that $U(\mathfrak{L}_{k-t-1,t+1}) \to \mathfrak{A}_{k-t-1,t+1}$ is an isomorphism of algebras. Consider the commutative diagram

$$\begin{array}{ccccccccc} 0 \to & U(\mathfrak{L}_{k-t-1,t+1}) & \to & U(\mathfrak{L}_{k-t,t}) & \to & U(\mathfrak{L}_{tt-1}) & \to 0 \\ & \downarrow & & \downarrow & & \downarrow & \\ 0 \to & \mathfrak{A}_{k-r-1,r+1} & \to & \mathfrak{A}_{k-r,r} & \to & \mathfrak{A}_{rr-1} & \to 0. \end{array}$$

Note that $\mathfrak{L}_{tt-1}$ is the free, graded Lie algebra generated by the set Ξ_{tt-1}. Hence, $\mathfrak{A}_{tt-1} \cong U(\mathfrak{L}_{tt-1})$. Next, observe that the horizontal lines in the diagram are exact and that the first and third vertical arrows are isomorphisms. By the five-lemma, the assertion follows. □

The following is a special case.

Corollary 3.1 *The morphism $\psi_k = \psi_{k,0} : \mathfrak{A}_k \to \mathfrak{P}_k$ is an isomorphism of Hopf algebras. Moreover, if $\mathfrak{P}_k$ is the space of primitives in the Hopf algebra $H_*(\Omega\mathbb{F}_k(\mathbb{R}^{n+1}),\mathbb{Q})$, then the natural morphism $U(\mathfrak{P}_k) \to H_*(\Omega\mathbb{F}_k(\mathbb{R}^{n+1}),\mathbb{Q})$ is an isomorphism of algebras.*

4 The Algebra $H_*(\Omega\mathbb{F}_{k+1}(S^{n+1}))$, $(n+1)$ Odd

We assume here, unless indicated otherwise, that homology is with integral coefficients. Now, we turn to the problem of describing the Hopf algebra $H_*(\Omega\mathbb{F}_{k+1}(S^{n+1}))$.

Consider the fibration

$$\mathrm{proj}_{k+1,1} : \mathbb{F}_{k+1}(S^{n+1}) \to S^{n+1},$$

where $(x_1, x_2, \cdots, x_{k+1}) \mapsto x_{k+1}$. The fiber at $-e = (-1, 0, \cdots)$ is identified with $\mathbb{F}_k(\mathbb{R}^{n+1})$ by means of the map $i_k : \mathbb{F}_k(\mathbb{R}^{n+1}) \to \mathbb{F}_{k+1}(S^{n+1})$, whereby $(x_1, \cdots, x_k) \mapsto (x_1, \cdots, x_k, -e)$ (see Chapter III, §1). Recall that, as $n+1$ is odd, there is a section $\delta'_{k+1} : S^{n+1} \to \mathbb{F}_{k+1}(S^{n+1})$ (see Chapter III, §3). This section, together with the imbedding i_k just described, induces a homotopy equivalence $\rho : \Omega\mathbb{F}_k(\mathbb{R}^{n+1}) \times \Omega(S^{n+1}) \simeq \Omega\mathbb{F}_{k+1}(S^{n+1})$ of spaces, where

$\rho(\omega) = \Omega(i_k)(\omega').\Omega(\delta')(\omega'')$, for $\omega' \in \Omega\mathbb{F}_k(\mathbb{R}^{n+1})$ and $\omega'' \in \Omega(S^{n+1})$ (see Chapter III, §3). Passing to homology with integral coefficients, and bearing in mind that the homology groups in question are torsion free, we obtain the following result.

Proposition 4.1 *There is a an isomorphism*

$$H_*(\rho) : H_*(\Omega\mathbb{F}_k(\mathbb{R}^{n+1})) \otimes H_*(\Omega(S^{n+1})) \xrightarrow{\cong} H_*(\Omega\mathbb{F}_{k+1}(S^{n+1}))$$

of coalgebras.

In fact, the isomorphism just described is one of Hopf algebras, as we shall see presently.

Theorem 4.1 *The isomorphism*

$$H_*(\rho) : H_*(\Omega\mathbb{F}_k(\mathbb{R}^{n+1})) \otimes H_*(\Omega(S^{n+1})) \xrightarrow{\cong} H_*(\Omega\mathbb{F}_{k+1}(S^{n+1}))$$

is one of Hopf algebras.

Proof. To prove the theorem, let $\{\bar{\delta}_{k+1}, \bar{\beta}_{rs}, \bar{\zeta}_{21}\}$ be the set of elements in $\pi_*(\Omega\mathbb{F}_{k+1}(S^{n+1}))$ adjoint to the corresponding elements $\{\delta_{k+1}, \beta_{rs}, \zeta_{21}\}$ in $\pi_*(\mathbb{F}_{k+1}(S^{n+1}))$, as defined in Chapter III, §§3 and 5 (cf. Theorem 5.1 and Proposition 5.1 of Chapter III). Also, denote by the same symbols the homology classes they define in $H_*(\Omega\mathbb{F}_{k+1}(S^{n+1}))$.

We need the following lemma.

Lemma 4.1 *For all admissible r, s, the relation $\bar{\delta}_{k+1}\bar{\beta}_{rs} \pm \bar{\beta}_{rs}\bar{\delta}_{k+1} = 0$ holds in $H_*(\Omega\mathbb{F}_{k+1}(S^{n+1}))$.*

Proof. The Samelson product $\bar{\delta}_{k+1}\bar{\beta}_{rs} \pm \bar{\beta}_{rs}\bar{\delta}_{k+1} = \pm\bar{\zeta}_{21}$ holds for all admissible r, s, according to Proposition 5.1 of Chapter III. Observe now that $\bar{\zeta}_{21}$ is in image$\{H_{2n-1}(\Omega(S^n)) \to H_{2n-1}(\Omega\mathbb{F}_{k+1})\}$. But, $H_{2n-1}(\Omega\mathbb{F}_k(S^n)) = 0$ for $n > 2$. Hence, $\bar{\zeta} = 0$, when $n > 2$. If $n = 2$, then (vii) of Theorem 5.2 of Chapter III implies that $\bar{\delta}_{k+1}$ commutes with each of the elements $\bar{\beta}_{rs}$. Thus the assertion is established. □

Now, to continue with the proof of the theorem, note that the set $\{\bar{\beta}_{rs}\}$ generates $H_*(\Omega\mathbb{F}_k(\mathbb{R}^{n+1}))$, as an algebra. Hence, the factors $H_*(\Omega(S^{n+1}))$ and $H_*(\Omega\mathbb{F}_k(\mathbb{R}^{n+1}))$ commute with each other. This proves the theorem. □

Definition 4.1 *Let Υ_k be a set of elements in one-to-one correspondence with the elements $\{\bar{\delta}_{k+1}, \bar{\beta}_{rs} \mid 1 \leq r < s \leq k\}$, and denote by $\mathfrak{A}'_k$ the associative, graded algebra generated by Υ_k. Also, let $\mathfrak{I}_k \subset A'_k$ denote the ideal generated by the Y-B relations of Theorems 5.1 and 5.2 of Chapter III transcribed in integral homology of $\Omega\mathbb{F}_{k+1}(S^{n+1})$.*

The algebra $\mathfrak{A}'_k$ admits a unique co-product $\mathfrak{A}'_k \to \mathfrak{A}'_k \otimes \mathfrak{A}'_k$, if the elements of Υ_k are required to be primitive. Note that the ideal $\mathfrak{J}_k$ generated by the Y-B relations is invariant under this coproduct. Thus $\mathfrak{A}'_k$ becomes a Hopf algebra.

The main theorem on the structure of the Pontryagin algebra is the following.

Theorem 4.2 *The morphism $\mathfrak{A}'_k \to H_*(\Omega\mathbb{F}_{k+1}(S^{n+1}))$ that sends an element in Υ_k to the corresponding element in the algebra $H_*(\Omega\mathbb{F}_{k+1}(S^{n+1}))$ induces an isomorphism of Hopf algebras.*

Proof. The proof of this theorem is similar to that of the analogous result for $H_*(\Omega\mathbb{F}_k(\mathbb{R}^{n+1}))$. On passing to based loops, we see that the fibration $\mathbb{F}_{k+1}(S^{n+1}) \to S^{n+1}$, the fiber of which is identified with, $\mathbb{F}_k(\mathbb{R}^{n+1})$, leads to a trivial fibration $\Omega\mathbb{F}_{k+1}(S^{n+1}) \to \Omega(S^{n+1})$, with fiber $\Omega\mathbb{F}_k(\mathbb{R}^{n+1})$. Next, let us consider the commutative diagram

$$\begin{array}{ccccccccc} 0 \to & H_*(\Omega_k)) & \to & H_*(\Omega_{k+1}) & \to & H_*(\Omega(S^{n+1})) & \to 0 \\ & \downarrow & & \downarrow & & \downarrow & \\ 0 \to & \mathfrak{A}_k & \to & \mathfrak{A}'_k & \to & \mathfrak{B} & \to 0, \end{array}$$

where, Ω_k and Ω_{k+1} denote, respectively, the based loop spaces of $\mathbb{F}_k(\mathbb{R}^{n+1})$ and $\mathbb{F}_{k+1}(S^{n+1})$, and $\mathfrak{B} = \mathfrak{A}'_k//\mathfrak{A}_k$. Here, homology is with integral coefficients. One can easily verify that $\mathfrak{B} = Z[\bar{\delta}_{k+1}]$. The upper and lower rows of the diagram are exact sequences of algebras, and the right and left vertical morphisms are isomorphisms. Hence, by the five-lemma, the middle vertical morphism is also an isomorphism, and this establishes the theorem. □

Denote the module of primitives of $H_*(\Omega\mathbb{F}_{k+1}(S^{n+1}))$ by $\mathfrak{P}_{k+1}$.

Corollary 4.1 *There is an isomorphism of $\mathfrak{P}_{k+1} \cong \mathfrak{P}_k \oplus \mathbb{Z}[\bar{\delta}_{k+1}]$, of Lie algebras, where $\mathfrak{P}_k$ is the module of primitives of $H_*(\Omega\mathbb{F}_k(\mathbb{R}^{n+1}))$, and $\mathbb{Z}[\bar{\delta}_{k+1}]$ is the trivial graded Lie algebra. Moreover, $H_*(\Omega\mathbb{F}_{k+1}(S^{n+1}); \mathbb{Q}) \cong U(\mathfrak{P}_{k+1} \otimes \mathbb{Q})$ is an isomorphism of Hopf algebras, where $U(\mathfrak{P}_{k+1} \otimes \mathbb{Q})$ is the universal enveloping algebra of $\mathfrak{P}_{k+1} \otimes$.*

5 The Algebra $H_*(\Omega\mathbb{F}_{k+1}(S^{n+1}))$, $(n+1)$ Even

Assume that $n+1 > 2$. Consider first the fibration

$$\mathrm{proj}_{k+1,1} : \mathbb{F}_{k+1}(S^{n+1}) \to S^{n+1},$$

where $(x_1, \cdots, x_{k+1})$ is mapped to x_{k+1}. The fiber is identified with $\mathbb{F}_k(\mathbb{R}^{n+1})$ by the map $i_k : \mathbb{F}_k(\mathbb{R}^{n+1}) \to \mathbb{F}_{k+1}(S^{n+1})$ (see Chapter III, §1). Let $\Delta_k(\mathbb{R}^{n+1})$ be the Dirac class in $\pi_n(\mathbb{F}_k(\mathbb{R}^{n+1})$. According to Theorem 6.1 of Chapter III, the image of the morphism $\partial_{n+1} : \pi_{n+1}(S^{n+1}) \to \pi_n(\mathbb{F}_k(\mathbb{R}^{n+1}))$ is the infinite cyclic group generated by $2\Delta_k$.

The map i_k induces a map $\Omega(i_k) : \Omega\mathbb{F}_k(\mathbb{R}^{n+1}) \to \Omega\mathbb{F}_{k+1}(S^{n+1})$. For brevity, put $\Omega_k = \Omega\mathbb{F}_k(\mathbb{R}^{n+1})$. Denote by $\bar{\Delta}_k \in \pi_{n-1}(\Omega_k)$ the element the suspension of which is Δ_k. Denote by the same symbol $\bar{\Delta}_k$ the corresponding homology class in $H_{n-1}(\Omega_k;\mathbb{Z})$. Put,

$$\mathfrak{D}_k = (2\bar{\Delta}_k)H_*(\Omega_k;\mathbb{Z}) \subset H_*(\Omega_k;\mathbb{Z}).$$

Here, $(2\bar{\Delta}_k)H_*(\Omega_k;\mathbb{Z})$ is the ideal in the Pontryagin algebra $H(\Omega_k;\mathbb{Z})$ generated by $2\bar{\Delta}_k$.

Theorem 5.1 *The map $\Omega(i_k)$ induces an exact sequence*

$$0 \to H_*(\Omega_k;\mathbb{Z})/\mathfrak{D}_k \to H_*(\Omega\mathbb{F}_{k+1}(S^{n+1});\mathbb{Z}) \to H_*(\Omega(S^{2n+1});\mathbb{Z}) \to 0$$

of algebras.

We shall give the proof in stages. To begin, let $i : S^n \to \Omega(S^{n+1})$ be the natural imbedding, and recall that there is a fiber sequence

$$S^n \xrightarrow{i} \Omega(S^{n+1}) \xrightarrow{\mathrm{p}} \Omega(S^{2n+1})$$

associated with the EHP sequence (see [84, Moore] or [110, Whitehead]). Consider the diagram

$$\begin{array}{ccccccc} \Omega_k & \to & \Omega(\mathrm{proj}_{k+1,1})^{-1}(S^n) & \to & S^n & \to & e \\ \downarrow & & \downarrow & & \downarrow & & \downarrow \\ \Omega_k & \to & \Omega\mathbb{F}_{k+1}(S^{n+1}) & \to & \Omega(S^{n+1}) & \xrightarrow{\mathrm{p}} & \Omega(S^{2n+1}) \end{array}$$

of fibrations, where the vertical arrows are the natural maps. Note that $\Omega(\mathrm{proj}_{k+1,1})^{-1}(S^n)$ is the fiber of

$$\mathrm{p} \circ \Omega(\mathrm{proj}_{k+1,1}) : \Omega\mathbb{F}_{k+1}(S^{n+1}) \to \Omega(S^{2n+1}).$$

Hence, the preceding diagram commutes, and, therefore,

$$\Omega(\mathrm{proj}_{k+1,1})^{-1}(S^n) \to S^n$$

is a fibration with fiber Ω_k.

Proposition 5.1 *There is an isomorphism*

$$H_*(\Omega(\mathrm{proj}_{k+1,1})^{-1}(S^n);\mathbb{Z}) \cong H_*(\Omega_k;\mathbb{Z})/\mathfrak{D}_k$$

of $H_*(\Omega_k;\mathbb{Z})$-modules.

To prove the proposition, we note that the E^2 term of the Serre spectral sequence of $\Omega(\mathrm{proj}_{k+1,1})|\Omega(\mathrm{proj}_{k+1,1})^{-1}(S^n) : \Omega(\mathrm{proj}_{k+1,1})^{-1}(S^n) \to S^n$ has the form $\mathrm{E}^2_{*,*} \cong H_*(S^n;\mathbb{Z}) \otimes H_*(\Omega_k;\mathbb{Z})$. Thus, for dimensional reasons, it follows that $\mathrm{d}^r : \mathrm{E}^r_{*,*} \to \mathrm{E}^r_{*,*}$ is trivial for $r < n$. Consequently,

$$\mathrm{E}^n_{*,*} \cong H_*(S^n;\mathbb{Z}) \otimes H_*(\Omega_k;\mathbb{Z}).$$

Next, regarding the fibration above as a principal Ω_k-fibration, we see that the differentials $\mathrm{d}^r : \mathrm{E}^r_{*,*} \to \mathrm{E}^r_{*,*}$ for all r commute with the action of the algebra $H_*(\Omega_k;\mathbb{Z})$ on itself by translation on the right. The proposition follows immediately from the following. Let u_n be the fundamental class of S^n.

Assertion 5.1 *The differential* $\mathrm{d}^n : \mathrm{E}^n_{n,m} \to \mathrm{E}^n_{0,(m+1)n-(m+1))}$ *takes* $u_n \otimes y$ *to* $2\bar{\Delta}_k(\mathbb{R}^{n+1}) \otimes y$.

Proof. Recall that

$$\mathrm{E}^n_{n,m(n-1)} \cong H_n(S^n) \otimes H_{m(n-1)}(\Omega_k),$$

and $\mathrm{E}^n_{0,m(n-1)} \cong H_{m(n-1)}(\Omega_k;\mathbb{Z})$. According to Theorem 6.1 of Chapter III, the $\partial : \pi_{n+1}(S^{n+1}) \to \pi_n(\mathbb{F}_k(\mathbb{R}^{n+1}))$ takes ι_{n+1} to $2\Delta_k(\mathbb{R}^{n+1})$, where ι_{n+1} is the fundamental class. On passing to the homotopy exact sequence of the associated loop space fibration, one sees that the desuspension $\iota_n \in \pi_n(\Omega(S^{n+1}))$ of ι_{n+1} goes to $2\bar{\Delta}_k(\mathbb{R}^{n+1})$. Passing to homology, one verifies the assertion. □

To finish the proof of Theorem 5.1, let us consider the fibration

$$\mathrm{p} \circ \Omega(\mathrm{proj}_{k+1,1}) : \Omega\mathbb{F}_{k+1}(S^{n+1}) \to \Omega(S^{2n+1}).$$

The E^2-term of the Serre spectral sequence is of the form

$$\mathrm{E}^2_{*,*} \cong H_*(\Omega(S^{2n+1};\mathbb{Z}) \otimes H_*(\Omega_k;\mathbb{Z})/\mathfrak{D}_k,$$

since, according to Proposition 5.1, $H_*(\Omega_k;\mathbb{Z})/\mathfrak{D}_k$ is the homology of the fiber. As n is odd, Theorem 3.1 implies that the algebra $H_*(\Omega_k;\mathbb{Z})$ is generated by even dimensional elements. Also, $H_*(\Omega(S^{2n+1};\mathbb{Z})$ is generated by a $2n$-dimensional elements. Hence, $\mathrm{E}^2_{*,*}$ consists entirely of even dimensional elements, and, therefore, $\mathrm{E}^2_{*,*} = \mathrm{E}^\infty_{*,*}$. This proves the theorem. □

Corollary 5.1 *There is an isomorphism*

$$H_*(\Omega(O_{n+2,2};\mathbb{Z}) \cong H_*(\Omega(S^n);\mathbb{Z})/2\mathfrak{D}_2 \otimes H_*(\Omega(S^{2n+1}))$$

of modules over $\mathbb{Z}$*, where* $\mathfrak{D}_2$ *is the ideal generated by the Dirac class* $\bar{\Delta}_2$.

There is another description of the algebra $H_*(\Omega\mathbb{F}_{k+1}(S^{n+1}),\mathbb{Z})$ based on the fibration $\mathrm{proj}_{k+1,3} : \mathbb{F}_{k+1}(S^{n+1}) \to \mathbb{F}_3(\mathbb{R}^{n+1}) \simeq O_{n+2,2}$ with $\mathbb{F}_{k-2,2}(\mathbb{R}^{n+1})$ as fiber. (See Chapter III, §1; also cf. Theorem 7.1 of Chapter V, §7).

Theorem 5.2 *There is an isomorphism*

$$\rho : H_*(\Omega(O_{n+2,2})) \otimes H_*(\Omega_{k-2,2}) \to H_*(\Omega\mathbb{F}_{k+1}(S^{n+1}),\mathbb{Z})$$

of algebras.

Proof. Let $\psi_{k+1} : O_{n+2,2} \to \mathbb{F}_{k+1}(S^{n+1})$ be the section defined in Chapter III, §1. Then, ψ_{k+1}, together with the imbedding $i_{k-2,2} : \mathbb{F}_{k-2,2}(\mathbb{R}^{n+1}) \to \mathbb{F}_{k+1}(S^{n+1})$ of $\mathbb{F}_{k-2,2}(\mathbb{R}^{n+1})$ as fiber, induces a homotopy equivalence

$$\Omega(\text{proj}_{k+1,3}) : \Omega(O_{n+2,2}) \times \Omega\mathbb{F}_{k-2,2} \to \Omega\mathbb{F}_{k+1}(S^{n+1}).$$

Thus, one obtains an isomorphism

$$\Phi_* : H_*(\Omega(O_{n+2,2}); \mathbb{Z}) \otimes H_*(\Omega\mathbb{F}_{k-2,2}; \mathbb{Z}) \to H_*(\Omega\mathbb{F}_{k+1}(S^{n+1}); \mathbb{Z})$$

of $H_*(\Omega\mathbb{F}_{k-2,2}; \mathbb{Z})$-modules, where $\mathbb{F}_{k-2,2}$ stands for $\mathbb{F}_{k-2,2}(\mathbb{R}^{n+1})$. Consider the affine maps

$$\psi_\omega : O_{n+2,2} \times S_1^n \to \mathbb{F}_{k+1}(S^{n+1})$$

for all $\beta_{rs} \in H_n(\Omega\mathbb{F}_{k-2,2}; \mathbb{Z})$, with $r > 2$. Observe that $\psi_\omega | O_{n+2,2} \simeq \psi_{k+1}$ and $\psi | S_1^n \simeq \omega$. Now, the induced maps

$$H_*(\Omega(\psi_\omega)) : H_*(\Omega(O_{n+2,2}); \mathbb{Z}) \times H_*(\Omega(S_1^n); \mathbb{Z}) \to H_*(\Omega\mathbb{F}_{k+1}(S^{n+1}); \mathbb{Z})$$

have the property that image$(H_*(\Omega(O_{n+2,2}); \mathbb{Z}) \to H_*(\Omega\mathbb{F}_{k+1}(S^{n+1}); \mathbb{Z})$ commutes with image$(H_*(\Omega(S_1^n); \mathbb{Z}) \to H_*(\Omega\mathbb{F}_{k+1}(S^{n+1}); \mathbb{Z})$. But, according to Theorem 3.1, the latter generate the algebra $H_*(\Omega\mathbb{F}_{k-2,2}; \mathbb{Z})$. Hence, $H_*(\Omega(O_{n+2,2}))$ commutes with the generators of the algebra $H_*(\Omega_{k-2,2})$ in $H_*(\Omega\mathbb{F}_{k+1}(S^{n+1}); \mathbb{Z})$. Thus, Φ_* is an isomorphism of algebras, and thus the theorem is established. □

6 Historical Remarks

The calculation of the algebra $H_*(\Omega\mathbb{F}_k(\mathbb{R}^{n+1}); \mathbb{Z})$ is given in [40, Fadell-Husseini]. The structure of the algebra $H_*(\Omega(\mathbb{F}_k(M)))$ for a general manifold M is the subject matter of [18, Cohen-Gitler]. Their results are based on the calculation of $H_*(\Omega(\mathbb{R}^{n+1}))$, and, although they were working independently and with a different point of view, the results they obtain overlap to some extent the results of this chapter. In [113, Xicoténcatl Merino], which is based on the methods of [18, Cohen-Gitler], one finds calculations of the Pontryagin algebras of the based loop spaces of orbit configuration spaces.

IX. RPT-Constructions

Our aim in this chapter is to represent the spaces of based and free loops $\Omega(X)$ and $\Lambda(X)$ as well as other associated spaces of a simply connected, countable CW-complex X with a single 0-vertex, e, as CW-complexes of a certain kind. In the case of $\Omega(X)$, the complex will have a nondegenerate product corresponding to the usual path multiplication. The first example of such constructions is the reduced-product complex X_∞ of [63, James]. The complexes we construct generalize his construction and will be called *constructions of the reduced-product type*, or RPT-complexes for short.

In §1 we begin by recalling some of the basic concepts and constructions of RPT-theory as given in [57, Husseini], [58, Husseini], [59, Husseini], and [60, Husseini]. We assume throughout that X is simply connected and countable. Following [63, James], such complexes will be called *special.* As X is simply connected, we assume that $X^{(0)} = X^{(1)} = \{e\}$. First we show how to find an RPT monoid $M(X)$ that models $\Omega(X)$, the space of closed paths based at e, and a right $M(X)$-complex, $E_\rho M(X)$that models $\mathcal{P}_1(X)$, the space of paths that end at e. The models fit in the diagram

$$\begin{array}{ccccc} M(X) & \to E_\rho M(X) & \to & E_\rho M(X)//M(X) = B\,M(X) \\ \downarrow \phi_F & \downarrow \phi_E & & \downarrow \phi_B \\ \Omega(X) \to & \mathcal{P}_1(X) \to & & X. \end{array}$$

The diagram commutes up to homotopy, ϕ_F is a homomorphism, ϕ_E is a map of a right $M(X)$ space to a right $\Omega(X)$ space, and all the vertical maps are homotopy equivalences.

Next, in §2 we show how to introduce a homotopy inverse, inv : $M(X) \to M(X)$. This is used in §3 to find a model for the fibration $\Lambda(X) \to X$, with the map being that which sends $\omega \in \Lambda(X)$ to $\omega(1)$. As a consequence, $\Lambda(X) \to X$ may be viewed as the *adjoint bundle* of the principal fibration $\mathcal{P}(X) \to X$.

In §3, more generally, we consider the associated space of paths,

$$\Lambda_\sigma = \{(\alpha, r) | \alpha(0) = x, \alpha(r) = \sigma(x)\},$$

where $\sigma : X \to X$ is a cellular map. We show how to construct an RPT-model for this fibration. This method provides an alternative construction for the cellular model for $\Lambda(X)$ given in §2.

With the questions regarding the Serre spectral sequence in mind (see Chapter X below), we consider in §5 the case when F is a CW-complex, and $M(X) \times F \to F$ is a cellular action of the RPT-complex $M(X)$ representing $\Omega(X)$. We define a spectral sequence, $\{\mathrm{E}^r_{*,*}, d^r\}$ of the Eilenberg-Moore type, where

$$\mathrm{E}^2_{*,*} \cong \mathrm{Tor}^{\mathrm{H}_*M}(\mathbb{K}, \mathrm{H}_*F),$$

which converges to $\mathrm{Tor}^{C_*M}(\mathbb{K}, C_*F)$, with C_*M and C_*F denoting the cellular chain complexes of M and F, respectively [28, Eilenberg-Moore].

1 RPT-Models for $\Omega(X)$

An RPT-monoid is, by definition, a special complex M with a monoid structure $M \times M \to M$, with the property that the product of two cells is again a cell. We call such a product *nondegenerate*. A *right* RPT-complex for an RPT-monoid M is a special complex N, with an action $N \times M \to N$, that is nondegenerate in the preceding sense. The *orbit space* is denoted by $N//M$. In [59, Husseini] it is proved that the natural projection, $\mathrm{proj} : N \to N//M$, is a quasi fibration. The first example of an RPT-complex is the James model X_∞ for the loop space $\Omega(S\#X)$ of the reduced suspension $S\#X$ of a special complex X. The model depends on the cellular decomposition of $S\#X$, but it is *minimal* in the sense that it is generated freely by the desuspensions of the cells of $S\#X$ or, equivalently, the cells of X.

Now, a general special complex X can be filtered,

$$X_0 \subset X_1 \subset \cdots \subset X_k \subset \cdots,$$

in such a way that $X_0 = \{e\}$ and, for each $k \geq 1$,

$$X_k = X_{k-1} \cup_{f_k} c(S\#L_k),$$

where $f_k : S\#L_k \to X_{k-1}$ and $c(L_k)$ is the reduced cone on the suspension of the CW-complex L_k of finite dimension. This filtration is the basis for an inductive construction of an RPT-model representing the space of loops $\Omega(X)$ based at e. The initial step of finding a model for the loop space $\Omega(X_1)$ is taken care of by means of the reduced product of [63, James]. Proceeding to $\Omega(X_2)$, one thinks of the process as being one of imposing *homotopy relations* on James's construction $(X_1)_\infty$, and so on.

In order to describe the process precisely, let $\Omega(X)$ and $\mathcal{P}_1(X)$ be, respectively, the Moore or (as they are called in [110, Whitehead, p.117]), *measured* loops and paths of X ending at e. The map that sends (α, r) to $\alpha(r)$ fibers $\mathcal{P}_1(X)$, with $\Omega(X)$ as fiber. Observe that the map

$$\mathcal{P}_1(X) \times \Omega(X) \to \mathcal{P}_1(X),$$

whereby $(\alpha, \omega) \mapsto \alpha * \omega$, with the convention that the composition of paths is from left to right, turns $\mathcal{P}_1(X)$ into a right $\Omega(X)$-space.

Theorem 1.1 *Given a special complex X, there is an* RPT*-complex $M(X)$, a contractible right* RPT*-complex $E_\rho M(X)$, and a homotopy commutative diagram*

$$\begin{array}{lll} M(X) \to & E_\rho M(X) \to & E_\rho M(X)//M(X) = BM(X) \\ \downarrow \phi_F & \downarrow \phi_E & \downarrow \phi_B \\ \Omega(X) \to & \mathcal{P}_1(X) \to & X, \end{array}$$

where ϕ_F is a homomorphism, ϕ_E is a map of an $M(X)$ space to an $\Omega(X)$ space, and all the vertical maps are homotopy equivalences.

Proof. To start, put

$$M_1 = (L_1)_\infty, \quad E_\rho M_1 = M_1 \cup (cL_1 \times M_1),$$

where cL_1 is the reduced cone on L_1, and the attaching map, $L_1 \times M_1 \to M_1$, is that induced by the multiplication in M_1. One can easily prove that $E_\rho M_1$ is a contractible M_1 complex and that the natural projection $EM_1 \to S\#L_1 = X_1$ is a quasi fibration. Moreover, using a path lifting function for the fibration $\mathcal{P}_1(X_1) \to X_1$, where $(\alpha, r) \mapsto \alpha(r)$, one obtains maps

$$L_1 \to \Omega(X_1), \quad M_1 \cup cL_1 \to \mathcal{P}_1(X_1)$$

such that the diagram

$$\begin{array}{lll} L^1 & \to M_1 \cup cL_1 & \to S\#L_1 = X_1 \\ \downarrow \phi_F & \quad \downarrow \phi_E & \quad \downarrow \phi_B \\ \Omega(X_1) \to & \mathcal{P}_1(X_1) \to & \quad X \end{array}$$

commutes up to homotopy. The vertical map ϕ_F can be uniquely extended to a homomorphism $\phi_F : M_1 \to \Omega(X_1)$ and the map ϕ_E, to a map $\phi_E : E_\rho M_1 = M_1 \cup (cL_1 \times M_1) \to \mathcal{P}_1(X_1)$ of an M_1-complex to an $\Omega(X_1)$ space. Then, the diagram

$$\begin{array}{lll} M_1 & \to E_\rho M_1 \to & EM_1//M_1 = BM_1 \\ \downarrow \phi_F & \downarrow \phi_E & \quad \downarrow \phi_B \\ \Omega(X_1) \to & \mathcal{P}_1(X_1) \to & \quad X \end{array}$$

commutes up to homotopy, and the vertical maps are homotopy equivalences.

The inductive step is treated similarly. Suppose we have an RPT-complex M_k, a contractible M_k-complex $E_\rho M_k$, and a homotopy commutative diagram

$$\begin{array}{ccccc} M_k & \to & E_\rho M_k & \to & E_\rho M_k//M_k = BM_k \\ \downarrow \phi_F & & \downarrow \phi_E & & \downarrow \phi_B \\ \Omega(X_k) & \to & \mathcal{P}_1(X_k)_1 & \to & X_k \end{array}$$

such that ϕ_F is a homomorphism and ϕ_E is a map of an M_k complex to an $\Omega(X_k)$-space, with the vertical maps being homotopy equivalences. Now consider

$$X_{k+1} = X_k \cup_{f_{k+1}} c(S\#L_{k+1}),$$

with L_{k+1} being a finite dimensional complex. The adjoint of the map $f_{k+1} : S\#L_{k+1} \to X_k$ can be deformed into a map

$$\bar{f}_{k+1} : L_{k+1} \to M_k.$$

Define M'_{k+1} to be $M_k \cup_{\bar{f}_{k+1}} cL_{k+1}$, and denote by M_{k+1} the universal RPT-complex containing M_k that is generated by M'_{k+1}. Put

$$E'_\rho M_{k+1} = E_\rho M_k \times_{M_k} M_{k+1}.$$

Write the $S\#L_{k+1} = c_-L_{k+1} \cup c_+L_{k+1}$, where $S\#L_{k+1}$ is the reduced suspension of L_{k+1}. There is a map $S\#L_{k+1} \to E'_\rho M_{k+1}$, where the left cone c_-L_{k+1} is mapped into $c_-L_{k+1} \to E_\rho M_k$ by a contraction of $L_{k+1} \to M_{k+1} \to EM_k$, while the right cone c_+L_{k+1} is mapped into M_{k+1} by the natural projection $c_+L_{k+1} \to M'_{k+1}$. Denote $S\#L_{k+1}$ by L'_{k+1}, and set

$$E_\rho M_{k+1} = EM_k \cup c(L'_{k+1}) \times M_{k+1}.$$

Then $E_\rho M_{k+1}$ is the desired contractible M_{k+1} complex that contains EM_k as an M_k-complex (see [59, Husseini]). Using an appropriate path lifting function, the morphisms ϕ_F and ϕ_E can be extended to yield a homotopy commutative diagram

$$\begin{array}{ccccc} M_{k+1} & \to & E_\rho M_{k+1} & \to & E_\rho M_{k+1}//M_{k+1} = BM_{k+1} \\ \downarrow \phi_F & & \downarrow \phi_E & & \downarrow \phi_B \\ \Omega(X_{k+1}) & \to & \mathcal{P}_1(X_{k+1}) & \to & X_{k+1}, \end{array}$$

subject to the conditions of the theorem. The details are straightforward and are given in [59, Husseini]. This establishes the induction step, and thus the theorem is proved. □

Let $\phi : M \to G$ be a homomorphism of the RPT-monoid M to the monoid G. Let $\mu : EM \times M \to EM$ be the action of M on the right RPT-complex EM of M. Assume that Y is a contractible space on which G acts, and denote the action by $\mu_G : Y \times G \to Y$. Choose a basepoint y_0 for Y.

Theorem 1.2 *There is a map $\Phi : EM \to Y$ that renders the diagram*

$$\begin{array}{ccc} EM \times M & \xrightarrow{\Phi \times \phi} & Y \times G \\ \downarrow \mu & & \downarrow \mu_G \\ EM & \xrightarrow{\mu} & Y \end{array}$$

commutative.

Proof. The proof proceeds along lines similar to that of Theorem 1.1. To start, let

$$M_1 \subset \cdots \subset M_k \subset M_{k+1} \subset \cdots \subset M$$

be a filtration of M by sub-RPT-monoids, such that

(*i*) $M_1 = (L_1)_\infty$, and
(*ii*) M_{k+1} is the universal RPT-monoid containing M_k as a submonoid generated by $M_k \cup_{f_k} cL_k$, where cL_k is the reduced cone on the CW-complex L_k and $f_k : L_k \to M_k$ is a cellular map.

It is easy to show (see the proof of Theorem 2.1 above or [59, Husseini]), that there is a companion filtration,

$$EM_1 \subset EM_2 \subset \cdots \subset EM_k \subset EM_{k+1} \subset \cdots \subset EM,$$

of EM by contractible subcomplexes EM_k, such that

(*i*) each EM_k is a right RPT-complex for M_k on which M_k acts freely;
(*ii*) $EM_1 = M_1 \cup_{g_1} cL_1 \times M_1$, where $M_1 = (L_1)_\infty$, $g_1 : L_1 \to M_1$ is the natural map; and
(*iii*) $EM_{k+1} = EM_k \times_{M_k} M_{k+1} \cup_{g_k} cL'_k \times M_{k+1}$ is the universal right RPT-complex for M_{k+1}, where $L'_k = S\#L_k = c_-L_k \cup c_+L_k$, and $g_k : L'_k \to EM_k \times_{M_k} M_{k+1}$ is a cellular map equal to a contraction of L_k in EM_k on the cone c_-L_k and to the natural map on c_+L_k.

To start the induction, note that the composite

$$L_1 \to M_1 \to G \to Y$$

is null homotopic, where the last map $G \to Y$ sends G to its orbit through the base point y_0 of Y. Hence, the composite above can be extended to a map

$$cL_1 \times M_1 \to Y \times G$$

of an M_1 space to a G space. The latter induces a map

$$\Phi_1 : E\,M_1 \times M_1 \to Y \times G$$

such that the diagram

$$\begin{array}{ccc} E\,M_1 \times M_1 & \overset{\Phi_1 \times \phi_1}{\to} & Y \times G \\ \downarrow \mu_1 & & \downarrow \mu_G \\ E\,M_1 & \overset{\Phi_1}{\to} & Y \end{array}$$

commutes, where $\mu_1 = \mu|(E\,M_1 \times M_1)$.

To establish the induction step, suppose that we have a map $\Phi_k : E\,M_k \to Y$ such that the diagram

$$\begin{array}{ccc} E\,M_k \times M_k & \overset{\Phi_k \times \phi_k}{\to} & Y \times G \\ \downarrow \mu_k & & \downarrow \mu_G \\ E\,M_k & \overset{\Phi_k}{\to} & Y \end{array}$$

commutes, where $\mu_k = \mu|(E\,M_k \times M_k)$. Observe that the maps $\Phi_k : E\,M_k \to Y$ and $\phi_{k+1} : M_{k+1} \to G$ induce uniquely a map

$$\Phi'_{k+1} : E\,M_k \times_{M_k} M_{k+1} \to Y.$$

In fact, one sets $\Phi'_{k+1}(x, \gamma) = \Phi_k(x).\phi_{k+1}(\gamma)$. Now, note that the composite

$$L'_k \overset{g_k}{\to} E\,M_k \times_{M_k} M_{k+1} \overset{\Phi'_{k+1}}{\to} Y$$

is null homotopic, since Y is contractible. Hence, the said composite induces a map

$$\Phi_{k+1} : E\,M_{k+1} = E\,M_k \times_{M_k} M_{k+1} \cup_{g_k} cL_k \times M_{k+1} \to Y$$

of an M_{k+1} complex to a G space. This establishes the induction step, and, therefore, the theorem is proved. □

The following result is a refinement of Theorem 1.2.

Theorem 1.3 *Let $\xi = (E_\xi, p_\xi, X, G_\xi, G_\xi)$ be a principal fiber bundle over the complex X. Then there is a diagram*

$$\begin{array}{ccccc} M(X) & \to & E_\rho M(X) & \to & E_\rho M(X)//M(X) = B\,M(X) \\ \downarrow \phi_F & & \downarrow \phi_E & & \downarrow \phi_B \\ G_\xi & \to & E_\xi & \to & X \end{array}$$

that commutes up to homotopy, where ϕ_F is a homomorphism and ϕ_E is a map of an $M(X)$ complex to a G_ξ space. Moreover, the map ϕ_B is a homotopy equivalence.

Proof. One proceeds by induction on the length of the filtration of X given above, in the manner of the proof of Theorem 1.2, taking the path lifting functions to be G_ξ-equivariant. Thus, one obtains at each stage a homomorphism $\phi_F : M_k \to G_\xi$ extending that of the previous stage. The homomorphism is readily extendable to a map $E_\rho M_k \to E_{\xi_k}$ of an M_k complex to a G_ξ space, where E_{ξ_k} is the restriction of the bundle ξ to X_k. The details are left to the reader. □

Assume now that B is a locally finite simplicial complex homotopy equivalent to X. Using simplicial paths, Milnor, in [82, Milnor], constructed a principal bundle

$$\xi_s = (EG_s, p_s, B, G_s, G_s),$$

where EG_s is contractible. Applying the previous theorem, we obtain the following corollary.

Corollary 1.1 *There is a homotopy commutative diagram*

$$\begin{array}{ccccc} M(X) & \to & E_\rho M(X) & \to & X \\ \downarrow \phi_F & & \downarrow \phi_E & & \downarrow \phi_B \\ G_s & \to & EG_s & \to & B, \end{array}$$

where ϕ_F is a homomorphism and ϕ_E is a map of an $M(X)$ complex to a G_s space, and all the vertical maps are homotopy equivalences.

Denote by $\mathcal{P}_0(X)$ the space of paths beginning at the basepoint, and put $\mathrm{p}_0 : \mathcal{P}_0(X) \to X$ equal to the map that sends a path to its endpoint. The space is a left $\Omega(X)$ space. Proceeding as in the proof of Theorem 1.1, one obtains and RPT-model for $\mathcal{P}_0(X)$. The details are left to the reader.

Theorem 1.4 *There is a contractible left* RPT*-complex $E_\lambda M(X)$ for $M(X)$ and a diagram*

$$\begin{array}{ccccc} M(X) & \to & E_\lambda M(X) & \to & E_\lambda M(X)//M(X) = B\,M(X) \\ \downarrow \phi_F & & \downarrow \phi_E & & \downarrow \phi_B \\ \Omega(X) & \to & \mathcal{P}_0(X) & \to & X, \end{array}$$

where ϕ_F is a homomorphism, ϕ_E is a map of an $M(X)$ space to an $\Omega(X)$ space, and all the vertical maps are homotopy equivalences.

2 Homotopy Inverse for $M(X)$

In order to construct RPT-models for some of the bundles associated with a principal bundle $EM \to X$, one needs an inverse map

$$\mathrm{inv} : M(X) \to M(X)$$

for the RPT-complex $M(X)$. It is well known that a homotopy inverse exists for homotopy associative H-spaces of the homotopy type of CW-complexes, but in general one does not know if the homotopy inverse is an anti-homomorphism. The result we give here assumes associativity, and the argument used in the proof depends in an essential manner on the RPT-structure.

We shall consider a more general problem. Let

$$\phi : M(X) \to G_s$$

be a homomorphism of $M(X)$ to the topological group G_s of simplicial loops, and denote by $\mathrm{inv}_s : G_s \to G_s$ the inverse-map in the group G_s.

Theorem 2.1 *Suppose that ϕ is a homotopy equivalence. Then there is a cellular anti-homomorphism* $\mathrm{inv} : M(X) \to M(X)$ *and a homotopy of anti-homomorphism* $\Phi : M(X) \times I \to G_s$ *such that*

$$\Phi|(M(X) \times \{0\}) = \phi \circ \mathrm{inv}, \quad \Phi|(M(X) \times \{1\} = \mathrm{inv}_s \circ \phi.$$

Proof. The proof is by induction on the length of the filtration

$$X_0 = \{e\} \subset X_1 \subset \cdots \subset X_k \subset \cdots \subset X,$$

where for each k

$$X_k = X_{k-1} \cup_{f_{k-1}} c(S\#L_{k-1}),$$

with L_{k-1} being a finite dimensional complex.

The theorem is certainly true for $k = 0$, for, in that case, $X_0 = \{e\}$ and $M(X_0)$ is just the two-sided unit e. Next, for the induction step, denote by M_{k-1} the RPT-complex of X_{k-1}, and assume that we have defined an anti-homomorphism $\mathrm{inv}_{k-1} : M_{k-1} \to M$ and a homotopy of antihomomorphisms $\Psi_{k-1} : M_{k-1} \times I \to G_s$ such that

$$\Psi_{k-1}|(M_{k-1} \times \{0\}) = \phi \circ \mathrm{inv}_{k-1}, \quad \Psi_{k-1}|(M_{k-1} \times \{1\}) = \mathrm{inv}_s \circ \phi.$$

Consider now the diagram

$$\begin{array}{ccccc} L_{k-1} & \overset{\bar{f}_{k-1}}{\to} & M_{k-1} & \subsetneq & M \\ & & \downarrow \phi & & \downarrow \phi \\ & & G_s & \overset{=}{\to} & G_s. \end{array}$$

As the composite of the horizontal morphisms of the upper row is null homotopic and ϕ is a homotopy equivalence, it follows that the composite

$$\phi \circ \bar{f}_{k-1} : L_{k-1} \to G_s$$

is also null homotopic.

Consider next the analogous diagram

$$\begin{array}{ccccc} L_{k-1} & \overset{\bar{f}_{k-1}}{\to} & M_{k-1} & \overset{\mathrm{inv}_{k-1}}{\to} & M \\ & & \downarrow \phi & & \downarrow \phi \\ & & G_s & \overset{\mathrm{inv}_s}{\to} & G_s, \end{array}$$

where the square is a diagram of antihomomorphisms. Note that it commutes up to homotopy. As $\phi \circ \bar{f}_{k-1}$ is null homotopic, it follows that the composite

$$\mathrm{inv}_{k-1} \circ \bar{f}_{k-1} : L_{k-1} \to M(X)$$

is also null homotopic. Hence, the map $\mathrm{inv}_{k-1} : M_{k-1} \to M(X)$ can be extended to a map

$$\mathrm{inv}'_k : M_{k-1} \cup_{\bar{f}_{k-1}} c(L_{k-1}) \to M(X)$$

such that, in the diagram

$$\begin{array}{ccccccc} L_{k-1} & \overset{\bar{f}_{k-1}}{\to} & M_{k-1} & \overset{\subsetneq}{\to} & M_{k-1} \cup L_{k-1} & \overset{\mathrm{inv}'_k}{\to} & M \\ & & \downarrow \phi & & \downarrow \phi & & \downarrow \\ & & G_s & \overset{=}{\to} & G_s & \overset{\mathrm{inv}_s}{\to} & G_s, \end{array}$$

the first square commutes, up to homotopy, relative to M_{k-1}, while the second square also commutes, up to homotopy, by a homotopy that is an extension of the inductively given homotopy $\Phi_{k-1} : M_{k-1} \times I \to G_s$. In a straightforward fashion, one extends the map $\mathrm{inv}'_{k-1} : M_{k-1} \cup cL_{k-1} \to M(X)$ to an antihomomorphism, $\mathrm{inv}_k : M_k \to M(X)$. Also, the homotopy of the antihomomorphism $\Psi_{k-1} : M_{k-1} \times I \to G_s$, inductively given, can be extended to a homotopy of antihomomorphisms $\Psi_k : M_k \times I \to G_s$, so as to make the diagram

$$\begin{array}{ccc} M_k & \overset{\mathrm{inv}_k}{\longrightarrow} & M(X) \\ \downarrow \phi & & \downarrow \phi \\ G_s & \overset{\mathrm{inv}_s}{\longrightarrow} & G_s \end{array}$$

homotopy commutative through antihomomorphisms. This finishes the proof of the inductive step. Thus the theorem is proved. □

The left $M(X)$ complex $E_\lambda M(X)$ can be obtained, up to equivalence, alternatively as follows. Observe that

$$M \times E_\rho M(X) \to E_\rho M(X),\ (\gamma, \alpha) \mapsto \alpha \,\mathrm{inv}(\gamma)$$

is a left action. Let $E'_\rho M(X)$ denote the space with this left action.

Theorem 2.2 *There is a map,* inv : $E_\lambda M(X) \to E'_\rho M(X)$, *of left* $M(X)$ *complexes such that the diagram*

$$\begin{array}{ccc} E_\lambda M(X) & \stackrel{\mathrm{inv}}{\to} & E'_\lambda M(X), \\ \downarrow & & \downarrow \\ X & \stackrel{\mathrm{id}}{\to} & X \end{array}$$

commutes, and inv *restricted to* $M(X)$ *is just the inverse map* inv.

The proof is similar to that of Theorem 2.1. The details are left to the reader.

The analogue of the map inv is the map

$$\mathrm{inv}_p : \mathcal{P}_0(X) \to \mathcal{P}_1(X),$$

where $\alpha \mapsto \alpha^{-1}$. Observe that the diagram

$$\begin{array}{ccc} \mathcal{P}_0(X) & \stackrel{\mathrm{inv}_p}{\to} & \mathcal{P}_1(X) \\ \downarrow & & \downarrow \\ X & \stackrel{\mathrm{id}}{\to} & X \end{array}$$

commutes. Also, proceeding in the same fashion, one obtains a map

$$\begin{array}{ccc} E_\lambda M & \stackrel{\Phi}{\to} & EG_s \\ \downarrow & & \downarrow \\ X & \stackrel{\simeq}{\to} & X' \end{array}$$

of an M-complex to G_s-spaces.

3 An RPT-Model for $\Lambda(X)$

We wish to find an RPT-model for $\Lambda(X)$. For the sake of brevity, put $M = M(X)$, where $M(X)$ is the RPT-model for $\Omega(X)$. First we need to recall some preparatory material.

Let $\phi : M \to G_s$ be a homomorphism of M into the group G_s of simplicial loops. For simplicity, write γ^{-1} for $\mathrm{inv}(\gamma)$. Suppose that

$$\lambda : M \times F \to F$$

is a left action of M on the space F. Denote by F_λ the space F with this action. In the usual manner, let M act on $E_\rho M \times F_\lambda$ according to the rule whereby $((y,z),\gamma) \mapsto (y.\gamma, \gamma^{-1}.z)$. This is certainly an M right action. Put

$$E_\rho M \times_M F_\lambda = (E_\rho M \times F_\lambda)//M.$$

We observe here that projection on the first factor turns the induced map $p : E_\rho M \times_M F_\lambda \to X$ into a quasi fibration.

Next, suppose that

$$\gamma_i : M \times F \to F \;\; i = 0,1$$

are two actions of M on the complex F. The actions γ_0 and γ_1 are said to be *homotopic* if, and only if, there is an action

$$\Gamma : (M \times I) \times F \to F$$

such that $\Gamma|M \times \{0\} = \gamma_0$ and $\Gamma|M \times \{1\} = \gamma_1$.

A special case of the preceding is the adjoint action $M \times M \to M$, whereby $(\omega,\gamma) \mapsto \gamma\omega\gamma^{-1})$. The induced action on $E_\rho M \times M$ is according to the rule

$$((\alpha,\omega),\gamma) \mapsto (\alpha\gamma, \gamma^{-1}\omega\gamma),$$

where $\alpha \in E_\rho M$ and $\omega, \gamma \in M$. This action turns $E_\rho M \times M$ into a right M-complex. Let us put

$$\mathrm{Ad}(E_\rho M) = E_\rho M \times_M M = (E_\rho M \times M)//M(X).$$

Theorem 3.1 *There is a fiber-preserving map*

$$\begin{array}{ccc} \mathrm{Ad}(E_\rho M) & \to & \Lambda(X) \\ \downarrow & & \downarrow \\ X & \xrightarrow{\mathrm{id}} & X \end{array}$$

that induces a homotopy equivalence.

The proof will be given in stages. Let us first describe $\Lambda(X)$ in terms of Milnor's simplicial model. Define

$$\mathrm{Ad}(EG_s) = EG_s \times_{G_s} G_s$$

to be the orbit space $(EG_s \times G_s)/G_s$, where G_s acts on the right on $EG_s \times G_s$, according to the rule $((e,g),h) \mapsto (e\,h, h^{-1}g\,h)$. Then projection on the first factor induces a map $\mathrm{Ad}(EG_s) \to X$ that fibers $\mathrm{Ad}(EG_s)$. It is, in fact, the bundle associated with the Milnor principal bundle with fiber G_s, with G_s acting on itself by the adjoint action.

Regarding EG_s as the space of simplicial paths ending at e, define the map

$$\varphi' : EG_s \times G_s \to \Lambda(X),\ (\alpha, \omega) \mapsto \alpha * \omega * \alpha^{-1}.$$

It is quite easy to verify that it induces a map $\varphi : EG_s \times_{G_s} G_s \to \Lambda(X)$.

Theorem 3.2 *The map* φ *induces a fiber homotopy equivalence*

$$\begin{array}{ccc} \mathrm{Ad}(EG_s) & \xrightarrow{\varphi} & \Lambda(X) \\ \downarrow & & \downarrow \\ X & \xrightarrow{\mathrm{id}} & X, \end{array}$$

where composition of paths is understood to be from left to right.

Proof. The proof is easily obtained. The map φ induces a morphism

$$\begin{array}{ccccc} \cdots \to & \pi_k(G_s) & \to \pi_k(\mathrm{Ad}(EG_s)) & \to \pi_k(X) \cdots \\ & \downarrow & \downarrow & \downarrow \\ \cdots \to & \pi_k(\Omega(X)) \to & \pi_k(\Lambda(X)) & \to \pi_k(X) \cdots \end{array}$$

of the homotopy exact sequence of the fibration $\mathrm{Ad}(EG_s) \to X$ to that of the fibration $\Lambda(X) \to X$. The induced morphisms on the homotopy groups of the fibers and base spaces are isomorphisms. Hence, the induced morphism on the homotopy groups of the total spaces is also an isomorphism. This means that the total space of one is weakly homotopy equivalent to that of the other. Hence, it is a homotopy equivalence, as required. □

According to Theorem 2.2 above, there is a map

$$\begin{array}{ccc} E_\rho M & \xrightarrow{\Phi} & EG_s \\ \downarrow & & \downarrow \\ X & \xrightarrow{\mathrm{id}} & X, \end{array}$$

where $\phi|M : M \to G_s$ is a homomorphism and Φ is a map of an M-complex to a G_s-space.

Next consider the quasi fibration $E_\rho M \times_M G_s \to X$ associated with the universal quasi fibration $E_\rho M \to X$, with M acting on G_s, according to the rule whereby $(g, \gamma) \mapsto (\phi(\gamma^{-1})g\phi(\gamma))$. Denote by $G_{s,1}$ the resulting M space with this action.

Lemma 3.1 *The map* $\mathrm{id} \times \phi : E_\rho M \times M \to EM \times G_s$ *induces a fiber homotopy equivalence*

$$\begin{array}{ccc} E_\rho M \times_M M & \to & E_\rho M \times_M G_{s,1} \\ \downarrow & & \downarrow \\ X & \xrightarrow{\mathrm{id}} & X. \end{array}$$

Proof. One verifies that the map

$$\mathrm{id} \times \phi : E_\rho M \times M \to E_\rho M \times G_{s,1},$$

where $(y, \omega) \mapsto (y, \phi(\omega))$, takes orbits to orbits. In fact, the composite

$$(E_\rho \times M) \times M \to (E_\rho \times M) \to (E_\rho M \times G_{s,1}),$$

which describes the action of M on the space on the left, is such that

$$((y, \omega), \gamma) \mapsto (y.\gamma, \gamma^{-1}.\omega.\gamma) \mapsto (y.\gamma, \phi(\gamma^{-1}).\phi(\omega).\phi(\gamma))$$

for $y \in E_\rho M, \omega \in M$, and $\gamma \in M$. Similarly, the composite

$$(E_\rho \times M) \times M \to (E_\rho \times G_s) \times M) \to (E_\rho M \times G_{s.1}),$$

which describes the action of M on the space on the right, is such that

$$((y, \omega), \gamma) \overset{\mathrm{id}\times\phi\times\mathrm{id}}{\mapsto} ((y, \phi(\omega)), \gamma) \mapsto (y.\omega, \phi(\gamma^{-1}).\phi(\omega).\phi(\gamma)).$$

Hence, $\mathrm{id} \times \phi$ induces a homotopy equivalence. □

Next, let $G_s \times M \to G_s$ be the action defined by the rule $(g, \gamma) \mapsto \phi(\gamma)^{-1}.g.\phi(\gamma)$. Denote by $G_{s,2}$ the space with the action just described. Recall that, according to Theorem 2.1, there is a homotopy

$$\Psi : M \times I \to G_s,$$

where each $\Psi \mid (M \times \{t\} : M \times \{t\} \to G_s$ is a homomorphism and

$$\Psi|M \times \{0\} = \mathrm{inv}_s \circ \phi, \; \Psi|(M \times \{1\} = \phi \circ \mathrm{inv},$$

with inv_s being the inverse map in G_s.

Lemma 3.2 *The homotopy Ψ induces a homotopy equivalence*

$$E_\rho M \times_M G_{s,1} \to E_\rho M \times_M G_{s,2}.$$

Proof. Note next that we have the quasi fibration

$$E_\rho M \times_M (G_s \times I) \to X,$$

where the action of M is by Ψ, which, as in the definition, contains the quasi fibrations

$$E_\rho M \times_M G_{s,1} \to X, \;\; E_\rho M \times_M G_{s,2} \to X.$$

Each is fiber homotopy equivalent to the ambient quasi fibration

$$E_\rho M \times_M (G_s \times I) \to X.$$

This implies the lemma. □

Consider next the map $\Phi \times \mathrm{id} : E_\rho M \times G_{s,2} \to E_\rho M \times G_s$.

Lemma 3.3 *The map* $\Phi \times \mathrm{id}$ *induces a homotopy equivalence*

$$E_\rho M \times_M G_{s,2} \to E_\rho M \times_{G_s} G_s.$$

Proof. Observe that the map $\Phi \times \mathrm{id} : E_\rho M \times G_{s,2} \to EG_s \times G_s$ is a map of an M space to a G_s space. In fact, the action of M on the space on the left,

$$(E_\rho M \times G_{s,2}) \times M \to E_\rho M \times G_{s,2},$$

is such that

$$((y,g),\gamma) \mapsto (y.\gamma, \phi(\gamma)^{-1}.g.\phi(\gamma)) \overset{\Phi}{\mapsto} \Phi(y).\phi(\gamma), \phi(\gamma)^{-1}.g.\phi(\gamma)).$$

Similarly, the action of G_s on the space on the right,

$$(E_\rho M \times G_{s,2}) \times G_s \to E_\rho M \times G_{s,2},$$

is such that

$$((y,g),\gamma) \overset{\Phi \times \mathrm{id} \times \phi}{\mapsto} (\Phi(y), g, \phi(\gamma)) \mapsto (\Phi(y).\phi(\gamma), \phi(\gamma)^{-1}.g.\phi(\gamma)).$$

This proves that $\Phi \times \mathrm{id}$ takes orbit to orbit and, hence, induces a map

$$\Phi \times \mathrm{id} : E_\rho M \times_M G_{s,2} \to EG_s \times_{G_s} G_{s,2}.$$

One sees easily that this latter map is a homotopy equivalence. □

Proof of Theorem 3.1

Lemmas 3.1 and 3.2 give us the sequence

$$\begin{array}{ccccc} E_\rho M \times_M & \to & E_\rho M \times_M G_{s,1} & \to & E_\rho M \times_M G_{s,2} \\ \downarrow & & \downarrow & & \downarrow \\ X & \overset{\mathrm{id}}{\to} & X & \overset{\mathrm{id}}{\to} & X \end{array}$$

of equivalences. Also, Lemma 3.3 and Theorem 3.2 give the sequence

$$\begin{array}{ccccc} E_\rho M \times_M G_{s,2} & \to & EG_s \times_{G_s} G_s & \to & \Lambda(X) \\ \downarrow & & \downarrow & & \downarrow \\ X & \overset{\mathrm{id}}{\to} & X & \overset{\mathrm{id}}{\to} & X \end{array}$$

of equivalences over X. This finishes the proof. □

4 An RPT-Model for $\Lambda_\sigma(X)$

Let $\sigma : X \to X$ be a map, and put

$$\Lambda_\sigma = \{(\alpha, r) | \alpha(r) = \sigma(\alpha(0))\}.$$

We wish to find an RPT-model for Λ_σ. Note that, if $\sigma = \mathrm{id}_X$, then $\Lambda_\sigma(X)$ is just $\Lambda(X)$. We offer here an alternative approach to the problem. It provides us with another construction for the RPT-model for $\Lambda(X)$ described in the previous section.

Now, let us consider the projection

$$\mathrm{proj} : X^{mI} \to X \times X,$$

where $(\alpha, r) \mapsto (\alpha(0), \alpha(r))$. It is a fibration with $\Omega(X)$ as fiber.

Assume that X is homotopy equivalent to a locally finite simplicial complex X', and let G_s be the Milnor group of simplicial paths of X'. Consider the action

$$(G_s \times G_s) \times G_s \to G_s, \ (\omega_1, \omega_2), \omega) \mapsto \omega_1.\omega.\omega_2^{-1}$$

for $(\omega_1, \omega_2) \in (G_s \times G_s)$ and $\omega \in G_s$. Observe that this action, when restricted to the diagonal, is just the adjoint action of G_s on itself.

Regarding $((\alpha_1, \alpha_2) \in (EG_s \times EG_s)$ and $x \in G_s$ as simplicial paths ending at the basepoint, note that the map

$$(EG_s \times EG_s) \times G_s) \to (X')^{mI}, \ ((\alpha_1, \alpha_2), \omega) \mapsto \alpha_1 \omega \alpha_2^{-1}$$

leads immediately to a map

$$\varphi : (EG_s \times EG_s) \times_{(G_s \times G_s)} G_s \to (X')^{mI}$$

that covers the identity map $X' \to X'$.

Theorem 4.1 *The map φ defines a fiber homotopy equivalence*

$$\begin{array}{ccc} (EG_s \times EG_s) \times_{(G_s \times G_s)} G_s \to & & (X')^{mI} \\ \downarrow & & \downarrow \mathrm{proj}_0 \times \mathrm{proj}_1 \\ X' \times X' & \overset{=}{\to} & X' \times X', \end{array}$$

where proj_0 *and* proj_1 *are, respectively, the projections that take a path to its initial and end points.*

Proof. Consider the commutative diagram

$$\begin{array}{ccccc} \cdots \to & \pi_k(G_s) & \to & \pi_k(E) & \to \pi_k(B) \cdots \\ & \downarrow & & \downarrow & \downarrow \\ \cdots \to & \pi_k(\Omega(X')) & \to & \pi_k((X')^{mI}) & \to \pi_k(B) \cdots \end{array}$$

of the homotopy exact sequences of the fibrations in question, where, for brevity

$$E = (EG_s \times EG_s) \times_{(G_s \times G_s)} G_s, \; B = X' \times X'.$$

The induced morphisms on the homotopy groups of the fibers and base-spaces are isomorphisms. Hence, the induced morphism on the homotopy groups of the total spaces is also an isomorphism. This means that the total space of one is weakly homotopy equivalent to that of the other. Hence, it is a homotopy equivalence, as required. □

Next, let us put

$$(EM_\rho \times EM_\lambda) \times_{(M \times M)} M = (EM_\rho \times EM_\lambda) \times M)//(M \times M),$$

where $M \times M$ acts on M according to the rule

$$((\alpha_1, \alpha_2), \omega) \mapsto \alpha_1.\omega.\alpha_2,$$

where $(\alpha_1, \alpha_2) \in (M \times M)$ and $\omega \in M$. (Recall that EM_λ is a left M space that is a model for the space of paths beginning at the base-point. See Theorem 2.2 and the remarks preceding it.)

Theorem 4.2 *There is a fiber homotopy equivalence*

$$\begin{array}{ccc} (EM_\rho \times EM_\lambda) \times_{(M \times M)} M & \rightarrow & X^{mI} \\ \downarrow & & \downarrow \\ X \times X & & = X \times X, \end{array}$$

where the vertical projections are those that send a path to its initial and its end point, respectively.

Proof. According to Corollary 1.1 above, there is a map

$$\begin{array}{ccc} E_\rho M & \xrightarrow{\Phi_\rho} & EG_s \\ \downarrow & & \downarrow \\ X & \xrightarrow{\text{id}} & X, \end{array}$$

where $\phi_\rho = \Phi_\rho|M : M \to G_s$ is a homomorphism and Φ_ρ is a map of a right M-complex to a right G_s space. Similarly, let G_s act on EG_s on the left according to the rule $(\omega, \alpha) \mapsto \omega\alpha^{-1}$, and apply Corollary 1.1 to obtain a map

$$\begin{array}{ccc} E_\lambda M & \xrightarrow{\Phi_\lambda} & EG_s \\ \downarrow & & \downarrow \\ X & \xrightarrow{\text{id}} & X, \end{array}$$

where $\phi_\lambda = \Phi_\lambda|M : M \to G_s$ is a homomorphism and Φ_λ is a map of a left M-complex to a left G_s space. The map $\Phi_\rho \times \Phi_\lambda$ induces a fiber homotopy equivalence

$$\begin{array}{ccc} (E_\rho M \times E_\lambda M) \times_{M\times M} M & \overset{\Phi_\rho\times\Phi_\lambda}{\to} & (EG_s \times EG_s) \times_{G_s\times G_s} G_s, \\ \downarrow & & \downarrow \\ X \times X & = & X \times X. \end{array}$$

Combining the homotopy equivalence induced by $\Phi_\rho \times \Phi_\lambda$ with Theorem 4.1, one proves the theorem. □

In preparation for the next result, let $M \times M \to M$ be the action whereby $(\omega, \alpha) \mapsto (\sigma\alpha)^{-1}\omega\alpha$. This action is called the *Reidemeister* action, and the space

$$(E_\rho M \times M)//M = E_\rho M \times_{M,\sigma} M \to X$$

is called the *Reidemeister* quasi fibration.

Now, let $\sigma : X \to X$ be a cellular map, and consider the space

$$\Lambda_\sigma(X) = \{(\alpha, r) \in X^{mI} | \sigma\alpha(0) = \alpha(r)\}.$$

The space, $\Lambda_\sigma(X)$, is just the space of Moore paths from x to $\sigma(x)$ for all $x \in X$.

Theorem 4.3 *There is a fiber homotopy equivalence*

$$\begin{array}{ccc} E_\rho M \times_{(M,\sigma)} M & \to & \Lambda_\sigma(X) \\ \downarrow & & \downarrow \mathrm{p} \\ X & \overset{=}{\to} & X, \end{array}$$

where $\mathrm{p}(\alpha) = \alpha(0)$.

Proof. First, we observe that the map $\sigma : X \to X$ induces a cellular homomorphism, which we denote by the same symbol, $\sigma : M(X) \to M(X)$. According to Theorem 1.2, it also induces map $E(\sigma) : E\rho M \to E_\rho M$ such that $E(\sigma)(x.\omega) = E(\sigma)(x).\sigma(\omega)$ for all x in $E_\rho M$ and ω in M. Put $E_\rho M \times_M M = (E_\rho M \times M)//M$, where M acts on itself according to the rule $(x, \omega) \mapsto \sigma(\omega^{-1}).x.\omega$.

The map $E_\rho M \to E_\rho M \times E_\rho M$, whereby $(x, \omega) \mapsto (E(\sigma)(x), x)$ for all x in $E_\rho M$, induces a map

$$\begin{array}{ccc} E_\rho M \times_{(M,\sigma)} M & \overset{\Phi_\sigma}{\to} & (E_\rho M \times E_\rho M) \times_{(M\times M)} M \\ \downarrow & & \downarrow \\ X & \to & X \times X, \end{array}$$

where the vertical maps are those induced by projection on the factors $E_\rho M$ and $E_\rho M \times E_\rho M$, respectively. On composing with the map

$$\begin{array}{ccc} (E_\rho M \times E_\rho M) \times_{(M\times M)} M & \to & X^{mI} \\ \downarrow & & \downarrow \\ X \times X & & = X \times X, \end{array}$$

where the vertical projections are those that send a path to its initial and end point, respectively, one establishes the theorem. □

5 A Cellular Spectral Sequence

Let M be an RPT-monoid and $EM \times_M F \to B$, a quasi fibration such as those representing $\Lambda(X)$ and $\Lambda_\sigma(X)$. Our aim here is to define a spectral sequence of the Eilenberg-Moore type, based on cellular chains, that converges to $H_*(EM \times_M F)$. This spectral sequence will play a key role in Chapter X, where we establish the stability of the Serre spectral sequence of the given fibration. Our approach is based on the construction of the universal quasi fibration of M given in [26, Dold-Lashof].

So,following [26, Dold-Lashof], put

$$\begin{aligned} &\mathcal{E}_0 M = M, \quad \cdots, \mathcal{E}_k M = E_{k-1} M \cup_{p_{k-1}} c\mathcal{E}_{k-1} M \times M, \cdots \\ &\mathcal{B}_0 M = \{*\}, \cdots, \mathcal{B}_k M = \mathcal{E}_k M // M, \qquad \cdots \end{aligned}$$

where $\mathcal{E}_k M \times M \to \mathcal{E}_k M$ is the action defined by the right action on the right factor and p_k is the natural projection. Observe that $\mathcal{E}_k M$ and $\mathcal{B}_k M$, for all $k \geq 0$, have a natural CW-structure and that $\mathrm{p}_k : \mathcal{E}_k M \to \mathcal{B}_k M$ is cellular. Put

$$\mathcal{E}_\infty M = \bigcup_{k\geq 0} \mathcal{E}_k M, \quad \mathcal{B}_\infty = \bigcup_{k\geq)} \mathcal{B}_k M,$$

and let $\mathrm{p}_\infty : E_\infty M \to B_\infty M$ be the natural projection. Observe that, if $M = M(X)$ for some special complex X and $EM(X) \to X$ is the universal quasi fibration given in §1 above, then we obtain the following homotopy commutative diagram

$$\begin{array}{ccccc} M(X) & \to & EM(X) & \to & X \\ \downarrow = & & \downarrow & & \downarrow \\ M & \to & \mathcal{E}_\infty M & \to & \mathcal{B}_\infty M. \end{array}$$

Next, let F be a CW-complex on which M acts cellularly. Denote the action by, $M \times F \to F$. Then

$$\mathcal{E}_\infty M \times_M F = (E_\infty M \times F)//M$$

is a CW-complex with a natural filtration

$$\mathcal{DL}(M): \left\{ \begin{array}{ccccccccc} M \times_M F = F & \subset \cdots \subset & \mathcal{E}_k M \times F & \subset \cdots \subset & \mathcal{E}_\infty M \times_M F \\ \downarrow & \downarrow & \downarrow & \downarrow & \downarrow \\ \{*\} & \subset \cdots \subset & \mathcal{B}_k M & \subset \cdots \subset & \mathcal{B}_\infty M. \end{array} \right.$$

Definition 5.1 *The cellular Eilenberg-Moore spectral sequence of the map*

$$\mathcal{E}_\infty M \times_M F \to \mathcal{B}_\infty M$$

is the spectral sequence $\{\mathrm{E}^r_{*,*}, \mathrm{d}^r\}$ *of the filtered cellular chain complex*

$$C_*F \subset \cdots \subset C_*(\mathcal{E}_k \times_M F) \subset \cdots \subset \mathcal{E}_\infty M \times_M F),$$

where $C_*(.)$ *is the cellular chain complex of* $(.)$. *We shall call it the E-M spectral sequence, for short.*

Note that $\mathcal{DL}(M)$ gives a geometric realization of the E-M spectral sequence. It provides an alternative approach to that of [95, Rothenberg-Steenrod].

Theorem 5.1 *The cellular E-M sequence* $\{\mathrm{E}^r, \mathrm{d}^r\}$ *has the following properties:*

$$(i)\ \mathrm{E}^2_{*,*} \cong \mathrm{Tor}^{H_*M}(\mathbb{K}, H_*F) \text{ and}$$
$$(ii)\ \mathrm{E}^\infty_{*,*} \cong \mathrm{Tor}^{C_*M}(\mathbb{K}, C_*F)$$
$$\cong E^0 H_*(\mathcal{E}_\infty M \times_M F; \mathbb{K}),$$

where $\mathbb{K}$ *is a PID and* H_*F *is* $\mathbb{K}$*-free.*

Proof. The proof proceeds naturally along the following lines. First note that

$$\begin{aligned} \mathrm{E}^0_{k,*} &= C_*((\mathcal{E}_k M \times_M F)/(\mathcal{E}_{k-1} M \times_M F)) \\ &\cong C_*(\mathcal{B}_k M/\mathcal{B}_{k-1} M) \otimes C_*F. \end{aligned}$$

Thus,

$$\mathrm{E}^1_{*,*} \cong C_*(\mathcal{B}_k M/\mathcal{B}_{k-1} M) \otimes H_*F,$$

and, on identifying $\mathrm{E}^1_{*,*}$ with $\sum_k C_*(\mathcal{B}_k M/\mathcal{B}_{k-1}M) \otimes H_*F$, we see that the differential $\mathrm{d}^1 : \mathrm{E}^1_{*,*} \to \mathrm{E}^1_{*,*}$ coincides with that induced by the differential on $C_*(\mathcal{B}_k M/\mathcal{B}_{k-1}M)$. Thus one deduces that $\mathrm{E}^2_{*,*} \cong \mathrm{Tor}^{H_*M}(\mathbb{K}, H_*F)$, if H_*F is $\mathbb{K}$-free.

Observe that $\mathrm{E}^2_{*,*} \Rightarrow \mathcal{E}^\infty_{*,*}$ because $\{\mathrm{E}^r_{*,*}, \mathrm{d}^r\}$ is a first quadrant sequence. Finally, the assertion that $\mathrm{E}^\infty \cong \mathrm{Tor}^{C_*M}(\mathbb{K}, C_*F)$ follows from the identification $H_*(\mathcal{E}^\infty M \times_M F) \cong \mathrm{Tor}^{C_*M}(\mathbb{K}, C_*F)$, which can be proved as follows.

Regard $\sum C_*(\mathcal{E}_k M/\mathcal{E}_{k-1}M)$ as a C_*M projective resolution of $\mathbb{K}$, and observe that $C_*(\mathcal{E}_k M \times_M F)/\mathcal{E}_{k-1}M \times_M F) \cong C_*(\mathcal{E}_k M/\mathcal{E}_{k-1}M) \otimes C_*(F)$. This allows one to regard $C_*(\mathcal{E}_\infty M \times_M F)$ as a C_*M-resolution of C_*F and, then, to deduce that $H_*(\mathcal{E}_\infty \times_M F) \cong \mathrm{Tor}^{C_*M}(\mathbb{K}, C_*F)$. Thus, the theorem is established. □

6 Historical Remarks

Adapting the combinatorial model of James, R. Cohen defines in [21] a combinatorial model for the free loop space of a suspension that incorporates the natural action of the circle group.

X. Cellular Chain Algebra Models

Our aim in this chapter is to present cellular chain algebra models for $C_*D_N(\Omega(\mathbb{F}_k(M)))$, the normalized chain algebra of singular cubes, where M is either $\mathbb{R}^{n+1}$ or S^{n+1}. The models are based on the RPT-models for $\Omega(\mathbb{F}_k(M))$ given in Chapter IX.

In §1 we prove that the cellular chain algebra $C_*(M(X))$ is isomorphic to the Adams-Hilton algebra, where $M(X)$ is an RPT- model for $\Omega(X()$ ([5, Adams-Hilton]). Thus, the RPT-model of $\Omega(X)$ is, in a manner of speaking, a geometric realization of the Adams-Hilton algebra.

In §2 we consider $X = S^{n_1} \times \cdots \times S^{n_m}$ as a CW-complex with the product cellular structure, where each S^{n_p} is a sphere of dimension n_p regarded as the CW- complex $e^0 \cup e^{n_p}$. We show that $C_*(M(X))$ is the cobar algebra of $C_*(X)$, where $M(X)$ is the RPT-model for the $\Omega(X)$. We also consider $O_{n+2,2}$ the Stiefel manifold of orthonormal 2-frames in $\mathbb{R}^{n+1}$ as a CW-complex, in the manner of Chapter III, and show that $C_*(M(O_{n+2,2}))$ is the cobar algebra of $C_*(O_{n+2,2})$.

In §§3 and 4 we prove that $C_*(M(X))$ is the cobar algebra of $C_*(X)$ when X is one of the complexes of Chapter VI homotopy equivalent to $\mathbb{F}_k(\mathbb{R}^{n+1})$ or $\mathbb{F}_{k+1}(S^{n+1})$ (cf. [1, Adams]). This result allows us to compute explicitly the differential in $C_*(M(X))$.

In §5, using the quadratic nature of the Y-B relations, as well as the filtration of $X \simeq \mathbb{F}_k(\mathbb{R}^{n+1})$ by the subcomplexes $X_{k-r,r}$ (see Chapter VI), we show that there is a homomorphism,

$$\phi : C_*(M(X)) \to H_*(\Omega(X))$$

of chain algebras, where $H_*(\Omega(X))$ has the trivial differential operator. This is a key result, which we use to prove that the Eilenberg-Moore spectral sequence of $\Lambda\mathbb{F}_k(M) \to M$ collapses when M is $\mathbb{R}^{n+1}$ or S^{n+1}.

1 The Adams-Hilton Algebra

Let X be a special CW-complex and $C_*(X)$, its cellular chain complex with integral coefficients. As $C_*(X)$ is free, we have a chain map

$$\phi^x : C_*(X) \to C_* D_N(X),$$

where $C_* D_N(X)$ is the normalized chain complex of singular cubes of X with integral coefficients. The chain map ϕ is actually a chain homotopy equivalence, because the chain complexes in question are free.

Let $M(X)$ be the RPT-complex generated by X, and denote its module of cellular chains by $\{C_*(M(X)), \partial\}$. Observe that the product

$$M(X) \times M(X) \to M(X)$$

is cellular, and, hence, it induces a morphism

$$C_*(M(X)) \otimes C_*(M(X)) \to C_*(M(X))$$

of chain groups, which turns $C_*(M(X))$ into a chain algebra.

Theorem 1.1 *There is a morphism*

$$\phi : C_*(M(X)) \to C_* D_N(\Omega(X))$$

of chain algebras that induces an isomorphism on homology.

Proof. The proof is by induction on $\dim X$. Assume that X is $(m-1)$-connected. Then we can assume, with no loss of generality, that $X^{(p)} = X^{(0)} = \{e\}$ for $p < m$ and that $X^{m)} = \bigvee_\alpha S_\alpha$, a bouquet of m-dimensional spheres S_α. Observing that $X^{(m)}$ is the suspension of a bouquet, say, $X^{(m)})'$, of $(m-1)$-dimensional spheres, one sees that $M(X^{(m)}$ is the James reduced product $((X^{(m)})')_\infty$. Moreover, its cellular chain algebra $C_*(M(X^m))$ is the free, associative, graded algebra generated by the classes corresponding to the cells $e^{(m-1}$ of $X^{(m-1)\prime}$, and the differential in $C_*(M(X^m))$ is trivial. Recall that $H_*(\Omega(X); \mathbb{Z})$ is the free, associative, graded algebra on $H_*((X^m)'; \mathbb{Z})$, since $X^{(m)}$ is a bouquet of spheres ([12, Bott-Samelson]). Now, mapping each $e^{(m-1)}$ to the cycle representing the corresponding homology class in $H_{(m-1)}(\Omega(X); \mathbb{Z})$, one obtains a morphism

$$\phi_m : C_*(M(X^{(m)}) \to C_* D_N(\Omega(X^{(m)}))$$

of algebras that induces an isomorphism in homology. Hence, ϕ_m is a homotopy equivalence. Thus the initial step is established.

The induction step is similar. Suppose that

$$\phi_{(n-1)} : C_*(M(X^{(n-1)})) \to C_* D_N(\Omega(X^{(n-1)}))$$

is a morphism of chain algebras that induces a homotopy equivalence. Put

$$X^{(n)} - X^{(n-1)} = \coprod_\alpha e^n_\alpha.$$

Recall that $M(X^{(n)})$ is the universal RPT-monoid generated by $M(X^{(n-1)})$ and the desuspension $\bar{e}^{(n-1)}\alpha$ of e^n_α. Let

$$f_\alpha : \partial D^{(n-1)} \to M(X^{(n-1)})$$

be the characteristic map of the cell $\bar{e}^{(n-1)}_\alpha$. Then, the composite

$$\partial D^{(n-1)} \stackrel{f_\alpha}{\to} M(X^{(n-1)}) \stackrel{\Phi^{(n-1)}}{\to} \Omega(X^{(n)}$$

is null homotopic. Therefore, the cycle ∂D^{n-1} is taken to a boundary ∂c_n in $C_n D_N(\Omega(X^{(n)})$ for some chain c_n of $C_n D_N(\Omega(X^{(n)}))$. Now, send $[\bar{e}_\alpha]$ to c_n. The morphism thus obtained, as $e^n\alpha$ ranges over the n dimensional cells of X, can be extended uniquely to a morphism

$$\phi_n : C_*(M(X^{(n)})) \to C_* D_N(\Omega(X^{(n-1)})),$$

which induces an isomorphism on homology with integral coefficients and, hence, is a homotopy equivalence. Thus the induction step is completed, and this finishes the proof of the theorem. □

Denote by $\{A(X), d, \theta\}$ the Adams-Hilton algebra of X. Recall that $A(X)$ is the tensor algebra of the desuspension of $C_*(X)$, $d : A(X) \to A(X)$, a suitably defined differential, and

$$\theta : A(X) \to C_* D_N(\Omega(X)),$$

a morphism of chain algebras that induces an isomorphism on homology, with $C_* D_N(\Omega(X))$ being the normalized chain algebra of singular cubes ([5, Adams-Hilton]). Observe that $C_*(M(X)) = A(X)$, as a graded algebra, but their differentials may not coincide, although they are both basically determined by the characteristic maps of the cells of X; for, in the process of defining them, one makes noncanonical choices. However, if one examines the proof of Theorem 2.1 of [5, Adams-Hilton], one readily sees that, if $C_*(M(X))$ and $A(X)$ are identified as graded algebras, then the differential $\partial : C_*(M(X)) \to C_*(M(X))$ and the morphism $\phi : C_*(M(X)) \to C_* D_N(\Omega(X))$ will do equally well for the Adams-Hilton algebra. Summing up, we have the following theorem.

Theorem 1.2 *There is a suitable differential* d : $A(X) \to A(X)$ *such that*

$$\begin{array}{ccc} C_*(M(X)) & \stackrel{\phi}{\to} & C_* D_N(\Omega(X)) \\ \iota \downarrow & & \mathrm{id} \downarrow \\ A(X) & \stackrel{\theta}{\to} & C_* D_N(\Omega(X)) \end{array}$$

is a commutative diagram of chain algebras, where ι, id *are the identity morphisms.*

In [5, Adams-Hilton] the authors define a free, contractible, twisted $A(X)$ chain complex, $C_*(X) \ltimes A(X)$ that models the space of paths $\mathcal{P}(X)$ that end at the base-point. One can show that $C_*(EM(X))$, where $EM(X)$ is the contractible RPT-complex constructed in Chapter VIII to model $\mathcal{P}(X)$, is the geometric realization of the contractible chain complex $A(X) \rtimes C_*(X)$, with an appropriate twisted differential (see [80, Mathews]).

2 An RPT-model for $\Omega(\prod_{i=1}^m S_i)$

Let S_i be an n-dimensional sphere, and regard it as the complex $S_i = e_i^0 \cup e^n$, where, respectively, e_i^0, e^n are cells of dimension 0, n. Assume throughout that $n \geq 2$ for all i. For brevity, put

$$X = S_{1,\cdots,m} = S_1 \times \cdots \times S_m,$$

and give it the product cell structure. We need a suitable RPT- model for the loop space, $\Omega(X)$.

Note that the boundary morphism $\partial : C_*(X) \to C_*(X))$ is trivial and, hence, $H_*(C(X)) = C_*(X)$. Choose a cellular approximation, $\nabla_X : X \to X \times X$, and let

$$C_*(\nabla_X) : C_*(X) \to C_*(X) \otimes C_*(X)$$

be the induced chain map. Observe that, as $H_*(C(X); \mathbb{Z}) = C_*(X)$, it follows that the coproduct $C_*(\nabla)$ is coassciative.

We wish to show that $F(C_*(X))$ is equivalent to the cellular chain algebra $C_*(M(X))$. For convenience, put

$$e_{i_1,\cdots,i_p} = e_{i_1}^n \times \cdots \times e_{i_p}^n,$$

for all $1 \leq p \leq m$, and denote its desuspension by $\bar{e}_{i_1,\cdots,i_p}$. Consider next the RPT-complex,

$$M = M(S_1 \times \cdots \times S_m).$$

Denote by the same symbol, $\bar{e}_{i_1,\cdots,i_p}$, the cellular chain that the cell defines, and observe that the cells $\{\bar{e}_{i_1,\cdots,i_p}\}$ are generators of the chain algebra $C_*(M(S_{1,\cdots,m}))$, as well as of the cobar algebra $F(C_*(S_{1,\cdots,m}))$. One would like to know how their differentials are related to each other.

Recall that, in defining the cellular structure of $M(X)$, one has to deform the desuspension of the characteristic maps of the cells of X to make them cellular. It is at this point where one has freedom in altering the choice for the boundary of the desuspended cell.

Theorem 2.1 *The cellular chain algebra $C_*(M(X))$ is isomorphic, as a chain algebra, to the cobar algebra $F(C_*(X))$.*

Before giving the proof, let us first prove the following lemma. The set of generating cells of $C_*(M(X))$, $\{\bar{e}| \subset X\}$, where e ranges over the cells of X, consists of cells of two kinds: cells e_i that correspond to the factors S_i and those that are products thereof. Let

$$\rho_m : C_*(M(X)) \to H_*(\Omega(X);\mathbb{Z}),$$

be the morphism that maps $\bar{e}$ to its homology class when e is indecomposable and to 0, otherwise. Then ρ is a morphism of chain algebras, provided that $H_*(\Omega(X)\ \mathbb{Z})$ is given the trivial boundary morphism.

Lemma 2.1 *The induced morphism,*

$$H_*(\rho) : H_*(C(M(X));\mathbb{Z}) \to H_*(\Omega(X);\mathbb{Z})$$

is an isomorphism of graded algebras.

Proof. One proceeds by induction on m. The case when $m = 1$ is obvious. The induction step depends on the commutative diagram

$$\begin{array}{ccccccccc} 0 \to & C_*(M(X_m)) & \to & C_*(M(X)) & \to & C_*(M(S_m)) & \to 0, \\ & \downarrow & & \downarrow & & \downarrow & \\ 0 \to & H(\Omega(X_m);\mathbb{Z}) & \to & H(\Omega(X);\mathbb{Z}) & \to & H(\Omega(S_m);\mathbb{Z}) & \to 0 \end{array}$$

of chain algebras, where $X_m = \{(x_1, \cdots, x_m) \mid x_m = e_m^0\}$. The vertical morphisms are those induced by ρ_m. Passing to homology, we obtain the commutative diagram

$$\begin{array}{ccccccccc} 0 \to & H_*(M(X_m)\ \mathbb{Z}) & \to & H_*(M(X);\mathbb{Z}) & \to & H_*(M(S_m);\mathbb{Z}) & \to 0, \\ & \downarrow & & \downarrow & & \downarrow & \\ 0 \to & H(\Omega(X_m);\mathbb{Z}) & \to & H(\Omega(X);\mathbb{Z}) & \to & H(\Omega(S_m);\mathbb{Z}) & \to 0 \end{array}$$

of graded algebras. Now, note that the vertical morphisms are isomorphisms by induction and that the lower sequence leads to an isomorphism

$$H_*(\Omega(X_m);\mathbb{Z}) \otimes H_*(\Omega(S_m);\mathbb{Z}) \cong H_*(\Omega(X);\mathbb{Z}),$$

while the upper leads to the isomorphism

$$H_*(M(X_m);\mathbb{Z}) \otimes H_*(M(S_m);\mathbb{Z}) \cong H_*(M(X);\mathbb{Z}).$$

This clearly implies that the morphism in the center is also an isomorphism. □

Proof of Theorem 2.1

Proceed by induction on m. The case when $m = 1$ is obvious. Suppose that the theorem is true for $(m-1)$, and let

$$\psi_{m-1} : F(C_*(X_m)) \to C_*(M(X_m))$$

be an isomorphism of chain algebras. Let $X_i = \{(x_1, \cdots, x_m) \mid x_i = e_i^0\}$, where e_i^0 is the vertex of S_i. Put $X' = \bigcup_{i=1}^m X_i$. Then $X = X' \cup e^n$, where $e^n = e^{n_1} \times \cdots \times e^{n_m}$ is the top dimensional cell of X. Note that $n = \sum_{i=1}^m n_i$.

As X' is the union of subcomplexes X_i, for $1 \leq i \leq m$, and each $X_i \cong S_{1,\cdots,m-1}$, it follows easily that there is an isomorphism

$$\psi'_m : F(C_*(X')) \to C_*(M(X'))$$

such that its restrictions to each X_i are given by the appropriate version of ψ_{m-1}. Consider now the commutative diagram

$$\begin{array}{ccccccccc} 0 \to & F(C_*(X')) & \to & F(C_*(X)) & \to & F(C_*(S^n)) & \to 0 \\ & \downarrow & & \downarrow & & \downarrow & \\ 0 \to & C_*(M(X')) & \to & C_*(M(X)) & \to & C_*(M(S^n)) & \to 0 \end{array}$$

of chain algebras and their morphisms, with the vertical morphisms being those induced by ψ_m. Here the horizontal morphisms on the right are those induced by the projection $p : X \to S^n$, which collapses X' to a point. Consider now the element

$$\omega = \sum_{(t,p-t)} \sum_{(\mu,\nu)} (-1)^{n^2\epsilon(\mu)} [\bar{e}_{\mu(1),\cdots,\mu(t)}][\bar{e}_{\nu(t+1),\cdots,\nu(p)}],$$

where (μ, ν) ranges over all $(t, p-t)$-shuffles of p elements of $F(C_*(X'))$. Here $\epsilon(\mu) = \sum_{i=1}^t \mu(i) - (i-1)$. It is the signature of the permutation that corresponds to the shuffle (μ, ν). It is easy to show that $d([\bar{e}^{n-1}] = \omega$ in $F(C_*(X)$. Hence, $d(\omega) = 0$ in $F(C_*(X'))$, which implies that $\partial(\psi'_m(\omega)) = 0$. Observe here that the projection $\rho_m : C_*(X) \to H_*(X; \mathbb{Z})$ takes ω to zero. Therefore, ω bounds in $C_*(M(X'))$.

Observe that the pair (X, X') is $(n-1)$-connected and that the characteristic map

$$f : (D^n, S^{n-1}) \to (X, X')$$

of the cell e^n generates the relative group $\pi_n(X, X', *)$. This group is infinite cyclic. Hence, its desuspension

$$\bar{f} : (D^{n-1}, S^{n-2}) \to (M(X), M(X'))$$

generates $\pi_{n-1}(X), M(X'))$. Therefore, the Hurewicz homomorphism

$$\pi_{n-1}(\Omega(X), \Omega(X') \to H_{n-1}((\Omega(X), \Omega(X'); \mathbb{Z})$$

is an isomorphism, and the groups in question are infinite cyclic generated by the class $[\bar{e}^{n-1}]$.

Also note that

$$H_{n-1}(F(C_*(X); \mathbb{Z}), F(C_*(X'))) \cong \mathbb{Z},$$

because (X, X') is $(n-1)$-connected, and that the latter group is generated by $[\bar{e}^{n-1}]$. Hence,

$$\omega = d(\lambda) + r\partial(\bar{e}^{n-1})$$

for some $\lambda \in C_*(M(X')) = F(C_*(X'))$ and an integer r. But none of the summands making up ω can appear in the boundary of any element in the cobar algebra $F(C_*(X'))$. Hence, $\lambda = 0$. Also, ω bounds $[\bar{e}^{n-1}]$ in $F(C_*(X))$. This implies that $\partial([e^{n-1}] = \omega$, in $C_*(M(X))$ and, therefore, ψ'_m can be extended to an isomorphism $F(C_*(X)) \to C_*(M(X))$. This finishes the induction step, and the theorem is thus established. □

Corollary 2.1 *The algebra $C_*(M(X))$ is the free, associative, chain algebra generated by the elements $[\bar{e}_{i_1,\cdots,i_p}]$, with $1 \le p \le m$. Moreover, the differential operator*

$$\partial : C_*(M(X)) \to C_*(M(X))$$

of $C_*(M(X))$ is given by rule

$$\partial(\bar{e}_{i_1,\cdots,i_p}) = \sum_{(t,p-t)} \sum_{(\mu,\nu)} (-1)^{n^2\epsilon(\mu)} \bar{e}_{\mu(i_1),\cdots,\mu(i_t)} \bar{e}_{\nu(i_{t+1}),\cdots,\nu(i_p)},$$

where (μ, ν) ranges over all $(t, p-t)$-shuffles of p elements.

Proof. It is easy to verify that the multifold cross product

$$\begin{aligned} C_*(S_1) \otimes \cdots \otimes C_*(S_m) &\to C_*(X) \\ &= C_*(S^{n_1} \times \cdots \times C_*(S^{n_m}), \end{aligned}$$

under which

$$1 \otimes \cdots \otimes 1 \otimes e_i^n \otimes 1 \otimes \cdots \otimes 1 \mapsto 1 \otimes \cdots \otimes 1 \otimes e_i^n \otimes 1 \otimes \cdots \otimes 1$$

for $i = 1, \cdots, m$, is an isomorphism of coalgebras. One sees easily that the differential is as stated, because the formula in question is just the transcription of how the coproduct in $C_*(S_{1,\cdots,m})$ determines the differential in the cobar algebra, $F(C_*(X))$. But the latter is isomorphic to the cellular chain algebra $C_*(M(X))$. □

There is an analogous result for $O_{n+2,2} \times S_2 \times \cdots \times S_m)$, where n is odd and each S_i for $2 \leq i \leq m$ is a sphere of dimension n. According to Proposition 5.1 of Chapter III, $O_{n+2,2}$ is homotopy equivalent to the CW-complex,

$$\begin{aligned} O_{n+2,2} &\simeq (S^{n+1} \vee S^n) \cup_\mu D^{2n+1} \\ &= e^0 \cup e^{n+1} \cup e^n \cup e^{2n+1}, \end{aligned}$$

where μ is in the homotopy class of $[\iota_{n+1}, \iota_n] + \zeta$. Now, choose a cellular approximation $\nabla_{n+2,2} : O_{n+2,2} \to O_{n+2,2} \times O_{n+2,2}$ to the diagonal map. Passing to cellular chains, one obtains the coproduct

$$C_*(\nabla_{O_{n+2,2}}) : C_*(O_{n+2,2}) \to C_*(O_{n+2,2}) \otimes C_*(O_{n+2,2}).$$

As $C_*(O_{n+2,2}) = H_*(C(O_{n+2,2}))$, one sees immediately that the coproduct $C_*(\nabla_{O_{n+2,2}})$ is coassociative. Similarly, choose a cellular approximation to the diagonal map

$$\nabla_{S_{2,\cdots,S_m}} : S_{2,\cdots,S_m} \to S_{2,\cdots,S_m} \times S_{2,\cdots,S_m},$$

where $S_{2,\cdots,m} = S_2 \times \cdots \times S_m$ has the product cell structure. Pass to the cellular chains to obtain the coproduct

$$C_*(\nabla_{S_{2,\cdots,S_m}}) : C_*(S_{2,\cdots,S_m}) \to C_*(S_{2,\cdots,S_m}) \otimes C_*(S_{2,\cdots,S_m}).$$

Again, this coproduct is coassociative, since the boundary homomorphism of $C_*(S_{2,\cdots,S_m})$ is trivial. Now give

$$C_*(O_{n+2,2}) \otimes C_*(S_{2,\cdots,S_m}) \cong C_*(O_{n+2,2} \times S_{2,\cdots,m})$$

the product coalgebra structure.

Theorem 2.2 *The chain algebra $C_*(M(O_{n+2,2} \times S_{2,\cdots,m}))$ of the* RPT-*monoid generated by $O_{n+2,2} \times S_{2,\cdots,m}$ is isomorphic as a chain algebra to the cobar algebra $F(C_*(O_{n+2,2} \times S_{2,\cdots,m}))$.*

Proof. The proof is by induction on m, the initial step being when $m = 2$. In this case, one needs to prove that $C_*(M(O_{n+2,2})) \overset{\text{id}}{\cong} F(C_*(O_{n+2,2}))$ is an isomorphism of chain algebras. The module $C_i(O_{n+2,2})$ is generated by $[e^n], [e^{n+1}]$ and $[e^{2n+1}]$ when $i = 1, n$ and $2n+1$, respectively. Now it is easy to see that coproduct in $C_*(O_{n+2,2})$ is such that

$$e^i \mapsto e^i \otimes 1 + 1 \otimes e^i$$

for $i = n$ and $i = n+1$. For $i = 2n+1$, the algebra structure of $H^*(O_{n+2,2}, \mathbb{Z})$ implies immediately that

$$e^{2n+1} \mapsto e^{2n+1} \otimes 1 + 1 \otimes e^{2n+1} + e^n \otimes e^{n+1} + (-1)^{n(n+1)} e^{n+1} \otimes e^n.$$

On the other hand, the desuspension of

$$\mu : (D^{2n+1}, S^{2n}) \to (O_{n+2,2}, S^{n+1} \vee S^n),$$

the characteristic map of the cell e^{2n+1}, is in the class of the elements $[\bar{\iota}_{n+1}, \iota_n] + \bar{\zeta}$ in $C_*(M(S^{n+1} \vee S^n))$. Note that $\bar{\zeta}$ is trivial in $C_*(M(S^{n+1}) \vee S^n))$ because ζ is of finite order. According to Samelson's theorem ([96, Samelson]), one has

$$\overline{[\iota_n, \iota_{n+1}]} = \pm(\bar{\iota}_{n+1}\bar{\iota}_n - (-1)^{(i+1)i}\bar{\iota}_n\bar{\iota}_{n+1})$$

in $C_*(M(S^{n+1} \vee S^n))$. Hence,

$$\partial(\bar{e}^{2n+1}) = \pm(\bar{e}^{n+1}\bar{e}^n - \bar{e}^n\bar{e}^{n+1})$$

in $C_*(M(O_{n+2,2}))$. This proves the initial step.

The rest of the proof, which is entirely similar to that of Theorem 2.1, is left to the reader. □

3 $C_*(M(X_{k-r,r}))$, $X_{k-r,r} \simeq \mathbb{F}_{k-r,r}$

Let $X_{k-r,r} \simeq \mathbb{F}_{k-r,r}(\mathbb{R}^{n+1})$ be the minimal CW- complex constructed in Chapter VI, §8. Assume that $n > 1$. Hence, $X_{k-r,r}$ is simply connected, and the boundary morphism of cellular chain $C_*(X_{k-r,r})$ is trivial. Denote by $M(X_{k-r,r})$ the RPT-complex generated by $X_{k-r,r}$. Choose a cellular approximation $\nabla : X_{k-r,r} \to X_{k-r,r} \times X_{k-r,r}$ for the diagonal map, and let

$$C_*(\nabla) : C_*(X_{k-r,r}) \to C_*(X_{k-r,r}) \otimes C_*(X_{k-r,r})$$

be the induced chain map. As the $C_*(X_{k-r,r}) = H_*(X_{k-r,r}; \mathbb{Z})$, it follows readily that the chain complex $C_*(X_{k-r,r})$ is a coassciative coalgebra.

Theorem 3.1 *The cellular chain algebra $C_*(M(X_{k-r,r}))$ of the* RPT- *complex $M(X_{k-r,r})$ is isomorphic, as a chain algebra, to the cobar algebra $F(C_*(X_{k-r,r}))$ of the coalgebra.*

Proof. Proceed by induction on the dimension of the generating (i.e., indecomposable) cells of $C_*(M(X_{k-r,r}))$. According to Chapter VII, §2, the indecomposable cells of $M(X_{k-r,r})$ are in one-one correspondence with the basic elements

$$\omega = \alpha_{\underline{r}\,\underline{s}} = \alpha_{r_1 s_1} \rtimes \cdots \rtimes \alpha_{r_m s_m},$$

as $(\underline{r}\,\underline{s})$ ranges over all indices such that $1 \le s_i < r_i$, and $r < r_1 < \cdots < r_m \le k$. Denote by $\bar{e}_\omega$ the indecomposable cell of $M(X_{k-r,r})$ that corresponds to

ω, and note that $\dim \bar{e}_\omega = (n-1) \times |\underline{r}|$ with ω being the basic element above. To prove the theorem, we must show that if

$$\omega \mapsto \omega \otimes 1 + 1 \otimes \omega + \sum_i \omega_i' \otimes \omega_i''$$

under the coproduct in $C_*(X_{k-r,r})$, then

$$\partial(\bar{e}_\omega) = \sum_i \bar{\omega}_i' \bar{\omega}_i''.$$

To do so, we consider the cellular map

$$\phi_\omega' : S_{1,\cdots,m} = S_1 \times \cdots S_m \to X_{k-r,r}$$

homotopic to the affine map

$$\phi_\omega : S_1 \times \cdots S_m \to \mathbb{F}_{k-r,r}(\mathbb{R}^{n+1}) \simeq X_{k-r,r}$$

of ω, as defined in Chapter VI, §2. The map, ϕ_ω', induces a cellular map

$$M(\phi_\omega') : M(S_{1,\cdots,m}) \to M(X_{k-r,r}),$$

where $M(S_{1,\cdots,m})$ is the RPT-complex generated by the usual product cellular structure of spheres on $S_{1,\cdots,m}$. According to Theorem 2.1, $C_*(M(X))$ is isomorphic to the cobar algebra $F(C_*(S_{1,\cdots,m}))$. For simplicity, denote by $e_{1,\cdots,m}$ the fundamental class of $S_{1,\cdots,m}$. Hence, the boundary of $\bar{e}_{1,\cdots,m}$ is given by the expression

$$\sum_{(t,m-t)} \sum_{(\mu,\nu)} (-1)^{n^2 \epsilon(\mu)} \bar{e}_{\nu(1),\cdots,\nu(t)} \otimes \bar{e}_{\mu(t+1),\cdots,\mu(m)},$$

where the sum is over all $(t, m-t)$-shuffles (μ, ν) of the set $\{1, \cdots, m\}$. To finish the proof, we need to compute the effect of the chain morphism $C_*(M(\phi_\omega'))$ induced by $M(\phi_\omega')$. This is achieved by looking at the factor table of ω. So, let λ_i be the ith-row of the factor table of $\omega = \alpha_{\underline{r}\,\underline{s}}$, and put

$$\lambda_{i_1,\cdots,i_p} = \lambda_{i_1} \rtimes \cdots \rtimes \lambda_{i_p}.$$

Denote by $\bar{\lambda}_{i_1,\cdots,i_p}$ its desuspension. (Recall that

$$\lambda_{i_1} \rtimes \cdots \rtimes \lambda_{i_p} = \sum_{\underline{r}} \pm \alpha_{\underline{r}\,\underline{s}},$$

where $(\underline{r}, \underline{s})$ ranges over a set of indices determined entirely by the rows λ_i of the factor table.) One easily verifies that, $\bar{\lambda}_{i_1,\cdots,i_p} = \sum_{\underline{r}} \pm \bar{\alpha}_{\underline{r}\,\underline{s}}$, where $(\underline{r}, \underline{s})$ ranges over the same set of indices as those that appear in the preceding expansion of $\lambda_{i_1,\cdots,i_p}$. Now, by naturality, one sees that

$$C_*(M(\phi'_\omega))(e_{1,\cdots,m}) = \bar{\lambda}_{1,\cdots,m}.$$

This implies that the expression for the boundary of the cell $\bar{e}_\omega = \bar{e}_{\alpha_{\underline{r}\,\underline{s}}}$ in $C_*(M(X))$ is

$$\bar{\alpha}_{\underline{r}\,\underline{s}} \mapsto \sum_{(t,p-t)} \sum_{(\mu,\nu)} (-1)^{n^2\epsilon(\mu)} \bar{\lambda}_{\mu(i_1),\cdots\mu(i_t)} \otimes \bar{\lambda}_{\nu(i_{t+1}),\cdots,\nu(i_m)},$$

with the summations being over all (μ,ν) $(t, m-t)$-shuffles of m elements. But this expression is the boundary of the cell $\bar{e}_\omega$ in the cobar algebra $F(C_*(X_{k-r,r}))$. This finishes the proof of the theorem. □

4 $C_*(M(Y_{k+1}), Y_{k+1} \simeq \mathbb{F}_{k+1}(S^{n+1})$, $(n+1)$ Odd

According to Theorem 9.1 of Chapter VI, there is a CW-complex $Y_{k+1} \simeq \mathbb{F}_{k+1}(S^{n+1})$ that contains $X_k \vee S^{n+1}$ as a subcomplex, where $X_k \simeq \mathbb{F}_k(\mathbb{R}^{n+1})$. Moreover, Y_{k+1} fits in a commutative diagram,

$$\begin{array}{ccccc} X_k & \to & Y_{k+1} & \to & S^{n+1} \\ \downarrow & & \downarrow & & \downarrow \\ \mathbb{F}_k(\mathbb{R}^{n+1}) & \to & \mathbb{F}_{k+1}(S^{n+1}) & \to & S^{n+1}, \end{array}$$

where the upper horizontal sequence is a quasi fibration and the lower, a fibration, while the vertical maps are homotopy equivalences. Note that X_k is a subcomplex of Y_{k+1}. For the diagonal map, choose a cellular approximation $Y_{k+1} \to Y_{k+1} \times Y_{k+1}$that extends the cellular approximation $X_k \to X_k \times X_k$ already obtained. Let

$$C_*(Y_{k+1}) \to C_*(Y_{k+1}) \otimes C_*(Y_{k+1})$$

be the induced coproduct. As $C(Y_{k+1}) = H_*(C(Y_{k+1}))$, it follows that this coproduct is coassociative, and $C_*(X_k)$ is a subcoalgebra.

Theorem 4.1 *The cellular chain algebra $C_*(M(Y_{k+1}))$ is isomorphic to the cobar algebra $F(C_*(Y_{k+1}))$ of the coalgebra $C_*(Y_{k+1})$.*

Proof. Recall that $C_*(M(Y_{k+1})) \cong F(C_*(Y_{k+1}))$, as a graded algebra. Thus, to prove the theorem, it is enough to verify that the differential operator of the cellular algebra $C_*(M(Y_{k+1}))$ is given by the coproduct in $C_*(Y_{k+1})$.

According to Theorem 9.1 of Chapter VI, $Y_{k+1} = \coprod e_\omega$, where ω ranges over the basis $\Gamma = \mathcal{B}^\wedge \cup \{\delta_{k+1}\} \times \mathcal{B}^\wedge$, with δ_{k+1} being the class of a section $S^{n+1} \to \mathbb{F}_{k+1}(S^{n+1})$, and $\mathcal{B}^\wedge = \{1, \beta_{\underline{r},\,\underline{s}}\}$, $(\underline{r},\, \underline{s}) \in (\mathcal{R}, \mathcal{S})$. Observe that the subcomplex $X_k = \coprod e_\omega$, where $\omega \in \mathcal{B}^\wedge$. According to Theorem 3.1, $C_*(M(X_k)) \cong F(C_*(X_k))$, as a chain algebra. This implies readily that the theorem is true for the cells $\bar{e}_\omega$, when $\omega = \beta_{\underline{r},\,\underline{s}}$.

To treat the remaining cases, consider first a cell e_ω, where

$$\begin{aligned}\omega &= \delta_{k+1} \rtimes \beta_{\underline{r},\underline{s}} \\ &= \delta_{k+1} \rtimes \beta_{21} \rtimes \beta r_2 s_2 \rtimes \cdots \beta_{r_m s_m}.\end{aligned}$$

One proceeds in a manner similar to that of the case of X_k. Let

$$\phi'_\omega : O_{n+2,2} \times S_{2,\cdots,m} \to Y_{k+1} \simeq \mathbb{F}_{k+1}(S^{n+1})$$

be a cellular map homotopic to the spherical map

$$\phi_\omega : O_{n+2,2} \times S_{2,\cdots,m} \to \mathbb{F}_{k+1}(S^{n+1})$$

of the cell e_ω (see Chapter VI, §9). Here, as above, we denote $S_2 \times \cdots \times S_m$ by $S_{2,\cdots,m}$. We observe here that the induced chain morphism

$$C_*(\phi'_\omega) : C_*(O_{n+2,2} \times S_{2,\cdots,m}) \to C_*(Y_{k+1})$$

is actually a morphism of coalgebras. In fact, it is, up to homotopy, a morphism of coalgebras. But the chain complexes in question have trivial boundary morphisms. Hence, the assertion.

By virtue of Chapter VI, §§6 and 9, we know that

$$e_\omega = C_*(\phi'_\omega)(e^{2n+1} \times e_2 \times e_m),$$

where e is the top cell of $O_{n+2,2}$, and e_i, for $2 \leq i \leq m$, is the top cell of the factor S_i. The preceding equation implies that

$$\bar{e}_\omega = C_*(M(\phi_\omega))(\bar{e}_{2n+1,1,2,\cdots,m}) = \bar{e}_\omega,$$

where $e_{2n+1,1,2,\cdots,m} = e^{2n+1} \times e_1 \times e_2 \times e_m$. Note that $e_{2n+1,1,2,\cdots,m}$ is, in fact, the top cell of $O_{n+2,2} \times S_{2,\cdots,m}$.

Next, let

$$\omega \mapsto \omega \otimes 1 + 1 \otimes \omega + \sum_i \omega'_i \otimes \omega'',$$

where $\dim \omega'_i$ and $\dim \omega''_i$ are both positive for all i, be the coproduct of ω in the coalgebra $C_*(Y_{k+1})$. Now, since Theorem 3.2 asserts that

$$C_*(M(O_{n+2,2} \times S_{2,\cdots,m})) \cong F(C_*(O_{n+2,2} \times S_{2,\cdots,m}),$$

it follows that

$$\partial(\bar{e}_{2n+1,1,2,\cdots,m}) = d(\bar{e}_{2n+1,1,2,\cdots,m}) = \sum_i \bar{\omega}'_i \bar{\omega}''_i.$$

By naturality, we obtain that $\partial(\bar{e}_\omega) = d(\bar{e}_\omega)$. This finishes the proof of the case under consideration and completes the proof of the theorem. □

It should be noted here that the images of the classes ω' and ω'' above are determined by the factor table of ϕ_ω. Therefore, the boundary $\partial(\bar{e}_\omega)$ is also determined by the said factor table. In fact, it is not hard to give an explicit expression for this boundary in terms of the rows λ_i of the factor table.

5 $C_*(M(Y))$, $Y \simeq \mathbb{F}_{k+1}(S^{n+1})$, $(n+1)$ Even

Assume that $(n+1)$ is even. Let $Y_{k+1} \simeq \mathbb{F}_{k+1}(S^{k+1})$ be the CW-complex constructed in Chapter VI, §9. Recall that Y_{k+1} contains

$$O_{n+2,2} \vee X_{k-2,2}$$

as subcomplex, where $O_{n+2,2} = S^n \cup_f D^{n+1} \times S^n$ is the usual cellular structure (see Theorem 6.3 of Chapter III).

Note that the cellular structure on Y_{k+1} can be described also as follows. Recall that the fibration

$$\mathrm{proj}_{k+1,1} : \mathbb{F}_{k+1}(S^{n+1}) \to S^{n+1}$$

that sends a tuple to its first entry has $\mathbb{F}_k(\mathbb{R}^{n+1})$ as fiber (see Chapter III, §1). This fibration yields the decomposition

$$\mathbb{F}_{k+1}(S^{n+1}) \cong (D^{n+1} \times \mathbb{F}_k(\mathbb{R}^{n+1})) \cup_g \mathbb{F}_k(\mathbb{R}^{n+1}),$$

where $g : \partial D^{n+1} \times \mathbb{F}_k(\mathbb{R}^{n+1}) \to \mathbb{F}_k(\mathbb{R}^{k+1})$ is a suitable action with the restriction $g|\partial D^{n+1} \to \mathbb{F}_k(\mathbb{R}^{n+1})$ in the Dirac class $\Delta_{k+1}(\mathbb{R}^{n+1})$. Replacing $\mathbb{F}_k(\mathbb{R}^{n+1})$ by the complex $X_k \simeq \mathbb{F}_{k+1}(\mathbb{R}^{n+1})$, one obtains, up to homotopy, the complex Y_{k+1} (see Theorem 9.3 of Chapter VI).

Observe that the projection $\mathbb{F}_{k+1}(S^{n+1}) \to \mathbb{F}_3(S^{n+1})$, whereby the tuple $(x_1, \cdots, x_{k+1})$ is mapped to (x_1, x_2, x_{k+1}), takes $X_{k-2,2}$ onto its natural image in X_k.

Choose a cellular approximation

$$\nabla : Y_{k+1} \to Y_{k+1} \times Y_{k+1}$$

such that $\nabla \mid X_k : X_k \to X_k \times X_k$ is a cellular approximation to the diagonal map of X_k. Passing to cellular chains, we obtain the coproduct

$$C_*(\nabla) : C_*(Y_{k+1}) \to C_*(Y_{k+1}) \otimes C_*(Y_{k+1}).$$

Proposition 5.1 *The coproduct $C_*(\nabla)$ is coassociative.*

Proof. For convenience, let us denote $C_*(Y_{k+1})$ by C. Consider now the diagram

$$\begin{array}{ccccc} C & \overset{\nabla}{\to} & C \otimes C & \overset{\nabla \otimes 1}{\to} & C \otimes C \otimes C \\ = \downarrow & & \downarrow & & = \downarrow \\ C & \overset{\nabla}{\to} & C \otimes C & \overset{1 \otimes \nabla}{\to} & C \otimes C \otimes C. \end{array}$$

To prove the proposition, we need only show that the diagram commutes. Clearly, it commutes up to chain homotopy. Let $e_\omega \in C$, and consider the map

$$\phi'_\omega : O_{n+2,2} \times S_2 \times \cdots \times S_m \to Y_{k+1}$$

homotopic to the multispherical map ϕ_ω, as in Chapter VI, §9. Recall that e_ω is the image of the fundamental class of $O_{n+2,2} \times S_2 \times \cdots \times S_m$. The coproduct is coassociative in the latter space. Hence, by naturality, it follows that

$$(\nabla \otimes 1) \circ \nabla(\omega) = (1 \otimes \nabla) \circ \nabla(\omega).$$

Therefore, the coproduct is coassociative in Y_{k+1}. □

Theorem 5.1 *There is a an isomorphism*

$$\rho : F(C_*(Y_{k+1})) \to C_*(M(Y_{k+1}))$$

of chain algebras, where $F(C_*(Y_{k+1}))$ is the cobar algebra of $C_*(Y_{k+1})$, and $C_*(M(Y_{k+1}))$ is the algebra of cellular chains of the RPT-complex $M(Y_{k+1})$ generated by Y_{k+1}.

Proof. The proof is by induction on k, with the initial step being when $k = 3$. Denote the classes in $C_*(O_{n+2,2}$, defined by the cells e^n, e^{n+1}, by the same symbols, and write $e^{2n1} = e^{n+1} \times e^n$ for the top dimensional cell. Keeping in mind that n is odd and greater than one, on computing the coproducts one obtains, for dimensional reasons, the following formulas.

$$e^n \mapsto e^n \otimes 1 + 1 \otimes e^n \quad \text{and}$$
$$e^{n+1} \mapsto e^{n+1} \otimes 1 + 1 \otimes e^{n+1}.$$

The cup product in $H^*(O_{n+2,2}, \mathbb{K})$, where $\mathbb{K} = \mathbb{Z}_2$ or $\mathbb{Z}_p$, with $p > 2$, implies that the coproduct of e^{2n+1} is given by the formula

$$e^{2n+1} \mapsto e^{2n+1} \otimes 1 + 1 \otimes e^{2n+1} + e^{n+1} \otimes e^n + (-1)^{n(n+1)} e^n \otimes e^{n+1}.$$

Next, one shows directly that $F(C_*(O_{n+2,2}) \cong C_*(M(O_{n+2,2}))$ (see Theorem 4.1 of [5, Adams-Hilton]).

The rest of the proof follows along the lines of that for Theorem 4.1 above. The details are left to the reader. □

6 The Eilenberg-Moore Spectral Sequence of $\Lambda(M)$

Let X stand for $X_{k-r,r} \simeq \mathbb{F}_{k-r,r}(\mathbb{R}^{n+1})$, or $Y_{k+1} \simeq \mathbb{F}_{k+1}(S^{n+1})$, with $n > 1$. Consider the fibration

$$\text{eval} : \Lambda(X) \to X,\ (\omega, r) \mapsto \omega(0) = \omega(r),$$

where $\Lambda(X)$ is the space of Moore closed loops in X.

Theorem 6.1 *The Eilenberg-Moore spectral sequence, over* $\mathbb{K}$, *of the fibration* eval : $\Lambda(X) \to X$ *is trivial where* $\mathbb{K}$ *is* $\mathbb{Z}_2$, *if* $X = Y_{k+1} \simeq \mathbb{F}_{k+1}(S^{n+1})$ *and* $n+1$ *is even, and is* $\mathbb{Z}$, *otherwise.*

We shall give the proof for the case when $X = X_{k-r,r} \simeq \mathbb{F}_{k-r,r}(\mathbb{R}^{n+1})$. The other cases are similar, and the details are left to the reader.

The key step is provided by the following theorem. The idea is to regard $H_*(\Omega(\mathbb{F}_{k-r,r};\mathbb{Z})$ as a chain algebra with the trivial boundary homomorphism and, then, to consider the morphism

$$\rho_{k-r,r} : C_*(M(X_{k-r,r})) \to H_*(\Omega(X_{k-r,r});\mathbb{Z})$$

that takes each cell $\bar{e}_\omega \in C_*(M(X_{k-r,r})$ to the homology class it defines in $H_*(\Omega(X_{k-r,r});\mathbb{Z})$ when $\omega = \alpha_{ts}$ for some r, s, and to 0 otherwise. Clearly, the morphism $\rho_{k-r,r}$ is a morphism of chain algebras.

Theorem 6.2 *The morphism*

$$\rho_{k-r,r} : C_*(M(X_{k-r,r})) \to H_*(\Omega(X_{k-r,r});\mathbb{Z})$$

of chain algebras, induces an isomorphism on homology.

Proof. The proof is similar to that of Lemma 2.1 above. One proceeds by induction on the filtration

$$X_{k-r,r} \supset X_{k-r-1,r+1} \supset \cdots \supset X_{1,k-1}$$

of Theorem 8.1 of Chapter VI. The initial step is to prove that

$$\rho_{1,k-1} : C_*(M(X_{1,k-1})) \to H_*(\Omega(X_{1,k-1});\mathbb{Z}).$$

This is immediate since $X_{1,k=1}$ is a bouquet of spheres, and the boundary morphism in $C_*(M(X_{1,k-1}))$ is, as a consequence, trivial. The inductive step is handled as in the proof of Lemma X.3.1. One uses the fibration

$$\mathrm{p}_t : X_{k-t,t} \to S_{t1} \vee \cdots \vee S_{tt-1},$$

of which the fiber is $X_{k-t-1,t+1}$. The details are left for the reader. □

Next let us consider the chain map

$$C_*(\mathrm{Ad}) : C_*(M(X_{k-r,r})) \otimes C_*(M(X_{k-r,r})) \to C_*(M(X_{k-r,r}))$$

induced by the adjoint action

$$\mathrm{Ad} : M(X_{k-r,r}) \times M(X_{k-r,r}) \to M(X_{k-r,r}),$$

where $(x, y) \mapsto xyx^{-1}$. The following assertion is easy to verify.

Assertion 6.1 *The diagram*

$$\begin{array}{ccc} C_*(M(X_{k-r,r})) \otimes C_*(M(X_{k-r,r})) & \stackrel{C_*(\mathrm{Ad})}{\rightarrow} & C_*(M(X_{k-r,r})) \\ \downarrow & & \downarrow \\ H_*(M(X_{k-r,r})) \otimes H_*(M(X_{k-r,r})) & \stackrel{H_*(\mathrm{Ad})}{\rightarrow} & H_*(M(X_{k-r,r})) \end{array}$$

commutes.

Proof of Theorem 6.1

According to the previous theorem, there is a homomorphism of chain algebras

$$\phi : C_*(M(X_{k-r,r})) \to H_*(M(X_{k-r,r}))$$

that induces an isomorphism on homology. The morphism ϕ induces in turn a morphism $\mathrm{E}^r_{*,*}(\phi) :' \mathrm{E}^r_{*,*} \to \mathrm{E}^r_{*,*}$ of the Eilenberg spectral sequence of $C_*(M(X_{k-r,r}))$ to that of $H_*(M(X_{k-r,r}))$ (see Theorem 5.1 of Chapter IX). But,

$$\begin{aligned} {}'\mathrm{E}^2 &\cong \mathrm{Tor}^{H_*(M(X_{k-r,r})}(\mathbb{Z}, \mathbb{Z}), \\ \mathrm{E}^2 &\cong \mathrm{Tor}^{H_*(M(X_{k-r,r}))}(\mathbb{Z}, \mathbb{Z}). \end{aligned}$$

Hence, $\mathrm{E}^r_{*,*}(\phi)$ is an isomorphism. This implies that $\mathrm{E}^r_{*,*}(\phi)$ is an isomorphism for all r. But $\mathrm{E}^r_{*,*}(\phi)$ is trivial, Hence, $'\mathrm{E}^r_{*,*}(\phi)$ is also trivial. □

It is worthwhile to note the following. Suppose that F is a CW-complex, and let

$$M(X) \times F \to F$$

be a cellular action.

Theorem 6.3 *Suppose that there is a morphism $C_*(F) \to H_*(F)$ of a $C_*(M(X))$-module to an $H_*(\Omega(X))$-module. Then the Eilenberg-Moore spectral sequence of the quasi fibration*

$$\mathrm{proj} : EM \times_M F \to X$$

collapses.

Proof. The morphism above induces a morphism of the cellular spectral sequence to that of homology. The latter collapses. The two agree at the E_2-level. Hence, the theorem. □

Corollary 6.1 *The Eilenberg-Moore spectral sequence of the fibration*

$$\mathrm{eval}_k : \Lambda(\mathbb{F}_k(\mathbb{R}^{n+1})) \to \mathbb{F}_k(\mathbb{R}^{n+1}),$$

where $\omega \mapsto \omega(0)$, collapses.

Proof. Note that there is an equivalence

$$\begin{array}{ccc} \Lambda(X) & \stackrel{\phi}{\to} & E\,M(X) \times_{M,Ad} M \\ \downarrow & & \downarrow \\ X & \stackrel{\mathrm{id}_X}{\to} & X. \end{array}$$

Moreover, the conditions of the preceding theorem are satisfied, with $F = M(X)$, and $M(X)$ acting on itself by the adjoint action (see Chapter VIII, §5). Now the corollary follows immediately. □

XI. The Serre Spectral Sequence

In this chapter we study the Serre spectral sequence $\{E^r_{*,*}(\mathrm{p}), d^r\}$ over $\mathbb{K} = \mathbb{Z}$, and $\mathbb{Z}_2$, of the path space fibration

$$\mathrm{p} : \mathcal{P}(M) \to M,$$

with $M = \mathbb{F}_k(\mathbb{R}^{n+1})$ or $\mathbb{F}_{k+1}(S^{n+1})$. Here the paths are based at an appropriate basepoint.

First, in §1 we take up the case of $M = \mathbb{R}^{n+1}$. We shall see that the spectral sequence stabilizes at the nth term, in the sense that

$$\mathrm{E}^{n+1}_{*,*}(\mathrm{p}) \cong \mathrm{E}^{\infty}_{*,*}(\mathrm{p}) \cong \mathbb{K}.$$

Consequently, regarding $H_*(\Omega(M); \mathbb{K})$ as a chain algebra, with the trivial differential and $\mathbb{K}$ as a trivial chain module over it, we interpret the $\mathrm{E}^n_{*,*}$ term of the spectral sequence as an acyclic, free resolution of $\mathbb{K}$ over $H_*(\Omega(M); \mathbb{K})$. This result, together with the fact that

$$H_*(\Lambda(M); \mathbb{K}) \cong \mathrm{Tor}^{H_*(\Omega(M);\mathbb{K})}(H_*(\Omega(M); \mathbb{K}), \mathbb{K})$$

(see §6 of Chapter IX), is the primary tool used in Chapter XII to study the module $H_*(\Lambda(M); \mathbb{Z}_2)$.

In §§2 and 3 we take up the case when $M = S^{n+1}$. The situation here is somewhat different,and the results of §1 are adapted suitably.

1 The Case of $\mathbb{F}_{k-r,r}$, $n > 1$

In keeping with the notation of Chapter II, put

$$\mathbb{F}_{r,k-r} = \mathbb{F}_r(\mathbb{R}^{n+1} - Q_{k-r}), \quad \mathbb{R}^{n+1}_{k-r} = \mathbb{R}^{n+1} - Q_{k-r},$$

and consider the path fibration

$$\mathrm{p}_{k-r,r} : \mathcal{P}(\mathbb{F}_{k-r,r}) \to \mathbb{F}_{k-r,r}$$

that sends a based Moore path (α, r) to its endpoint $\alpha(r)$. Throughout this section, we assume that the homology and cohomology groups are with integral coefficients.

Theorem 1.1 *The Serre spectral sequence* $\{E^t_{*,*}(p_{k-r,r}), d^t\}$ *of the path fibration* $p_{k-r,r} : \mathcal{P}(\mathbb{F}_{k-r,r}) \to \mathbb{F}_{k-r,r}$ *has the following properties:*

$$\begin{aligned}
&(i)\ E^2_{*,*}(p_{k-r,r}) \cong H_*(\mathbb{F}_{k-r,r}) \otimes H_*(\Omega(\mathbb{F}_{k-r,r})),\\
&(ii)\ E^2_{*,*}(p_{k-r,r}) \cong E^r_{*,*}(p_{k-r,r}),\ \text{for } r \leq n, \quad \text{and}\\
&(iii)\ E^{n+1}_{*,*}(p_{k-r,r}) \cong E^{\infty}_{*,*}(p_{k-r,r}) = \mathbb{Z},
\end{aligned}$$

where homology is with integral coefficients.

To start, consider the following segment of the fundamental fiber sequence $\mathcal{F}_k$ of $\mathbb{F}_k(\mathbb{R}^{n+1})$ (see Chapter II, §1):

$$\begin{array}{cccc}
\cdots \leftarrow & \mathbb{F}_{k-t,t} & \leftarrow \mathbb{F}_{k-t-1,t+1} & \leftarrow \cdots \\
\downarrow & \downarrow p_{k-t,t} & \downarrow p_{k-t-1,t+2} & \downarrow \\
\cdots & \mathbb{R}^{n+1}_t & \mathbb{R}^{n+1}_{t+1} & \cdots .
\end{array}$$

Recall that the vertical maps are the projections on the first nonconstant factor. Observe that

$$\mathbb{R}^{n+1}_t \simeq (\vee^t_{j=1} S_{t+1j}) \cong (S^n)^{\vee t}$$

for $r \leq t < k$, where the notation is that of Chapter II (see Proposition 1.1 of Chapter II).

The proof of Theorem 1.1 will be by induction, but first we need some preparatory work. To simplify the notation, put

$$E = \mathbb{F}_{k-t,t},\ B = \mathbb{R}^{n+1}_t,\ F = \mathbb{F}_{k-t-1,t+1},\ p = p_{k-t,t},$$

and consider the fibration

$$p : E \to B, \ (q_1, \cdots, q_t, x_1, \cdots, x_{k-t}) \mapsto x_1,$$

with fiber F. The map $i : F \to E$ that imbeds F as the fiber at the basepoint induces the commutative diagram

$$\mathcal{D}_{k-t,t} : \left\{ \begin{array}{ccccc}
\Omega(F) & \to & \mathcal{P}(F) & \to & F \\
\downarrow \Omega(i) & & \downarrow & & \downarrow = \\
\Omega(E) & \to & \Omega(E,F) & \to & F \\
\downarrow = & & \downarrow & & \downarrow \\
\Omega(E) & \to & \mathcal{P}(E) & \to & E
\end{array} \right.$$

of path fibrations, where $\Omega(E,F) = \{(x,\alpha) \mid \alpha(r) = x \in F\} = i^*\mathcal{P}(E)$. The vertical maps are the natural maps induced by i or by the identity.

Recall that there is a section $\rho : \mathbb{R}^{n+1}_t \to \mathbb{F}_{k-t,t}$, with restrictions to the spheres $\{S_{t+1j} | 1 \leq j \leq t\}$ that realize the elements $\{\alpha_{t+1s} | 1 \leq s < t+1\}$

(see the proof of Theorem 4.3 or Proposition 1.1 of Chapter II). The section ρ, in turn, leads to a twisted product

$$\begin{array}{ccc} E & \simeq & B \rtimes F, \\ \downarrow = & & \downarrow = \\ \mathbb{F}_{k-t,t} & \simeq & \mathbb{R}_t^{n+1} \rtimes \mathbb{F}_{k-t-1,t+1} \end{array}$$

(see Chapter II, §5). This twisted product itself induces an isomorphism

$$H_*(E) \cong H_*(B) \otimes H_*(F),\ H_*(\Omega(E)) \cong H_*(\Omega(B) \otimes H_*(\Omega(F))$$

of modules. Observe that these isomorphisms define twisted-product structures of coalgebras on the former and of Pontryagin algebras on the latter (see Chapter VIII). Denote by $\mathrm{p}_F : \mathcal{P}(F) \to F$ and $\mathrm{p}_E : \mathcal{P}(E) \to E$ the projections that send a path to its endpoint. Also, denote by

$$\mathrm{p}_{(E,F)} : \Omega(E,F) \to F$$

the map where $(\alpha, r) \mapsto \alpha(r)$. Note that it is the pull-back $i^*\mathcal{P}(E)$ of $\mathrm{p}_E : \mathcal{P}(E) \to E$ by $i : F \to E$ and, therefore, the right action of $\Omega(E)$ on $\mathcal{P}(E)$ induces a right action of $\Omega(E)$ on $\Omega(E,F)$. The section $\rho : B \to E$ induces an imbedding $\Omega(\rho) : \Omega(B) \to \Omega(E)$. The action of $\Omega(E)$ induces, by restriction, an action

$$\gamma : \Omega(B) \times \mathcal{P}(F) \to \Omega(E,F),$$

where $\gamma(\omega, \alpha) = \omega.\alpha$.

Proposition 1.1 *The map γ induces a fiber homotopy equivalence*

$$\mathcal{D}'_{k-t,t} : \left\{ \begin{array}{ccccc} \Omega(B) \times \Omega(F) & \to & \Omega(B) \times \mathcal{P}(F) & \overset{\mathrm{proj}}{\to} & F \\ \downarrow \gamma_0 & & \downarrow \gamma & & \downarrow = \\ \Omega(E) & \to & \Omega(E,F) & \overset{\mathrm{p}_{(E,F)}}{\to} & F, \end{array} \right.$$

where proj *is the composite of the projection on $\mathcal{P}(F)$ and* p_F.

Proof. The proof is immediate: just observe that γ_0 and γ are homotopy equivalences. □

Next, denote by

$$\begin{array}{l} \{\mathrm{E}^m(\mathrm{p}_F), \mathrm{d}^m(\mathrm{p}_F), m \geq 2\}, \\ \{\mathrm{E}^m(\mathrm{p}_E), \mathrm{d}^m(\mathrm{p}_E), m \geq 2\}, \text{ and} \\ \{\mathrm{E}^m(\mathrm{p}_{(E,F)}), \mathrm{d}^m(\mathrm{p}_{(E,F)}), m \geq 2\} \end{array}$$

the Serre spectral sequences of, respectively, the maps $\mathrm{p}_F, \mathrm{p}_E$, and $\mathrm{p}_{(E,F)}$.

Lemma 1.1 *The spectral sequence* $\{\mathrm{E}^m(\mathrm{p}_{(E,F)}), \mathrm{d}^m(\mathrm{p}_{(E,F)}), m \geq 2\}$ *has the following properties:*

$$\begin{array}{ll}(i)\ \mathrm{E}^m_{p,q}(\mathrm{p}_{(E,F)}) = \sum_{q=q'+u} \mathrm{E}^m_{p,q'}(\mathrm{p}_F) \otimes H_u(\Omega(B)) & \textit{and} \\ (ii)\ \mathrm{d}^m(\mathrm{p}_{(E,F)}) \ = \mathrm{d}^m(\mathrm{p}_F) \otimes 1, & \textit{for } 2 \leq m < \infty,\end{array}$$

for all $p, q \geq 0$.

Proof. Proposition 1.1 implies that there is a morphism

$$\phi^r(\gamma) : \mathrm{E}^r_{*,*}(\mathrm{proj}) \to \mathrm{E}^r_{*,*}(\mathrm{p}_{(E,F)}).$$

Observe that

$$\mathrm{E}^r_{p,q}(\mathrm{proj}) = \sum_{q=q'+u} \mathrm{E}^r_{p,q'}(\mathrm{p}_F) \otimes H_u(\Omega(B)),$$

and $\mathrm{d}^r(\mathrm{proj}) = \mathrm{d}^r(\mathrm{p}_F) \otimes 1$, for all $r \geq 2$. Note that $H_*(B) = H_*(\vee_{j=1}^t S_{t+1j})$. Hence, $H_*(\Omega(B))$ is the free associative algebra on the desuspensions $\{\bar{\alpha}_{ts} \mid r < t \leq k\}$ (see Chapter VII, §1). Thus, $H_*(\Omega(B))$ is torsion free, and the universal coefficient applies. This proves that

$$\mathrm{E}^{n+1}_{*,q}(\mathrm{proj}) = \sum_{q=q'+u} \mathrm{E}^r_{p,q'}(\mathrm{proj}) \otimes H_u(\Omega(B))$$

implies the desired conclusions. □

Lemma 1.2 *The spectral sequences*

$$\{\mathrm{E}^m(\mathrm{p}_F), \mathrm{d}^m(\mathrm{p}_F), m \geq 2\},\ \{\mathrm{E}^m(\mathrm{p}_{(E,F)}), \mathrm{d}^m(\mathrm{p}_{(E,F)}), m \geq 2\}$$

have the following properties:

$$\begin{array}{ll}(i)\ \mathrm{E}^2_{*,*}(\mathrm{p}_F) & = E^n_{*,*}(\mathrm{p}_F) \ \textit{and} \\ (ii)\ \mathrm{E}^2_{*,*}(\mathrm{p}_{(E,F)}) & = \mathrm{E}^n_{*,*}(\mathrm{p}_{(E,F)}).\end{array}$$

Proof. We prove (i) only, as the other is similar. By the Leray-Serre Theorem, we have

$$\mathrm{E}^2_{*,*}(\mathrm{p}_F) = H_*(B) \otimes H_*(\Omega(F)).$$

Note that $H_i(F) = 0$ and $i \neq 0 \mod n$ and $i > (k-t-1)n$ (see the remark preceding Theorem 5.2 of Chapter VI). Now the assertion follows immediately. □

Lemma 1.3 *Assume that* $\mathrm{E}^{n+1}_{*,*}(\mathrm{p}_F) = \mathbb{K}$. *Then,*

$$\begin{array}{l}(i)\ \mathrm{E}^{n+1}_{0,*}(\mathrm{p}_{(E,F)}) = H_*(\Omega(B)), \\ (ii)\ \mathrm{E}^{n+1}_{p,*}(\mathrm{p}_{(E,F)}) = 0, \ \textit{if } p > 0, \ \textit{and} \\ (iii)\ \mathrm{E}^{n+1}_{*,*}(\mathrm{p}_{(E,F)}) = \mathrm{E}^{\infty}_{*,*}(\mathrm{p}_{(E,F)}).\end{array}$$

Proof. The statements of Lemma 1.3 are an immediate consequence of Lemma 1.1. □

Lemma 1.4 *The map $i : F \to E$ induces an injective morphism*

$$\phi^n(i) : \mathrm{E}^n_{*,*}(\mathrm{p}_F) \to \mathrm{E}^n_{*,*}(\mathrm{p}_E).$$

Proof. Arguing as above, one proves that

$$\mathrm{E}^n_{*,*(\mathrm{p}_E)} = \mathrm{E}^2_{*,*}(\mathrm{p}_E) = H_*(E) \otimes H_*(\Omega).$$

Note that $H_*(i) : H_*(F) \to H_*(E)$ and $H_*(\Omega(i)) : H_*(\Omega(F)) \to H_*(\Omega(E))$ and that the modules in question are free, implying thus the assertion. □

Definition 1.1 *Denote by $\mathcal{C}_*$ the* DG*-module*

$$\mathrm{E}^n_{*,*}(\mathrm{p}_E)/\mathrm{image}(\phi^n : \mathrm{E}^n_{*,*}(\mathrm{p}_F) \to \mathrm{E}^n_{*,*}(\mathrm{p}_E),$$

where the differential is that induced by

$$\mathrm{d}^n(\mathrm{p}_{(E,F)})) : \mathrm{E}^n_{*,*}(\mathrm{p}_E) \to \mathrm{E}^n_{*,*}(\mathrm{p}_E),$$

while the grading is induced by the total degree in $\mathrm{E}^n_{*,*}(\mathrm{p}_E)$.

Next, let $\mathbb{Z} \to H_*(B)$ be the coaugmentation of the coalgebra $H_*(B)$ induced by the imbedding of the basepoint in $B \simeq \vee^t_{j=1} S_{t+1j}$, and put $\bar{H}_*(B) = H_*(B)/\mathbb{Z}$. The exact sequence of modules

$$0 \to \mathbb{Z} \to H_*(B) \to \bar{H}_*(B) \to 0$$

is split-exact. On multiplying by $H_*(F) \otimes H_*(\Omega(E))$, we obtain the split-exact sequence

$$\begin{aligned} 0 &\to H_*(F) \otimes H_*(\Omega(E)) &&= \mathrm{E}^n(\mathrm{p}_E \circ i) \\ &\to H_*(B) \otimes H_*(F) \otimes H_*(\Omega(E)) &&= \mathrm{E}^n(\mathrm{p}_E) \\ &\to (\bar{H}_*(B) \otimes H_*(F) \otimes H_*(\Omega(E)) &&= \mathcal{C}_* \\ &\to 0, \end{aligned}$$

of DG-modules, where the differentials are induced by $\mathrm{d}^n(\mathrm{p}_E) : \mathrm{E}^n_{*,*}(\mathrm{p}_E) \to \mathrm{E}^n_{*,*}(\mathrm{p}_E)$. Observe here that

$$\mathcal{C}_* \cong (H_*(E)/H_*(F)) \otimes H_*(\Omega(E)),$$

where $H_*(F)$ is identified with its image under the morphism induced by $i : F \to E$.

It is worthwhile to note that $\mathcal{C}_*$ can be identified with the $\mathrm{E}^n_{*,*}$-term of the relative Leray-Serre spectral sequence of the fiber pair

$$\mathrm{p} : (\mathcal{P}(E), \Omega(E, F)) \to (E, F),$$

where p is map induced by p_E. The details are left to the reader.

Denote by $d : \mathcal{C}_* \to \mathcal{C}_*$ the induced differential. Filter $\mathcal{C}_*$ by putting

$$\mathcal{F}(\mathcal{C}) : F_p(\mathcal{C}) = \sum_{m=0}^{p} \bar{H}_*(B) \otimes H_{mn}(F) \otimes H_*(\Omega(E)).$$

For brevity, denote by F_p the submodule just defined.

Assertion 1.1 *The submodules $F_p \subset \mathcal{C}$ of the filtration are* DG*-submodules.*

Proof. Let $\omega \in \bar{H}_n(B) \otimes H_{pn}(F) \subset E^n_{*,*}(p_E)$. Since $H_*(E) \cong H_*(B) \otimes H_*(F)$ and $H_*(B) \cong \sum_{s=1}^{t} H_*(S_{t+1s})$, where the sum is a direct sum, it follows that

$$E^n_{t,*}(p_E) = 0, \text{if } t \neq 0 \mod n, \text{ and } t > 0.$$

Therefore, the morphism

$$\begin{aligned} d^n(p_E) : & E^n_{*,*}(p_E) \cong H_*(E) \otimes H_*(\Omega(E)) \\ \to & E^n_{*,*}(p_E) \cong H_*(E) \otimes H_*(\Omega(E)) \end{aligned}$$

takes $H_n(B) \otimes \sum_{m=0}^{p} H_{mn}(F)$ to the sum

$$\begin{aligned} &\sum_{m=0}^{p} H_{mn}(F) \otimes H_*(\Omega(E)) + \\ &H_n(B) \otimes \sum_{m=0}^{p-1} H_{n(m-1)}(F) \otimes H_*(\Omega(E)). \end{aligned}$$

Passing to $\mathcal{C} = H_*(E)/H_*(F) \otimes H_*(\Omega(E))$, we see that the first summand goes to zero. This proves that the submodules $F_p(\mathcal{C})$ are invariant under the differential of $d(\mathcal{C})$. □

The spectral sequence $\{E^r_{*,*}(\mathcal{C})\}$ has the properties described in Assertions 1.2, 1.3, and 1.4.

Assertion 1.2 $E^0_{*,*}(\mathcal{C}) = E^1_{*,*}(\mathcal{C})$.

Proof. Clearly,

$$F_p/F_{p-1} = H_n(B) \otimes H_{pn}(F) \otimes H_*(\Omega(E)).$$

The calculation given in Assertion 1.1 shows readily that the differential

$$d^0 : E^0_* = \sum_p F_p/F_{p-1} \to E^0_* = \sum_p F_p/F_{p-1}$$

is trivial. Hence, $E^0_{*,*} = E^1_{*,*}$. □

Assertion 1.3 *There is an isomorphism*

$$\begin{array}{ccc} E^1_{p,*}(\mathcal{C}) & \overset{d^1(\mathcal{C})}{\to} & E^1_{p-1,*}(\mathcal{C}) \\ \downarrow & & \downarrow \\ E^n_{n,*}(p_B) \otimes E^n_{np,*}(p_F) & \overset{1 \otimes d^n(p_F)}{\to} & E^n_{n,*}(p_B) \otimes E^n_{n(p-1),*}(p_F) \end{array}$$

of DG*-modules.*

Proof. The module $H_n(B) \otimes H_{np}(F) \cong H_{n(p+1)}(E)$ is generated by elements of the form $\omega = \omega_B \rtimes \omega_F$, where $\omega_B = \alpha_{t+1,s} \in H_n(B)$ and $\omega_F = \alpha_{\underline{u}\,\underline{s}} \in H_{np}(F)$ are basic elements. Recall that, according to [32, Fadell-Hurewicz], the differential

$$\mathrm{d}^n(p_E) : \mathrm{E}^n_{*,*}(\mathrm{p}_E) \to \mathrm{E}^n_{*,*}(\mathrm{p}_E)$$

of the spectral sequence of $\mathrm{E}^*_{*,*}(\mathrm{p}_E)$ is determined by the coproduct $H_*(\nabla) : H_*(E) \to H_*(E) \otimes H_*(E)$, and $H_*(\nabla)(\omega) \in H_*(E)$, by the factor table

$$\mathcal{T}(\phi_{\underline{r}\,\underline{s}}) : \begin{cases} \alpha_{t+1 s_1} & + \sum_{m=2}^{p} \lambda_{1m}(\omega_m) \\ \cdots & + \cdots \\ \alpha_{r_j s_j} & + \sum_{m=j+1}^{p} \lambda_{jm}(\omega_m) \\ \cdots & + \cdots \\ \alpha_{r_{p+1} s_{p+1}} & \end{cases}$$

of ω (see Theorem 3.1 of Chapter VIII). Recall that the rows $\lambda_i(\omega)$ are the restrictions of the affine map

$$\phi_\omega : S_1 \times \cdots \times S_{p+1} \to \mathbb{F}_{k-t,t}$$

of ω to the factors S_i (see Chapter VI, §3). Now, according to Theorem 2.2 of Chapter VII, the coproduct

$$H(\nabla) : H_*(E) \to H_*(E) \otimes H_*(E)$$

is such that

$$H(\nabla)(\omega) = \sum \pm \lambda_{1,\cdots,\hat{i},\cdots,p+1} \otimes \lambda_i + \varpi,$$

where ϖ is a sum of elements in $\sum H_j(E) \otimes H_i(E)$, with $i \geq 2$, $i + j = p+1$, while

$$\lambda_i(\omega) = \alpha_{r_i s_i} + \sum_{m=i+1}^{p+1} \lambda_{im}(\omega_m)$$

is the ith row of the factor table of ω, and $\lambda_{1,\cdots,\hat{i},\cdots,p+1}$ is the twisted product of the rows of the factor table, excluding $\lambda_i(\omega)$. Denote by $\bar{\lambda}_i(\omega)$ the expression obtained from $\lambda_i(\omega)$ by replacing every occurrence of α by $\bar{\alpha}$. Hence, the morphism,

$$\begin{aligned} \mathrm{d}^n(\mathrm{p}_E) : \mathrm{E}^n_{*,*}(\mathrm{p}_E) \cong H_*(E) \otimes H_*(\Omega(E)) \to \\ \mathrm{E}^n_{*,*}(\mathrm{p}_E) \cong H_*(E) \otimes H_*(\Omega(E)) \end{aligned}$$

is such that

$$\text{(I)} \qquad \mathrm{d}^n(\mathrm{p}_E)(\omega \otimes \xi) = \sum_{j=1}^{p+1} \pm \lambda_{1,\cdots,\hat{j},\cdots,p+1} \otimes \bar{\lambda}_j(\omega)\xi.$$

Now, recall that the factor table of

$$\begin{aligned}\omega &= \omega_B \rtimes \omega_F \\ &= \alpha_{t+1,s} \rtimes \alpha_{\underline{u}\,\underline{s}}\end{aligned}$$

is obtained from that of ω_F by adjoining to the factor table of ω_F the row

$$\lambda_1(\omega) = \alpha_{t+1s_1} + \sum_{m=2}^{p+1} \lambda_{1m}(\omega_m)$$

in the first position. The row λ_1 consists of the sum of $\omega_B = \alpha_{t+1s}$, with its perturbations λ_{1m} by the rows of ω_F.
Note that the differential $\mathrm{d}(\mathcal{C}) : \mathcal{C} \to \mathcal{C}$ induces the boundary morphism

$$\partial : H(F_p/F_{p-1}) \to H(F_{p-1}/F_{p-2})$$

of the triple (F_p, F_{p-1}, F_{p-2}), and the latter coincides with the differential

$$\mathrm{d}^1(\mathcal{C}) : \mathrm{E}^1_{*,*}(\mathcal{C}) \to \mathrm{E}^1_{*,*}(\mathcal{C}).$$

Now, as indicated above, we have

$$H(F_p/F_{p-1}) = F_p/F_{p-1} = H_n(B) \otimes H_{np}(F) \otimes H_*(\Omega(E)),$$

as well as

$$H(F_{p-1}/F_{p-2}) = F_{p-1}/F_{p-2} = H_n(B) \otimes H_{(p-1)n}(F) \otimes H_*(\Omega(E)).$$

Consider the expression

$$\mathrm{d}(\omega \otimes \xi) = \sum_{j=1}^{p} \pm\lambda_{1,\cdots,\hat{\jmath},\cdots,p} \otimes \bar{\lambda}_j(\omega)\xi$$

given in (I) above. We see easily that

$$\lambda_{\hat{1},2,\cdots,p+1} \otimes \bar{\lambda}_1(\omega)\xi = \omega_F \otimes \bar{\lambda}_1(\omega)\xi = \alpha_{\underline{u}\,\underline{s}} \otimes \bar{\lambda}_1\xi.$$

However, this sum is equal to zero in $\mathcal{C}$. Moreover,

$$\alpha_{t+1s_1} \otimes \sum_{j=2}^{p+1} \pm\lambda_{1,\cdots,\hat{\jmath},\cdots,p+1}\xi = \omega_B \otimes \sum_{j=2}^{p+1} \pm\lambda_{2,\cdots,\hat{\jmath},\cdots,p} \otimes \bar{\lambda}_j(\omega)\xi,$$

where $\lambda_{2,\cdots,\hat{\jmath},\cdots,p+1}$ is the twisted product of the rows $\lambda_j(\omega_F)$ of the factor table of $\omega_F = \alpha_{\underline{u}\,\underline{s}}$ and $\bar{\lambda}_j(\omega_F)$ is the element of $H_*(\Omega(E))$ obtained therefrom. This implies that immediately that $\mathrm{E}^1_{*,*}(\mathcal{C})$ is isomorphic to the DG-module $(H_n(B) \otimes H_*(\Omega(B))) \otimes (H_*(F) \otimes H_*(\Omega(F))$, with $\mathrm{d}^1(\mathcal{C})$ corresponding to $1 \otimes d^n(\mathrm{p}_F)$. □

Assertion 1.4 *If* $\mathrm{E}^{n+1}_{*,*}(\mathrm{p}_F) = \mathrm{E}^{\infty}_{*,*}(\mathrm{p}_F)$, *then*

$$\begin{aligned}\mathrm{E}^2_{p,*}(\mathcal{C}) &= 0, \quad \textit{for } p > 0, \quad \textit{and} \\ \mathrm{E}^2_{0,*}(\mathcal{C}) &= \bar{H}_n(B) \otimes H_*(\Omega(B)).\end{aligned}$$

Proof. Denote the term $E^n_{*,*}(\mathrm{p}_F)$ by $\mathcal{C}'_*$. Then, the differential

$$\mathrm{d}^n(\mathrm{p}_F) : \mathrm{E}^n_{*,*}(\mathrm{p}_F) \to \mathrm{E}^n_{*,*}(\mathrm{p}_F)$$

turns $\mathcal{C}'_*$ into a DG-module. Since $\mathrm{E}^{n+1}_{*,*}(\mathrm{p}_F) = \mathrm{E}^{\infty}_{*,*}(\mathrm{p}_F)$ and $H_*(\mathcal{P}(F); \mathbb{K}) \cong \mathbb{K}$, we can regard the DG-module $\mathcal{C}'_*$ as a an acyclic resolution of $\mathbb{K}$. Applying Assertion 1.3 and the Universal Coefficient Theorem, we obtain the desired result. □

Assertion 1.5 *If* $\mathrm{E}^{n+1}_{*,*}(\mathrm{p}_F) = \mathrm{E}^{\infty}_{*,*}(\mathrm{p}_F)$, *then* $\mathrm{E}^2_{*,*}(\mathcal{C}) = \cdots = \mathrm{E}^{\infty}_{*,*}(\mathcal{C})$.

Proof. This is an immediate consequence of Assertion 1.4. □

Proof of Theorem 1.1

Proof. We proceed by descending induction on t. The induction begins with $t = k - 1$. Here, the fibration in question is

$$\mathrm{p}_{1,k-1} : \mathbb{F}_1(\mathbb{R}^{n+1}_{k-1}) \overset{\simeq}{\to} (\mathbb{R}^{n+1}_{k-1}) \simeq (\vee_{j=1}^{k-1} S_{kj}).$$

As each S_{kj} is an n-sphere, and $n > 1$ by assumption, we conclude that $\mathbb{F}_{k-t,t}$ is simply connected. The proof of Statement (i) of the theorem follows from general principles. On the other hand, Statements (ii) and (iii) can easily be seen to follow for dimensional reasons. Thus the initial step of the induction is proved.

Assume next that the theorem is true for the path fibration

$$\mathrm{p}_{k-t-1,t+1} : \mathcal{P}(\mathbb{F}_{k-t-1,t+1}) \to \mathbb{F}_{k-t-1,t+1}.$$

The base space is simply connected, and, therefore, Statement (i) of the theorem holds, while Statement (ii) is true for dimensional reasons.

To prove Statement (iii), consider the exact sequence

$$0 \to \mathbb{K} \to H_*(B) \to \bar{H}_*(B) \to 0$$

of $\mathbb{K}$-modules. Multiplying by $H_*(F) \otimes H_*(\Omega(E))$, we obtain the exact sequence

$$0 \to \mathcal{C}^F_* \to \mathcal{C}^E_* \to \mathcal{C}_* \to 0,$$

where

$$\begin{aligned} \mathcal{C}_*^F &= H_*(F) \otimes H_*(\Omega(E)), \\ \mathcal{C}_*^E &= H_*(E) \otimes H_*(\Omega(E)), \text{ and} \\ \mathcal{C}_* &= H_*(E)/H_*(F) \otimes H_*(\Omega(E)). \end{aligned}$$

As above, we give each term in the sequence the differential induced by that of the Serre spectral sequence of the fibration $p_E : \mathcal{P}(E) \to E$. Passing to homology, the preceding exact sequence of DG-modules yields the long exact sequence

$$\cdots \to H_*(\mathcal{C}) \xrightarrow{\partial} H(\mathcal{C}^F) \to H_*(\mathcal{C}^E) \to \cdots .$$

By the induction assumption and Lemma 1.1, we see that $H_*(\mathcal{C}^F) = H_*(\Omega B))$; also, by the induction assumption and the preceding Assertions, we have $H_*(\mathcal{C}) = \bar{H}_*(B) \otimes H_*(\Omega(B))$. Thus we obtain the long exact sequence

$$\cdots \to \bar{H}_*(B) \otimes H_*(\Omega(B)) \xrightarrow{\partial} H_*(\Omega(B)) \to H_*(\mathcal{C}^E) \to \cdots .$$

It is easy to show that the morphism,

$$\bar{H}_*(B) \otimes H_*(\Omega(B)) \xrightarrow{\partial} H_*(\Omega(B))$$

is just the differential d^n of the Serre spectral sequence of the fibration $p_B : \mathcal{P}(B) \to B$. From the long exact sequence, we deduce that $H_*(\mathcal{C}^E) = \mathbb{K}$. However,

$$\mathcal{C}_*^E = \mathrm{E}^n_{*,*}(p_E),$$

where $\mathrm{E}^n_{*,*}(p_E)$ is the E^n-term of the spectral sequence of the fibration $p_E : \mathcal{P}(E) \to E$. This finishes the proof of Statement (iii). □

Corollary 1.1 *The Serre spectral sequence of $\Lambda(\mathbb{F}_{k-r,r}) \to \mathbb{F}_{k-r,r}$ is such that $\mathrm{E}^t_{*,*} \cong \mathrm{E}^\infty_{*,*}$, for $t \geq (n+1)$.*

Proof. The $\{\mathrm{E}^n_{*,*}(p_{k-r}, \mathrm{d}(p_{k-r}\}$ term of the Serre spectral sequence of $p_{k-r} : \mathcal{P}(\mathbb{F}_{k-r,r}) \to \mathbb{F}_{k-r,r}$ is an acyclic free resolution of $\mathbb{K}$ over $H_*(\Omega(\mathbb{F}_{k-r,r}))$. Hence, $\mathrm{E}^{n+1}_{*,*}(p_{k-r})$ is isomorphic to the $\mathrm{E}^2_{*,*}$ term of the Eilenberg-Moore spectral sequence. But the latter collapses. Hence, $\mathrm{E}^{n+1} \cong \mathrm{E}^\infty_{*,*}$. □

2 The Case of $\mathbb{F}_{k+1}(S^{(n+1)})$, $(n+1)$ Odd

In this section, homology is with integral coefficients. Our aim here is to show that the homology Serre spectral sequence of the path fibration

$$\mathrm{p} : \mathcal{P}(\mathbb{F}_{k+1}(S^{n+1})) \to \mathbb{F}_{k+1}(S^{n+1}),$$

where $(\alpha, r) \mapsto \alpha(0)$, stabilizes in the sense of Theorem 1.1.

First, we need to describe the Serre spectral sequence of an intermediate fibration. Recall that the projection

$$\mathrm{proj}_{k+1,1} : \mathbb{F}_{k+1}(S^{n+1}) \to S^{n+1}$$

such that $(x_1, \cdots, x_n) \mapsto x_1$is a fibration, the fiber of which is homeomorphic to $\mathbb{F}_k(\mathbb{R}^{n+1})$ (see Chapter III, §2). Following the conventions of the previous section, put

$$F = \mathbb{F}_k(\mathbb{R}^{n+1}),\ E = \mathbb{F}_{k+1}(S^{n+1}),\ B = S^{n+1},$$

and denote by $i : F \to E$ the imbedding defined in Chapter III, §2. As we showed there, there is a section $\delta'_{k+1} : B \to E$, and this leads to a twisted product representation $E = B \rtimes F$. Consider the path fibrations

$$\mathrm{p}_E : \mathcal{P}(E) \to E,\ \mathrm{p}_F : \mathcal{F}(F) \to F,$$

where $(\alpha, r) \mapsto \alpha(0)$. Put

$$\Omega(E, F) = \{(\alpha, r) \mid \alpha(r) \in F\},$$

and denote by

$$\mathrm{p}_{(E,F)} : \Omega(E, F) \to F$$

the fibration where the projection sends a path to its endpoint. Observe that the fibration $\mathrm{p}_{(E,F)} : \Omega(E, F) \to F$ is the pull-back of $\mathrm{p}_E : \mathcal{P}(\mathbb{F}_{k+1}(S^{n+1})) \to S^{n+1}$ by the map $i : F \to E$.

Theorem 2.1 *The Serre spectral sequence $\{\mathrm{E}^m_{*,*}(\mathrm{p}_{(E,F)}), \mathrm{d}^m(\mathrm{p}_{(E,F)})\}$ of the fibration $\mathrm{p}_{(E,F)} : \Omega(E, F) \to F$ has the following properties:*

(i) $\mathrm{E}^2_{*,*}(\mathrm{p}_{(E,F)}) \cong \cdots \cong \mathrm{E}^n_{*,*}(\mathrm{p}_{(E,F)})$;
(ii) $\mathrm{E}^{n+1}_{p,*} \cong 0$, *if* $p \neq 0$, *and* $H_*(\Omega(B))$, *if* $p = 0$; *and*
(iii) $\mathrm{E}^m_{*,*}(\mathrm{p}_{(E,F)}) \cong \mathrm{E}^\infty_{*,*}(\mathrm{p}_{(E,F)})$ *for* $m \geq (n+1)$.

Proof. To prove Statement (i), recall that

$$\mathrm{E}^2_{*,*}(\mathrm{p}_{(E,F)}) \cong H_*(F) \otimes H_*(\Omega(E, F)).$$

The twisted-product structure on $F = \mathbb{F}_k(\mathbb{R}^{n+1})$, given in Chapter II, §4, leads to an additive isomorphism

$$H_*(F) \cong \bigotimes_{m=2}^{k} H_*(S_{m1} \vee \cdots \vee S_{mm-1}).$$

Recall that each S_{mj} is an n-sphere and, hence, $H_i(F) = 0$, if $i \neq 0 \mod n$. This proves Statement (i).

To prove Statement (ii), consider the commutative diagram

$$\begin{array}{ccccc} \Omega(F) & \to & \mathcal{P}(F) & \overset{p_F}{\to} & F \\ \downarrow & & \downarrow & & \downarrow \\ \Omega(E) & \to & \Omega(E,F) & \overset{p_{(E,F)}}{\to} & F, \end{array}$$

where the rows are the path fibrations and the vertical maps are those induced by $i : F \to E$. The latter induces a morphism

$$\phi^r : \mathrm{E}^r_{*,*}(\mathrm{p}_F) \to \mathrm{E}^r_{*,*}$$

of spectral sequences. But $\mathrm{p}_F : \mathcal{P}(F) \to F$ and $p_{(E,F)} : \Omega(E,F) \to F$ are, respectively, principal $\Omega(F)$ and $\Omega(E)$ fibrations. Hence, the Serre spectral sequences

$$\{\mathrm{E}^*_{*,*}(\mathrm{p}_F), \mathrm{d}(\mathrm{p}_F)\}, \ \ \{\mathrm{E}^*_{*,*}(\mathrm{p}_{(E,F)}), \mathrm{d}(\mathrm{p}_{(E,F)})\}$$

are DG-sequences of, respectively, $H_*(\Omega(F))$ and $H_*(\Omega(E))$ modules. Moreover, the morphism ϕ^r respects these actions. Hence, it induces, in turn, the morphism

$$\phi'^r : \mathrm{E}^r_{*,*}(p_F) \otimes_{H_*(\Omega(F))} H_*(\Omega(E)) \to \mathrm{E}^r_{*,*}(p_{(E,F)}).$$

As the differentials of the spectral sequences commute with the actions of the Pontryagin algebras of the relevant loop space, the morphism ${\phi'}_r$ can easily be seen to be a morphism of spectral sequences, where the differential of $\mathrm{E}^r_{*,*}(p_F) \otimes_{H_*(\Omega(F))} H_*(\Omega(E))$ is $\mathrm{d}^r(\mathrm{p}_F) \otimes 1$.

The twisted product representation $E = B \rtimes F$ leads to the homotopy equivalence

$$\psi : \Omega(B) \times \Omega(F) \to \Omega(E),$$

where $((\beta, r_\beta), (\alpha, r_\alpha))$ goes to the loop composition $\beta \circ s * \alpha \circ i$, with $s : B \to E$ being the section giving the twisted product representation. One sees easily that ψ is a map of a right $\Omega(F)$ space to a right $\Omega(E)$ space. Hence,

$$\mathrm{E}^r_{*,*}(\mathrm{p}_F) \otimes_{H_*(\Omega(F))} H_*(\Omega(E)) \cong H_*(F) \otimes H_*(\Omega(F)) \otimes H_*(\Omega(B)),$$

for $2 \leq r \leq n$, and the differential is just $\mathrm{d}^r(\mathrm{p}_F) \otimes 1$. This immediately implies that

$$\mathrm{E}^{n+1}_{p,*}(\mathrm{p}_{(E,F)}) \cong \begin{cases} 0, & \text{if } p \neq 0, \\ H_*(\Omega(B)) & \text{if } p = 0, \end{cases}$$

as required. Therefore,

$$\mathrm{E}^{n+1}_{p,*}(\mathrm{p}_{(E,F)}) \cong \mathrm{E}^{n+1}_{p,*}(\mathrm{p}_F) \otimes H_*(\Omega(B))$$

for all $p \geq 0$. But $\mathrm{E}^{n+1}_{*,*}(\mathrm{p}_F) \cong \mathbb{K}$, according to Theorem 1.1. This finishes the proof of Statement (ii). Statement (iii) is an immediate corollary. □

In order to prove that the Serre spectral sequence of $p_E : \mathcal{P}(E) \to E$ stabilizes in the same manner, we proceed as in the proof of Theorem 1.1. So, consider the exact sequence

$$0 \to \mathbb{K} \to H_*(B) \to \bar{H}_*(B) \to 0$$

of $\mathbb{K}$-modules. Multiplying by $H_*(F) \otimes H_*(\Omega(E))$, we obtain the exact sequence

$$0 \to \mathcal{C}_*^F \to \mathcal{C}_*^E \to \mathcal{C}_*^{(E,F)} \to 0,$$

where

$$\begin{aligned} \mathcal{C}_*^F &= H_*(F) \otimes H_*(\Omega(E)), \\ \mathcal{C}_*^E &= H_*(E) \otimes H_*(\Omega(E)), \text{ and} \\ \mathcal{C}_*^{(E,F)} &= H_*(E)/H_*(F) \otimes H_*(\Omega(E)). \end{aligned}$$

Observe here that $\mathcal{C}_*^E$ can be identified with $\mathrm{E}^n_{*,*}(p_E)$, the $\mathrm{E}^n_{*,*}$ term of the Serre homology spectral sequence of p_E, and $\mathcal{C}_*^F$, with $\mathrm{E}^n_{*,*}(p_{(E,F)})$, the $\mathrm{E}^n_{*,*}$ term of the spectral sequence of $p_{(E,F)}$. We give $\mathcal{C}_*^E$ and $\mathcal{C}_*^F$ the differentials $\mathrm{d}^n(p_E)$ and $\mathrm{d}^n(p_F)$, respectively. Then, $\mathcal{C}_*^F$ becomes a DG-submodule of $\mathcal{C}_*^E$, as can easily be verified. Hence, the exact sequence

$$0 \to \mathcal{C}_*^F \to \mathcal{C}_*^E \to \mathcal{C}_*^{(E,F)} \to 0$$

is one of DG-modules.

Filter $\mathcal{C}_*^{(E,F)}$ by putting

$$\mathcal{F}(\mathcal{C}^{(E,F)}) : F_p(\mathcal{C}^{(E,F)}) = \textstyle\sum_{m=0}^{p} \bar{H}_*(B) \otimes H_m(F) \otimes H_*(\Omega(E)).$$

For brevity, denote by F_p the submodule just defined. The spectral sequence $\{\mathrm{E}^r_{*,*}(\mathcal{C}^{(E,F)}), \mathrm{d}^r(\mathcal{C}^{(E,F)})\}$ has the following properties.

Assertion 2.1 $\mathrm{E}^0_{*,*}(\mathcal{C}^{(E,F)}) = \mathrm{E}^1_{*,*}(\mathcal{C}^{(E,F)})$.

Assertion 2.2 *There is an isomorphism*

$$\begin{array}{ccc} \mathrm{E}^1_{p,*}(\mathcal{C}^{(E,F)})) & \overset{\mathrm{d}^1(\mathcal{C}^{(E,F)}))}{\rightarrow} & \mathrm{E}^1_{p-1,*}(\mathcal{C}^{(E,F)})) \\ \downarrow & & \downarrow \\ \mathrm{E}^{n+1}_{n+1,*}(p_B) \otimes \mathrm{E}^n_{np,*}(p_F) & \overset{1\otimes \mathrm{d}^n(p_F)}{\rightarrow} & \mathrm{E}^{n+1}_{n+1,*}(p_B) \otimes \mathrm{E}^n_{n(p-1),*}(p_F) \end{array}$$

of DG-*modules.*

Assertion 2.3

$$\begin{aligned} &\mathrm{E}^2_{p,*}(\mathcal{C}) = 0, \quad \textit{for } p > 0 \quad \textit{and} \\ &\mathrm{E}^2_{0,*}(\mathcal{C}) = \bar{H}_{n+1}(B) \otimes H_*(\Omega(B)). \end{aligned}$$

Assertion 2.4 $\mathrm{E}^2_{*,*}(\mathcal{C}) = \cdots = \mathrm{E}^{\infty}_{*,*}(\mathcal{C})$.

The proofs are similar to those of Assertions 1.2, 1.3, and 1.4 and are left to the reader.

Theorem 2.2 *The Serre spectral sequence* $\{\mathrm{E}^r_{*,*}(\mathrm{p}_E), \mathrm{d}^r(\mathrm{p}_E)\}$ *has the following properties:*

(*i*) $\mathrm{E}^2_{*,*}(\mathrm{p}_E) \cong \mathrm{E}^r_{*,*}(\mathrm{p}_E)$, *for* $2 \leq r \leq n$;
(*ii*) *the projection* $p \circ \mathrm{p}_E : \mathcal{P}(E) \to B$, *induces an isomorphism* $\mathrm{E}^r_{*,*}(\mathrm{p}_E) \cong \mathrm{E}^r_{*,*}(\mathrm{p}_B)$ *of* DG-*modules, for* $r \geq (n+1)$; *and*
(*iii*) $\mathrm{E}^r_{*,*}(\mathrm{p}_E) \cong \mathrm{E}^{\infty}_{*,*} \cong \mathbb{K}$, *for* $r \geq (n+2)$.

Proof. To prove Statement (i), note that $H_*(E) \cong H_*(B) \otimes H_*(F)$. But $B = S^{n+1}$ and $F = \mathbb{F}_k(\mathbb{R}^{n+1})$. Thus, the assertion follows for dimensional reasons.

To prove Statement (ii), we need to show that $\mathrm{E}^{n+1}_{*,*}(\mathrm{p}_E)$ is isomorphic to $H_*(B) \otimes H_*(\Omega(B))$. To do so, we consider the exact sequence of DG-modules

$$0 \to \mathcal{C}^F_* \to \mathcal{C}^E_* \to \mathcal{C}^{(E,F)}_* \to 0.$$

Passing to homology, one obtains the long exact sequence

$$\cdots \to H_*(\mathcal{C}^{(E,F)}) \to H_*(\mathcal{C}^F) \to H_*(\mathcal{C}^E) \to H_*(\mathcal{C}^{(E,F)}) \to \cdots .$$

According to Theorem 1.1, $H_*(\mathcal{C}^F) \cong H_*(\Omega(B))$, while Assertions 2.1, 2.2, and 2.3 imply that $H_*(\mathcal{C}^{(E,F)}) \cong \bar{H}_*(B) \otimes H_*(\Omega(B))$. Moreover, the connecting homomorphism $H_*(\mathcal{C}^{(E,F)}) \to H_*(\mathcal{C}^F)$ is easily seen to be trivial, since $\mathrm{d}^n(\mathrm{p}_E) : \mathrm{E}^n_{*,*}(\mathrm{p}_E) \to \mathrm{E}^n_{*,*}(\mathrm{p}_E)$ is trivial on $\mathcal{C}^{(E,F)}$. Therefore, $H_*(\mathcal{C}^E) \cong H_*(B) \otimes H_*(\Omega(B))$. But $H_*(\mathcal{C}^E) = \mathrm{E}^{n+1}_{*,*}(\mathrm{p}_E)$. Now, Statement (ii) and the rest of the theorem follow easily. □

Recall that Corollary 1.1 implies that the computation of the homology of $\Lambda(F) = \Lambda(\mathbb{F}_k(\mathbb{R}^{n+1})$ reduces to a single stage: namely, that of computing the homology of the DG-module

$$\mathrm{d}^n(\mathrm{eval}_F) : \mathrm{E}^n_{*,*}(\mathrm{eval}_F) \to \mathrm{E}^n_{*,*}(\mathrm{eval}_F),$$

where $\mathrm{eval}_F : \Lambda(F) \to F$ is the evaluation map. The analogous problem for $\Lambda(\mathbb{F}_{k+1}(S^{n+1})$ requires two stages.

One way to proceed is to compute both the $\mathrm{E}^n_{*,*}$ and $\mathrm{E}^{n+1}_{*,*}$ terms of the Serre spectral sequence of the evaluation map $\mathrm{eval}_E : \Lambda(E) \to E$. Alternatively, we consider the composite map

$$\mathrm{proj}_{k+1,1} \circ \mathrm{eval}_E : \Lambda(E) \to B.$$

The latter is a fibration the fiber of which is

$$\Lambda(E,F) = \{\omega \in \Lambda(\mathbb{F}_{k+1}(S^{n+1}) | \omega(1) \in F\}.$$

Let $\mathrm{eval}_{(E,F)} : \Lambda(E,F) \to F$ be the evaluation map where $\omega \mapsto \omega(1)$.

Corollary 2.1 *The Serre spectral sequence* $\{\mathrm{E}^r_{*,*}(\mathrm{eval}_{(E,F)}), \mathrm{d}^r(\mathrm{eval}_{(E,F)}))\}$ *has the following properties:*

$$\begin{aligned}
&(i)\ \mathrm{E}^2_{*,*}(\mathrm{eval}_{(E,F)}) \cong \cdots \cong \mathrm{E}^n_{*,*}(\mathrm{eval}_{(E,F)});\\
&(ii)\ \mathrm{E}^{n+1}_{*,*}(\mathrm{eval}_{(E,F)}) \cong H_*(\Lambda(F)) \otimes H_*(\Omega(B));\ and\\
&(iii)\ \mathrm{E}^{n+1}_{p,*}(\mathrm{eval}_{(E,F)}) = \mathrm{E}^{\infty}_{*,*}(\mathrm{eval}_{(E,F)}).
\end{aligned}$$

Proof. Statement (i) follows for dimensional reasons. As to Statement (ii), we have

$$\begin{aligned}
\mathrm{E}^n_{*,*}(\mathrm{eval}_{(E,F)}) &\cong H_*(F) \otimes H_*(\Omega(E))\\
&\cong H_*(F) \otimes H_*(\Omega(F)) \otimes H_*(\Omega(B)).
\end{aligned}$$

Moreover, $\mathrm{d}^n(\Lambda(\mathrm{eval}_{(E,F)})) = \mathrm{d}^n(\mathrm{eval}_F) \otimes 1$, as can be easily seen (cf. Theorem 4.1 of Chapter VIII). This proves Statement (ii).

To prove Statement (iii), note that Theorem 2.1 implies that $E^n_{*,*}(\mathrm{p}_F) \cong H_*(F) \otimes H_*(\Omega(F))$ is an $H_*(\Omega(F))$ acyclic free resolution of $\mathbb{K}$. Hence,

$$\mathrm{E}^{n+1}_{*,*}(\mathrm{eval}_{(E,F)}) \cong \mathrm{Tor}^{H_*(\Omega(F)}(H_*(\Omega(E)), \mathbb{K}),$$

where the term on the right is the E^2 term of the Eilenberg-Moore spectral sequence. But the latter sequence collapses. This proves Statement (iii). □

Next, denote the map

$$\mathrm{proj}_{k+1,1} \circ \mathrm{eval}_E : \Lambda(E) \to B$$

by $\mathrm{p}_{k+1,1}$.

Corollary 2.2 *The Serre spectral sequence* $\{\mathrm{E}^r_{*,*}(\mathrm{p}_{k+1,1}), \mathrm{d}^r(\mathrm{p}_{k+1,1})\}$ *has the following properties:*

$$\begin{aligned}
&(i)\ \mathrm{E}^2_{*,*}(\mathrm{p}_{k+1,1}) \cong \cdots \cong \mathrm{E}^n_{*,*}(\mathrm{p}_{k+1,1});\\
&(ii)\ \mathrm{E}^{n+1}_{*,*}(\mathrm{p}_{k+1,1}) \cong H_*(B) \otimes H_*(\Lambda(E,F));\ and\\
&(iii)\ \mathrm{E}^t_{*,*} \cong \mathrm{E}^{\infty}_{*,*},\ for\ t \geq (n+2).
\end{aligned}$$

Putting the two corollaries together, we obtain the following. First, consider the map $\Lambda(\mathrm{proj}_{k+1,1}) : \Lambda(\mathbb{F}_{k+1}(S^{n+1})) \to \Lambda(S^{n+1})$ induced by the projection $\mathrm{proj}_{k+1,1} : \mathbb{F}_{k+1}(S^{n+1}) \to S^{n+1}$ that sends a tuple to its first entry.

Corollary 2.3 *The natural injection* $\Lambda(\mathbb{F}_k(\mathbb{R}^{n+1})) \to \Lambda(\mathbb{F}_{k+1}(S^{n+1}))$ *induces an isomorphism*

$$H_*(\Lambda\mathbb{F}_k(\mathbb{R}^{n+1})) \otimes H_*(\Lambda(S^{n+1})) \to H_*(\Lambda\mathbb{F}_{k+1}(S^{n+1}))$$

of coalgebras.

3 The Case of $\mathbb{F}_{k+1}(S^{n+1})$, $(n+1)$ Even

Assume that $n > 1$. In keeping with the notation of the previous sections, put $E = \mathbb{F}_{k+1}(S^{n+1})$, $F = \mathbb{F}_{k+1}(\mathbb{R}^{n+1})$, and $B = S^{n+1}$. Then, the projection $\mathrm{p} : E \to B$ that sends a tuple to its first entry is a fibration the fiber of which at the basepoint can be naturally identified with F.

Assume here that homology is with coefficients in $\mathbb{Z}_2$. Denote by $\mathrm{p}_E : \mathcal{P}(E) \to E$, $\mathrm{p}_F : \mathcal{P}(F) \to F$ and $\mathrm{p}_B : \mathcal{P}(B) \to B$ the projection that sends a path to its endpoint for E, F, and B, respectively. Put

$$\Omega(E,F) = \{\omega \in \mathcal{P}(E) | \omega(r) \in F\},$$

and let $\mathrm{p}_{(E,F)} : \Omega(E,F) \to F$ be the map where $\omega \mapsto \omega(r)$.

Theorem 3.1 *The Serre homology spectral sequence of* p_E *has the following properties:*

(i) $\mathrm{E}^r_{*,*}(\mathrm{p}_E) \cong H_*(E) \otimes H_*(\Omega(E))$, *for* $2 \leq r \leq n$;
(ii) *the map* $\mathrm{p} : E \to B$, *induces an isomorphism,*
$\mathrm{E}^r_{*,*}(\mathrm{p}) : \mathrm{E}^r_{*,*}(\mathrm{p}_E) \to \mathrm{E}^r_{*,*}(\mathrm{p}_B)$, *for* $r \geq (n+1)$;
and, hence,
(iii) $\mathrm{E}^r_{*,*}(\mathrm{p}_E) \cong \mathbb{Z}_2$, *for* $r > (n+1)$.

Proof. The proof is essentially the same as that for Theorem 2.2. First, one shows that the mod 2 Serre spectral sequence of the projection $\mathrm{p}_{(E,F)} : \Omega(E,F) \to F$, whereby a path is sent to its endpoint, has the following properties (cf. Theorem 2.1):

(i) $\mathrm{E}^r_{*,*}(\mathrm{p}_{(E,F)}) \cong H_*(F) \otimes H_*(\Omega(E))$, for $2 \leq r \leq n$;
(ii) $\mathrm{E}^{n+1}_{*,*}(\mathrm{p}_{(E,F)}) \cong H_*(\Omega(B))$; and
(iii) $\mathrm{E}^r_{*,*}(\mathrm{p}_{(E,F)}) \cong \mathrm{E}^{n+1}_{*,*}(\mathrm{p}_{(E,F)})$, for $r \geq (n+1)$.

The proof amounts to comparing the path fibrations

$$\mathrm{p}_F : \mathcal{P}(F) \to F, \ \mathrm{p}_{(E,F)} : \Omega(E,F) \to F$$

with each other, noting that they are both principal and, hence, that the differential operators of the spectral sequences commute with the action of the respective Pontryagin algebras.

Next, one considers the $\mathrm{E}^n_{*,*}$ terms of the spectral sequences

$$\{\mathrm{E}^r_{*,*}(\mathrm{p}_{(E,F)}), \mathrm{d}^r(\mathrm{p}_{(E,F)})\}, \ \{\mathrm{E}^r_{*,*}(\mathrm{p}_E), \mathrm{d}^r(\mathrm{p}_E)\}.$$

Put

$$\begin{aligned} C^F_* &= \mathrm{E}^n_{*,*}(\mathrm{p}_{(E,F)}) \\ &= H_*(F) \otimes H_*(\Omega(E)) \\ C^E_* &= \mathrm{E}^r_{*,*}(\mathrm{p}_E) \\ &= H_*(E) \otimes H_*(\Omega(E)) \end{aligned}$$

and their quotient

$$\mathcal{C}_*^{(E,F)} = H_*(E)/H_*(F) \otimes H_*(\Omega(E))$$

as DG-modules, with the differentials being those induced by the differential $\mathrm{d}^n(\mathrm{p}_E)$ of the spectral sequence of $\mathrm{p}_E : \mathcal{P}(E) \to E$. As in §2 above, one proves that $H_*(\mathcal{C}^{(E,F)}) \cong \bar{H}_*(B) \otimes H_*(\Omega(B)$, obtaining Statement (ii). Statement (iii) follows as in the case when $(n+1)$ is even. The details of these steps are left to the reader. □

The procedure for computing the mod 2 homology groups of $\Lambda(E)$ is also similar to the case when $(n+1)$ is odd. One computes directly both the $\mathrm{E}^n_{*,*}$ and $\mathrm{E}^{n+1}_{*,*}$ terms of the Serre spectral sequence of the evaluation map $\mathrm{eval}_E : \Lambda(E) \to E$, or, alternatively, one considers the composite map

$$\mathrm{proj}_{k+1,1} \circ \mathrm{p}_{\Lambda(E)} : \Lambda(E) \to B.$$

The latter is a fibration the fiber of which is

$$\Lambda(E,F) = \{\omega \in \Lambda(\mathbb{F}_{k+1}(S^{n+1}) | \omega(1) \in F\}.$$

Let $\mathrm{eval}_{(E,F)} : \Lambda(E,F) \to F$ be the evaluation map where $\omega \mapsto \omega(1)$.

The proofs of the following are, with the necessary modifications, like those of Corollaries 2.1, 2.2, and 2.3. The details are left to the reader.

Corollary 3.1 *The Serre spectral sequence* $\{\mathrm{E}^r_{*,*}(\mathrm{eval}_{(E,F)}), \mathrm{d}^r(\mathrm{eval}_{(E,F)})\}$ *has the following properties:*

(i) $\mathrm{E}^2_{*,*}(\mathrm{eval}_{(E,F)}) \cong \cdots \cong \mathrm{E}^n_{*,*}(\mathrm{eval}_{(E,F)})$;
(ii) $\mathrm{E}^{n+1}_{*,*}(\mathrm{eval}_{(E,F)}) \cong H_*(\Lambda(F)) \otimes H_*(\Omega(B))$; *and*
(iii) $\mathrm{E}^{n+1}_{p,*}(\mathrm{eval}_{(E,F)})) = \mathrm{E}^\infty_{*,*}(\mathrm{eval}_{(E,F)})$.

Next, denote the map

$$\mathrm{proj}_{k+1,1} \circ \mathrm{p}_{\Lambda(E)} : \Lambda(E) \to B$$

by $\mathrm{p}_{k+1,1}$.

Corollary 3.2 *The Serre spectral sequence* $\{\mathrm{E}^r_{*,*}(\mathrm{p}_{k+1,1}), \mathrm{d}^r(\mathrm{p}_{k+1,1})\}$ *has the following properties:*

(i) $\mathrm{E}^2_{*,*}(\mathrm{p}_{k+1,1}) \cong \cdots \cong \mathrm{E}^n_{*,*}(\mathrm{p}_{k+1,1})$;
(ii) $\mathrm{E}^{n+1}_{*,*}(\mathrm{p}_{k+1,1}) \cong H_*(B) \otimes H_*(\Lambda(E,F))$; *and*
(iii) $\mathrm{E}^t_{*,*} \cong \mathrm{E}^\infty_{*,*}$, *for* $t \geq (n+2)$.

Let $\Lambda(\mathrm{proj}_{k+1,1}) : \Lambda(\mathbb{F}_{k+1}(S^{n+1})) \to \Lambda(S^{n+1})$ be the map induced by the projection $\mathrm{proj}_{k+1,1} : \mathbb{F}_{k+1}(S^{n+1}) \to S^{n+1}$ that sends a tuple to its first entry.

Corollary 3.3 *The natural injection* $\Lambda(\mathbb{F}_k(\mathbb{R}^{n+1})) \to \Lambda(\mathbb{F}_{k+1}(S^{n+1}))$ *induces an isomorphism*

$$H_*(\Lambda(\mathbb{F}_k(\mathbb{R}^{n+1}))) \otimes H_*(\Lambda(S^{n+1})) \to H_*(\Lambda(\mathbb{F}_{k+1}(S^{n+1})))$$

of coalgebras.

Let us consider the case for rational homology.

Theorem 3.2 *The Serre rational homology spectral sequence of* p_E *has the following properties:*

(*i*) $\mathrm{E}^r_{*,*}(\mathrm{p}_E) \cong H_*(E) \otimes H_*(\Omega(E))$, *for* $2 \leq r \leq n$;
(*ii*) *the map* $\mathrm{p} : E \to B$, *induces an isomorphism* $\mathrm{E}^r_{*,*}(\mathrm{p}) : \mathrm{E}^r_{*,*}(\mathrm{p}_E) \to \mathrm{E}^r_{*,*}(\mathrm{p}_B)$, *of spectral sequences, for* $r \geq (n+1)$; *and, consequently,*
(*iii*) $\mathrm{E}^r_{*,*}(\mathrm{p}_E) \cong \mathbb{Q}$, *for* $r > (n+1)$.

The proof is based on the twisted product representation

$$\mathbb{F}_{k+1}(S^{n+1}) \simeq O_{n+2,2} \rtimes \mathbb{F}_{k-2,2}(\mathbb{R}^{n+1})$$

(see Chapter III, §1). It follows along the same lines as that for Theorems 2.1 and 2.2. The details are left to the reader.

XII. Computation of $H_*(\Lambda(M))$

We have seen in the previous chapters that the space $\mathbb{F}_k(M)$ can be described as a twisted product of simpler spaces when M is $\mathbb{R}^{n+1}$ or S^{n+1}. The simpler spaces are bouquets of n-dimensional spheres when $M = \mathbb{R}^{n+1}$; when $M = S^{n+1}$, they include the Stiefel manifold $O_{n+2,2}$ of orthonormal 2-frames in $\mathbb{R}^{n+2}$, as well. We have also seen that the space $\Omega\mathbb{F}_k(M)$ of based loops splits as a product of the loop spaces of the split factors as spaces, but not as loop spaces. A natural question to ask is whether the space of free loops $\Lambda\mathbb{F}_k(M)$ splits, at the homology level, as a tensor product of the homology of the split factors of $\mathbb{F}_k(M)$. We shall see in this chapter that this is the exception: it is true for $k = 3$, but not in general.

In §1 we show that, for mod 2 homology, in the special case $k = 3$,

$$\begin{aligned} H_*(\Lambda\mathbb{F}_3(\mathbb{R}^{n+1}) &\cong H_*(\Lambda\mathbb{F}_2(\mathbb{R}_1^{n+1}) \\ &\cong H_*(\Lambda(\mathbb{R}_1^{n+1})) \otimes H_*(\Lambda(\mathbb{R}_2^{n+1})) \end{aligned}$$

additively. This is done by noting that the natural imbedding of $\mathbb{F}_2(\mathbb{R}_1^{n+1})$ into $\mathbb{F}_3(\mathbb{R}^{n+1})$ is a homotopy equivalence and then considering the fibration

$$\Lambda(\mathrm{p}_1) : \Lambda\mathbb{F}_2(\mathbb{R}_1^{n+1}) \to \Lambda(\mathbb{R}_1^{n+1})$$

the fiber of which is $\Lambda(\mathbb{R}_2^{n+1})$. We proceed to show that the fiber is totally nonhomologous to zero. The argument amounts to computing the Serre spectral sequence of the fibrations in the commutative diagram

$$\begin{array}{ccccc} \Omega(\mathbb{R}_2^{n+1}) & \to & \Lambda(\mathbb{R}_2^{n+1}) & \to & \mathbb{R}_2^{n+1} \\ \downarrow & & \downarrow & & \downarrow \\ \Omega\mathbb{F}_2(\mathbb{R}_1^{n+1}) & \to & \Lambda\mathbb{F}_2(\mathbb{R}_1^{n+1}) & \to & \mathbb{F}_2(\mathbb{R}_1^{n+1}). \end{array}$$

The horizontal maps are the evaluation maps whereby $\omega \mapsto \omega(1)$, while the vertical maps are those induced by the natural inclusions. Using the results of Chapter IX, §3 on the structure of the free loop space as well as Corollary 1.1 of Chapter XI, we are able to establish the required result.

In §2 we show that the fiber $\Lambda(\mathbb{R}_{k-1}^{n+1})$ of the fibration

$$\Lambda(\mathrm{proj}_{k,k-1}) : \Lambda\mathbb{F}_k(\mathbb{R}^{n+1}) \to \Lambda\mathbb{F}_{k-1}(\mathbb{R}^{n+1}),$$

where $\text{proj}_{k,k-1}$ is the projection that takes $(x_1, \cdots, x_k)$ to $x_1, \cdots, x_{k-1})$, is not totally non-homologous to zero, if $k > 3$. We do this by considering the spectral sequence of the subfibration

$$\Lambda(\mathbb{R}^{n+1}_{k-1}) \to \Lambda(\mathbb{F}_{2,k-2}) \to \Lambda(\mathbb{R}^{n+1}_{k-2}).$$

In §3 we compute the mod 2 Poincaré series of $\Lambda(\mathbb{R}^{n+1}_r)$ and show that the Betti numbers grow exponentially. This result is used in §4 to obtain a lower estimate of the homology mod 2 of $\Lambda(\mathbb{F}_k(\mathbb{R}^{n+1}))$, which does not arise from $\Lambda(\mathbb{F}_r(\mathbb{R}^{n+1}))$, for $r < k$.

In the remaining section we study cup length in $\Lambda\mathbb{F}_k(\mathbb{R}^{n+1})$ is infinite and that the space contains compact subsets of arbitrarily high LS-category.

1 Splitting of $H_*(\Lambda\mathbb{F}_k(\mathbb{R}^{n+1}); \mathbb{Z}_2)$

Recall that, although $\mathbb{F}_k(\mathbb{R}^{n+1}) \simeq \mathbb{F}_r(\mathbb{R}^{n+1}) \ltimes \mathbb{F}_{k-r,r}(\mathbb{R}^{n+1})$, the fibration $\text{proj}_{k,r} : \mathbb{F}_k(\mathbb{R}^{n+1}) \to \mathbb{F}_r(\mathbb{R}^{n+1})$ is not fiber homotopically trivial except for special cases (see Theorem 5.1 of Chapter II). Passing to the associated free loop spaces, we ask whether

$$\Lambda(\text{proj}_{k,k-1}) : \Lambda\mathbb{F}_k(\mathbb{R}^{n+1}) \to \Lambda\mathbb{F}_{k-1}(\mathbb{R}^{n+1})$$

is trivial in any sense at all.

Theorem 1.1 *The fiber of the fibration*

$$\Lambda(\text{proj}_{k,k-1}) : \Lambda\mathbb{F}_k(\mathbb{R}^{n+1}) \to \Lambda\mathbb{F}_{k-1}(\mathbb{R}^{n+1})$$

is totally nonhomologous to 0 *over* $\mathbb{Z}_2$ *if, and only if,* $k = 3$.

Proof of ($\Leftarrow$)

Assume that $k = 3$, and consider the diagram

$$\mathcal{D} : \left\{ \begin{array}{ccccc} \mathbb{F}_3(\mathbb{R}^{n+1}) & \leftarrow & \mathbb{F}_{2,1}(\mathbb{R}^{n+1}) & \leftarrow & \mathbb{F}_{1,2}(\mathbb{R}^{n+1}) \\ \downarrow & & \downarrow & & \\ \mathbb{R}^{n+1} & & \mathbb{R}^{n+1}_1 & & \end{array} \right.$$

of fibrations (see Chapter II, §1). Recall that $\mathbb{F}_{2,1} \to \mathbb{F}_3(\mathbb{R}^{n+1})$ is a homotopy equivalence. Hence, to compute the homology $H_*(\Lambda\mathbb{F}_3(\mathbb{R}^{n+1}))$, it is enough to consider the fibration

$$\mathrm{p}_1 : \mathbb{F}_{2,1}(\mathbb{R}^{n+1}) \to \mathbb{R}^{n+1}_1, \ (q_1, x_2, x_3) \mapsto (q_1, x_2),$$

the fiber of which is $\mathbb{F}_{1,2}(\mathbb{R}^{n+1})$. Let us denote by

$$i : \mathbb{F}_{1,2} \to \mathbb{F}_{2,1}(\mathbb{R}^{n+1})$$

the imbedding where $x \mapsto (q_1, q_2, x)$. To prove the theorem, it is enough to prove that the morphism

$$\text{(I)} \quad H_*(\Lambda(i)) : H_*(\Lambda(\mathbb{R}_2^{n+1});\mathbb{Z}_2) \to H_*(\Lambda\mathbb{F}_2(\mathbb{R}_1^{n+1});\mathbb{Z}_2)$$

is injective. So, consider the diagram

$$\mathcal{D} : \left\{ \begin{array}{ccccc} \Omega(\mathbb{R}_2^{n+1}) & \to & \Lambda(\mathbb{R}_2^{n+1}) & \to & \mathbb{R}_2^{n+1} \\ \downarrow & & \downarrow & & \downarrow \\ \Omega\mathbb{F}_2(\mathbb{R}_1^{n+1}) & \to & \Lambda\mathbb{F}_2(\mathbb{R}_1^{n+1}) & \to & \mathbb{F}_2(\mathbb{R}_1^{n+1}), \end{array} \right.$$

where the second horizontal maps are the evaluation maps that send ω to $\omega(1)$ and the vertical maps are those induced by the natural inclusion $i : \mathbb{R}_2^{n+1} \to \mathbb{F}_2(\mathbb{R}_1^{n+1})$. Note that each of the horizontal lines is a fibration and that the diagram is a map of one fibration to the other. Therefore, the diagram induces a morphism

$$\phi^r : {}'\mathrm{E}^r_{*,*} \to \mathrm{E}^r_{*,*}$$

of the integral Serre spectral sequence of the first row to that of the second. But, according to Corollary 1.1 of Chapter XI, the Serre spectral sequences stabilize at the $(n+1)^{\mathrm{st}}$ term. Therefore, to prove that $\Lambda(\mathbb{R}_2^{n+1}) \to \Lambda\mathbb{F}_2(\mathbb{R}_1^{n+1})$ is injective, it is sufficient to prove that the morphism of spectral sequences

$$\text{(II)} \quad \phi^{n+1} : {}'\mathrm{E}^{n+1}_{*,*} \otimes \mathbb{Z}_2 \to \mathrm{E}^{n+1}_{*,*} \otimes \mathbb{Z}_2$$

is injective.

The proof amounts to a careful analysis of the spectral sequences and their differential operators. To begin, let us collect what we know about them.

Assertion 1.1 *Assuming the coefficient ring to be* $\mathbb{Z}$, *we have*

$$\begin{array}{ll} (i)\ {}'\mathrm{E}^2_{*,*} & \cong H_*(\mathbb{R}_2^{n+1}) \otimes H_*(\Omega(\mathbb{R}_2^{n+1})); \\ (ii)\ {}'\mathrm{E}^2_{*,*} & \cong {}'\mathrm{E}^n_{*,*};\ \textit{and} \\ (iii)\ {}'\mathrm{E}^{(n+1)}_{*,*} & \cong {}'\mathrm{E}^\infty_{*,*}. \end{array}$$

Proof. Keeping the notation of the preceding chapters, recall that

$$H_*(\mathbb{R}_2^{n+1};\mathbb{Z}) \cong \mathbb{Z}\{\alpha_{31}, \alpha_{32}\},$$

since $\mathbb{R}_2^{n+1} \simeq (S_{31} \vee S_{32})$, where the term on the right is the free module generated by $\{\alpha_{31}, \alpha_{32}\}$. As $n > 1$, it follows that the base space $\mathbb{R}_2^{n+1}$ is simply connected, and, hence, Statement (i) is a consequence of the structure theorem of the Serre spectral sequence of a fibration ([97, Serre]). Statement (ii) follows for dimensional reasons: the homology of $\mathbb{R}_2^{n+1}$ in positive dimensions is concentrated in dimension n. Finally, Statement (iii) follows because, according to Corollary 1.1 of Chapter XI, the Serre spectral sequence $\{{}'\mathrm{E}^r_{*,*}, {}'\mathrm{d}^n\}$ stabilizes at the $(n+1)$-st stage. □

Consequently, we see that

$$\begin{aligned}'\mathrm{E}^n_{*,*} &\cong H_*(\mathbb{R}^{n+1}_2;\mathbb{Z}) \otimes H_*(\Omega(\mathbb{R}^{n+1}_2);\mathbb{Z}) \\ &\cong \mathbb{Z}\{\alpha_{31},\alpha_{32}\} \otimes H_*(\Omega(\mathbb{R}^{n+1}_2);\mathbb{Z}).\end{aligned}$$

For simplicity, let us put

$$\mathrm{ad}(\mu)(\omega) = [\mu,\omega],\ \mu,\omega \in H_*(\Omega(\mathbb{R}^{n+1});\mathbb{Z}),$$

where square brackets stand for the Samelson product in homology ([96, Samelson]).

Assertion 1.2 *The differential*

$$'\mathrm{d}^n : {'\mathrm{E}^n_{*,*}} \to {'\mathrm{E}^n_{0,*}}$$

is such that

$$\begin{aligned}'\mathrm{d}^n(\alpha_{31}\otimes\omega) &= \mathrm{ad}(\bar\alpha_{31})(\omega) \text{ and} \\ '\mathrm{d}^n(\alpha_{32}\otimes\omega) &= \quad \mathrm{ad}(\bar\alpha_{32})(\omega)\end{aligned}$$

for all $\omega \in H_*(\Omega(\mathbb{R}^{n+1}_2);\mathbb{Z})$.

Proof. The differential $'\mathrm{d}^n$ depends on the action of $\Omega(\mathbb{R}^{n+1}_2)$, the loop space of the base, on the fiber (see [32, Fadell-Hurewicz]). The fiber here is $\Omega(\mathbb{R}^{n+1}_2)$ itself, and the action is the adjoint action

$$\Omega(\mathbb{R}^{n+1}_2) \times \Omega(\mathbb{R}^{n+1}_2) \to \Omega(\mathbb{R}^{n+1}_2),$$

where $(g,h) \mapsto g^{-1}hg$ for g a loop in the base and h an element in the fiber (see Chapter IX, §4). Now, note that this action is the composite of the coproduct

$$\Omega(\nabla) : \Omega_2 \to \Omega_2 \times \Omega_2$$

with the map

$$\Omega_2 \times \Omega_2 \times \Omega_2 \to \Omega_2,\ (g,x,h) \mapsto g^{-1}x\,h,$$

where $\Omega_2 = \Omega(\mathbb{R}^{n+1}_2)$. Recall that the induced coproduct

$$H_*(\mathbb{R}^{n+1}_2;\mathbb{Z}) \to H_*(\mathbb{R}^{n+1}_2;\mathbb{Z}) \otimes H_*(\mathbb{R}^{n+1}_2;\mathbb{Z})$$

is such that $\alpha_{rs} \mapsto \alpha_{rs}\otimes 1 + 1\otimes\alpha_{rs}$, where $(r,s) = (3,1)$ or $(3,2)$. Also, recall that $H_*(\Omega_2) \cong \mathbb{Z}(\bar\alpha_{31},\bar\alpha_{32})$, where $\bar\alpha_{rs}$ is the desuspension of α_{rs} (see Chapter VIII, §3). Hence, passing to homology, we see that the induced adjoint action

$$H_*(\Omega_2) \otimes H_*(\Omega_2) \to H_*(\Omega_2)$$

is such that $(\bar\alpha_{rs},\omega,\bar\alpha_{rs})$ goes to the Samelson product $[\bar\alpha_{rs},\omega]$ (cf. [32, Fadell-Hurewicz]). □

Let us turn next to the spectral sequence $\{\mathrm{E}^r_{*,*}, \mathrm{d}^r\}$ of the fibration

$$\Omega(\mathbb{F}_2(\mathbb{R}^{n+1}_1)) \to \Lambda(\mathbb{F}_2(\mathbb{R}^{n+1}_1)) \to \mathbb{F}_2(\mathbb{R}^{n+1}_1).$$

Assertion 1.3

$$\begin{aligned} &(i)\ \mathrm{E}^2_{*,*} \cong H_*(\mathbb{F}_2(\mathbb{R}^{n+1}_1);\mathbb{Z}) \otimes H_*(\Omega(\mathbb{F}_2(\mathbb{R}^{n+1}_1));\mathbb{Z});\\ &(ii)\ \mathrm{E}^2_{*,*} \cong \mathrm{E}^n_{*,*};\ \textit{and}\\ &(iii)\ \mathrm{E}^{n+1}_{*,*} \cong \mathrm{E}^\infty_{*,*}.\end{aligned}$$

Proof. The proof is similar to that of Assertion 1.1 given above. To begin, recall that, according to Chapter VI, §1,

$$H_n(\mathbb{F}_2(\mathbb{R}^{n+1}_1);\mathbb{Z}) \cong \mathbb{Z}\{\alpha_{21}, \alpha_{31}, \alpha_{32}\},$$

while

$$H_{2n}(\mathbb{F}_2(\mathbb{R}^{n+1}_1);\mathbb{Z}) \cong \mathbb{Z}\{\alpha_{21} \rtimes \alpha_{31}, \alpha_{21} \rtimes \alpha_{32}\}.$$

Now, the the proof proceeds along the lines of that for Assertion 1.1. □

As above, we see that

$$\begin{aligned}\mathrm{E}^n_{n,*} &\cong H_n(\mathbb{F}_{2,1};\mathbb{Z}) \otimes H_*(\Omega(\mathbb{F}_{2,1});\mathbb{Z})\\ &\cong \mathbb{Z}\{\alpha_{21}, \alpha_{21}, \alpha_{21}\} \otimes H_*(\Omega(\mathbb{F}_{2,1});\mathbb{Z}).\end{aligned}$$

Assertion 1.4 *The differential* $\mathrm{d}^n : \mathrm{E}^n_{n,*} \to \mathrm{E}^n_{0,*}$ *is given by the formulas*

$$\begin{aligned}\mathrm{d}^n(\alpha_{21} \otimes \omega) &= \mathrm{ad}(\bar{\alpha}_{21})(\omega),\\ \mathrm{d}^n(\alpha_{31} \otimes \omega) &= \mathrm{ad}\alpha_{31}(\omega),\ \textit{and}\\ \mathrm{d}^n(\alpha_{32} \otimes \omega) &= \mathrm{ad}\alpha_{32}(\omega)\end{aligned}$$

for all $\omega \in H_*(\Omega(\mathbb{R}^{n+1}_2);\mathbb{Z})$.

The proof is similar to that of Assertion 1.2 above and is left to the reader.

Next, we turn to the differential in filtration index $2n$. Consider

$$\begin{aligned}\mathrm{E}^n_{2n,*} &\cong H_{2n}(\mathbb{F}_{2,1};\mathbb{Z}) \otimes H_*(\Omega(\mathbb{F}_{2,1});\mathbb{Z})\\ &\cong \mathbb{Z}\{\alpha_{21} \rtimes \alpha_{31}, \alpha_{21} \rtimes \alpha_{32}\} \otimes H_*(\Omega(\mathbb{F}_{2,1});\mathbb{Z}).\end{aligned}$$

Assertion 1.5 *Assume that the coefficient ring is* $\mathbb{Z}$. *Then the differential* $\mathrm{d}^n : \mathrm{E}^n_{2n,*} \to \mathrm{E}^n_{n,*}$ *is given by the formulas*

$$\begin{aligned}\mathrm{d}^n(\alpha_{21} \rtimes \alpha_{31} \otimes \omega) &= \alpha_{31} \otimes \mathrm{ad}(\alpha_{21} + (-1)^{n^2}\bar{\alpha}_{32})(\omega) +\\ &\qquad (\alpha_{21} + \alpha_{31}) \otimes \mathrm{ad}(\alpha_{32})(\omega)\ \ \textit{and}\\ \mathrm{d}^n(\alpha_{21} \rtimes \alpha_{32} \otimes \omega) &= \alpha_{32} \otimes \mathrm{ad}(\alpha_{21} + \bar{\alpha}_{31})(\omega) +\\ &\qquad (\alpha_{21} + \alpha_{32}) \otimes \mathrm{ad}(\alpha_{31})(\omega)\end{aligned}$$

for all ω *in the algebra* $H_*(\Omega(\mathbb{R}^{n+1}_2);\mathbb{Z})$.

Proof. As in the proof of Assertion 1.2 above, we need to describe the coproduct

$$H_*(\nabla) : H_*(\mathbb{F}_2(\mathbb{R}_1^{n+1}); \mathbb{Z}) \to H_*(\mathbb{F}_2(\mathbb{R}_1^{n+1}); \mathbb{Z}) \otimes H_*(\mathbb{F}_2(\mathbb{R}_1^{n+1}); \mathbb{Z}).$$

Now, recall that, according Theorem 2.2 of Chapter VII,

$$\begin{aligned} &\alpha_{21} \rtimes \alpha_{31} \to \alpha_{21} \rtimes \alpha_{31} \otimes 1 + 1 \otimes \alpha_{21} \rtimes \alpha_{31} + \\ &\lambda_1 \otimes \lambda_2 + (-1)^{n^2} \lambda_2 \otimes \lambda_1, \end{aligned}$$

where $\lambda_1 = \alpha_{21} + (-1)^{n+1}\alpha_{32}$ and $\lambda_2 = \alpha_{31}$. The rest of the argument follows along lines similar to those for Assertions 1.2 and 1.4. □

Let us define

$$\begin{aligned} A' &= H_*(\Omega(\mathbb{R}_2^{n+1}); \mathbb{Z}_2) \\ &\cong H_*(\Omega(\mathbb{R}_2^{n+1}); \mathbb{Z}) \otimes \mathbb{Z}_2 \\ &\cong \mathbb{Z}_2(\bar{\alpha}_{31}, \bar{\alpha}_{32}). \end{aligned}$$

The isomorphisms are consequences of the facts (see Chapter VIII, §3)

$$H_*(\Omega(\mathbb{R}_2^{n+1}); \mathbb{Z}) \cong H_*(\Omega(S_{31} \vee S_{32}); \mathbb{Z}) \cong \mathbb{Z}(\bar{\alpha}_{31}, \bar{\alpha}_{32}),$$

where $\bar{\alpha}_{31}, \bar{\alpha}_{32}$ stand for the desuspensions of α_{31}, α_{32}, respectively, and $\mathbb{Z}(\bar{\alpha}_{31}, \bar{\alpha}_{32})$ is the free associative algebra generated by these elements. Similarly, put

$$\begin{aligned} A &= H_*(\Omega(\mathbb{F}_3(\mathbb{R}^{n+1}); \mathbb{Z}_2) \\ &\cong H_*(\Omega(\mathbb{F}_3(\mathbb{R}^{n+1}); \mathbb{Z}) \otimes \mathbb{Z}_2 \\ &\cong \mathbb{Z}_2(\bar{\alpha}_{21}, \bar{\alpha}_{31}, \bar{\alpha}_{32})/I, \end{aligned}$$

since, according to Chapter VIII, §3,

$$\begin{aligned} A &\cong H_*(\Omega(\mathbb{F}_3(\mathbb{R}^{n+1}); \mathbb{Z}) \otimes \mathbb{Z}_2 \\ &\cong H_*(\Omega(S_{21} \rtimes (S_{31} \vee S_{32}); z) \otimes \mathbb{Z}_2 \\ &\cong (\mathbb{Z}(\bar{\alpha}_{21}, \bar{\alpha}_{31}, \bar{\alpha}_{32})/I) \otimes \mathbb{Z}_2, \end{aligned}$$

where I is the ideal generated by the Y-B relations.
The morphism $'\mathrm{E}^{n+1}_{*,*} \otimes \mathbb{Z}_2 \to \mathrm{E}^{n+1}_{*,*} \otimes \mathbb{Z}_2$ gives rise to the commutative diagram

$$\mathcal{E} : \begin{cases} 0 \leftarrow {'\mathrm{E}^n_{0,*}} \leftarrow {'\mathrm{E}^n_{n,*}} \leftarrow 0 \\ \qquad\quad \downarrow \qquad\quad \downarrow \qquad \downarrow \\ 0 \leftarrow \mathrm{E}^n_{0,*} \leftarrow \mathrm{E}^n_{2n,*} \leftarrow \mathrm{E}^n_{2n,*} \end{cases}$$

of DG-modules, where the vertical morphisms are those induced by ϕ, while the horizontal morphism lines are those induced by $'\mathrm{d}^n$ and d^n, respectively.
A key step in the proof that ϕ is injective is the following.

Assertion 1.6 *For all ω in the algebra $H_*(\Omega(\mathbb{F}_{k-r}(\mathbb{R}^{n+1}_r);\mathbb{Z}_2)$, the relation $[\omega,\omega]=0$ holds.*

Proof. Let $\omega \in H_*(\Omega(\mathbb{F}_{k-r}(\mathbb{R}^{n+1}_r);\mathbb{Z}_2)$. Then ω is the reduction mod 2 of an integral class ω' in $H_*(\Omega(\mathbb{F}_{k-r}(\mathbb{R}^{n+1}_r);\mathbb{Z})$. Note that, if ω' is of odd dimension, then $[\omega',\omega']=0$, and the assertion is certainly true. If, on the other hand, ω' is of even dimension, then

$$[\omega',\omega'] = 2\omega'\omega'$$

in $H_*(\Omega(\mathbb{F}_{k-r}(\mathbb{R}^{n+1}_r);\mathbb{Z})$. But this algebra is torsion-free. Hence,

$$[\omega',\omega'] = 0, \text{ mod } 2.$$

This proves the Assertion. □

Assertion 1.7 *The derivation*

$$\mathrm{ad}(\bar{\alpha}_{21}+\bar{\alpha}_{31}+\bar{\alpha}_{32}) : A \to A$$

is trivial.

Proof. As the algebra A is generated by $\{\bar{\alpha}_{rs} \mid 1 \le s < r \le 3\}$, it is enough to show that $\mathrm{ad}(\bar{\alpha}_{21}+\bar{\alpha}_{31}+\bar{\alpha}_{32})$ annihilates the generators. Now note that

$$\mathrm{ad}(\bar{\alpha}_{21}+\bar{\alpha}_{31}+\bar{\alpha}_{32})(\alpha_{21}) = \\ \mathrm{ad}(\bar{\alpha}_{21})(\bar{\alpha}_{21}) + \mathrm{ad}(\bar{\alpha}_{21})(\bar{\alpha}_{31}+\bar{\alpha}_{32}) = 0+0,$$

where the two terms vanish, according to the previous Assertion and to the Y-B relations, respectively.
Similarly, one shows that $\mathrm{ad}(\bar{\alpha}_{21}+\bar{\alpha}_{31}+\bar{\alpha}_{32})$ annihilates the other remaining generators. □

Define ξ to be $(\bar{\alpha}_{21}+\bar{\alpha}_{31}+\bar{\alpha}_{32})$. We immediately have the following.

Assertion 1.8 *The natural imbedding $A' \to A$ induces an isomorphism*

$$\psi : A' \otimes \mathbb{Z}_2(\xi) \to A,$$

of algebras, where $\xi \mapsto \sum \bar{\alpha}_{rs}$.

Now one easily obtains the next assertion.

Assertion 1.9 *The morphism ψ induces, in turn, the morphism*

$$\bar{\psi} : ({}'\mathrm{E}^n_{*,*} \otimes \mathbb{Z}_2) \otimes \mathbb{Z}_2(\xi) \to \mathrm{E}^n_{*,*} \otimes \mathbb{Z}_2$$

of DG- *modules.*

Assertion 1.9 is the key to the computation of $E^{n+1}_{0,*}$.

Assertion 1.10 *The morphism $\bar{\psi}$ induces an isomorphism*

$$'\mathrm{E}^{n+1}_{0,*} \otimes \mathbb{Z}_2 \otimes \mathbb{Z}_2(\xi) \overset{\cong}{\to} \mathrm{E}^{n+1}_{0,*} \otimes \mathbb{Z}_2.$$

Proof. Note that

$$'\mathrm{E}^{n+1}_{0,*} \otimes \mathbb{Z}_2(\xi) \cong \mathrm{coker}('\mathrm{d}^n :' \mathrm{E}^n_{n,*} \to' \mathrm{E}^n_{0,*}) \otimes \mathbb{Z}_2(\xi),$$

since the vector space $\mathbb{Z}_2(\xi)$ is clearly free over $\mathbb{Z}_2$. Next, observe that

$$\mathrm{image}('\mathrm{d}^n :' \mathrm{E}^n_{n,*} \otimes \mathbb{Z}_2(\xi) \to' \mathrm{E}^n_{0,*}) = \mathrm{image}(\mathrm{d}^n : \mathrm{E}^n_{n,0} \to \mathrm{E}^n_{0,*}),$$

by virtue of Assertion 1.9. Thus we obtain the isomorphism

$$\mathrm{coker}('\mathrm{d}^n :' \mathrm{E}^n_{n,*} \to' \mathrm{E}^n_{0,*}) \otimes \mathbb{Z}_2(\xi) \cong \mathrm{coker}(d^n : \mathrm{E}^n_{n,*} \to \mathrm{E}^n_{0,*}).$$

This proves the Assertion. □

Assertion 1.11 *The morphism*

$$\phi^{n+1} : '\mathrm{E}^{n+1}_{n,*} \otimes \mathbb{Z}_2 \to \mathrm{E}^{n+1}_{n,*} \otimes \mathbb{Z}_2$$

is injective.

Proof. It is enough to prove that

$$\begin{array}{l} \mathrm{image}(\phi^n : '\mathrm{E}^n_{n,*} \otimes \mathbb{Z}_2 \to \mathrm{E}^n_{n,*}) \otimes \mathbb{Z}_2 \\ \cap\, \mathrm{image}(\mathrm{d}^n : \mathrm{E}^n_{2n,*} \otimes \mathbb{Z}_2 \to \mathrm{E}^n_{n,*}) \otimes \mathbb{Z}_2 = 0. \end{array}$$

To do this, recall that the differential $\mathrm{d}^n : \mathrm{E}^n_{2n,*} \otimes \mathbb{Z}_2 \to \mathrm{E}^n_{2n,*} \otimes \mathbb{Z}_2$ is given by the expressions

$$\begin{array}{l} (\alpha_{21} \rtimes \alpha_{31}) \otimes \omega_1 + \alpha_{31} \otimes [\bar{\alpha}_{21} + \bar{\alpha}_{32}, \omega_1] + (\alpha_{21} + \alpha_{32}) \otimes [\bar{\alpha}_{31}, \omega_1] \\ (\alpha_{21} \rtimes \alpha_{32}) \otimes \omega_2 + \alpha_{32} \otimes [\bar{\alpha}_{21} + \bar{\alpha}_{31}, \omega_2] + (\alpha_{21} + \alpha_{31}) \otimes [\bar{\alpha}_{32}, \omega_2]. \end{array}$$

Observe now that $\mathrm{ad}(\bar{\alpha}_{21} + \bar{\alpha}_{32} + \bar{\alpha}_{31}) = 0$ or, equivalently, $\mathrm{ad}(\bar{\alpha}_{21} + \bar{\alpha}_{32}) = \mathrm{ad}(\bar{\alpha}_{31})$. Similarly, one shows that $\mathrm{ad}(\bar{\alpha}_{21} + \bar{\alpha}_{31}) = \mathrm{ad}(\bar{\alpha}_{32})$. Then, the preceding expressions for d^n become

$$\begin{array}{l} (\alpha_{21} \rtimes \alpha_{31}) \otimes \omega_1 + \alpha_{31} \otimes [\bar{\alpha}_{31}, \omega_1] + (\alpha_{21} + \alpha_{32}) \otimes [\bar{\alpha}_{31}, \omega_1] \\ (\alpha_{21} \rtimes \alpha_{32}) \otimes \omega_2 + \alpha_{32} \otimes [\bar{\alpha}_{32}, \omega_2] + (\alpha_{21} + \alpha_{31}) \otimes [\bar{\alpha}_{32}, \omega_2]. \end{array}$$

Now, let $\zeta \in \mathrm{E}^n_{2n,*} \otimes \mathbb{Z}_2$. Then,

$$\zeta = \begin{cases} (\alpha_{21} \rtimes \alpha_{31}) \rtimes \omega_1, \\ (\alpha_{21} \rtimes \alpha_{32}) \rtimes \omega_2, \\ (\alpha_{21} \rtimes \alpha_{31}) \rtimes \omega_1 + (\alpha_{21} \rtimes \alpha_{32}) \rtimes \omega_2. \end{cases}$$

We shall prove the Assertion for the third case only. The other cases are similarly proved and are left for the reader. If $\mathrm{d}^n(\zeta)$ is in the image of

$$\phi^n : {}'\mathrm{E}^n_{n,*} \otimes \mathbb{Z}_2 \to \mathrm{E}^n_{n,*} \otimes \mathbb{Z}_2,$$

where

$$\zeta = (\alpha_{21} \rtimes \alpha_{31}) \rtimes \omega_1 + (\alpha_{21} \rtimes \alpha_{32}) \rtimes \omega_2,$$

then it follows that

$$\alpha_{21} \otimes ([\bar{\alpha}_{31}, \omega_1] + [\bar{\alpha}_{32}, \omega_2]) = 0,$$

which, in turn, implies that

$$[\bar{\alpha}_{31}, \omega_1] + [\bar{\alpha}_{32}, \omega_2] = 0.$$

Now, note that $[\bar{\alpha}_{31}, \omega_1] + [\bar{\alpha}_{32}, \omega_2]$ is the coefficient of α_{31} as well as of α_{32} in the expression for $\mathrm{d}^n(\zeta)$ given above. Therefore,

$$\mathrm{d}^n((\alpha_{21} \rtimes \alpha_{31}) \otimes \omega_1 + (\alpha_{21} \rtimes \alpha_{32}) \otimes \omega_2) = 0,$$

which is the statement asserted at the beginning of the proof. □

The following is an easy result.

Corollary 1.1 *The module $H_*(\Lambda(\mathbb{F}_3(\mathbb{R}^{n+1}));\mathbb{Z}_2)$ contains*

$$\bar{H}_*(\Lambda(S_{21});\mathbb{Z}_2) \oplus \bar{H}_*(\Lambda(S_{31});\mathbb{Z}_2) \oplus \bar{H}_*(\Lambda(S_{32});\mathbb{Z}_2)$$

as a submodule.

Let $\mu = \alpha_{31} + \alpha_{32}$, and let $S_\mu \to (S_{31} \vee S_{32})$ be a map representing the homotopy class μ.

Corollary 1.2 *The morphism*

$$H_*(\Lambda(S_\mu);\mathbb{Z}_2) \subset H_*(\Lambda(\mathbb{F}_3(\mathbb{R}^{n+1}));\mathbb{Z}_2)$$

induced by μ is injective.

Proof. Note that the map $S_\mu \to (S_{31} \vee S_{32})$ can be regarded as a homotopy section for the projection

$$\mathbb{F}_3(\mathbb{R}^{n+1}) \to \mathbb{F}_2(\mathbb{R}^{n+1}),\ (x_1, x_2, x_3). \mapsto (x_2, x_3).$$

This proves that the induced morphism

$$H_*(\Lambda(S_\mu);\mathbb{Z}_2) \to H_*(\Lambda(\mathbb{F}_3(\mathbb{R}^{n+1}));\mathbb{Z}_2)$$

is injective. □

Thus Statement (II) is proved and the implication ($\Leftarrow$) is established.

Proof of ($\Rightarrow$)

Consider the fibration

$$\Lambda(\text{proj}_{k,k-1}) : \Lambda\mathbb{F}_k(\mathbb{R}^{n+1}) \to \Lambda\mathbb{F}_{k-1}(\mathbb{R}^{n+1})$$

induced by the projection $(x_1, x_2, \cdots, x_k) \mapsto (x_1, x_2, \cdots, x_{k-1})$. Denote by $\mathbb{R}^{n+1}_{k-1}$ the fiber at the point $(q_1, \cdots, q_{k-1})$, and let

$$i_{k-1} : \mathbb{R}^{n+1}_{k-1} \to \mathbb{F}_k(\mathbb{R}^{n+1}),\ x \mapsto (q_1, \cdots, q_{k-1}, x)$$

be the injection of the fiber. We need to show that the induced morphism

$$\text{(III)} \qquad H_*(\Lambda(i_{k-1})) : H_*(\Lambda(\mathbb{R}^{n+1}_{k-1})) \to H_*(\Lambda(\mathbb{F}_k(\mathbb{R}^{n+1})))$$

is not injective.

Observe that the fibration

$$(\text{proj}_k|\mathbb{F}_{2,k-2}) : \mathbb{F}_{2,k-2} = \mathbb{F}_2(\mathbb{R}^{n+1}_{k-2}) \to \mathbb{R}^{n+1}_{k-2},\ (x_1, x_2) \mapsto x_2$$

can be regarded as a subspace of $\mathbb{F}_k(\mathbb{R}^{n+1})$, by virtue of the morphism

$$i_{2,k-2} : \mathbb{F}_{2,k-2} \to \mathbb{F}_k(\mathbb{R}^{n+1}),\ (x_1, x_2) \mapsto (q_1, q_2, \cdots, q_{k-2}, x_1, x_2).$$

In fact, it is the imbedding of the subfibration in the fibration in question. The imbedding $i_{2,k-2} : \mathbb{F}_{2,k-2} \to \mathbb{F}_k(\mathbb{R}^{n+1})$ leads to the commutative diagram

$$\mathcal{F}_1 : \begin{cases} \mathbb{R}^{n+1}_{k-1} \to \mathbb{F}_{2,k-2}(\mathbb{R}^{n+1}) \to \mathbb{R}^{n+1}_{k-1} \\ \downarrow \qquad\qquad \downarrow \qquad\qquad\qquad \downarrow \\ \mathbb{R}^{n+1}_{k-1} \to \mathbb{F}_k(\mathbb{R}^{n+1}) \qquad \to \mathbb{F}_{k-1}(\mathbb{R}^{n+1}), \end{cases}$$

the rows of which are fibrations. Passing to the space of free loops, we obtain the similar diagram,

$$\Lambda(\mathcal{F}_1) : \begin{cases} \Lambda(\mathbb{R}^{n+1}_{k-1}) \to \Lambda\mathbb{F}_{2,k-2}(\mathbb{R}^{n+1}) \to \Lambda(\mathbb{R}^{n+1}_{k-1}) \\ \downarrow \qquad\qquad \downarrow \qquad\qquad\qquad \downarrow \\ \Lambda(\mathbb{R}^{n+1}_{k-1}) \to \Lambda\mathbb{F}_k(\mathbb{R}^{n+1}) \qquad \to \Lambda\mathbb{F}_{k-1}(\mathbb{R}^{n+1}). \end{cases}$$

Thus, to prove Statement (III), it is enough to prove that the induced morphism

$$\text{(IV)} \qquad H_*(\Lambda(\mathbb{R}^{n+1}_{k-1})) \to H_*(\Lambda\mathbb{F}_{2,k-2}(\mathbb{R}^{n+1}))$$

is not injective. The morphism (IV) can be factored through the total space of another fibration that is easier to handle: namely, the space

$$\Lambda(\mathbb{F}_{2,k-2}, \mathbb{R}^{n+1}_{k-1}) = \{\omega \in \Lambda\mathbb{F}_{2,k-2} | \omega(1) \in \mathbb{R}^{n+1}_{k-1}\}.$$

The natural projection $\omega \mapsto \omega(1)$ yields a fibration

$$\rho_{k-2,2} = \text{eval}_{2,k-2}|\Lambda(\mathbb{F}_{2,k-2}, \mathbb{R}^{n+1}_{k-1}) : \Lambda(\mathbb{F}_{2,k-2}, \mathbb{R}^{n+1}_{k-1}) \to \mathbb{R}^{n+1}_{k-1},$$

with $\Omega\mathbb{F}_{2,k-2}(\mathbb{R}^{n+1})$ as fiber. Thus, to prove Statement (II), it is enough to show that the morphism

$$\text{(V)} \qquad H_*(\Lambda(\mathbb{R}^{n+1}_{k-1})) \to H_*(\Lambda(\mathbb{F}_{2,k-2}, \mathbb{R}^{n+1}_{k-1}))$$

is not injective.

Consider $\{\mathrm{E}^r(\rho_{k-2,2}), \mathrm{d}^r(\rho_{k-2,2})\}$, the Serre spectral sequence of $\rho_{k-2,2}$. Then, we have

$$\begin{aligned} (i)\ \mathrm{E}^2_{*,*}(\rho_{k-2,2}) &\cong H_*(\mathbb{R}^{n+1}_{k-1}) \otimes H_*(\Omega\mathbb{F}_{2,k-2}); \\ (ii)\ \mathrm{E}^2_{*,*}(\rho_{k-2,2}) &\cong \mathrm{E}^n_{*,*}(\rho_{k-2,2}); \text{ and} \\ (iii)\ \mathrm{E}^{n+1}_{*,*}(\rho_{k-2,2}) &\cong \mathrm{E}^\infty_{*,*}(\rho_{k-2,2}). \end{aligned}$$

Recall that, in the notation of Chapter II,

$$\begin{aligned} \mathbb{R}^{n+1}_{k-1} &\simeq (S_{k1} \vee \cdots \vee S_{kk-1}), \\ \mathbb{R}^{n+1}_{k-2} &\simeq (S_{k-11} \vee \cdots \vee S_{k-1k-2}), \text{and} \\ \mathbb{F}_{2,k-2} &\simeq (S_{k1} \vee \cdots \vee S_{k-1k-2}) \ltimes (S_{k-11} \vee \cdots \vee S_{k-1k-2}). \end{aligned}$$

Passing to the homology of based loop spaces, we see that, according to Chapter VIII, §3, we have

$$\begin{aligned} H_*(\Omega(\mathbb{R}^{n+1}_{k-1})) &\cong \mathbb{Z}_2(\bar{\alpha}_{k1}, \cdots, \bar{\alpha}_{kk-1}), \\ H_*(\Omega(\mathbb{R}^{n+1}_{2})) &\cong \mathbb{Z}_2(\bar{\alpha}_{k-11}, \cdots, \bar{\alpha}_{k-1k-2}), \text{ and} \\ H_*(\Omega(\mathbb{F}_{2,k-2})) &\cong \mathbb{Z}_2(\Xi_{k,k-2})/I, \end{aligned}$$

as Pontryagin algebras, where

$$\Xi_{k,k-2} = \{\bar{\alpha}_{k1}, \cdots, \bar{\alpha}_{kk-1}, \bar{\alpha}_{k-11}, \cdots, \bar{\alpha}_{k-1k-2}\},$$

and I is the ideal generated by the Y-B relations. The differential operator $\mathrm{d}^n : \mathrm{E}^n_{n,*} \to \mathrm{E}^n_{0,*}$ is such that

$$\mathrm{d}^n(\alpha_{ks} \otimes \omega_s) = [\bar{\alpha}_{ks}, \omega_s],\ s = 1, 2, \cdots, k-1,$$

where $\omega_s \in \mathbb{Z}_2(\Xi k, k-2/I$.

To compute the E^{n+1}-term of the Serre spectral sequence of the fibration $\rho_{k-2,2} : \Lambda(\mathbb{F}_{2,k-2}, \mathbb{R}^{n+1}_{k-1}) \to \mathbb{R}^{n+1}_{k-1}$, it is convenient to identify $\mathrm{E}^n_{n,*}(\rho_{k-2,2})$ with $\mathcal{A}^{\oplus(k-1)}$ and $\mathrm{E}^n_{0,*}(\rho_{k-2,2})$ with $\mathcal{A}$, where $\mathcal{A} = \mathbb{Z}_2(\Xi_{k,k-2}/I$. Summing up, we have the following.

Proposition 1.1 *The differential* $\mathrm{d}^n : \mathrm{E}^n_{n,*}(\rho_{k-2,2}) \to \mathrm{E}^n_{0,*}(\rho_{k-2,2})$ *is the morphism, where* $(\omega_1, \cdots, \omega_{k-l}) \mapsto \sum_{i=1}^{k-1}[\bar{\alpha}_{ki}, \omega_i]$, *and* $\mathrm{E}^{n+1}_{0,*}(\rho_{k-2,2})$ *is isomorphic to* $\text{coker}(\mathcal{A}^{\oplus(k-1)} \to \mathcal{A})$.

Let $\{'\mathrm{E}^r, '\mathrm{d}^r\}$ be the mod 2 Serre spectral sequence of the fibration

$$\mathrm{eval}: \Lambda(\mathbb{R}^{n+1}_{k-1}) \to \mathbb{R}^{n+1}_{k-1}, \; \omega \mapsto \omega(1).$$

Identify $'\mathrm{E}^n_{n,*}$ with $\mathcal{A}'^{\oplus(k-2)}$, and $'\mathrm{E}^n_{0,*}$ with $\mathcal{A}'$), where

$$\mathcal{A}' = \mathbb{Z}_2(\bar{\alpha}_{k-1k-2}, \cdots, \bar{\alpha}_{k-1k-2}).$$

Proposition 1.2 *With the identifications above,* $'\mathrm{d}^n : {'\mathrm{E}^n_{n,*}} \to {'\mathrm{E}^n_{0,*}}$ *is the morphism where* $(\omega_1, \cdots, \omega_{k-2}) \mapsto \sum_{i=1}^{k-2} [\bar{\alpha}_{k-1i}, \omega_1]$ *and* $'\mathrm{E}^{n+1}_{0,*}$ *is isomorphic to* $\mathrm{coker}(\mathcal{A}'^{\oplus(k-2)} \to \mathcal{A}')$.

The proof is similar to that of Proposition 1.1 and, therefore, will be omitted.

Proof of Statement (V)

Note that

$$'\mathrm{E}^n_{n,2(n-1)} \cong H_n(\mathbb{R}^{n+1}_{k-1}) \otimes H_{2(n_1)}(\Omega(\mathbb{R}^{n+1}_{k-1})),$$

and that $H_{2(n_1)}(\Omega(\mathbb{R}^{n+1}_{k-1}))$ is isomorphic to the module generated by

$$\{\bar{\alpha}_{kj}\bar{\alpha}_{kt} | 1 \le j, t < k\}.$$

Hence, image$('\mathrm{E}^n_{n,2(n-1)} \to {'\mathrm{E}^n_{0,3(n-1)}})$ is generated by

$$\{[\bar{\alpha}_{ki}, \bar{\alpha}_{kj}\bar{\alpha}_{kt}] | 1 \le i < k, \; 1 \le j, t < k\}.$$

As $'\mathrm{E}^{n+1}_{0,3(n-1)} \cong$ image$('\mathrm{E}^n_{n,2(n-1)} \to {'\mathrm{E}^n_{0,3(n-1)}})$, we conclude that the image of $\bar{\alpha}_{k2}[\bar{\alpha}_{k1}, \bar{\alpha}_{kk-1}]$ is $\neq 0$ in $'\mathrm{E}^{n+1}_{0,3(n-1)}$. But

$$[\bar{\alpha}_{k2}, \bar{\alpha}_{k1}\bar{\alpha}_{k-11}] + [\bar{\alpha}_{k1}, \bar{\alpha}_{k2}\bar{\alpha}_{k-11}] = \bar{\alpha}_{k2}[\bar{\alpha}_{k1}, \bar{\alpha}_{kk-1}].$$

Hence, the morphism

$$'\mathrm{E}^{n+1}_{0,3(n-1)} \to \mathrm{E}^{n+1}_{0,3(n-1)}$$

is not injective. This implies Statement (V). □

2 Coproducts in $H_*(\Lambda\mathbb{F}_3(\mathbb{R}^{n+1}); \mathbb{Z}_2)$

We wish to show that the coproduct length in the mod 2 homology of $(\Lambda\mathbb{F}_3(\mathbb{R}^{n+1}), E)$ is infinite, where E is a suitable subspace the homology of which is like that of the union of the factors $\Lambda(S_{rs})$.

Let us consider now the affine map

$$\phi_\omega : S_1 \times S_2 \to \mathbb{F}_k(\mathbb{R}^{n+1}), \quad (\xi_1, \xi_2) \mapsto (q_1, q_1 + \xi_1, q_1 + 2\xi_2),$$

where ξ_1, ξ_2 are unit vectors. This yields the Y-B relation

$$[\alpha_{21}, \alpha_{31} + \alpha_{32}] = 0.$$

Arguing as in Chapter VI, §§1 and 4, one can easily show that the fundamental class of $S_1 \times S_2$ is taken by $H_{2n}(\phi_\omega)$ to $\alpha_{21} \rtimes \alpha_{31} + \alpha_{21} \rtimes \alpha_{32}$. Denote by

$$\Lambda(\phi_\omega) : \Lambda(S_1 \times S_2) \to \Lambda(\mathbb{F}_3(\mathbb{R}^{n+1})$$

the map induced by ϕ_ω.

Theorem 2.1 *The induced morphism*

$$H_*(\Lambda(\phi_\omega) : H_*(\Lambda(S_1 \times S_2);\mathbb{Z}_2) \to H_*(\Lambda\mathbb{F}_3(\mathbb{R}^{n+1});\mathbb{Z}_2)$$

is injective.

Proof. Observe, first of all, that the homotopy class of

$$\phi_\omega|S_1 : S_1 = S^n \to \mathbb{F}_3(\mathbb{R}^{n+1}),\ \xi_1 \mapsto (q_1, q_1 + \xi_1, q_1 + \xi_1 + 2)$$

is α_{21} and that

$$\mathrm{proj}_{21} : \mathbb{F}_3(\mathbb{R}^{n+1}) \to \mathbb{F}_2(\mathbb{R}^{n+1}),\ (x_1, x_2, x_3) \mapsto (x_1, x_2)$$

takes $[\phi_\omega|S_1]$ to α_{21}. Hence, $\phi_\omega|S_1$ is a homotopy section for proj_{21}, and, therefore, the induced morphism

$$H_*(\Lambda(\phi_\omega|S_1)) : H_*(\Lambda(S_1);\mathbb{Z}_2) \to H_*(\mathbb{F}_3(\mathbb{R}^{n+1});\mathbb{Z}_2)$$

is injective. Similarly, one proves that

$$H_*(\Lambda(\phi_\omega|S_2)) : H_*(\Lambda(S_2);\mathbb{Z}_2) \to H_*(\Lambda\mathbb{F}_3(\mathbb{R}^{n+1});\mathbb{Z}_2)$$

is injective. On passing to cohomology mod 2, we see immediately that

$$H^*(\Lambda(\phi_\omega)) : H^*(\Lambda\mathbb{F}_3(\mathbb{R}^{n+1}));\mathbb{Z}_2) \to H^*(\Lambda(S_1 \times S_2);\mathbb{Z}_2)$$

is surjective, since the algebras $H^*(\Lambda(S_1;\mathbb{Z}_2)$ and $H^*(\Lambda(S_2);\mathbb{Z}_2)$ generate $H^*(\Lambda(S_1 \times S_2);\mathbb{Z}_2)$ as an algebra. In turn, the latter implies that the induced homology mod 2 morphism is injective. □

Let us put

$$E = \Lambda(S_{21}) \vee \Lambda(S_{31}) \vee \Lambda(S_{32}).$$

Corollary 2.1 *Suppose that E' is a subspace of $\Lambda\mathbb{F}_3(\mathbb{R}^{n+1})$ containing E and is such that $H^m(E', E; \mathbb{Z}_2) = 0$ for m sufficiently large. Then, the coproduct length mod 2 of $(\Lambda\mathbb{F}_3(\mathbb{R}^{n+1}), E')$ is infinite.*

Proof. To identify

$$\text{image}(H_*(\Lambda(S_1 \times S_2); \mathbb{Z}_2) \to H_*(\Lambda\mathbb{F}_3(\mathbb{R}^{n+1}); \mathbb{Z}_2),$$

proceed as follows. One easily proves that

$$\bar{H}_*(\Lambda(S_{21}); \mathbb{Z}_2) \oplus \bar{H}_*(\Lambda(S_{31}); \mathbb{Z}_2) \oplus \bar{H}_*(\Lambda(S_{32}); \mathbb{Z}_2) \cong \bar{H}_*(E; \mathbb{Z}_2).$$

Observe that the cross product

$$H_*(\Lambda(S_1); \mathbb{Z}_2) \otimes H_*(\Lambda(S_2); \mathbb{Z}_2) \to H_*(\Lambda(S_1 \times S_2); \mathbb{Z}_2)$$

is an isomorphism of coalgebras. Next, note that the homotopy class of

$$\phi_\omega | S_2 : S_2 \to \mathbb{F}_3(\mathbb{R}^{n+1})$$

is $\alpha_{31} + \alpha_{32}$. According to Corollary 1.1, the isomorphism

$$H_*(\Lambda(S_{21}); \mathbb{Z}_2) \otimes H_*(\Lambda(S_{31} \vee S_{32}); \mathbb{Z}_2) \to H_*(\Lambda(\mathbb{F}_3(\mathbb{R}^{n+1})); \mathbb{Z}_2)$$

induces an injection

$$\bar{H}_*(\Lambda(S_{21}); \mathbb{Z}_2) \oplus \bar{H}_*(\Lambda(S_{31}); \mathbb{Z}_2) \oplus \bar{H}_*(\Lambda(S_{32}); \mathbb{Z}_2) \to H_*(\Lambda(S_{31} \vee S_{32}); \mathbb{Z}_2).$$

Therefore,

$$\begin{gathered}\text{image}(H_*(E; \mathbb{Z}_2) \to H_*(\Lambda(\mathbb{F}_3(\mathbb{R}^{n+1})); \mathbb{Z}_2)) \cap \\ \text{image}(\bar{H}_*(\Lambda(S_1); \mathbb{Z}_2) \otimes \bar{H}_*(\Lambda(S_2); \mathbb{Z}_2) \to H_*(\Lambda\mathbb{F}_3(\mathbb{R}^{n+1}); \mathbb{Z}_2)) = 0,\end{gathered}$$

as one can easily see, because, on projecting by proj_{21}, proj_{31}, and proj_{32}, the module

$$\text{image}(\bar{H}_*(\Lambda(S_1); \mathbb{Z}_2) \otimes \bar{H}_*(\Lambda(S_2); \mathbb{Z}_2) \to H_*(\Lambda\mathbb{F}_3(\mathbb{R}^{n+1}); \mathbb{Z}_2))$$

goes to 0. As the submodule $H_*(\Omega(S_1 \times S_2); \mathbb{Z}_2)$ of $H_*(\Lambda(S_1 \times S_2); \mathbb{Z}_2)$ contains elements of arbitrarily high coproduct length, the same is true of the module $H_*(\Lambda(S_1 \times S_2); \mathbb{Z}_2)$. These elements ξ, viewed as lying in $H_*(\Lambda\mathbb{F}_3(\mathbb{R}^{n+1}); \mathbb{Z}_2)$, provide one with elements of arbitrarily high coproduct length. Observe that the natural morphism

$$H_*(\Lambda\mathbb{F}_3(\mathbb{R}^{n+1}); \mathbb{Z}_2) \to H_*(\Lambda\mathbb{F}_3(\mathbb{R}^{n+1}), E; \mathbb{Z}_2)$$

takes these elements ξ to elements $\xi' \neq 0$. By naturality, the coproduct length of ξ is the same as that of ξ'.

The preceding shows that the coalgebra $H_*(\Lambda\mathbb{F}_3(\mathbb{R}^{n+1}), E; \mathbb{Z}_2)$ has infinite coproduct length. This in turn implies that the coproduct length of the coalgebra $H_*(\Lambda\mathbb{F}_3(\mathbb{R}^{n+1}), E'; \mathbb{Z}_2)$ is also infinite. □

3 The Growth of $H_*(\Lambda(\mathbb{R}^{n+1}_{k-1})$

Our aim in this chapter is to compute the The mod 2 Poincaré series of $\Lambda(\mathbb{R}^{n+1}_{k-1})$. We proceed in stages.

Consider the Serre spectral sequence $\{\mathrm{E}^r_{*,*}, \mathrm{d}^r\}$ of the fibration

$$\mathrm{eval}_{k-1} : \Lambda(\mathbb{R}^{n+1}_{k-1}) \to \mathbb{R}^{n+1}_{k-1}, \quad \omega \mapsto \omega(1).$$

Assertion 3.1 *$H_i(\Lambda(\mathbb{R}^{n+1}_{k-1})) = 0$, for $i \neq m(n-1), m(n-1)+1$.*

Proof. Recall that

$$\mathrm{E}^2_{*,*} = \mathrm{E}^n_{*,*} \cong H_*(\mathbb{R}^{n+1}_{k-1}) \otimes H_*(\Omega(\mathbb{R}^{n+1}_{k-1})),$$

where homology is mod 2 homology. According to Chapter V, §1,

$$H_n(\mathbb{R}^{n+1}_{k-1}; \mathbb{Z}_2) \cong \sum_{i=1}^{k-1} \mathbb{Z}_2[\alpha_{ki}],$$

where $\mathbb{Z}_2[\alpha_{ki}]$ is the graded vector space generated by the element α_{ki} of dimension n. Also, according to Chapter VIII, §4, the Pontryagin algebra

$$H_*(\Omega(\mathbb{R}^{n+1}_{k-1})) \cong \mathbb{Z}_2(\bar{\alpha}_{k1}, \cdots, \bar{\alpha}_{kk-1}),$$

as algebras, where the right hand side is the associative graded algebra generated by the desuspensions of the elements $\{\alpha_{rs} | 1 \leq s < r \leq k\}$. Hence, $\mathrm{E}^n_{n,q} = 0$ for $q \neq (m-1)(n-1)$, and $\mathrm{E}^n_{0,q} = 0$ for $q \neq m(n-1)$. As $\mathrm{E}^{n+1}_{*,*} = \mathrm{E}^{\infty}_{*,*}$, we obtain the desired conclusion. □

Assertion 3.2 *For all $m \geq 1$, it follows that*

$$\dim H_{m(n-1)+1}(\Lambda(\mathbb{R}^{n+1}_{k-1})) = \dim H_{m(n-1)}(\Lambda(\mathbb{R}^{n+1}_{k+1})).$$

Proof. Consider the differential

$$\mathrm{d}^n : \mathrm{E}^n_{n,(m-1)(n-1)} \to \mathrm{E}^n_{0,m(n-1)}.$$

Note that

$$\begin{aligned} \ker(d^n) &= E^{n+1}_{0,m(n-1)} \cong H_{m(n-1)+1}(\Lambda(\mathbb{R}^{n+1}_{k-1})) \quad \text{and} \\ \mathrm{coker}(\mathrm{d}^n) &= \mathrm{E}^{n+1}_{0,m(n-1)} \cong H_{m(n-1)}(\Lambda(\mathbb{R}^{n+1}_{k-1})). \end{aligned}$$

As a consequence, we see that

$$\begin{aligned} &\dim \mathrm{E}^n_{n,(m-1)(n-1)} - \dim H_{m(n-1)+1}(\Lambda(\mathbb{R}^{n+1}_{k-1})) = \\ &\dim \mathrm{E}^n_{0,m(n-1)} \quad\;\; - \dim H_{m(n-1)}(\Lambda(\mathbb{R}^{n+1}_{k-1})). \end{aligned}$$

But, $\dim \mathrm{E}^n_{n,(m-1)(n-1)} = (k-1)^m = \dim \mathrm{E}^n_{0,m(n-1)}$. □

By definition, let $a_m = \dim H_{m(n-1)+1}(\Lambda(\mathbb{R}^{n+1}_{k-1}))$.

Assertion 3.3 *The mod 2 Poincaré series of* $H_*(\Lambda(\mathbb{R}^{n+1}_{k-1}))$ *is*

$$P(\Lambda(\mathbb{R}^{n+1}_{k-1})) = 1 + (1+T)(\sum_{m\geq 1} a_m T^{m(n-1)}).$$

This is an easy consequence of Assertion 3.2.

Next, we turn to the problem of finding an expression for a_m in terms of m and k. Put, for the sake of simplicity,

$$A_m = H_{m(n-1)}(\Omega(\mathbb{R}^{n+1}_{k-1})),\ m \geq 0$$

and $A_* = \sum_m A_m$, and denote the generators $\bar{\alpha}_{ki}$ by ω_i for $1 \leq i \leq (k-1)$. Denote by Γ the cyclic group $\{\gamma^i | \gamma^m = 1\}$ of order m, and let it act on A_m according to the rule

$$(\gamma, \omega_{i_1} \cdots \omega_{i_m}) = \omega_{i_2} \cdots \omega_{i_m} \omega_{i_1},$$

where $\gamma \in \Gamma$ is the generator.

Assertion 3.4 *The module* $\ker \mathrm{d}^n : \mathrm{E}^n_{n,(m-1)(n-1)} \to \mathrm{E}^n_{0,m(n-1}$ *is naturally isomorphic to* A^Γ_m.

Proof. Identifying $\mathrm{E}^n_{*,*}$ with $H_*(\mathbb{R}^{n+1}_{k-1}) \otimes H_*(\Omega(\mathbb{R}^{n+1}_{k-1}))$, we see that the differential

$$\mathrm{d}^n : H_n(\mathbb{R}^{n+1}_{k-1}) \otimes H_{m(n-1)}(\Omega(\mathbb{R}^{n+1}_{k-1})) \to H_{(n-1)(m+1)}(\Omega(\mathbb{R}^{n+1}_{k-1})),\ m \geq 0$$

operates according to the rule

$$\begin{aligned} \mathrm{d}^n(\alpha_{ks} \otimes \omega) &= [\bar{\alpha}_{ks}, \omega] \\ &= \bar{\alpha}_{ks}\omega + \omega\bar{\alpha}_{ks} \end{aligned}$$

for $1 \leq s < k$. Put $\omega = \omega_{i_2} \cdots \omega_{i_m}$, and observe that the formula for d^n becomes

$$\begin{aligned} &\mathrm{d}^n(\alpha_{ks} \otimes \omega_{i_2} \cdots \omega_{i_m}) = \\ &\quad \omega_s\omega_{i_2} \cdots \omega_{i_m} + \omega_{i_2} \cdots \omega_{i_m}\omega_s \\ &\quad = (\mathrm{id} + \gamma)(\omega_s\omega_{i_2} \cdots \omega_{i_m}). \end{aligned}$$

Note that $H_n(\mathbb{R}^{n+1}_{k-1}) \cong A_1$. Identify $H_n(\mathbb{R}^{n+1}_{k-1}) \otimes A_{m-1}$ with A_m to establish the assertion. □

Consider next

$$X_m = \{\omega_{i_1} \cdots \omega_{i_m} | 1 \leq i_1, \cdots, i_m < k\},$$

where m is fixed. The group Γ acts on X where the generator γ permutes the elements of X circularly.

Assertion 3.5 *The orbits of Γ in X form a basis for the module A^Γ_m. In particular, $a_m = |X/\Gamma|$.*

Proof. Let $\xi \in A^\Gamma_m$, and express it as $\sum_i \xi_i$, where each $\xi_i = \omega_{i_1} \cdots \omega_{i_m}$. As $\gamma(\xi) = \xi$ and the elements of X are a basis for A_m, we see that the terms ξ_i are permuted among themselves by γ and its powers. Hence,

$$\xi = \sum_{u=1}^{p}(\sum_{i=1}^{m} \gamma^i(\xi_{j_u})),$$

where ξ_{j_u} are among the terms appearing in the expression for ξ above.

Conversely, if $\xi \in X$, then $\sum_{i=1}^m \gamma^i(\xi)$ is in A^Γ_m. This proves the assertion. □

Let $\mathcal{K} = \{K||K| \text{ divides } m\}$ be the lattice of subgroups of Γ. As usual, put $\Gamma_x = \{\gamma \in \Gamma|\gamma(x) = x\}$ for $x \in X$.

Assertion 3.6 *Given $K \in \mathcal{K}$, there is an $x \in X$ such that $\Gamma_x = K$.*

Proof. Let $K \in \mathcal{K}$. Then K is a cyclic subgroup generated by γ^s, where $s|K| = m$. Choose x_s to be the element of length s defined by the formula

$$x_s = \omega_1\omega_2 \cdots \omega_2\omega_1,$$

where $\omega_1 \neq \omega_2$. Put $x = x_s \cdots x_s$, where x_s is repeated m/s times. Then $x \in X$ is such that $x = \gamma^s(x)$ and $x \neq \gamma^r(x)$ for $1 < r < s$. It is easy to verify that $\Gamma_x = K$. □

Let $K \in \mathcal{K}$, and define

$$f(K) = |\{x \in X|K \subset \Gamma_x\}|, \; g(K) = |\{x \in X|K = \Gamma_x\}|.$$

It is easily seen that

$$f(K) = \sum_{L \supset K} g(L).$$

Now, it follows directly from the Moebius Inversion Formula that

$$\text{(I)}: \qquad g(K) = \sum_{L \supset K} \mu(|L|/|K|)f(L),$$

where μ is the Moebius function (see the Appendix of Chapter II of [14, Bourbaki]).

Now we are able to complete the computation of the Poincaré series of $H_*(\Lambda(\mathbb{R}^{n+1}_{k-1}); \mathbb{Z}_2)$ partially given in Assertion 3.3.

Theorem 3.1 *The mod 2 Poincaré series of $H_*(\Lambda(\mathbb{R}^{n+1}_{k-1}))$ is*

$$P(\Lambda(\mathbb{R}^{n+1}_{k-1})) = 1 + (1+T)(\sum_{m\geq 1} a_m T^{m(n-1)}),$$

where $a_m = \sum_{r|m} \frac{1}{r}(\sum_{s|r} \mu(\frac{r}{s})(k-1)^s$.

Proof. Put $r = |K|$ and $s = |L|$. Then s runs through the divisors of m that are multiples of r, as L ranges through the subgroups of Γ that contain K. Moreover, $|X^K| = (k-1)^r$, as one can verify easily. Hence, Formula (I) above immediately implies that

$$g(K) = \sum_{r|s} \mu(\frac{r}{s})(k-1)^s.$$

But the number of distinct orbits of X, the isotropy subgroup of which is exactly K, is just $\frac{1}{r}g(K)$. Hence,

$$a_m = \sum_{r|m} \frac{1}{r}(\sum_{s|r} \mu(\frac{r}{s})(k-1)^s),$$

r running through the divisors of m as K ranges over the subgroups of Γ. Thus, Theorem 3.1 is proved. □

Corollary 3.1 $a_m = \frac{1}{m}\sum_{g|m} \phi(m/g)(k-1)^g$.

Proof. Consider the identity $m' = \sum_{s'|m'} \phi(s')$. To prove the identity, note first that it is true if m' is a prime number. Then, note that both sides are homomorphisms of $\mathbb{N}^* \to \mathbb{N}^*$.) Apply the Moebius Inversion Formula to deduce the identity

$$\begin{aligned}\phi(m') &= \textstyle\sum_{s'|m'} s' \mu(\frac{m'}{s'}) \\ &= \textstyle\sum_{r'|m'} \frac{m'}{r'} \mu(r').\end{aligned}$$

On the other hand, the coefficient of $(k-1)^s$ in the expression for ma_m is

$$\begin{aligned} m\textstyle\sum_{r|m} \frac{1}{r} \sum_{s|r} \mu(\frac{r}{s}) &= \textstyle\sum_{r'|m'} \frac{m'}{r'} \sum_{r'} \mu(r') \\ &= \textstyle\sum_{r'|m'} \frac{m'}{r'} \mu(r') \\ &= \phi(m'). \end{aligned}$$

This finishes the proof. □

Corollary 3.2 *If m is a prime number p, then*

$$a_p = \frac{1}{p}((k-1)^{p-1} + (p-1))(k-1)).$$

In particular, it follows that the coefficients of the mod 2 Poincaré series of $\Lambda(\mathbb{R}^{n+1}_{k-1})$ grow exponentially.

4 The Growth of $H_*(\Lambda\mathbb{F}_k(\mathbb{R}^{n+1}))$

In this section we compare the growth of the Betti numbers $H_*(\Lambda\mathbb{F}_k(\mathbb{R}^{n+1}))$ with those of $H_*(\Lambda\mathbb{F}_{k-1}(\mathbb{R}^{n+1}))$. The problem is complicated by the fact that $H_*(\Lambda\mathbb{F}_k(\mathbb{R}^{n+1}))$ need not split as a product of

$$H_*(\Lambda\mathbb{F}_{k-1}(\mathbb{R}^{n+1})) \otimes H_*(\Lambda\mathbb{R}^{n+1}_{k-1}).$$

However, since there are sections $\mathbb{F}_{k-1}(\mathbb{R}^{n+1}) \to \mathbb{F}_k(\mathbb{R}^{n+1})$, $H_*(\Lambda\mathbb{F}_k(\mathbb{R}^{n+1}))$ contains submodules isomorphic to $H_*(\Lambda\mathbb{F}_{k-1}(\mathbb{R}^{n+1}))$. We are interested in the homology that does not arise from such sections.

Precisely, let

$$\mathrm{proj}_\rho : \mathbb{F}_k(\mathbb{R}^{n+1}) \to \mathbb{F}_{k-1}(\mathbb{R}^{n+1})$$

be the projection that deletes the indices not in ρ, where ρ is a subset of $\{1, \cdots, k\}$ of $k-1$ elements. We wish to estimate the rank of the submodule

$$\cap_\rho \ker(H_m(\Lambda(\mathrm{proj}_\rho)) : H_m(\Lambda\mathbb{F}_k(\mathbb{R}^{+1})) \to H_m(\Lambda\mathbb{F}_{k-1}(\mathbb{R}^{n+1})),$$

where ρ ranges over all subsets of $k-1$ elements. So, let $\{E^{n+1}_{*,*}, \mathrm{d}^{n+1}\}$ be the $(n+1)$-term of the Serre spectral sequence of the fibration

$$\mathrm{eval}_k : \Lambda\mathbb{F}_k(\mathbb{R}^{n+1}) \to \mathbb{F}_k(\mathbb{R}^{n+1})$$

that sends a loop ω to $\omega(1)$. As the differential operators d^r for $r > n$ are trivial, it follows that $\mathrm{E}^{n+1}_{*,*}$ survives to $\mathrm{E}^{\infty}_{*,*}$ and, therefore, the natural morphism

$$\mathrm{E}^{n+1}_{(k-1)n,*} \to \mathrm{E}^{\infty}_{(k-1)n,*} \cong \mathrm{E}^{0}_{(k-1)n,*}H_*(\Lambda\mathbb{F}_k(\mathbb{R}^{n+1}))$$

is an isomorphism. As above, let $H_{(k-1)n,*}(\Lambda\mathbb{F}_k(\mathbb{R}^{n+1}))$ be the corresponding submodule of $H_*(\Lambda\mathbb{F}_k(\mathbb{R}^{n+1}))$.

Proposition 4.1 *The module*

$$\cap_\rho \ker(H_m(\Lambda(\mathrm{proj}_\rho)) : H_m(\Lambda\mathbb{F}_k(\mathbb{R}^{n+1})) \to H_m(\Lambda\mathbb{F}_{k-1}(\mathbb{R}^{n+1}))$$

contains $H_{(k-1)n,*}(\Lambda\mathbb{F}_k(\mathbb{R}^{n+1}))$.

Proof. The projections

$$H_*(\Lambda(\mathrm{proj}_\rho)) : H_*(\Lambda\mathbb{F}_k(\mathbb{R}^{n+1})) \to H_*(\Lambda\mathbb{F}_{k-1}(\mathbb{R}^{n+1}))$$

annihilate the basic elements $\alpha_{\underline{r},\underline{s}}$, with $|\underline{r}| = k-1$ of $H_{(k-1)n}(\mathbb{F}_k(\mathbb{R}^{n+1}))$, for all subsets $\rho \subset \{1, \cdots, k\}$ of cardinality $(k-1)$. This implies that the induced morphisms

$$H_*(\Lambda(\mathrm{proj}_\rho)) : H_*(\Lambda\mathbb{F}_k(\mathbb{R}^{n+1})) \to H_*(\Lambda\mathbb{F}_{k-1}(\mathbb{R}^{n+1}))$$

take the submodule $H_{(k-1)n,*}(\Lambda\mathbb{F}_k(\mathbb{R}^{n+1}))$ to zero. □

The next result is the first step in the estimate of the growth of the Betti numbers of the homology groups $H_*(\Lambda\mathbb{F}_k(\mathbb{R}^{n+1}))$ as a function of k.

Theorem 4.1 *The inequality*

$$\operatorname{rank} H_{(k-1)n,m}(\Lambda\mathbb{F}_k(\mathbb{R}^{n+1})) \geq \frac{1}{2}(k-2)b_m \operatorname{rank} H_{2n,m}(\Lambda\mathbb{F}_3(\mathbb{R}^{n+1}))$$

holds for an infinite number of integers m, where $b_m = \sum_{r=0}^{m+1} a_r$ and

$$a_r = \sum_{r|m} \frac{1}{r}(\sum_{s|r} \mu(\frac{r}{s})2^s = \frac{1}{r}\sum_{g|r}\phi(r/g)2^g,$$

with μ and ϕ being, respectively, the Moebius and Euler functions.

Proof. Consider the isomorphism

$$\mathrm{E}^n_{(k-1)n,*} \cong H_{(k-1)n}(\mathbb{F}_k(\mathbb{R}^{n+1})) \otimes H_*(\Omega\mathbb{F}_k(\mathbb{R}^{n+1})),$$

where $\mathrm{E}^n_{(k-1)n,*}$ is the nth-term of the Serre spectral sequence of the evaluation map $\mathrm{eval}_k : \Lambda\mathbb{F}_k(\mathbb{R}^{n+1}) \to \mathbb{F}_k(\mathbb{R}^{n+1})$. The module $H_{(k-1)n}(\mathbb{F}_k(\mathbb{R}^{n+1}))$ is generated by the basic $(k-1)$-fold twisted products

$$\{\alpha_{r_1 s_1} \rtimes \cdots \rtimes \alpha_{r_k s_k}\},$$

where $r_1 = 2 < \cdots < r_k = k$ and $1 \leq s_j < r_j$ for $j = 1, \cdots, k$, and the algebra $H_*(\Omega\mathbb{F}_k(\mathbb{R}^{n+1}))$ is generated by the set

$$\{\bar{\alpha}_{21}; \bar{\alpha}_{31}, \bar{\alpha}_{32}; \cdots; \bar{\alpha}_{k1}, \cdots, \bar{\alpha}_{kk-1}\},$$

where the elements $\bar{\alpha}$ are the desuspensions of the elements α. Let $\mathcal{A}$ be the subalgebra of $H_*(\Omega\mathbb{F}_k(\mathbb{R}^{n+1}))$ generated by the set $\{\bar{\alpha}_{k1}, \bar{\alpha}_{k-11}, \bar{\alpha}_{kk-1}\}$. Put

$$\gamma_1 = \alpha_{r_1,s_1} \rtimes \cdots \rtimes \alpha_{k-2 s_{k-2}} \rtimes \alpha_{k-11} \rtimes \alpha_{k1}$$

and

$$\gamma_2 = \alpha_{r_1 s_1} \rtimes \cdots \alpha_{k-2 s_{k-2}} \rtimes \alpha_{k-11} \rtimes \alpha_{kk-1},$$

and consider elements of the form $\gamma_1 \otimes \omega_1, \gamma_2 \otimes \omega_2 \in \mathrm{E}^n_{(k-1)n,*}$, where ω_1, ω_2 are in $\mathcal{A}$. Recall that the differential $\mathrm{d}^n : \mathrm{E}^n_{(k-1)n,*} \to \mathrm{E}^n_{(k-2)n,*}$ is that where, in the notation of Chapter VIII, §5 above,

$$\gamma_1 \otimes \omega_1 = \alpha_{r_1 s_1} \rtimes \cdots \rtimes \alpha_{k-11} \otimes \omega_1$$

maps to

$$\sum \lambda_1 \rtimes \cdots \rtimes \hat{\lambda}_j \rtimes \cdots \rtimes \lambda_k \otimes [\bar{\lambda}_j, \omega_1],$$

where λ_j is the jth row in the factor table

$$\mathcal{T}_1(\phi_{\underline{r}\,\underline{s}}) : \begin{cases} \alpha_{r_1 s_1} & + \sum_{m=2}^{p} \lambda_{1m} \\ \cdots & + \cdots \\ \alpha_{r_j s_j} & + \sum_{m=j+1}^{p} \lambda_{jm} \\ \cdots & + \cdots \\ \alpha_{k-11} + \alpha_{kk-1} & \\ \alpha_{k1} & \end{cases}$$

of the basic element γ_1, and where $\bar{\lambda}$ is its desuspension. In the same manner, the differential $\mathrm{d}^n : \mathrm{E}^n_{(k-1)n,*} \to \mathrm{E}^n_{(k-2)n,*}$ takes $\gamma_2 \otimes \omega_2$ to

$$\sum \lambda'_1 \rtimes \cdots \rtimes \hat{\lambda}'_j \rtimes \cdots \rtimes \lambda'_k \otimes [\bar{\lambda}'_j, \omega_2],$$

where λ'_j is the jth row of the factor table

$$\mathcal{T}_2(\phi_{\underline{r}\,\underline{s}}) : \begin{cases} \alpha_{r_1 s_1} & + \sum_{m=2}^{p} \lambda_{1m} \\ \cdots & + \cdots \\ \alpha_{r_j s_j} & + \sum_{m=j+1}^{p} \lambda_{jm} \\ \cdots & + \cdots \\ \alpha_{k-11} + \alpha_{k1} & \\ \alpha_{kk-1} & \end{cases}$$

of the basic element γ_2 and $\bar{\lambda}'_j$ is its desuspension. Observe now that

$$\begin{cases} \lambda_j & = \lambda'_j, \quad \text{for } 1 \leq j \leq (k-2) \\ \lambda_{k-1} & = \alpha_{k-11} + \alpha_{kk-1} \\ \lambda_k & = \alpha_{k1} \\ \lambda'_{k-1} & = \alpha_{k-11} + \alpha_{k1} \\ \lambda'_k & = \alpha_{kk-1}. \end{cases}$$

The mod 2 Y-B relations imply that

$$\begin{aligned} \mathrm{ad}(\bar{\alpha}_{k1})(\omega) &= \mathrm{ad}(\bar{\lambda}_{k-1})(\omega) \\ &= \mathrm{ad}(\bar{\alpha}_{k-11} + \bar{\alpha}_{kk-1})(\omega) \end{aligned}$$

for $\omega \in \{\bar{\alpha}_{k1}, \bar{\alpha}_{k-11}, \bar{\alpha}_{kk-1}\}$. Similarly, one shows that

$$\begin{aligned} \mathrm{ad}(\bar{\alpha}_{kk-1})(\omega) &= \mathrm{ad}(\bar{\lambda}'_{k-1})(\omega) \\ &= \mathrm{ad}(\bar{\alpha}_{k-11} + \bar{\alpha}_{k1})(\omega) \end{aligned}$$

for $\omega \in \{\bar{\alpha}_{k1}, \bar{\alpha}_{k-11}, \bar{\alpha}_{kk-1}\}$. With λ standing for any of the first $(k-2)$-rows of the factor table of either γ_1 or γ_2, we note that $\mathrm{ad}(\bar{\lambda})$ annihilates $\mathcal{A}$, since it annihilates the generators of $\mathcal{A}$. Therefore,

$$\begin{aligned} d^n(\gamma_1 \otimes \omega_1) &= \beta \rtimes \alpha_{k1} \otimes [\bar{\alpha}_{k1}, \omega_1] + \beta \rtimes \lambda_{k-1} \otimes [\bar{\alpha}_{k1}, \omega_1] \quad \text{and} \\ d^n(\gamma_2 \otimes \omega_2) &= \beta \rtimes \alpha_{kk-1} \otimes [\bar{\alpha}_{kk-1}, \omega_2] + \beta \rtimes \lambda'_{k-1} \otimes [\bar{\alpha}_{kk-1}, \omega_2], \end{aligned}$$

where $\beta = \alpha_{r_1 s_1} \rtimes \cdots \rtimes \alpha_{r_{k-2} s_{k-2}}$. On decomposing the twisted products $\beta \rtimes \lambda_{k-1}$ and $\beta \rtimes \lambda'_{k-1}$ in terms of the twisted products $\alpha_{r_i s_i} \rtimes \alpha_{r_j s_j}$, one easily sees that

$$\beta \rtimes (\alpha_{k1} + \lambda_{k-1}) = \beta \rtimes (\alpha_{kk-1} + \lambda'_{k-1})$$

(see also Chapter VI, §5). Hence,

$$\text{(I)} \qquad [\bar{\alpha}_{k1}, \omega_1] + [\bar{\alpha}_{kk-1}, \omega_2] = 0$$

in $\mathcal{A}$. By definition, let

$$\mathcal{V}_m = \{\omega_1, \omega_2 \in \mathcal{A}_{m-1} | \text{ (I) is satisfied}\}$$

and $\mathcal{V} = \bigoplus_m \mathcal{V}_m$.

To compute the dimension of $\mathcal{V}$, note that the correspondence

$$\bar{\alpha}_{k1} \mapsto \bar{\alpha}_{31}, \ \bar{\alpha}_{kk-1} \mapsto \bar{\alpha}_{32}, \ \bar{\alpha}_{k-11} \mapsto \bar{\alpha}_{21}$$

induces an isomorphism $\mathcal{A} \cong H_*(\Omega\mathbb{F}_3(\mathbb{R}^{n+1}))$ of algebras. Thus, we can regard $\{\bar{\alpha}_{21}, \bar{\alpha}_{31}, \bar{\alpha}_{32}\}$ as the generators of $\mathcal{A}$, and what remains to be done is to determine the dimension of $\mathcal{V}_m$.

Put

$$\Omega(\mathbb{F}_3(\mathbb{R}^{n+1}), \mathbb{R}_2^{n+1}) = \{\omega \in \Lambda\mathbb{F}_3(\mathbb{R}^{n+1}) | \omega(1) \in \mathbb{R}_2^{n+1}\},$$

where $\mathbb{R}_2^{n+1}$ is the fiber of $\text{proj}_{21} : \mathbb{F}_3(\mathbb{R}^{n+1}) \to \mathbb{F}_2(\mathbb{R}^{n+1})$. Denote by $\{'\mathrm{E}^*_{*,*}, '\mathrm{d}^*\}$ the Serre spectral sequence of the fibration

$$\mathrm{p}_2 : \Omega(\mathbb{F}_3(\mathbb{R}^{n+1}), \mathbb{R}_2^{n+1}) \to \mathbb{R}_2^{n+1}, \ \omega \mapsto \omega(1).$$

Put $\zeta = \bar{\alpha}_{31} + \bar{\alpha}_{32} + \bar{\alpha}_{21}$, and denote by $\{''\mathrm{E}^n_{*,*}, ''\mathrm{d}^*\}$ the Serre spectral sequence of the fibration $\Lambda(\mathbb{R}_2^{n+1}) \to \mathbb{R}_2^{n+1}$. Note that the natural imbedding $\Lambda(\mathbb{R}_2^{n+1}) \subset \Omega(\mathbb{F}_3(\mathbb{R}^{n+1}), \mathbb{R}_2^{n+1})$ induces a morphism $\phi^* : '' \mathrm{E}^n_{*,*} \to ' \mathrm{E}^*_{*,*}$ of spectral sequences.

Assertion 4.1 *The morphism ϕ^* induces an isomorphism*

$$''\mathrm{E}^n_{*,*} \otimes \mathbb{Z}_2(\zeta) \to ' \mathrm{E}^n_{*,*}$$

of spectral sequences.

The proof is similar to that of Theorem 1.1 and is left to the reader.

With this Assertion established, we note that

$$\mathcal{V}_m \cong ' \mathrm{E}^n_{n,m(n-1)},$$

and, by virtue of Theorem 3.1, $\dim '' \mathrm{E}^{n+1}_{n,(m-1)(n-1)} = a_m$. Hence, the isomorphism $''\mathrm{E}^{n+1}_{n,*} \otimes \mathbb{Z}_2(\zeta) \to ' \mathrm{E}^{n+1}_{n,*}$ implies that $\dim \mathcal{V}_m = \sum_{i=0}^m a_i$. This completes the proof Theorem 4.1. □

5 Cup Length in $H^*(\Lambda(\mathbb{F}_{k-r,r});\mathbb{Z}_2)$

We wish to show that $\mathbb{F}_{k-r,r}$ has arbitrarily large mod 2 cup length. As before, write $\Omega_{k-r,r}$ for $\Omega\mathbb{F}_{k-r,r}$, and $\Lambda_{k-r,r}$, for $\Lambda\mathbb{F}_{k-r,r}$.

Recall that $H_*(\Omega(S^n);\mathbb{Z}) \cong T(x, n-1)$, where $T(x, n-1)$ is the associative algebra generated by the $(n-1)$-dimensional element x, and the isomorphism is one of Hopf algebras. To compute the dual Hopf algebra, we need the following concept. Recall that the *divided polynomial algebra*, $\Gamma[z^*, m]$, generated by the m-dimensional element z^*, m being even, is the algebra generated by $\{\gamma_r(z^*) \mid 0 \le r\}$, where

$$\gamma_0(z^*) = 1,\ \gamma_1(z^*) = z^*, \text{ and } \gamma_r(z^*)\gamma_s(z^*) = \frac{(r+s)!}{r!s!}\gamma_{r+s}(z^*).$$

Hence,

$$H^*(\Omega(S^n);\mathbb{Z}) \cong \begin{cases} \Gamma[(y^*, n-1],\ n \,\text{odd} \\ E(x^*, n-1) \otimes \Gamma[y^*, 2n-2],\ n \,\text{even}, \end{cases}$$

where $E(x^*, n-1)$ is the exterior algebra generated by the $(n-1)$-dimensional element x^*, with $(n-1)$ being odd. Therefore,

$$H^*(\Omega(S^n);\mathbb{Z}_2) \cong \bigotimes_i Q_2[\gamma_{2^i}(y^*)],$$

where $Q_2[\gamma_{2^i}(y^*)]$ is the truncated polynomial $\mathbb{Z}_2[\gamma_{2^i}(y^*)]/(\gamma^2_{2^i}(y^*))$ generated by the element $\gamma^2_{2^i}(y^*)$, with $i \ge 1$.

Recall (see Chapter II, §1) that $\Omega(\mathbb{F}_{k-r,r})$ splits

$$\Omega(\mathbb{F}_{k-r,r}) \simeq \prod_{t=r+1}^{k} \Omega(S_{t1} \vee \cdots \vee S_{tt-1}).$$

Also, for each bouquet we have the splitting

$$\Omega(S_{t1} \vee \cdots \vee S_{tt-1}) \simeq \prod_w \Omega(S^{|\omega|}),$$

where $\omega \in \mathcal{W}_{k-r,r}$ ranges over the basic Whitehead products in

$$\{\alpha_{ts} \mid 1 \le s < t \le k, \text{ and } (r+1) \le t\}.$$

Let us put

$$\mathcal{T}_{k-r,r} = \{\gamma^*_{2^u}(y^*_w) | w \in \mathcal{W}_{k-r,r}\},$$

where $y^*_\omega \in H^*(\Omega(S^{|\omega|});\mathbb{Z}_2)$ is the generator defined by $\omega \in \mathcal{W}_{k-r,r}$.

Theorem 5.1 *Given $\omega \in \mathcal{T}_{k-r,r}$, there is an element ζ^*_ω in the algebra $H^*(\Lambda(S_{t1} \vee \cdots \vee S_{tt-1}); \mathbb{Z}_2)$ such that*

$$(\zeta^*_\omega y^*_\omega) \in \text{image}(H^*(\Lambda_{k-r,r}; \mathbb{Z}_2) \to H^*(\Omega_{k-r,r}; \mathbb{Z}_2))$$

is not equal to zero.

Proof. Consider the mod 2 Serre cohomology spectral sequence $\{\mathrm{E}^{*,*}_m, \mathrm{d}_m\}$ of the fibration

$$\Lambda_{k-r,r} = \Lambda(\mathbb{F}_{k-r,r}) \to \mathbb{F}_{k-r,r},$$

defined by the evaluation map. Let $\omega \in \mathcal{W}_{k-r,r}$. Then,

$$\mathrm{d}_n(y^*_\omega) = \sum \alpha^*_{rs} \otimes \bar{\lambda}_{tu}.$$

Define ζ^*_ω to be the product of all $\tau^* \in \mathcal{T}_{k-r,r}$ of dimension less than that of y^*_ω. One easily determines inductively that ζ^*_ω, and $(\zeta^*_\omega)y^*_\omega$ are cocycles. This is enough to establish the assertion. □

Let us consider next the case when

$$\omega = [\alpha_{31}, \alpha_{32}] \in \text{image}(\pi_n(S_{31} \vee S_{32}) \to \pi_n(\mathbb{F}_3(\mathbb{R}^{n+1}))).$$

Let $\zeta^*_\omega \in H^*(\Lambda\mathbb{F}_3(\mathbb{R}^{n+1}); \mathbb{Z}_2)$ be the multiplier corresponding to $y^*_\omega \in H^*(\Omega\mathbb{F}_3(\mathbb{R}^{n+1}); \mathbb{Z}_2)$ and $z^*_\omega \in H^*(\Lambda\mathbb{F}_3(\mathbb{R}^{n+1}); \mathbb{Z}_2)$ such that

$$z^*_\omega \mid \Omega(\mathbb{F}_3(\mathbb{R}^{n+1}) = \zeta^*_\omega y^*_\omega.$$

Corollary 5.1 *The element*

$$\gamma_{2^i}(\alpha^*_{21})\gamma_{2^j}(\alpha^*_{31})\gamma(\alpha^*_{32})z^*_\omega, \; i,j \geq 1$$

in an element in $H^*(\Lambda\mathbb{F}_3(\mathbb{R}^{n+1}); \mathbb{Z}_2)$ the restriction of which to $\Omega\mathbb{F}_3(\mathbb{R}^{n+1})$ is nontrivial.

Proof. It is easy to see that the multiplier corresponding to ω, ζ^*_ω, is $\alpha^*_{21}\alpha^*_{31}\alpha^*_{32}$. Hence,

$$z^*_\omega | \Omega(\mathbb{F}_3(\mathbb{R}^{n+1}) = \alpha^*_{21}\alpha^*_{31}\alpha^*_{32}y^*_\omega.$$

Note that

$$(\gamma_{2^i}(\alpha^*_{21})\gamma_{2^j}(\alpha^*_{31})\gamma(\alpha^*_{32})z^*\omega) \mid \Omega(\mathbb{F}_3(\mathbb{R}^{n+1})) = \\ (\gamma_{2^i}(\alpha^*_{21})\gamma_{2^j}(\alpha^*_{31})\gamma(\alpha^*_{32})y^*_\omega).$$

But the latter product in $H^*(\Omega\mathbb{F}_3(\mathbb{R}^{n+1}); \mathbb{Z}_2)$ is nontrivial. □

Corollary 5.2 *The cup length of* $(\Lambda\mathbb{F}_3(\mathbb{R}^{n+1}), E)$ *is infinite, where* $E \simeq \Lambda_{21} \vee \Lambda_{31} \vee \Lambda_{32}$.

Proof. Consider the word

$$\lambda^* = \gamma_2(\alpha_{21}^*)\gamma_2(\alpha_{31}^*)\gamma_2(\alpha_{32}^*)z^*\omega,$$

as in Corollary 5.1. Then the words

$$\prod_{r>1} \gamma_{2^r}(\alpha_{21}^*) \prod_{s>1} \gamma_{2^s}(\alpha_{31}^*) \prod_{t>1} \gamma_{2^t}(\alpha_{32}^*)\lambda^*$$

are nontrivial in $(\Lambda\mathbb{F}_3(\mathbb{R}^{n+1}), E)$ and vanish when restricted to E. Hence, $H^*(\Lambda\mathbb{F}_3(\mathbb{R}^{n+1}), E; \mathbb{Z}_2)$ has words of arbitrarily high cup length. □

Corollary 5.3 *The cup-length of* $\Lambda\mathbb{F}_{k-r,r}$ *is infinite.*

Proof. The words

$$\prod_{r>1} \gamma_{2^r}(\alpha_{21}^*) \prod_{s>1} \gamma_{2^s}(\alpha_{31}^*) \prod_{t>1} \gamma_{2^t}(\alpha_{32}^*)\lambda^*$$

of Corollary 5.2 are nontrivial in $H^*(\Lambda\mathbb{F}_{k-1,1}; \mathbb{Z}_2)$ for $k \geq 3$. By applying an appropriate permutation one obtains similar words that are nontrivial in $H^*(\Lambda\mathbb{F}_{k-r,r}; \mathbb{Z}_2)$. □

6 Historical Remarks

Theorem 1 of [39, Fadell-Husseini] is true only if $k = 3$; in fact, Theorem 1.1 above is the corrected version. Furthermore, Theorem 4.1 replaces inequality (3) of [39, Fadell-Husseini] regarding the growth of the Poincaré series of $\mathbb{F}_k(\mathbb{R}^{n+1})$ in comparison with that of $\mathbb{F}_r(\mathbb{R}^{n+1})$. A sketch of Theorem 5.1 appears as the Proposition of §4 of [39, Fadell-Husseini].

The role of the cyclic action in the rational homology of the free loop spaces of bouquets of suspensions is recognized in [21, R. Cohen], and there it is used to obtain a decomposition of the homology of the free loop space. Theorem 3.1 is announced without proof in [67, Kallel], and a similar result for rational coefficients is given in [94, Roos]. The latter approach is based on the connection between the homology of the free loop space and cyclic homology (see [70, Loday-Quillen]). The problem is reduced to the study of the cyclic representation of the cyclic group, as above, but the calculations are within the framework of representation theory.

Lower estimates on the rank of $H_*(\Lambda(X); \mathbb{Q})$, when X is a bouquet of spheres, can be found in [109, Vigué-Poirrier].

XIII. Γ-Category and Ends

In this chapter we prepare the ground for our applications to problems of the k-body type, to be presented in Chapter XIV. We are chiefly motivated by the notion of *neighborhoods of infinity*, introduced in [8, Bahri-Rabinowitz] in their study of 3-body problems. Intuitively speaking, the neighborhoods of infinity consist of configurations of three bodies that separate into simpler clusters moving away from each other. Another approach that deals with this is that of admissible and non-admissible sets, introduced in [75, 76, 77, Majer-Terracini]. In an attempt to bring the two points of view together, we offer the concept of *ends*. This concept makes it possible to adopt a variational approach that depends on the use of a category theory of the relative type.

In §1, anticipating the precise notions of topological and analytic ends in §2, we present a theory of relative category for ANR metric spaces that is subadditive. (See the Historical Remarks below.)

In §2, as in [8, Bahri-Rabinowitz], we restrict ourselves to the 3-body case and introduce the notion of ends in the Sobolev space $W_T^{1,2}(\mathbb{R}^{3(n+1)})$, which consists of triples $\{u = (u_1, u_2, u_3)\}$, where each $u_i : \mathbb{R} \to \mathbb{R}^{n+1}$ is T-periodic. The analytic analogue of $\Lambda(\mathbb{F}_3(\mathbb{R}^{n+1}))$ is the space

$$\Lambda' = \{u \in W_T^{1,2}(\mathbb{R}^{3(n+1)}) | u_i(t) \neq u_j(t),\, i \neq j,\, t \in \mathbb{R}\}.$$

We define *topological ends* E^t in Λ' in terms of the average values $[u_i]$ and the L_2-norm values $|\dot{u}_i|_2$, and we are able to show that the pair (Λ', E^t) has infinite cup length. We then refine the notion of ends to define a smaller class of smaller ends, B, which we call *analytic ends*. The latter involves a functional $f : \Lambda' \to \mathbb{R}$ and, using the fact that $B \subset E^t$, we show that $\text{cat}_{\Lambda'}(\Lambda', B)$ is infinite. This prepares the way for the application in Chapter XIV of the min-max method to problems of the 3-body type, using analytic ends.

The category of the ends E^t and B in §2, as well as their relative category, is infinite. The Bahri-Rabinowitz approach depends on homology, and there one meets a similar situation: the homology of the neighborhoods of infinity and, as well, the relative homology of the space modulo these neighborhoods are of infinite dimension.

An essential feature of [75, Mayer-Terracini] is a novel description of sets consisting of clusters that cannot be separated into subclusters without introducing collision. They call such sets *admissible*; the neighborhoods of infinity of [8, Bahri-Rabinowitz] are then analogous to these nonadmissible sets. As categorical sets are generally regarded as being negligible: the view taken in [40, Fadell-Husseini] is that categorical sets in $\Lambda(\mathbb{F}_k(\mathbb{R}^{n+1}))$ should be sets that are contractible in the spaces

$$\Lambda_{ij} = \{u = (u_1, \cdots, u_k) \in W_T^{1,2}(\mathbb{R}^{k(n+1)})\},$$

where the orbits u_i and u_j never collide. This idea led us to the concept of Γ-category. This theory, following [40, Fadell-Husseini], is presented in §3. It gives us an alternative to the collision index of [75, Mayer-Terracini], that is not only of greater simplicity and naturality but also makes the necessary computations more tractable.

The general Γ-category amounts to specifying a finite collection Γ of open sets

$$\{\Gamma_i \subset W_T^{1,2}(\mathbb{R}^{3(n+1)})\}.$$

Then, given $A \subset \cap_i \Gamma_i$, we say that $\Gamma \mathrm{cat} A \leq p$ if, and only if, for any decomposition $A = \cup A_i$, $A_i \subset \Gamma_i$, the inequalities $\mathrm{cat}_{\Gamma_i}(A) \leq p$ hold. In §4 we let Γ be the open collection $\{\Lambda_{ij}\}$, and $A \subset \Lambda'$. Then, it turns out that nonadmissible sets are analogous to sets of $\Gamma\mathrm{cat} \leq 2$, while admissible sets correspond to sets of $\Gamma\mathrm{cat} > 2$. In addition, we establish the existence in Λ' of compact subsets with Γ-category that is arbitrarily high. The tools are now in place for an alternative approach in Chapter XIV to the approach of [77, Majer-Terracini] to problems of the k-body type.

1 Relative Category

Let Y be a metric, path-connected ANR and $A \subset Y$, a fixed closed neighborhood retract in Y.

Definition 1.1 *A subset $C \subset Y$ is said to be categorical in Y relative to A, if, and only if, there is a homotopy $h : C \times I \to Y$ such that*

(i) $h(u, 0) = u$ for all $u \in C$;
(ii) $h(u, t) = u$, for all $u \in A \cap C$ and $t \in I$; and
(iii) $h(u, 1) \in A$ for $u \in C$.

Since Y is an ANR, property (ii) of Definition 1.1 can be replaced by the following equivalent condition.

(*ii*)′ $h(u, t) \in A$, for all $u \in A \cap C$ and t with $t \in I$.

The homotopy h above is also called a *compression* of C into A relative to A. As Y is an ANR and C is closed in Y, the homotopy h is extendable to a homotopy $H : Y \times I \to Y$, so that $H(u,t) = u$ for all $u \in A$.

Definition 1.2 *Let X be a closed subset of Y. Put* $\mathrm{cat}_Y(X;A) = n < \infty$ *if, and only if, there is a closed cover $\{C_1, \cdots, C_n\}$ in Y of X such that each C_i is categorical in Y relative to A and n is minimal with respect to this property. If no such cover exists, we set* $\mathrm{cat}_Y(X;A) = \infty$.

Observe that the sets C_i may be taken to be open, since A is assumed to be a neighborhood retract in Y.

The following proposition states the basic properties of the relative category needed here. The proofs are straightforward and are left for the reader.

Proposition 1.1 $\mathrm{cat}_Y(X;A)$ *has the following properties:*

(*i*) *(monotoneity)* $\mathrm{cat}_Y(X_1;A) \le \mathrm{cat}_Y(X_2;A)$ *if* $X_1 \subset X_2$;
(*ii*) *(subadditivity)* $\mathrm{cat}_Y(X_1 \cup X_2;A) \le \mathrm{cat}_Y(X_1,A) + \mathrm{cat}_Y(X_2;A)$;
(*iii*) *(continuity) if A is a neighborhood retract in Y, then there is an open set $U \supset X$ such that* $\mathrm{cat}_Y(\bar{U};A) = \mathrm{cat}_Y(X;A)$; *and*
(*iv*) *(finiteness) if X is compact, then* $\mathrm{cat}_Y(X;A)$ *is finite.*

Equally easy to prove is the following lemma.

Lemma 1.1 *Suppose that X is closed in Y. Then*

$$\mathrm{cat}_Y(X;A) = \mathrm{cat}_Y(X \cup A;A).$$

Thus, one may assume, for computational purposes and without loss of generality, that $X \supset A$.

Our relative category, $\mathrm{cat}_Y(X;A)$, with $A \subset X$, can be described in the manner of G.Whitehead's characterization of the classical Lusternik-Schnirelman category ([110, Whitehead]). To do so, proceed as follows. Put

$$W^m(Y,A) = \{(y_1, \cdots, y_m) \mid y_i \in A, \text{ for some } i \le m\},$$

and let

$$\Delta^m(X,A) : (X;A) \to (Y,A)^{\times m} = (Y^{\times m}, W^m(Y,A))$$

be the map induced by the diagonal map, where $x \mapsto (x_1, \cdots, x_m)$ with $x_i = x$ for all $i \le m$. ($W^m(Y,A)$ is the so-called fat wedge when A is just the basepoint of X.) We shall say that $\Delta^m(X,A)$ is *compressible* in $W^m(Y,A)$ if, and only if, there is a homotopy $H : X \times I \to Y^{\times m}$ such that

$$H_0 = \Delta^m(X,A), \; H_1(X) \in W^m(X,A)$$

and $\mathrm{proj}_i \circ H(a,t) = a$, $a \in A$, $0 \le t \le 1$, where $\mathrm{proj}_i : Y^{\times m} \to Y$ is the projection on the ith factor.

Proposition 1.2 *Let (Y, A) be an ANR(metric) pair, with $A \subset Y$ being a closed subspace and $A \subset X \subset Y$ as above. Then, $\text{cat}_Y(X; A) = m < \infty$ if, and only if,*

$$\Delta^m(X, A) : (X, A) \to (Y, A)^{\times m}$$

is compressible to $W^m(Y, A)$ and m is minimal with respect to this property.

Proof. Proceed as in the proof of the absolute case. As (Y, A) is an ANR pair, the homotopies involved can be assumed to be globally defined. The various homotopies contracting the C_i's to A induce a homotopy that yields the required compression of

$$\Delta^m(X, A) : (X, A) \to (Y, A)^{\times m}$$

into the relative fat wedge $W^m(Y, A)$. Conversely, using such a compressing homotopy composed with the projections on the various factors, one defines C_i's as the subsets mapped into A. The contracting homotopies of the C_i's into A are those induced by the compressing homotopy. □

Proposition 1.2 provides an easy proof of the behaviour of $\text{cat}_Y(X, A)$ under a deformation of Y.

Proposition 1.3 *Let $\eta : (Y, A) \times I \to (Y, A)$ denote a map such that*

$$\eta_0 = \text{id}, \text{ and } \eta_t(A) \subset A, \text{ for all } t \in [0, 1].$$

Then, if $Y \supset X \supset A$ and $X, \eta_1(A)$ and are closed in Y, it follows that $\text{cat}_Y(\eta_1(X), A) \geq \text{cat}_Y(X, A)$.

Proof. It suffices to show that $\text{cat}_Y(\eta_1(X) \cup A, A) \geq \text{cat}_Y(X, A)$. Assume, without loss of generality, that $\text{cat}_Y(\eta_1(X) \cup A, A) = m < \infty$. Then,

$$\Delta^m(\eta_1(X) \cup A, A) : (\eta_1(X) \cup A, A) \to (Y, A)^{\times m}$$

is compressible in $W^m(Y, A)$ by means of a compression, say, H. Note now that the map $K : X \to Y^{\times m}$ defined by the rule

$$K(x, t) = \begin{cases} (\eta(x, t), \cdots, \eta(x, t)) & 0 \leq t \leq 1/2 \\ H(\eta_1(x), 2t - 1), \cdots, H(\eta_1(x), 2t - 1) & 1/2 \leq t \leq 1 \end{cases}$$

is a compression of (X, A) in $W^m(Y, A)$. This proves the assertion. □

Cohomology cup products and their duals, homology coproducts, provide one with a lower bound $\text{cat}_Y(X; A)$ with $X \supset A$ ([36, 38, Fadell-Husseini]). As such, they are important tools for estimating the relative coproduct length, $\text{copl}_Y(X, A)$. The latter is defined as follows. Let $A \subset X \subset Y$. Observe that the exterior product or, equivalently, the cross product (see [25, Dold], [101, Spanier]) induces the morphism

$$\bigotimes_1^m H_*(Y,A) \stackrel{\cong}{\to} H_*((Y,A)^{\times m}),$$

where homology is singular homology with coefficients in a field $\mathbb{K}$. Recall that these products induce isomorphisms on homology, if (Y, A) is an ANR(metric) pair or when A is open in Y. Composing this with the morphism induced by $\Delta^m(X, A)$, we obtain the morphism

$$H_*(\Delta^m(X,A)) : H_*(X,A) \to \bigotimes_1^m H_*(Y,A).$$

Definition 1.3 *Let $A \subset X \subset Y$ as above, and let $i : (X, \emptyset) \to (X, A)$ be the natural inclusion. Then, the relative coproduct length,* $\mathrm{copl}_Y(X, A)$, *is defined by the rule*

$$\mathrm{copl}_Y(X,A) = \min\{m | H_*(\Delta^m(X,A)) \circ H_*(i) = 0\},$$

if $H_*(\Delta^m(X,A)) \circ H_*(i) = 0$, for some m. Otherwise, set $\mathrm{copl}_Y(X,A) = \infty$.

Proposition 1.4 *Let (Y, A) be an ANR(metric) space with A being closed in Y. Then, for any closed subset X of Y,*

$$\mathrm{cat}_Y(X,A) = \mathrm{cat}_Y(X \cup A, A) \geq \mathrm{copl}_Y(X \cup A, A).$$

The proof is straightforward and is left for the reader.

Recall that, if (Y, A) is an ANR pair or A is an open subset of Y, then the exterior product, or cross product, induces a morphism

$$\bigotimes_1^m H^*(Y,A) \stackrel{\cong}{\to} H^*((Y,A)^{\times m}).$$

Composing with the morphism induced by $\Delta^m(X, A)$, we obtain the morphism

$$H^*(\Delta^m(X,A)) : \bigotimes_1^m H^m(Y,A) \to H^*(X,A).$$

Note that

$$H^*(\Delta^m(X,A))(y_1 \otimes \cdots \otimes y_m) = H^*(j)(y_1 \cup \cdots \cup y_m),$$

where $j : (X, A) \to (Y, A)$ is the morphism induced by the natural imbedding $X \to Y$.

Definition 1.4 *Let $A \subset X \subset Y$ be as above. Then the relative cup length,* $\mathrm{cl}_Y(X, A)$, *is defined by the rule*

$$\mathrm{cl}_Y(X,A) = \min\{m \mid H^*(\Delta^m(X,A)) = 0\},$$

if $H^*(\Delta^m(X, A) = 0$ for some m. Otherwise, we set $\mathrm{cl}(X, A) = \infty$.

Proposition 1.5 *Suppose that (Y, A) is an ANR(metric) pair, with A closed in Y. Then, for any closed subset X of Y,*

$$\mathrm{cat}_Y(X; A) = \mathrm{cat}_Y(X \cup A; A) \geq \mathrm{cl}_Y(X \cup A, A).$$

The proof follows the lines of the analogous assertion in the absolute case. It is left for the reader.

It is worthwhile to note here that, when Y is paracompact and $X \supset A$ are closed subsets of Y, then the relative cup length $\mathrm{cl}_Y(X, A)$ can be defined using the Alexander-Spanier, or the Čech, cohomology. Following [101, Spanier], denote by $\bar{H}^*$ the Alexander-Spanier cohomology. Again we assume that the coefficients are in a field $\mathbb{K}$. Put

$$\mathrm{cl}_Y(X, A) = \min\{m \mid \bar{H}^*(j)(y_1 \cdots y_m) = 0\},$$

where $y_1 \cdots y_m$ is the m-fold product in $\bar{H}^*(X, A)$ and $\bar{H}^*(j)$ is the morphism induced by the inclusion $j : (X, A) \to (Y, A)$. Thus, if we drop the requirement that A is an ANR, the definition $\mathrm{cat}_Y(X, A)$ makes sense in a limited way, and $\mathrm{cat}_Y(Y, A) \geq \mathrm{cl}_Y(Y, A)$.

A nontrivial product $y_1 \cdots y_m \in \bar{H}^*(Y)$, the restriction of which to A is trivial, is often useful in comparing $\mathrm{cl}_Y(Y, A)$ with $\mathrm{cl}_Y(X, B)$, where $B \subset A$.

Proposition 1.6 *Suppose that Y is paracompact, and let $A, B \subset Y$ be closed subsets with $B \subset A$. Suppose also that there is a nonzero $m-fold$ product $y_1 \cdots y_m \in \bar{H}^*(Y)$ such that $\bar{H}^*(i_A)(y_i) = 0$ for each i, where $i_A : A \to Y$ is the natural inclusion. Then, $\mathrm{cl}_Y(Y, B) > m$.*

Proof. Since each y_i pulls back to some $u_i \in \bar{H}^*(Y, A)$, and the product $y_1 \cdots y_m \neq 0 \bar{H}^*(Y)$ by assumption, it follows readily that the product $u = u_1 \cdots u_m \neq 0 \in \bar{H}^*(Y, A)$. As

$$\bar{H}^*(i_B) \circ \bar{H}^*(j)(u_1 \cdots u_m) = y_1 \cdots y_m \neq 0,$$

where $i_B : (Y, \emptyset) \to (Y, B)$ is the natural morphism, we easily see that $\bar{H}^*(j)(u_1 \cdots u_m) \neq 0 \in \bar{H}^*(Y, B)$, where $j : (Y, B) \to (Y, A)$ is the natural imbedding. □

2 Ends in $W_T^{1,2}(\mathbb{R}^{3(n+1)})$

In this and the next section (as well as in Chapter XIV), we shall require the following setting. If we use the representation $\mathbb{R}^{k(n+1)} = \mathbb{R}^{n+1} \times \cdots \times \mathbb{R}^{n+1}$, $k-times$, then the Sobolev space $W_T^{1,2}(\mathbb{R}^{k(n+1)} = \mathcal{H}$ ([81, Mawhin-Willem]), has a typical element $u = (u_1, \cdots, u_k)$, where u_i is a T-periodic map from $\mathbb{R}$ to $\mathbb{R}^{n+1}$. The inner product in $W^{1,2}(\mathbb{R}^{k(n+1)})$ is given by

$$u.v = \int_0^T (\dot{u}(t)\dot{v}(t) + u(t)v(t))dt,$$

with the corresponding norm denoted by $\|u\|$.

There is an equivalent norm $\|u\|'$ (see [8, Bahri-Rabinowitz])

$$\|u\|' = ((\sum_{i=1}^{3} \int_0^T \dot{u_i}^2\, dt) + [u]^2)^{\frac{1}{2}},$$

where $[u]$ is the average value $[u] = \frac{1}{T}\int_0^T u\, dt$.

We also employ the norms defined by

$$\|u\|_2 = (\int_0^T |u|^2\, dt)^{\frac{1}{2}}, \quad \|u\|_\infty = \sup_{t\in[0,T]} |u(t)|.$$

Recall that the Sobolev inequality asserts that

$$\|u_i - [u_i]\|_\infty \leq \frac{T}{12}\|\dot{u}_i\|_2^2,$$

where $1 \leq i \leq 3$, and $u = (u_1, \cdots, u_3) \in W^{1,2}(\mathbb{R}^{3(n+1)})$ (see Proposition 1.3 of [81, Mawhin-Willem]). Furthermore, set

$$\sigma(u) = \sum_{i=1}^{3} \sqrt{\frac{T}{12}}\|\dot{u}_i\|_2.$$

It is convenient to define the notion of ends as either closed or open subsets of Λ'.

Definition 2.1 *The (closed) topological end $E_c^t(\Lambda', \epsilon)$ is defined by*

$$E_c^t(\Lambda', \epsilon) = \{u \in \Lambda' \mid |[u_k] - \tfrac{1}{2}[u_i + u_j]| \geq \tfrac{1}{2}|[u_i - u_j]| + \sigma(u) + \epsilon\}$$

for some permutation (i, j, k) of $(1, 2, 3)$, where $i < j$ and $\epsilon > 0$.

Definition 2.2 *The (open) topological end $E_o^t(\Lambda', \epsilon)$ is defined by*

$$E_o^t(\Lambda', \epsilon) = \{u \in \Lambda' \mid |[u_k] - \tfrac{1}{2}[u_i + u_j]| > \tfrac{1}{2}|[u_i - u_j]| + \sigma(u) + \epsilon\}$$

for some permutation (i, j, k) of $(1, 2, 3)$, where $i < j$ and $\epsilon > 0$.

We note here that $E_c^t(\Lambda', \epsilon)$ is appropriate for the study of the relative category of $(\Lambda', E_c^t(\Lambda', \epsilon))$, while the open end $E_o^t(\Lambda', \epsilon)$ is appropriate for the relative singular homology of $(\Lambda', E_o^t(\Lambda', \epsilon))$. Also note that

$$E_o^t(\Lambda', \epsilon) \subset E_c^t(\Lambda', \epsilon) \subset E_o^t(\Lambda', \epsilon'),$$

if $\epsilon' < \epsilon$. We now study the singular homology of $E_o^t(\Lambda', \epsilon)$. For the sake of simplicity, write E^t for $E_o^t(\Lambda', \epsilon)$. Note that E^t admits a natural partition, $E^t = \bigcup_{i,j} E^t(i,j)$, into three subsets, where

$$E^t(i,j) = \{u \in \Lambda' \mid \|[u_k] - \frac{1}{2}[u_i + u_j]\| > \frac{1}{2}\|[u_i - u_j]\| + \sigma(u) + \epsilon\},$$

$i < j$, and (i,j,k) is a permutation of $(1,2,3)$.

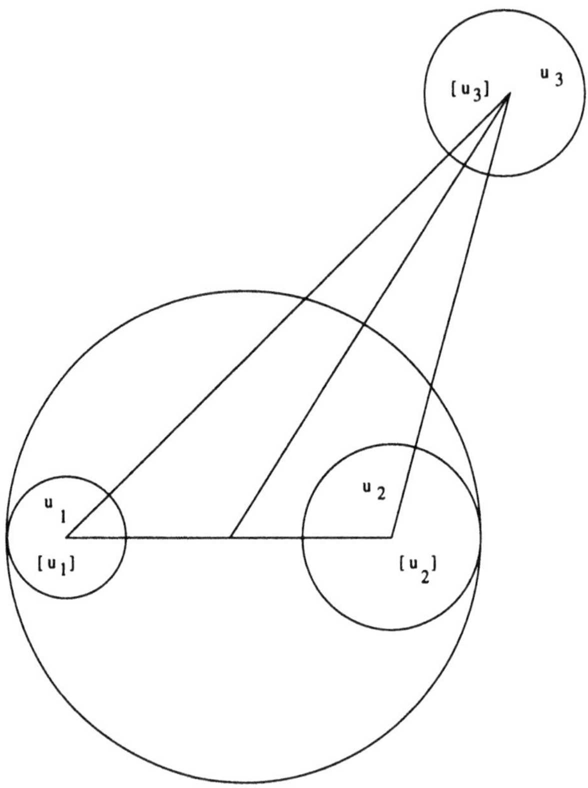

Fig. XIII.1. The Subset $E^t(1,2)$

We wish to determine the homotopy type of $E^t(i,j)$. So consider the space

$$E^t(1,2) = \{u \in \Lambda' \mid \|[u_3] - \frac{1}{2}[u_1 + u_2]\| > \frac{1}{2}\|[u_1 - u_2]\| + \sigma(u) + \epsilon\}.$$

Fig. XIII. 1 illustrates how each $u \in E^t$ carries along with it a triangle the vertices of which are $[u_1], [u_2]$, and $[u_3]$, with Sobolev neighborhoods

$N([u_i], r_i)$ centered at $[u_i]$ and with $r_i = \sqrt{\frac{T}{12}}\|\dot{u}_i\|_2$ as radii. The larger circle has radius $\frac{1}{2}|[u_1 - u_2]| + r_1 + r_2 + \epsilon$. When $u \in E^t(1,2)$, the Sobolev neighborhoods $N([u_1], r_1)$ and $N([u_2], r_2)$ are disjoint from $N([u_3], r_3)$.

Now consider the projection

$$\mathrm{proj}_{1,2} : E^t(1,2) \to \Lambda'\mathbb{F}_2(\mathbb{R}^{n+1}),\ u = (u_1, u_2, u_3) \mapsto (u_1, u_2).$$

Denote by $E^t_*(1,2)$ the subset of $E^t(1,2)$, where u_3 is a constant path.

Lemma 2.1 *$E^t_*(1,2)$ is a strong deformation retract of $E^t(1,2)$.*

Proof. Consider the homotopy

$$h : E^t(1,2) \times I \to E^t(1,2),\ (u_1, u_2, u_3, s) \mapsto (u_1, u_2, (1-s)u_3 + s[u_3]).$$

One sees that h is a deformation retraction of $E^t(1,2)$ onto $E^t_*(1,2)$. □

Observe that the deformation h of the preceding lemma induces a homotopy equivalence of $E^t(1,2)$ with $\Lambda'(\mathbb{F}_2(\mathbb{R}^{n+1}))$.

Lemma 2.2 *The projection*

$$\mathrm{proj}_{1,2} : E^t_*(1,2) \to \Lambda'\mathbb{F}_2(\mathbb{R}^{n+1})$$

is a fibration.

Proof. Define a lifting function λ as follows. Let $(u_1, u_2, x) \in E^t_*(1,2)$, and choose a path (w_1, w_2) in $\Lambda'\mathbb{F}_2(\mathbb{R}^{n+1})$, where $(w_1, w_2)(s) = (w_1(s), w_2(s))$, with $0 \le s \le 1$, and $w_1(s), w_2(s)$ are paths in $\mathbb{R}^{n+1}$ such that $(w_1(0), w_2(0)) = (u_1, u_2)$. Also, let $v(s) = (1/2)[w_1(s) + w_2(s)]$, so that $v(0) = (1/2)[u_1 + u_2]$ and

$$|x - v(0)| > (1/2)|[u_1 - u_2]| + \sigma(u) + \epsilon.$$

Set $\alpha = |x - v(0)| - (1/2)|[u_1 - u_2]| + \sigma(u) + \epsilon > 0$, and note that $\sigma(u) = r_1 + r_2$, where $r_i = \sqrt{(T/12)}\|\dot{u}\|_2$. Set

$$\lambda((u_1, u_2, x), w_1, w_2)(s) = (w_1(s), w_2(s), w_3(s)),\ 0 \le s \le 1,$$

where $w_3(s) = v(s) + M(s)(x - v(0))/|x - v(0)|$, with

$$M(s) =$$
$$1/2|[w_1(s) + w_2(s)]| + \sqrt{(T/12)}(\|\dot{w}_1(s)\|_2 + \|\dot{w}_2(s)\|_2 + \epsilon + \alpha.$$

It is easily verifiable that λ is the required lifting function. □

Lemma 2.3 *The fibrations*

$$\begin{aligned} \mathrm{proj}_{1,2} &: E^t_*(1,2) &&\to \Lambda'\mathbb{F}_2(\mathbb{R}^{n+1}) \\ \mathrm{proj}_1 &: \Lambda'\mathbb{F}_2(\mathbb{R}^{n+1}) \times A &&\to \Lambda'\mathbb{F}_2(\mathbb{R}^{n+1}) \end{aligned}$$

are fiber homotopically equivalent. Here, proj_1 *is projection on the first factor, and* $A = \{x \in \mathbb{R}^{n+1} \mid |x| > \epsilon\}$.

Proof. Let (q_1, q_2) be the basepoint in $\mathbb{F}_2(\mathbb{R}^{n+1})$ (see Chapter II, §1), and identify it with the corresponding pair of constant loops in $\Lambda'\mathbb{F}_2(\mathbb{R}^{n+1})$. Then, the fiber at the basepoint is $\{x \mid |x| > 1 + \epsilon\}$. Consider the map

$$\begin{array}{ccc} E_*^t(1,2) & \xrightarrow{\phi} & \Lambda'\mathbb{F}_2(\mathbb{R}^{n+1}) \times A \\ \downarrow & & \downarrow \\ \Lambda'\mathbb{F}_2(\mathbb{R}^{n+1}) & \xrightarrow{\mathrm{id}} & \Lambda'\mathbb{F}_2(\mathbb{R}^{n+1}), \end{array}$$

where $\phi(u_1, u_2, u_3) = (u_1, u_2, |[u_3] - (1/2)[u_1 + u_2]|)$. Note that ϕ is a map over $\Lambda'\mathbb{F}_2(\mathbb{R}^{n+1})$ and that its restriction on the fiber at (q_1, q_2) is a homotopy equivalence. According to [24, Dold], we can conclude that ϕ is, indeed, a fiber homotopy equivalence. □

The following corollaries are immediate, where homology (as is the case here and until the end of the section) has coefficients in $\mathbb{Z}_2$. Recall, first, the twisted product representation

$$\Lambda'\mathbb{F}_2(\mathbb{R}^{n+1}) \simeq S^n \rtimes \Omega(\mathbb{F}_2(\mathbb{R}^{n+1})).$$

Corollary 2.1 *$E^t(1,2)$ is homotopy equivalent to $\Lambda'\mathbb{F}_2(\mathbb{R}^{n+1}) \times S^n$. Furthermore, the generator in $H_n(S^)$ corresponds to $\alpha_{31} + \alpha_{32}$ (see Chapter VI, §1) in $H_*(\Lambda\mathbb{F}_3(\mathbb{R}^{n+1}))$, while the generators in $H_*(\Lambda'\mathbb{F}_2(\mathbb{R}^{n+1}))$ correspond to α_{21} and $\bar{\alpha}_{21}$ (see Chapter II, §1 and Chapter VIII, §1).*

Corollary 2.2 *$E^t(i,j)$, where $1 \le i < j \le 3$ is homotopy equivalent to $\Lambda'\mathbb{F}_2(\mathbb{R}^{n+1}) \times S^n$. Furthermore, the generators in $H_*(E^t(i,j))$ correspond to*

$$\alpha_{ki} + \alpha_{kj},\ \alpha_{ij},\ \text{and}\ \bar{\alpha}_{ij}$$

in $H_*(\Lambda\mathbb{F}_3(\mathbb{R}^{n+1}))$, where $i, j, k \in \{1,2,3\}$.

Studying the homology of the end $E^t(\Lambda')$ requires the study of the intersections

$$E^t(1,2) \cap E^t(1,3),\ E^t(1,2) \cap E^t(1,3) \cap E^t(2,3).$$

Proposition 2.1 *Let $E^t(12,13) = E^t(1,2) \cap E^t(1,3)$ and $E^t(1,2,3) = E^t(1,2) \cap E^t(1,3) \cap E^t(2,3)$. Then,*

(i) *the sets $E^t(12,13)$ and $E^t(1,2,3)$ are of the homotopy type of open sets in $\mathbb{F}_3(\mathbb{R}^{n+1})$, and*
(ii) *$H^m(E^t(12,13))$ and $H^m(E^t(1,2,3))$ are trivial for $m \ge 3(n+1)$.*

Proof. We consider the case $E^t(12,13)$ only, since the case of the triple intersection has a similar argument. If $u \in E^t(12,13)$, then it follows that the three Sobolev neighborhoods $N([u_i], r_i)$ for $i = 1,2,3$ are mutually disjoint and, hence, the homotopy

$$h : E^t(12,13) \times I \to E^t(12,13),$$

where $(u_1, u_2, u_3, s) \mapsto (1-s)(u_1, u_2, u_3) + s([u_1], [u_2], [u_3])$, yields a deformation retraction of $E^t(12,13)$ onto an open subset of $\mathbb{F}_3(\mathbb{R}^{n+1})$.

The second assertion follows from the general fact that the homology of open sets in any $\mathbb{R}^m$ vanishes in dimensions $\geq m$ (see [25, Dold]). □

The study of the homology of $E^t(\Lambda')$ leads to the set

$$A^t(\Lambda', \epsilon) = \\ \{u \in \Lambda' \mid \|[u_k] - (1/2)[u_i + u_j]\| > (3/2)\|[u_i - u_j]\| + \sigma(u) + \epsilon\}$$

for some permutation (i, j, k) of $(1, 2, 3)$, with $i < j$ and $\epsilon > 0$. The set $A^t(\Lambda')$ has a natural partition

$$A^t(\Lambda', \epsilon) = A^t(1,2) \cup A^t(1,3) \cup A^t(2,3),$$

where, for fixed $i < j$,

$$A^t(i,j) = \\ \{u \in \Lambda' \mid \|[u_k] - (1/2)[u_i + u_j]\| > (3/2)\|[u_i - u_j]\| + \sigma(u) + \epsilon\}.$$

Observe that

(*i*) $A^t(i,j) \subset E^t(i,j)$, and the inclusion is a homotopy equivalence;
(*ii*) the three sets $A^t(i,j)$, $i < j$, are mutually disjoint; and
(*iii*) Corollary 3.2 above remains valid when $A^t(i,j)$ is substituted for $E^t(i,j)$.

Proposition 2.2 *The inclusion* $j : A^t(\Lambda') \to E^t(\Lambda', \epsilon)$ *induces an isomorphism*

$$H_m(j) : H_m(A^t(\Lambda', \epsilon)) \to H_m(E^t(\Lambda', \epsilon))$$

for $m \geq 3(n+1)$.

Proof. Consider the morphism of Mayer-Vietoris sequences,

$$\begin{array}{ccc} 0 & & 0 \\ \downarrow & & \downarrow \\ H_m(A^t(1,2)) \oplus H_m(A^t(1,3)) & \longrightarrow & H_m(E^t(1,2)) \oplus H_m(E^t(1,3)) \\ \downarrow & & \downarrow \\ H_m(A^t(1,2) \cup A^t(1,3)) & \xrightarrow{H_m(\alpha)} & H_m(E^t(1,2) \cup E^t(1,3)), \\ \downarrow & & \downarrow \\ 0 & & 0 \end{array}$$

where $m \geq 3(n+1)$. One sees immediately that $H_m(\alpha)$ is an isomorphism. The final step is to consider the morphism

$$\begin{array}{ccc} 0 & & 0 \\ \downarrow & & \downarrow \\ H_m(A((12,13)) \oplus H_m(A(2,3)) & \rightarrowtail & H_m(E(12,13)) \oplus H_m(E(2,3)) \\ \downarrow & & \downarrow \\ H_m(A(12,13,23)) & \xrightarrow{H_m(\alpha)} & H_m(E(12,13,23)) \\ \downarrow & & \downarrow \\ 0 & & 0 \end{array}$$

between the pairs

$$(A^t(1,2) \cup A^t(1,3), A^t(2,3)), \text{ and } (E^t(1,2) \cup E^t(1,3), E^t(2,3)).$$

Here

$$\begin{array}{ll} A((ij,ik)) & = A^t(i,j) \cup A^t(i,k), \\ E((ij,ik)) & = E^t(i,j) \cup E^t(i,k), \\ A((12,13,23)) & = A^t(1,2) \cup A^t(1,3) \cup A^t(2,3), \text{ and} \\ E((12,13,23)) & = E^t(1,2) \cup E^t(1,3) \cup E^t(2,3). \end{array}$$

One easily sees that $H_m(j)$ is an isomorphism for $m \geq 3(n+1)$. □

In our applications in Chapter XIV, we make use of a control function $g : \Lambda' \to \mathbb{R}$ given by

$$g(u) = (1/6)(([u_1 - u_2])^2 + ([u_1 - u_3])^2 + ([u_2 - u_3])^2)$$

for $u \in \Lambda'\mathbb{F}_3(\mathbb{R}^{n+1})$. The function g plays the role of a distance function and also defines closed neighborhoods of infinity, using sets $\{u \mid g(u) \geq \phi(u)\}$ for appropriate ϕ's. The next result shows that some such neighborhoods of infinity are contained in topological ends $E^t(\Lambda', \epsilon)$.

Proposition 2.3 *Assume that* $\alpha(u) \geq (1/2)(\sum_{i=1}^3 r_i(u) + \epsilon)$. *Then,*

$$B(\alpha, \epsilon) = \{u \in \Lambda' \mid g(u) \geq 2(\sqrt{3}+1)^2(\alpha(u))^2\} \subset E_c^t(\Lambda', \epsilon).$$

The proof depends on a simple geometric lemma, which we consider first.

As illustrated in Fig. XII. 2, let $A = (-a, 0)$, $B = (a, 0)$ and $C = (x, y)$ be a given triangle $\triangle ABC$, with AB the shortest side. Define g by

$$6g = (x+a)^2 + y^2 + (x-a)^2 + y^2 + 4a^2.$$

Then, for a fixed g, $3g = x^2 + y^2 + 3a^2$ describes a circle that intersects the vertical axis in $D = (0, v)$, and AB remains the shortest side of $\triangle ABD$. Choose $b \geq a$ so that $3(2b)^2 = 6g$ and v' to make $\triangle EFG$ equilateral, with $E = (-b, 0)$, $F = (b, 0)$ and $G = (0, v')$. Hence, $v' \leq v$. Thus, $v' \leq v$ implies that $\sqrt{3}b \leq \sqrt{x^2 + y^2}$.

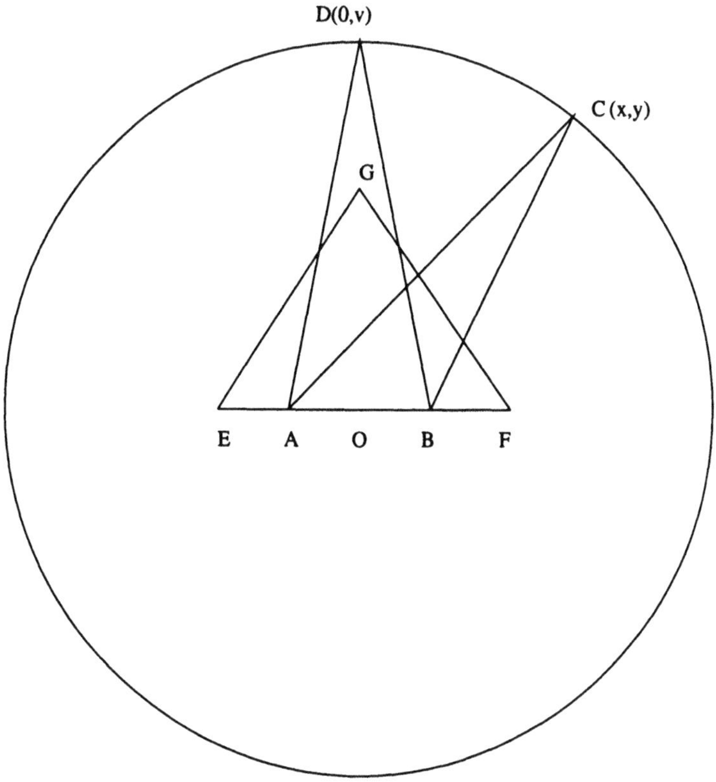

Fig. XIII.2. A Basic Triangle

Lemma 2.4 *Suppose that* $g \geq 2(\sqrt{3}+1)^2\alpha^2$. *Then,* $\sqrt{x^2+y^2} \geq 2\alpha + a$.

Proof. Observe that

$$x^2 + y^2 \geq 3b^2 = (3/2)g \geq 3(\sqrt{3}+1)^2\alpha^2.$$

Hence, $b \geq (\sqrt{3}+1)\alpha$, and the latter implies that $\sqrt{3}b \geq 2\alpha + b$. Thus, $\sqrt{x^2+y^2} \geq 2\alpha + a$, since $a \leq b$. □

Proof of Proposition 2.3

Take $u = (u_1, u_2, u_3)$ so that $g(u) \geq 2(\sqrt{3}+1)^2\alpha(u)^2$. Identify the triangle $([u_1],[u_2],[u_3])$ with ΔABC (see Fig. XIII. 2). We may assume that $|[u_1] - [u_2]|$ is the shortest side of ΔABC. Applying Lemma 2.4, we see that

$$\begin{aligned} |[u_3] - (1/2)[u_1 + u_2]| &\geq 2\alpha(u) + (1/2)|[u_1 - u_2]| \\ &\geq \textstyle\sum_{i=1}^3 r_i(u) + \epsilon + (1/2)|[u_1 - u_2]|. \end{aligned}$$

This establishes the proposition.

Observe here that $\alpha(u)$, being sufficiently smooth, implies that $B(\alpha,\epsilon)$ is a neighborhood retract in Λ'; hence, $(\Lambda', B(\alpha,\epsilon))$ is an ANR pair, as, for example, when

$$B(\alpha,\epsilon) = \{u | g(u) - 2(\sqrt{3}+1)^2(\alpha(u))^2 \leq 0\}$$

is a manifold the boundary of which admits a non-zero normal vector field.

Theorem 2.1 *Suppose that* $\alpha(u) \geq (1/2)((\sum_{i=1}^{3} r_i(u) + \epsilon))$. *Then*

$$\operatorname{cat}_{\Lambda'}(\Lambda', B(\alpha,\epsilon)) = \infty,$$

and Λ' *contains compact subsets of arbitrarily large* $\operatorname{cat}_{\Lambda'}(\Lambda', B(\alpha,\epsilon))$.

Proof. First we observe that $B(\alpha,\epsilon)$ is contained in the closed end $E_c^t(\Lambda',\epsilon)$. On the other hand,

$$E_c^t(\Lambda',\epsilon) = \bigcap_m U_m,$$

where U_m is the open end $E_o^t(\Lambda', \epsilon - 1/m)$. Hence, $E_c^t(\Lambda',\epsilon)$ is the inverse limit of the open ends U_m, and, therefore,

$$\lim_{m\to\infty} H^*(U_m) \cong \bar{H}^*(E_c^t(\Lambda',\epsilon)).$$

Observe that the connecting morphisms are bijective in dimensions $\geq 3(n+1)$. In a similar manner, we see that

$$\lim_{m\to\infty} H^*(\Lambda', U_m) \cong \bar{H}^*(\Lambda', E_c^t(\Lambda',\epsilon)).$$

By virtue of the results of Corollaries 1.1 and 5.2 of Chapter XII, it now follows that there exist in the diagram

$$\begin{array}{ccc} \bar{H}^*(\Lambda', E_c^t(\Lambda',\epsilon)) & \to & \bar{H}^*(\Lambda') \\ \downarrow & & \mathrm{id}\downarrow \\ \bar{H}^*(\Lambda', B(\alpha,\epsilon)) & \to & \bar{H}^*(\Lambda') \end{array}$$

nonzero q-fold products that pull back to $\bar{H}^*(\Lambda', E_c^t(\Lambda',\epsilon))$ for every q and, hence, to $\bar{H}^*(\Lambda', B(\alpha,\epsilon))$. Thus, $\operatorname{cl}_{\Lambda'}(\Lambda', E_c^t(\Lambda',\epsilon))$ is infinite and, therefore, $\operatorname{cat}_{\Lambda'}(\Lambda', B(\alpha,\epsilon))$ is also infinite (see Proposition 2.2). Now, note that the Alexander-Spanier cohomology $\bar{H}^*(\Lambda', B(\alpha,\epsilon))$ is isomorphic to the singular cohomology $H^*(\Lambda', B(\alpha,\epsilon))$. As singular homology has compact supports, the proof is complete. □

The sets $B(\alpha,\epsilon)$ in Theorem 2.1 are the precursors of the analytic ends, defined in the next chapter, where $\alpha(u)$ is expressed in terms of a given functional $f : \Lambda' \to \mathbb{R}$. We shall call $B(\alpha,\epsilon)$ *pre-analytic ends.*

3 Γ-category

We consider here a simple extension of the classical LS-category that is based on a given finite open collection $\Gamma = \{\Gamma_i\}$ of sets in the ambient space in question. It provides a suitable substitution for the Majer-Terracini collision index (see [40, Fadell-Husseini]).

The general setting for Γ-category is the following. Let $\Gamma = \{\Gamma_1, \cdots, \Gamma_m\}$ be a fixed family of open sets in an ANR-space Y. Let $A \subset \cap_{i=1}^m \Gamma_i$, be a closed subset.

Definition 3.1 *$\Gamma\text{cat}(A) \leq p$ if, and only if, there is a decomposition $A = \cup_{i=1}^m A_i$ where $A_i \subset \Gamma_i$ is a closed subset and $\text{cat}_{\Gamma_i}(A_i) \leq p$ for all i. Set $\Gamma\text{cat}(A) = p$, if $\Gamma\text{cat}(A) \leq p$ for some p and p is minimal with $\Gamma\text{cat} \leq p$. If no such p exists, we put $\Gamma\text{cat}(A) = \infty$.*

Γcat satisfies the usual basic properties.

Proposition 3.1 *Γcat satisfies the following properties:*

(i) *(monotonicity) if $A \subset B$, then $\Gamma\text{cat}(A) \leq \Gamma\text{cat}(B)$;*
(ii) *(continuity) $\Gamma\text{cat}(A) = p$, implies that there is an open set U such that $\Gamma\text{cat}(\bar{U}) = p$;*
(iii) *(subadditivity) $\Gamma\text{cat}(A \cup B) \leq \Gamma\text{cat}(A) + \Gamma\text{cat}(B)$;*
(iv) *(compactness) if A is compact, then $\Gamma\text{cat}(A) < \infty$;*
(v) *(supinvariance) suppose that $A \subset X$, $X \subset \cap_i \Gamma_i$, and let $H : X \times I \to X$ be a deformation with $H_0 = \text{id}$; then $\Gamma\text{cat}(A) \leq \Gamma\text{cat}(\overline{H_1(A)})$.*
(vi) *(refinement) suppose that $\Gamma = \{\Gamma_i\}$ is a refinement of $\Gamma' = \{\Gamma'_j\}$; then, $m\Gamma\text{cat}(A) \geq \Gamma'\text{cat}(A)$, where m is the cardinality of Γ.*

Proof. (i)-(v) follow easily from the analogous properties of the LS-category. To illustrate, let us prove (v). Suppose that $A = \cup_i A_i$ is a decomposition of A into closed subsets. Then,

$$\text{cat}_{\Gamma_i}(A_i) \leq \text{cat}_{\Gamma_i}(\overline{H_1(A_i)}).$$

To prove (vi), assume, without loss of generality, that $\Gamma\text{cat}(A) = p < \infty$. Then, $A = \cup_{i=1}^m A_i$, with $A_i \subset \Gamma_i$ and $\text{cat}_{\Gamma_i}(A_i) \leq p$. For each j, let B_j denote the union of all $A_i \subset \Gamma_i \subset \Gamma'_j$. Then, $\cup_j B_j = A$. Hence, $\text{cat}_{\Gamma_j}(B_j) \leq mp$. □

For an alternate way to view Γ-category, recall that a subset $B \subset Y$ is called *categorical in* Y if, and only if, B is contractible in Y to a point. Proceeding similarly, we introduce the following concept.

Definition 3.2 *A closed set $A \subset \cap_i \Gamma_i$ is called Γ-categorical if there is a decomposition $A = \cup_i A_i$ into closed sets, where each $A_i \subset \Gamma_i$ is categorical in Γ_i in the usual sense.*

Observe that if $\Gamma\text{cat}(A) = p$,one has the decompositions

$$\begin{aligned} A &= \cup_{i=1}^m A_i \\ &= \cup_{i=1}^m (\cup_{k=1}^p A_{i,k}) \\ &= \cup_{k=1}^p (\cup_i A_{i,k}), \end{aligned}$$

where each $A_{i,k}$ is contractible in Γ_i. Thus, we see that $\Gamma\text{cat}(A) \leq p$ if, and only if, $A = \cup_{k=1}^p B_k$ with each B_k being Γ-categorical. In particular, to say that $\Gamma\text{cat}(A) = 1$ is just to say that A is Γ-categorical.

Let us also observe that the condition that A_i be closed may be replaced by the condition that there be a cover $A = \cup_i A_i$, where A_i is an open subset of Γ_i.

As in the case of the classical LS-category, cup length in cohomology provides a lower bound for Γ-category. Let us recall how the cup length is defined. Suppose that X is connected and $A \subset X$ is closed. Then $\text{cl}_X(A)$ is defined to be the minimal integer k for which $H^*(i)(\alpha_1 \cdots \alpha_k) = 0$, for all k-fold products $\alpha_1 \cdots \alpha_k \in H^*(X;\mathbb{K})$, where each $\dim \alpha_j \geq 1$. Here $i : A \to X$ is the natural imbedding, and $H^*(X;\mathbb{K})$ is the Alexander-Spanier cohomology with coefficients in $\mathbb{K}$ ([101, Spanier]). Then, one shows in the standard manner that $\text{cl}_X(A) \leq \text{cat}_X(A)$, where $\text{cat}_X(A)$ is the classical LS-category.

Definition 3.3 *Let A be a closed subset of X. Then, say that $\text{cl}_\Gamma(A) \leq p$ if, and only if, there is a decomposition $A = \cup_i A_i$, where each A_i is closed in Γ_i, such that for all i $\text{cl}_{\Gamma_i}(A_i) \leq p$. Put $\text{cl}_\Gamma(A) = p$, if $\text{cl}_\Gamma(A) \leq p$ for some p, and p is minimal with respect to this property. As usual, put $\text{cl}_\Gamma(A) = \infty$ otherwise.*

Proposition 3.2 *$\text{cl}_\Gamma(A)$ is a lower bound for $\text{cat}_\Gamma(A)$.*

4 Strongly Admissible Sets

Our aim in this section is to characterize a class of admissible sets, as defined in [75, Majer-Terracini], by means of a suitable Γcat. To start, let M be a metric ANR, and put $Y = M^{\times k}$. Then, let $X = \mathbb{F}_k(M)$ be the k-configuration space of M, and define

$$\mathbb{F}_k(i,j) = \{x = (x_1, \cdots, x_k) | x_i \neq x_j\},$$

where i, j range over all pairs of indices such that $1 \leq i < j \leq k$. Note that these spaces are intermediate between $\mathbb{F}_k(M)$ and $M^{\times k}$ and that $\mathbb{F}_k(M) = \cap_{i<j} \mathbb{F}_k(i,j)$. The set $\{\mathbb{F}_k(i,j) | i < j\}$ is a Γ-family of open sets, as in §3.

Similarly, we can pass to the associated spaces of based and free loops: put,

$$\Omega\Gamma = \{\Omega\mathbb{F}_k(i,j) = \Omega_{ij}\},\ \Lambda\Gamma = \{\Lambda\mathbb{F}_k(i,j) = \Lambda_{ij}\}.$$

Clearly, $\Omega\mathbb{F}_k(M) = \cap_{i<j}\mathbb{F}_k(i,j)$ and $\Lambda\mathbb{F}_k(M) = \cap_{i<j}\Lambda\mathbb{F}_k(i,j)$.

Lemma 4.1 *Let $A \subset \Omega\mathbb{F}_k(M)$ be a closed subset. Then the inequality $\Lambda\Gamma\mathrm{cat}(A) = \Omega\Gamma\mathrm{cat}(A)$ holds.*

Proof. Let $A = \cup A_{ij}$, where each $A_{ij} \subset \Omega(i,j) = \Omega(\mathbb{F}_k(i,j)$. Consider now the fibration $\Lambda\mathbb{F}_2(i,j) \to \mathbb{F}_2(i,j)$, with fiber $\Omega\mathbb{F}_2(i,j)$. As it has a section, the trivial section, for example, we conclude, by virtue of [35, Fadell-Husseini], that

$$\mathrm{cat}_{\Lambda_{ij}}(A) = \mathrm{cat}_{\Omega_{ij}}(A)$$

holds for all $i < j$. □

The analogue of $\Lambda\mathbb{F}_k(i,j)$ in $\mathcal{H}$ is the space

$$\Lambda'_{ij} = \{u \in \mathcal{H} | u_i(t) \neq u_j(t),\ \forall t \in \mathbb{R}\},$$

and the corresponding family open Γ collection is the family $\Lambda'\Gamma = \{\Lambda'_{ij}\}$. According to [88, Palais], the natural imbedding

$$\phi : \mathcal{H} \to C^0_T(\mathbb{R}, \mathbb{R}^{k(n+1)}) = C^0(S^1, \mathbb{R}^{k(n+1)})$$

induces a homotopy equivalence $\phi_0 : \Lambda'\mathbb{F}_k(\mathbb{R}^{n+1})) \to \Lambda(\mathbb{F}_k(\mathbb{R}^{n+1}))$. Note that ϕ induces a homotopy equivalence $\phi_{ij} : \Lambda'_{ij} \to \Lambda_{ij}$, where $\Lambda' = \Lambda'\mathbb{F}_k(\mathbb{R}^{n+1}) = \cap_{i<j}\Lambda'_{ij}$. Also, denote by $D(\mathcal{H})$ the set of deformations

$$h : \mathcal{H} \times I \to \mathcal{H}$$

from $h_0 = \mathrm{id}$ to $h_1 \in \mathbb{R}^{k(n+1)}$, the space of constant functions.

The importance of the next lemma we explained in the introduction to this chapter. Recall that, broadly speaking, the work in [8, Bahri-Rabinowitz] leads one to expect that the solutions to problems of the 3-body type consist of those configurations of bodies that interact with each other and do not separate into subclusters. In the space of free loops Λ', these subclusters consist of interlocking loops that cannot be separated without causing collisions to occur. Thus, the negligible sets are those where the bodies can be separated without causing collision. The following lemma shows that such sets are, in fact, negligible, if measured by the $\Lambda'\Gamma$ category.

Lemma 4.2 *Suppose that $A \subset \Lambda'\mathbb{F}_k(\mathbb{R}^{n+1})$ is a closed set for which there is an*

$$h : \mathcal{H} \times I \to \mathcal{H}$$

in $D(\mathcal{H})$ with the property that, for every $u \in A$, there is a pair i, j, $i < j$ depending on u such that, for every $s \in [0,1]$ and $t \in \mathbb{R}$, one has $h_i(u,s)(t) \neq h_j(u,s)(t)$. Then $\Lambda'\Gamma\mathrm{cat}(A) \leq 2$.

Proof. Let

$$V_{ij} = \{u \in \mathcal{H} | h_i(u,s)(t) \neq h_j(u,s)(t), \ \forall s,t\}.$$

Observe that V_{ij} is an open set deformable into $\mathbb{F}_k(i,j)$. But the latter is of the homotopy type of $\mathbb{F}_2(\mathbb{R}^{n+1})$, which, in turn, is homotopy equivalent to S^n. As $\text{cat}(S^n) = 2$, it follows that each V_{ij} is the union of two open categorical sets in Λ_{ij}. Hence, $\Lambda'\Gamma\text{cat}(A) \leq 2$. □

Lemma 4.2 suggests the following concept.

Definition 4.1 *A subset $A \subset \Lambda'\mathbb{F}_k(\mathbb{R}^{n+1})$ will be called* strongly admissible *if, and only if, for every $h \in D(\mathcal{H})$, there is a $u \in A$ with the following property: for every $i < j$, there exists an $s_{ij} \in I = [0,1]$ and a $t_{ij} \in \mathbb{R}$ such that $h_i(u, s_{ij}(t_{ij}) = h_j(u, s_{ij})(t_{ij})$, where $h = (h_1, \cdots, h_k)$.*

See Remark (5.1) of [75, Majer-Terracini] regarding the preceding notion. Strongly admissible sets are, roughly speaking, those in which any attempt to separate the loops by means of homotopies in $D(\mathcal{H})$ results in a collision somewhere, someplace. The concept is more restrictive than that of admissible sets in [75, Majer-Terracini] (see §5 below).

Corollary 4.1 *Suppose that $\Lambda\Gamma'\text{cat}(A) > 2$, where $A \subset \Lambda'\mathbb{F}_k(\mathbb{R}^{n+1})$. Then, A is strongly admissible.*

Proof. Assume otherwise. Then there is a homotopy $h \in D(\mathcal{H})$ such that, for every $u \in A$, there is a pair $i, j \in \{1, \cdots, k\}$ with the property that, for every $s \in [0,1]$ and $t \in \mathbb{R}$,

$$h_i(u,s)(t) \neq h_j(u,s)(t).$$

According to the previous lemma, $\Lambda\Gamma'\text{cat}(A) \leq 2$, which contradicts the assumption on A. □

Note also that, by virtue of Lemma 4.2, sets that are not strongly admissible and, in particular, nonadmissible sets have Γ-category ≤ 2. Hence, sets of large Γ-category are admissible.

We now apply the above corollary to the following calculation, which is basic to the Majer-Terracini method. Recall that, for $u \in \mathcal{H}$, $\|u\|_2$ is the L_2-norm, while $\|u\|_\infty = \sup_t\{|u_i(t)|\}$.

Proposition 4.1 *Let*

$$E = \{u \in \mathbb{F}_k(\mathbb{R}^{n+1}) | \frac{1}{2k} \sum_{i \neq j} [u_i - u_j]^2 \geq \frac{T}{6} \|\dot{u}\|_2^2\}.$$

Then, $\Lambda'\Gamma\text{cat}(E) \leq 2$.

Proof. Suppose, to the contrary, that $\Lambda'\text{cat}(E) > 2$, and consider the deformation $h = (h_1, \cdots, h_k) \in D(\mathcal{H})$, defined by the rule

$$h_i(u,s)(t) = (1-s)u_i(t) + s[u_i].$$

Note that h deforms linearly the loop u_i to the constant loop $[u_i]$. Then, according the preceding corollary, there is a $u \in E$ with the property that, for every $i < j$, there is an s_{ij} and t_{ij} such that

$$(1-s_{ij})u_i(t_{ij}) + s_{ij}[u_i] = (1-s_{ij})u_j(t_{ij}) + s_{ij}[u_j].$$

Rewriting the equation, we obtain

$$(1-s_{ij})(u_i(t_{ij}) - u_j(t_{ij}) - [u_i - u_j]) = [u_i - u_j].$$

Using the Sobolev inequality and the latter equation, we see that

$$|[u_i - u_j]|^2 < \|u_i - u_j - [u_i - u_j]\|_\infty^2 \le \frac{T}{12}\|\dot{u}_i - \dot{u}_j\|_2^2.$$

Summing over all pairs $i \neq j$, we conclude that

$$\sum_{i\neq j}[u_i - u_j]^2 < \frac{T}{12}\sum_{i\neq j} 2(\|u_i\|_2^2 + \|u_j\|_2^2) \le \frac{T}{6}2(k-1)\|\dot{u}\|_2^2.$$

Hence, $(1/2k)\sum_{i\neq j}[u_i - u_j]^2 < (T/6)\|\dot{u}\|_2^2$. But this is a contradiction. □

The expression appearing in the last line of the proof above, defines an important control function,

$$g : \Lambda'(\mathbb{F}_k(\mathbb{R}^{n+1})) \to \mathbb{R}, \quad u \mapsto (1/2k)\sum_{i\neq j}[u_i - u_j]^2.$$

Using the function g, we obtain the following restatement of the preceding result.

Corollary 4.2 *Let*

$$E = \{u \in \Lambda'(\mathbb{F}_k(\mathbb{R}^{n+1})) | g(u) \ge (T/6)\|\dot{u}\|_2^2\}.$$

Then $\Lambda'\Gamma\text{cat}(E) \le 2$.

The sets defined in the corollary play the role of neighborhoods of infinity.

Now, consider a C^1-functional $f : \Lambda'(\mathbb{F}_k(\mathbb{R}^{n+1})) \to \mathbb{R}$ of the form

$$f(u) = \frac{1}{2}\int_0^T |\dot{u}|^2 dt - \int_0^T V(u,t)dt,$$

where $V(u,t) \le 0$ for all u and t, with $0 \le t \le T$.

Corollary 4.3 *Let*

$$E_f = \{u \in \Lambda'(\mathbb{F}_k(\mathbb{R}^{n+1})) | g(u) \geq \frac{T}{3} f(u)\}.$$

Then, $\Lambda'\Gamma\mathrm{cat}(E_f) \leq 2$.

Proof. Observe that $E_f \subset E$. Appealing to the monotone property of $\Lambda'\Gamma\mathrm{cat}$, we verify the assertion. □

There is another set of interest associated with E_f : the intersection of a lower and an upper level set. Namely,

$$E_{c,b} = \{u \in \Lambda'(\mathbb{F}_k(\mathbb{R}^{n+1})) | f(u) \leq c,\ g(u) \geq b\}.$$

Corollary 4.4 *If* $c > 0$ *and* $b \geq (T/3)c$, *then* $\Lambda'\Gamma\mathrm{cat}(E_{c,b}) \leq 2$.

Proof. Observe that $g(u) \geq b \geq (T/3)c \geq (T/3)f(u)$. Hence, $E_{c,b} \subset E_f$, and the previous corollary implies the rest. □

It is important to know that the loop spaces, based or not, contain compact subsets of arbitrarily high Γ-category. This question is taken care of by the following. Consider the fibration

$$\mathrm{proj}_{k,2} : \mathbb{F}_k(\mathbb{R}^{n+1}) \to \mathbb{F}_2(\mathbb{R}^{n+1}),$$

where $(x_1, \cdots, x_k) \mapsto (x_1, x_2)$. Let σ_k be the section for $\mathrm{proj}_{k,2}$ given by

$$\sigma_k : \mathbb{F}_2(\mathbb{R}^{n+1}) \to \mathbb{F}_k(\mathbb{R}^{n+1}),\quad (x_1, x_2) \mapsto (x_1, x_2, x_3', \cdots, x_k'),$$

with $x_j' = x_1 + 4(j-1)(x_2 - x_1)$, for $j \geq 3$. Note that $\sigma_k(x_1, x_2) \in \mathbb{F}_k(L(x_1, x_2))$,where $L(x_1, x_2)$, is the half line from x_1 to x_2. Put $\mathbb{F}_2^*(\mathbb{R}^{n+1}) = \sigma_k(\mathbb{F}_2(\mathbb{R}^{n+1}))$. Observe here that σ_k is an imbedding with $\mathrm{proj}_{k,2}$ as its inverse.

Lemma 4.3 *Let* $C^* \subset \mathbb{F}_2^*(\mathbb{R}^{n+1})$, *and suppose that* C^* *is contractible in* $\mathbb{F}_k(i,j)$. *Then* $C = \sigma_k^{-1}(C^*)$ *is contractible in* $\mathbb{F}_2(\mathbb{R}^{n+1})$.

Proof. Denote by $\mathrm{proj}_{i,j} : \mathbb{F}_k(i,j) \to \mathbb{F}_2(\mathbb{R}^{n+1})$ the projection where the tuple $(x_1, \cdots, x_k)$ goes to (x_i, x_j). Note that $\mathrm{proj}_{i,j} \circ \sigma_k : \mathbb{F}_2(\mathbb{R}^{n+1}) \to \mathbb{F}_2(\mathbb{R}^{n+1})$ is homotopic to the identity. For, writing

$$\sigma_k(x_1, x_2) = (x_1, x_2, x_3', \cdots, x_k'),$$

where $x_3', \cdots, x_k' \in L(x_1, x_2)$, one sees that a simple homotopy along $L(x_1, x_2)$ will slide (x_1, x_2) to (x_i, x_j). Next, let $G : C^* \times I \to \mathbb{F}_k(i,j)$ be a homotopy contracting C^* to a point, and define $\hat{G}$ by

$$\hat{G} : C \times I \to \mathbb{F}_2(\mathbb{R}^{n+1}),\ ((x_1, x_2), t) \mapsto \mathrm{proj}_{i,j}(G(\sigma_k(x_1, x_2), t)).$$

Then, $\hat{G}_0 = \mathrm{proj}_{i,j} \circ \sigma_k$ and $\hat{G}_1$ is constant; but, there is a homotopy from the identity on C to $\mathrm{proj}_{i,j} \circ \sigma_k$. Combining the latter homotopy with $\hat{G}$, one obtains a contraction of C. □

The following proposition describes the relation of LS-category to the Γcat.

Proposition 4.2 *Let* $\Gamma = \{\mathbb{F}_k(i,j) \mid 1 \le i < j \le k\}$. *Then*

$$\mathrm{cat}_{\mathbb{F}_2(\mathbb{R}^{n+1})}(A) \le \frac{k(k-1)}{2} \Gamma\mathrm{cat}(A^*)$$

where $A \subset \mathbb{F}_2(\mathbb{R}^{n+1})$, *and* $A^* = \sigma_k(A)$.

Proof. Suppose $\Gamma\mathrm{cat}(A^*) = t < \infty$, and let $A^* = \cup_{i<j} B^*_{ij}$ be a decomposition into closed sets, where, for each $i < j$, $\mathrm{cat}_{\mathbb{F}_2(i,j)}(B_{ij}) \le t$. Now, each B^*_{ij} can be decomposed as the union, $\cup_{m=1}^{t} B^*_{ij}(m)$, of sets $B^*_{ij}(m)$ contractible in $\mathbb{F}_k(i,j)$. Lemma 4.2 implies that the set $B_{ij}(m) = \sigma_k^{-1}(B^*_{i,j}(m))$ is contractible in $\mathbb{F}_2(\mathbb{R}^{n+1})$. But $A = \cup_{i,j,m} B_{ij}(m)$. So the proposition follows. □

Recall that we are assuming throughout Part Three that $n > 2$. Thus $\mathbb{F}_k(\mathbb{R}^{n+1})$ is simply connected, and, therefore, the associated based and free loop spaces $\Omega(\mathbb{F}_k(\mathbb{R}^{n+1}))$ and $\Lambda(\mathbb{F}_k(\mathbb{R}^{n+1}))$ are path connected. This implies clearly that the point to which the space is contracted may be any point in the space.

The section $\sigma_k : \mathbb{F}_2(\mathbb{R}^{n+1}) \to \mathbb{F}_k(\mathbb{R}^{n+1})$ induces sections

$$\begin{aligned} \Omega(\sigma_k) &: \Omega(\mathbb{F}_2(\mathbb{R}^{n+1})) \to \Omega(\mathbb{F}_k(\mathbb{R}^{n+1})) \\ \Omega'(\sigma_k) &: \Omega(\mathbb{F}_2(\mathbb{R}^{n+1})) \to \Omega'(\mathbb{F}_k(\mathbb{R}^{n+1})) \\ \Lambda(\sigma_k) &: \Lambda(\mathbb{F}_2(\mathbb{R}^{n+1})) \to \Lambda(\mathbb{F}_k(\mathbb{R}^{n+1})) \\ \Lambda'(\sigma_k) &: \Lambda'(\mathbb{F}_2(\mathbb{R}^{n+1})) \to \Lambda'(\mathbb{F}_k(\mathbb{R}^{n+1})), \end{aligned}$$

where Ω' is the subset of Λ' consisting of the loops based at a particular point. Put

$$\begin{aligned} \Lambda\Gamma &= \{\Lambda'_{ij} = \Lambda\mathbb{F}_k(i,j)\} \\ \Lambda'\Gamma &= \{\Lambda'_{ij} = \Lambda'\mathbb{F}_k(i,j)\}. \end{aligned}$$

Proposition 4.3 *Let* $A \subset \Lambda(\mathbb{F}_2(\mathbb{R}^{n+1}))$ *be a closed subset, and denote by* $A^* \subset \Lambda(\mathbb{F}_k(\mathbb{R}^{n+1}))$ *its image under the section* $\Lambda(\sigma_k) : \Lambda(\mathbb{F}_2(\mathbb{R}^{n+1})) \to \Lambda(\mathbb{F}_k(\mathbb{R}^{n+1}))$. *Then,*

$$\mathrm{cat}_{\Lambda(\mathbb{F}_2(\mathbb{R}^{n+1}))}(A) \le \frac{k(k-1)}{2} \Lambda\Gamma\mathrm{cat}(A^*),$$

where $A^* = \sigma_k(A)$. *Moreover, the statement above remains valid if* Λ *is replaced by* Λ', Ω, *and* Ω' *throughout.*

The proof is entirely similar to that of the preceding proposition and is left for the reader.

Corollary 4.5 *The spaces* $\Lambda\mathbb{F}_k(\mathbb{R}^{n+1})$ *and* $\Lambda'\mathbb{F}_k(\mathbb{R}^{n+1})$ *contain compact subsets of arbitrarily high* $\Lambda\Gamma$- *and* $\Lambda'\Gamma$-*category.*

Proof. The space $\Lambda\mathbb{F}_k(\mathbb{R}^{n+1})$ contains compact subsets A of arbitrarily high LS-category, since it has infinite cup length over $\mathbb{Z}_2$ ([37, Fadell-Husseini]). The same is certainly true for $\Lambda'\mathbb{F}_k(\mathbb{R}^{n+1})$, since they are of the same homotopy type. The rest follows from the preceding proposition.

Alternatively, observe that $\Omega(\mathbb{F}_2(\mathbb{R}^{n+1}))$ has compact subsets A of arbitrarily high category. Since

$$\mathrm{cat}_{\Omega(\mathbb{F}_2(\mathbb{R}^{n+1}))}(A) = \mathrm{cat}_{\Lambda(\mathbb{F}_2(\mathbb{R}^{n+1}))}(A),$$

([35, Fadell-Husseini]), the rest follows. □

5 Historical Notes

The definition of neighborhoods of infinity in [8, Bahri-Rabinowitz], takes into account the level of the functional $f : \mathcal{H} \to \mathbb{R}$ as well as the distance of a particle u from its average $[u]$. The former depends on both potential and kinetic energies, while the latter is determined by kinetic energy only. Their neighborhoods of infinity thus resemble our sets E_f of §4. Our definition of topological ends restricts the kinetic energy to a degree that would allow one of the three bodies to break off from the other two, but it says nothing directly about the potential energy. Our definition of analytic ends requires that the distances between the various pairs of bodies be sufficiently large that, in a manner of speaking, the gradient of f will increase g. This is very important for the arguments in Chapter XIV showing the existence of critical values.

The point of view of [8, Bahri-Rabinowitz] is that also of [91, 92, 93, Riahi] in the study of problems of the 4-body type and higher. Our concept of ends, both topological and analytic, can be extended to these cases. We elected not to carry out such a program here: it would have taken us outside the scope of this monograph, and we just wanted to show how some of our results can be usefully applied.

Lusternik-Schnirelman category is a topological tool that specifies certain subsets of a manifold M, analogous to the unstable manifolds of critical points defined by a Morse function, for the purpose of finding the critical values and estimating the number of critical points of a functional $f : M \to \mathbb{R}$. The first relative form of this appears in [90, Reeken]. Other versions related to it are given in [40, Fadell-Husseini], as well as in [47, Fournier-Willem] and [106, Szulkin]. These fail to be subadditive in the usual sense, in contrast to the version given here, but they satisfy an alternative version that involves both relative and absolute category.

The form given here is based on a notion that first appears in §11 of [31, Fadell] and is studied further in [89, Ramsey]. To explain how it is different, let us recall the definition in this chapter: given a pair of spaces (Y, A) and a closed subset $X \subset Y$, a set C is said to be *categorical in* Y *relative to* A if C can be deformed into A relative to $A \cap C$; then, $\text{cat}_Y(X; A) = n$ if there is a closed cover $\{C_1, \cdots, C_n\}$ of X such that each C_i is categorical in Y relative to A. The variants of relative category introduced in [90, Reeken], [47, Fournier-Willem], [106, Szulkin], and [40, Fadell-Husseini] require only that C_1 be categorical in Y relative to A, while the other C_i's, for $i \geq 2$, are assumed to be contractible to a point in Y. All of these concepts, as well as their equivariant and cohomological versions, are now referred to as index theories; see [42, Fadell-Husseini-Rabinowitz] and [33, 34, Fadell-Husseini].

The concept of Γ-category is motivated by the notion of the collision index introduced in [76, Majer-Terracini]. This collision index uses a generalized notion of collision, as well as other technicalities, that make it difficult to handle. Precisely, given a pair $(h, u) \in D(\mathcal{H}) \times \Lambda'$, define a relation on the set of indices $\{1, \cdots, k\}$ of $h = (h_1, \cdots, h_k) \in D(\mathcal{H})$ according to the rule,

$$(*) \qquad i r_{h,u} j \equiv h_i(u, s_{ij})(t_{ij}) = h_j(u, s_{ij})(t_{ij})$$

for some $s_{ij} \in [0, 1]$ and some $t_{ij} \in \mathbb{R}$. Our *strongly admissible* sets are precisely those that consist of $u \in \Lambda'$ for which there is an $h \in D(\mathcal{H})$ such that the relation (*) is satisfied for every pair i, j. In [76, Majer-Terracini], the relation (*) is used generate a transitive relation denoted by $R_{h,u}$. Then they call a closed set $A \subset \Lambda'$ *admissible* if, and only if, for every $h \in D(\mathcal{H})$, there is a $u \in A$ such that $i R_{h,u} j$ for every pair $i < j$. With this concept in hand, they define the collision index $\text{i}(A)$ as the minimal integer p for which A is the union of p nonadmissible closed sets of Λ'. Thus the nonadmissible sets are the analogue of categorical sets.

In [78, Majer-Terracini], one finds an approach using the other relative category theory of [90, Reeken], [47, Fournier-Willem], [106, Szulkin], and [40, Fadell-Husseini]. It is based on the concept of an alternative admissible set that does not allow collisions with the first body.

XIV. Problems of k-body Type

Our basic objective here is to illustrate how the results of Chapter XIII can be used in the study of problems of the k-body type. More specifically, we discuss two examples to show how our results relate to the studies of the 3-body type in [8, Bahri-Rabinowitz] and of the k-body type in [75, 76, 77, 78, Majer-Terracicni].

With the notation being as in the previous chapters, and using variational methods, we consider functionals $f : \Lambda' \to \mathbb{R}$ of the form

$$f(u) = \frac{1}{2}\int_0^T |\dot{u}|^2 dt - \int_0^T \sum_{i \neq j} V_{ij}(u_i - u_j, t)dt$$

associated with dynamical systems of k-body type $\ddot{u} + V'(x) = 0$, with potential V of the form

$$V = \sum_{i \neq j} V_{ij}(x_i - x_j), \ \ 1 \leq i, j \leq k.$$

In the works cited, above it is shown that the functional f has critical points, but with the conditions on the components V_{ij} that differ somewhat in the different works. For our illustrations, we have chosen to assume the conditions in [77, 78, Majer-Terracini].

In the first example, we use our concept of topological end, $E_c^t(\Lambda', \epsilon)$, which resembles the notion of neighborhood of infinity in [8, Bahri-Rabinowitz] but, as we have seen, has the advantage that $\mathrm{cat}_{\Lambda'}(\Lambda', E_c^t(\Lambda', \epsilon))$ is computable through cohomology. Here, the conditions on the V_{ij} are as in [78, Majer-Terracicni] and $k = 3$ as in [8, Bahri-Rabinowitz]. In the second example, we follow [77, Majer-Terracini], but we illustrate how to replace the concept of collision index by the more natural concept of Γ-category. Both examples are presented in outline form, with suitable references for the details.

We have seen in Chapter XIII that any reasonable candidate E for an end in $\Lambda' = \Lambda'(\mathbb{F}_3(\mathbb{R}^{n+1}))$ is a subspace that is not only rich in homology but also has infinite cup product length and, hence, is of infinite category. To deduce that the relative category $\mathrm{cat}_{\Lambda'}(\Lambda'; E)$ remains infinite requires a close analysis of the homology of Λ'. Also, in a certain sense, $E_c^t(\Lambda', \epsilon)$ is, in fact, a universal end, for one easily sees that the proof of Theorem 2.1 in Chapter XIII implies that $\mathrm{cat}_{\Lambda'}(\Lambda', B) = \infty$ for any closed ANR, $B \subset \Lambda'$ such that $B \subset E_c^t(\Lambda', \epsilon)$, where $E_c^t(\Lambda', \epsilon)$ is a topological end (see §2 of Chapter XIII). Thus, we see that Λ' has a multitude of compact subsets, which are the support of nonzero cohomology products that, in turn, vanish on $E_c^t(\Lambda', \epsilon)$.

In our first example from [75, 76, 77, 78, Majer-Terracini], using our results of Chapter XII on the homology and cohomology of $\Lambda'(\mathbb{F}_3(\mathbb{R}^{n+1}))$, we prove that $\mathrm{cat}(\Lambda', B) = \infty$ for an appropriate subset B of the topological end $E_c^t(\Lambda', \epsilon)$. We note here that in [78, Majer-Terracini] this fact is established by a clever argument that requires only that the category of $\Lambda'(\mathbb{F}_2(\mathbb{R}^{n+1}))$ is infinite but avoids a deeper study of the cohomology of $\Lambda'(\mathbb{F}_3(\mathbb{R}^{n+1}))$ and of the end $E_c^t(\Lambda', \epsilon)$. We hope that our alternative approach may prove useful in future problems.

1 Analytic Ends

In this chapter we keep the notation of Chapter XIII. We are interested in sets of the form

$$B(\phi) = \{u \in \Lambda' | g(u) \geq \phi(\|\dot{u}\|_2^2)\},$$

where $\phi : \mathbb{R}^+ \to \mathbb{R}^+$ is sufficiently smooth, as introduced in [78, Majer-Terracini]. Observe that $B(\phi)$ is a manifold with boundary.

First, we wish to modify the pre-analytic ends of §2 of Chapter XIII to accommodate the analytic needs of a suitable deformation lemma. So, let

$$\rho(u) = \frac{1}{2}\sqrt{T}(\sum_{i=1}^{3} \|\dot{u}\|_2^2)^{1/2},$$

where $u \in \Lambda'$.

Assertion 1.1 *The inequality $\rho(u) \geq \sigma(u)$ holds.*

Proof. Recall that

$$\sigma(u) = \sqrt{\frac{T}{12}} \sum_{i=1}^{3} \|\dot{u}\|_2.$$

A simple argument yields the desired inequality. □

Next, put

$$E_c^t(\Lambda', \epsilon, \rho) = \{u \in \Lambda' | \, |u_k - \frac{1}{2}|[u_i - u_j]| \geq \frac{1}{2}|[u_i - u_j]| + \rho(u) + \epsilon\}$$

for some permutation (i, j, k) of $(1, 2, 3)$, where $i < j$ and $\epsilon > 0$. The following proposition is similar to Proposition 2.3 of Chapter XIII, and its proof is along the same lines.

Proposition 1.1 *If* $\alpha(u) \geq \frac{1}{2}(\rho(u)+\epsilon)$, *then*

$$B(\alpha,\epsilon) \subset E_c^t(\Lambda',\epsilon,\rho) \subset E_c^t(\Lambda,\epsilon),$$

where

$$B(\alpha,\epsilon) = \{u \in \Lambda' | g(u) \geq 2(\sqrt{3}+1)^2(\alpha(u))^2\}.$$

Corollary 1.1 *Let*

$$\phi_1(s) = \frac{1}{2}(\sqrt{3}+1)^2(\frac{T}{4}s + \sqrt{T}\epsilon\sqrt{s} + \epsilon^2).$$

If $\phi : \mathbb{R}^+ \to \mathbb{R}^+$ *is a function with* $\phi \geq \phi_1$, *then,*

$$B(\phi,\epsilon) = \{u \in \Lambda' | g(u) \geq \phi(\|\dot{u}\|_2^2)\} \subset E_c^t(\Lambda',\epsilon).$$

The corollary follows easily: note that $\epsilon > 0$ is arbitrary.

Corollary 1.2 *If* $B(\phi,\epsilon) = \{u | \Phi(u) = g(u) - \phi(\|\dot{u}\|_2^2) \geq 0\}$ *denotes an ANR (manifold with boundary), where* $\phi \geq \phi_1$, *then* Λ' *contains subsets* $A \subset \Lambda'$ *of arbitrarily high category* $\mathrm{cat}_\Lambda(A, B(\phi,\epsilon))$.

Definition 1.1 *If* $\phi : \mathbb{R}^+ \to \mathbb{R}^+$, *the set* $B(\phi) = \{u | g(u) \geq \phi(u)\}$ *is called an analytic end.*

As an additional example, suppose that $f : \Lambda' \to \mathbb{R}$ is a functional of the form

$$f(u) = \frac{1}{2}\int_0^T \dot{u}|^2 dt - \int_0^T V(u,t)dt,$$

where $V(u,t) \leq 0$. It is easy to show that, if

$$\phi(u) = 2(\sqrt{3}+1)^2(\sqrt{(T/2)f(u)} + \epsilon)^2, \quad \epsilon > 0$$

and $B(\phi) = \{u | g(u) \geq \phi(u)\}$, then $B(\phi)$ is an analytic end and $B(\phi) \subset E_c^t(\Lambda',\epsilon)$.

Now, speaking more generally, let $\Lambda' = \Lambda'(\mathbb{F}_k(\mathbb{R}^{n+1}))$. Both of the examples we give in this chapter involve functionals f of the above form, where, more specifically, the potential function $V(u,t)$ has the form

$$V(u,t) = \sum_{i \neq j} V_{ij}(u_i - u_j, t), \; 1 \leq i \neq j \leq k.$$

In these cases the functional is invariant under translations in $\mathbb{R}^{n+1}$, i.e., $f(u+a) = f(u_1+a, \cdots, u_k+a) = f(u)$ for all $a \in \mathbb{R}^{n+1}$. Moreover, analytic ends of the form $\{u | g(u) \geq \phi(|\dot{u}|_2^2)\}$ or $\{u | g(u) \geq \phi(f(u))\}$ are also invariant under this action. Furthermore, several of the results required in the proofs are valid modulo translations or, equivalently, in $\Lambda'/\mathbb{R}^{n+1}$. This orbit space may be identified as follows. Let $\mathcal{H}_0 \subset \mathcal{H} = W_T^{1,2}(\mathbb{R}^{(n+1)k})$ denote the subspace defined by

$$\mathcal{H}_0 = \{u \in \mathcal{H} | \sum_{i=1}^{k} [u_i] = 0\}$$

and $\pi : \mathcal{H} \to \mathcal{H}_0$, the orthogonal projection given by $\pi(u) = u = \bar{x}$, where

$$(\bar{x})_i = x = \frac{1}{k} \sum_{i=1}^{k} [u_i], \quad 1 \leq i \leq k.$$

Observe that the map π induces a bijection $\mathcal{H}/\mathbb{R}^{n+1} \to \mathcal{H}_0$. Restricting the map to Λ', we obtain the identification

$$\Lambda'/\mathbb{R}^{n+1} \to \Lambda'_0 = \mathcal{H}_0 \cap \Lambda'.$$

Note next that the projection π induces a retraction $\pi_0 = \pi/\Lambda' : \Lambda' \to \Lambda'_0$ that is, in fact, a strong deformation retract of Λ' onto Λ'_0. Therefore, in the applications to follow, we consider $f|\Lambda'_0 : \Lambda'_0 \to \mathbb{R}$ rather than f itself. Of course, one needs to check that the results for Λ'_0 carry over from Λ'. Here is an example of this in the case where $\Lambda' = \Lambda'(\mathbb{F}_3(\mathbb{R}^{n+1}))$.

Proposition 1.2 *Let $B(\phi) = \{u | g(u) \geq \phi(u)\}$ denote an analytic end, where ϕ is invariant under translation in $\mathbb{R}^{n+1}$. Then,*

$$\mathrm{cat}_{\Lambda'(\mathbb{F}_3(\mathbb{R}^{n+1}))}(A; B(\phi)) = \mathrm{cat}_{\Lambda'_0(\mathbb{F}_3(\mathbb{R}^{n+1}))}(\pi_0(A); \pi_0(B(\phi))).$$

Consequently, when $B(\phi)$ is an ANR and $B(\phi) \subset E_c^t(\Lambda', \epsilon)$, it follows that $\Lambda'_0(\mathbb{F}_3(\mathbb{R}^{n+1}))$ contains compact subsets A of arbitrarily high relative category $\mathrm{cat}_{\Lambda'_0(\mathbb{F}_3(\mathbb{R}^{n+1}))}(\pi_0(A); \pi_0(B(\phi)))$.

With this being understood, by abuse of notation, we shall hereon use Λ' for Λ'_0 and f for $f|\Lambda'_0$ and, in general, keep the same notation on passing to the orbit space.

2 The First Example

As mentioned above, our first example is concerned with the case of three bodies and is meant to illustrate the advantage of using the notion of topological end, $E_c^t(\Lambda', \epsilon)$, introduced in Chapter XIII. The example is representative of several theorems that assert the existence of critical points of functionals of the form

$$f(u) = \frac{1}{2}\int_0^T |\dot{u}|^2 dt - \int_0^T \sum_{i \neq j} V_{ij}(u_i - u_j, t)dt,$$

subject to various sets of conditions on the potential V. This particular example is one of the results of [78, Majer-Terracini].

Let $V_{ij} : \mathbb{R}^{n+1} - \{0\} \times \mathbb{R} \to \mathbb{R}$, where $1 \leq i \neq j \leq 3$, be C^1-functions which satisfy the following assumptions:

(V_i) for all $x \in \mathbb{R}^{n+1} - \{0\}$ and $t \in \mathbb{R}$, $V_{ij}(x,t) \leq 0$,
(V_2) there exists $\rho_0 > 0$ and a C^1-function $U : \mathbb{R}^{n+1} - \{0\} \to \mathbb{R}$ such that $\lim_{x \to 0} U(x) = -\lim_{x \to 0} V_{ij}(x,t) = +\infty$, and $-V_{ij}(x) \geq |\nabla U(x)|^2$ for $0 < |x| < \rho_0$, and
(V_3) as $|x| \to +\infty$, $(\nabla V_{ij}.x)/(|\nabla V_{ij}||x|^1/2)$ approaches $+\infty$ uniformly in t, where $\nabla V_{ij} = (\partial V_{ij}/\partial x_1, \partial V_{ij}/\partial x_2, \partial V_{ij}/\partial x_3)$.

Theorem 2.1 *Assume that* $(V_1), (V_2)$, *and* (V_3) *hold, and let* $f : \Lambda' = \Lambda'(\mathbb{F}_3(\mathbb{R}^{n+1})) \to \mathbb{R}$, $n > 1$ *be the function given by*

$$f(u) = \frac{1}{2}\int_0^T |\dot{u}|^2 dt - \int_0^T \sum_{1 \leq i \neq j \leq 3} V_{ij}(u_i - u_j, t)dt.$$

Then, for each $T > 0$, f *possesses an unbounded sequence of critical values.*

We content ourselves with a sketch of the proof in [78, Majer-Terracicni], suitably modified to highlight the role of the topological end $E_c^t(\Lambda, \epsilon)$.

Our first objective is to find an appropriate analytic end (in our sense) of the form $B(\phi) = \{u | g(u) \geq \phi(\|\dot{u}\|_2^2)\}$ to satisfy certain conditions. For example, if we let

$$M(\phi) = \Lambda' - B(\phi) = \{u | g(u) < \phi(\|\dot{u}\|_2^2)\},$$

then $M(\phi)$ is required to be a smooth manifold with boundary $\partial M(\phi)$, the latter with a unit exterior normal ν_ϕ. This is accomplished simply by assuming that $\phi > 0$ and is at least of class C^2.

Here is a consequence of V_1 and V_3.

Proposition 2.1 *There exists a* C^2*-function* $\phi_0 : \mathbb{R}^+ \to \mathbb{R}^+$ *such that for every increasing* $\phi \in C^2(\mathbb{R}^+, \mathbb{R}^+)$ *with* $\phi \geq \phi_0$, *the equation*

$$\nabla f(u) = \lambda \nu_\phi(u),\ \lambda > 0$$

has no solution u *on* $\partial M(\phi)$.

For the reader's convenience, we now collect in the following theorem many of the results of [74, Majer] and [78, Majer-Terracicni].

Put $f^a = \{u \in \Lambda' | f(u) \leq a\}$, and denote by $K_c = \{u \in \Lambda' | f(u) = c, f'(u) = 0\}$, the critical points of f at level $c \in \mathbb{R}$. Also, let $\phi : \mathbb{R}^+ \to \mathbb{R}^+$ be a C^2-function with the following properties:

(*i*) ϕ is bounded away from 0;
(*ii*) $\phi' > 0$; and
(*iii*) $\phi \geq \phi_0$,

where ϕ_0 is as in the Proposition 2.1 above. Put

$$B = B(\phi) = \{u | g(u) \geq \phi(\|\dot{u}\|_2^2)\}, M = M(\phi) = \{u | g(u) < \phi(\|\dot{u}\|_2^2)\}.$$

Theorem 2.2 *If $K_c \cap M = \emptyset$, then there is an $\epsilon_0 > 0$ and $\epsilon \in (0, \epsilon_0]$ and a continuous deformation $\eta : \Lambda' \times [0,1] \to \Lambda'$ with the following properties:*

(*i*) *$\eta(u, 0) = u$, for $u \in \Lambda'$;*
(*ii*) *$\eta(u, t) = u$ for $t \in [0,1]$ if $|f(u) - c| \geq \epsilon_0$;*
(*iii*) *$(\partial f / \partial t)(\eta(u,t)) \leq 0$;*
(*iv*) *$B = \Lambda' - M$ is invariant under η; and*
(*v*) *$\eta(f^{c+\epsilon}, 1) \subset f^{c-\epsilon} \cup B$.*

Next, let, for $m \geq 2$,

$$\mathcal{A}_m = \{A \subset \Lambda' | A \text{ is compact and } \mathrm{cat}_{\Lambda'}(A, B(\phi)) \geq m\}.$$

Recall that, if $\phi \geq \phi_1$, where ϕ_1 is that given in Corollary 1.2 above, then there exist compact subsets A with relative category $\mathrm{cat}_{\Lambda'}(A, B(\phi))$ that is arbitrarily large, since $B(\phi)$ is contained in the topological end $E_c^t(\Lambda', \epsilon)$. This is precisely the alternative to the [78, Majer-Terracini] argument that uses only $\Lambda'(\mathbb{F}_2(\mathbb{R}^{n+1}))$ to obtain such compact sets, ignoring the richer homology structure of $\Lambda'(\mathbb{F}_3(\mathbb{R}^{n+1}))$. In any case, the family $\mathcal{A}_m$ is non-empty when $\phi \geq \phi_1$, and the numbers

$$c_m = \inf_{A \in \mathcal{A}_m} \sup_A f(u)$$

are well-defined.

Suppose that $\phi : \mathbb{R}^+ \to \mathbb{R}^+$ satisfies the conditions given in the paragraph before Theorem 2.2.

Proposition 2.2 *If $\phi \geq \phi_1$, then $K_{c_m} \cap M \neq \emptyset$, for all m.*

Proof. Put $c = c_m$, and suppose, to the contrary, that $K_c \cap M = \emptyset$. Then, there is an ϵ_0, ϵ and a deformation $\eta : \Lambda' \times I \to \Lambda'$ that leaves $B(\phi)$ invariant and takes $f^{c+\epsilon}$ into $f^{c-\epsilon} \cup B(\phi)$. There is a compact set $A \subset \Lambda'$ such that $\sup_A f \leq c + \epsilon$ and $\mathrm{cat}_{\Lambda'}(A, B(\phi)) \geq m$. On setting $\eta(A, 1) = A'$, one sees that $\mathrm{cat}_{\Lambda'}(A', B(\phi)) \geq m$ $\sup_A f \leq c - \epsilon$. But this a contradiction of the fact that $c = c_m$. □

Next, the final step in the proof of the theorem depends on the following result, which is also of independent interest.

Theorem 2.3 *Assume that V_1 and V_2 hold for the functional f. Then, for every open subset M of Λ' and every $c \in \mathbb{R}$, $\mathrm{cat}_{\Lambda'}(f^c, \Lambda - M) < \infty$.*

We refer the reader to [78, Majer-Terracini] for the proof. Note, however that the relative category used in the that article, which is different from the one given in Chapter XIII, is called cat* in [31, Fadell]. It is cat*, however, that dominates our relative category.

Corollary 2.1 *The sequence $\{c_m\}$ above is unbounded.*

Proof. Suppose, to the contrary, that the sequence $\{c_m\}$ is bounded. As the sequence is monotone nondecreasing, it follows that $\lim_m c_m = c < \infty$. Hence, $\mathrm{cat}_{\Lambda'}(f^{c+\epsilon}, B(\phi)) = +\infty$, which contradicts the preceding theorem. Thus the theorem is established. □

Observe that the proof of Proposition 2.2 shows that, if $\phi \geq \phi_1$, in addition to conditions (i)-(iii) in the paragraph preceding Theorem 2.2, there are critical points u_m for f in the complement of the analytic end $B(\phi)$ such that $\lim_m f(u_m) = +\infty$. If we extend the notion of topological end to read

$$E_c^t(\Lambda', \epsilon, \beta(\dot{u})) = \{u \in \Lambda' || [u_k - \frac{1}{2}[u_i + u_j]|| \geq |\Lambda' \frac{1}{2}[u_i + u_j]|| + \beta(\dot{u}) + \epsilon\}$$

for some function $\beta(\dot{u})$, where $i < j$ and (i, j, k) is a permutation of $(1, 2, 3)$, then, by a simple argument one can show both that

$$\beta(\dot{u}) \geq 3/2\sqrt{\phi(\|\dot{u}\|_2^2)},$$

$E_c^t(\Lambda', \epsilon, \beta(\dot{u})) \subset B(\phi)$ and also that the solutions $u_m \notin E_c^t(\Lambda', \epsilon, \beta(\dot{u}))$.

3 The Second Example

In this section we are again concerned with a C^2-functional $f' : \Lambda' = \Lambda'(\mathbb{F}_k(\mathbb{R}^{n+1})) \to \mathbb{R}$, with $k \geq 3$, of the form

$$(1) \qquad f(u) = \frac{1}{2}\int_0^T |\dot{u}|^2 dt - \int_0^T \sum_{i \neq j} V_{ij}(u_i - u_j, t)dt,$$

as considered in [75, 77, Majer-Terracini], with the following assumptions on the C^1-functions V_{ij}, where $1 \leq i \neq j \leq k$:

(V_1) $V_{ij}(x,t) \le 0, \;\; x \in \mathbb{R}^{n+1} - 0, \;\; t \in \mathbb{R}$;
(V_2) $\exists \rho_0 > 0$, and $U \in C^1(\mathbb{R}^{n+1} - 0, \mathbb{R})$ such that
$-V_{ij}(x,t) \ge |\nabla U(x)|^2 + U(x), \forall x, 0 < |x| < \rho_0, t \in \mathbb{R}$,
and $\lim_{x\to 0} U(x) = +\infty$; and
(V_3) $\exists \rho > 0$ and θ, $0 \le \theta < \pi/2$ such that
$\arccos(\nabla V_{ij}(x,t), x)/|V_{ij}(x,t)||x| \le \theta, \forall |x| > \rho$.

Note that V_2 is a version of V_2 of §2, while V_3 is a more geometric version of V_3 of §2.

The main result of [77, Majer-Terracini] is as in the following theorem.

Theorem 3.1 *For each $T > 0$, f possesses an unbounded sequence of critical values.*

Just as in the first example, we do not give a complete proof of the theorem: we shall concentrate on those aspects of the proof that involve the use of Γ-category instead of the collision index. (See the Historical Remarks at the end of Chapter XIII.) To give the reader a little insight into the study of such problems, we state, without proof, the necessary analytic results. The part of the proof requiring Γ-category is given in some detail.

One of the first steps in the proof is to investigate whether a restricted Palais-Smale condition, which is usually called the PS-condition for short, is satisfied by our functional. Recall that it requires of the C^1-functional $f : \Lambda' \to \mathbb{R}$ that every sequence $\{u_i\}$ in Λ' such that $f(u_i) \to c$ and $f'(u_i) \to 0$ have a convergent subsequence. But our functional $f : \Lambda' \to \mathbb{R}$ is invariant under translation by $\mathbb{R}^{n+1}$ and, hence, cannot satisfy the PS-condition. To overcome this difficulty, we work in the orbit space $\Lambda'/\mathbb{R}^{n+1}$, and ask that the $\{u_i\}$ be bounded in the sense that $g(u_j) \le b$, for all j, where $g : \Lambda' \to \mathbb{R}$ is the function defined by the equation

$$g(u) = (1/2k) \sum_{i \ne j} [u_i - u_j]^2.$$

To simplify the notation, we shall write Λ' for $\Lambda'/\mathbb{R}^{n+1}$ and f, g for the induced functionals. The conditions V_1, V_2 and V_3 continue to obtain.

The proofs of the following propositions are consequences of V_1, V_2, and V_3. They are given in [75, Majer-Terracini].

Proposition 3.1 *If $u_i \to v \in \partial\Lambda'$ and $g(u_i)$ is a bounded sequence, then $f(u_i) \to +\infty$.*

Proposition 3.2 *Let $\{u_i\}$ be a sequence in Λ', and $c \in \mathbb{R}$, such that $f(u_i) \to c$, $f'(u_i) \to 0$ and $g(u_i) \le b$ for some $b > 0$ and all i. Then $\{u_i\}$ has a convergent subsequence in Λ'.*

Proposition 3.3 *Given $c \in \mathbb{R}$ and $b > 0$, every sequence $\{u_i\}$ in Λ' such that $f(u_i) \to c$, $g(u_i) \to b$ and $f'(u_i) - \lambda_i g(u_i) \to 0$, where $\lambda_i \ge 0$, has a convergent subsequence.*

Proposition 3.4 *There is an increasing function $b(c) > 0$, $c > 0$ such that the nonlinear eigenvalue problem $f'(u) = \lambda g(u)$, $\lambda > 0$ has no solution with $f(u) \leq c$, $g(u) > b$.*

We note that an explicit formula for $b(c)$, derived in [75, Majer-Terracini], leads to a priori bounds for the critical points of the functional f.

The preceding propositions are used in the proof of the following deformation lemma. First, let us recall the following standard notation. Given $h \in C^1(\Lambda', \mathbb{R})$, put

$$\begin{aligned} K_c &= \{u \in \Lambda' | h(u) = c, h'(u) = 0\}, \\ h^b &= \{u \in \Lambda' | h(u) \leq b\}, \text{ and} \\ h_b &= \{u \in \Lambda' | h(u) \geq b\}. \end{aligned}$$

Theorem 3.2 *(Deformation Theorem) Given $c < \bar{c}$, let $b = b(c)$ be the increasing function in Proposition 4.4. Then, for $\bar{\epsilon} > 0$ and every neighborhood $N \subset \{f^{c+\bar{\epsilon}} \cap g^{b(\bar{c})}$ of $K_c \cap g^{b(\bar{c})}$, there is a continuous deformation $\eta : \Lambda' \times I \to \Lambda'$ and an ϵ, $0 < \epsilon < \bar{\epsilon}$; such that*

(i) $\eta(u, s) = u$ *if* $s = 0$, $|f(u) - c| \geq \bar{\epsilon}$, *or* $g(u) \geq b(\bar{c})$;
(ii) $\frac{\partial}{\partial s} f(\eta(x, s)) \leq 0$, *i.e.* $f(x, s)$ *is nonincreasing for every fixed* x;
(iii) $\eta(f^{c+\epsilon}) \cap g^{b(\bar{c})} - N) \subset f^{c-\epsilon} \cap g^{b(\bar{c})}$; *and*
(iv) if $K_c = \emptyset$, *then* $\eta(f^{c+\epsilon}) \cap g^{b(\bar{c})} \subset f^{c-\epsilon} \cap g^{b(\bar{c})}$.

Proof of Theorem: Final Step

Let $\{\Lambda'_{ij}\}$ be the open cover of $\Lambda'(\mathbb{F}_k(\mathbb{R}^{n+1}))$ introduced in §4 of Chapter XIII. Recall that

$$\Lambda'_{ij} = \{u \in \mathcal{H} | u_i(t) \neq u_j(t), \forall t \in \mathbb{R}\}.$$

Observe that Λ'_{ij} is invariant under translations by elements of $\mathbb{R}^{n+1}$. The family $\Lambda'\Gamma = \{\Lambda'_{ij}\}$ gives rise to a similar family in the orbit space, and the $\Lambda'\Gamma$-category of a set A in $\Lambda'(\mathbb{F}_k(\mathbb{R}^{n+1}))$ is the same as that of its image in the orbit space $\Lambda'(\mathbb{F}_k(\mathbb{R}^{n+1}))/\mathbb{R}^{n+1}$. Moreover, according to Proposition 4.5 of Chapter XIII, the image of Λ' contains compact subsets of arbitrarily high $\Lambda'\Gamma$-category. Following our convention mentioned above, by abuse of notation, we write $\mathcal{H}$ for $\mathcal{H}/\mathbb{R}^{n+1}$ and $\Lambda' = \Lambda'(\mathbb{F}_k(\mathbb{R}^{n+1}))$ for its image in the orbit space $\mathcal{H}/\mathbb{R}^{n+1}$. Similarly, we write $\Lambda'\Gamma = \{\Lambda'_{ij}\}$ for the open cover of its image. Observe also that, according to Corollary 4.4 of Chapter XIII, if $b \geq \frac{T}{3}c$, then it follows that $\Lambda'\Gamma\text{cat}(E_{c,b}) \leq 2$, where

$$E_{c,b} = \{u \in \Lambda' | f(u) \leq c, \text{ and } g(u) \geq b\}.$$

Now, put

$$\mathcal{A}_m = \{A \subset \Lambda' | A \text{ is compact and } \Lambda'\Gamma\text{cat}(A) \geq 2m\},$$

and let

$$c_m^* = \inf_{A \in \mathcal{A}_m} \sup_A f(u).$$

Note that, because $\mathcal{A}_m$ is non-empty, it follows that c_m^* is well-defined and $c_m^* \leq c_{m+1}^*$. Also, choose $\bar{c}_m > c_m^*$.

Now, let $b(c)$ be the function in Proposition 3.4, and assume further that $b(c) \geq \frac{T}{3}c$, if necessary. Define c_m by

$$c_m = \inf\{s | f^s \cup (f^{\bar{c}_m} \cap g_{b(\bar{c}_m)} \supset A, \text{for some } A \in \mathcal{A}_m\}.$$

Then, $c_m \leq c_m^* < \bar{c}_m$. Also, note that, for $\epsilon > 0$,

$$\Lambda' \Gamma \text{cat}(f^{c_m+\epsilon} \cup (f^{\bar{c}_m} \cap g_{b(\bar{c}_m)} \geq 2m,$$

and, hence, $\Lambda'\Gamma\text{cat}(f^{c_m+\epsilon}) \geq 2m - 2$, which in turn implies that $c_{m-1}^* \leq c_m \leq c_m^* < \bar{c}_m$.

The next three propositions complete the proof.

Proposition 3.5 *For each $m \geq 2$, the critical set $K_{c_m} \cap g_{b(\bar{c}_m)} \neq \emptyset$.*

Proof. Suppose, to the contrary, that $K_{c_m} \cap g_{b(\bar{c}_m)} = \emptyset$. Now, using the notation of the Deformation Theorem, put $c = c_m$ and $\bar{c} = \bar{c}_m$. Choose $\epsilon < \bar{c} - c$. Then,

$$\Lambda' \Gamma \text{cat}(f^{c+\epsilon} \cup (f^{\bar{c}} \cap g_{b(\bar{c})})) \geq 2m,$$

and

$$\eta(f^{c+\epsilon} \cup (f^{\bar{c}} \cap g_{b(\bar{c})}), 1) \subset f^{c-\epsilon} \cup (f^{\bar{c}} \cap g_{b(\bar{c})})).$$

This contradicts the definition of c. □

For the next proposition, suppose that $m \geq 2$. Then, as above, $c_m \leq c_m^* < \bar{c}_m$. Suppose further that

$$c_m = c_{m+1} = \cdots = c_{m+s} = c, \; s \geq 1,$$

where we assume, without loss of generality, that $\bar{c}_{m+s} \geq \bar{c}_{m+j}$, for $j = 0, \cdots, s$.

Proposition 3.6 *With the assumptions above,*

$$\Lambda' \Gamma \text{cat}(K_c \cap g^{b(\bar{c}_{m+s})}) \geq 2s + 1.$$

Proof. Let $B = K_c \cap g^{b(\bar{c}_{m+s})}$, and note that B is compact. Next, choose an open neighborhood N such that

$$\bar{N} \supset N \supset B$$

and

$$\Lambda'\Gamma\text{cat}(\bar{N}) = \Lambda'\Gamma\text{cat}(B) \le 2s < \infty.$$

Let $\bar{\epsilon} = 1$, and apply the Deformation Theorem to obtain an ϵ and a deformation η. Choose a compact set $A \in \mathcal{A}_{m+s}$ such that $A \subset f^{c+\epsilon} \cup (f^{\bar{c}_{m+s}} \cap g_{b(\bar{c}_{m+s})})$, with $\Lambda'\Gamma\text{cat}(A) \ge 2(m+s)$. Suppose that $\Lambda'\Gamma\text{cat}(\bar{N}) \le 2s$; then $\Lambda'\Gamma\text{cat}(A - N) \ge 2m$, and

$$\eta(f^{c+\epsilon} \cup (f^{\bar{c}_{m+s}} \cap g_{b(\bar{c}_{m+s})}) - N \subset f^{c-\epsilon} \cup (f^{\bar{c}_{m+s}} \cap g_{b(\bar{c}_{m+s})}).$$

But this contradicts the definition of c_m. □

Proposition 3.7 *The sequence* $\{c_m\}$ *is unbounded.*

Proof. Suppose that the sequence $\{c_m\}$ is bounded. As $c^*_{m-1} \le c_m \le c^*_m < \bar{c}_m$, it follows that the sequence $\{c^*_m\}$ is also bounded. Let $c = \lim_{m\to\infty} c^*_m$, and choose $\bar{c} > c$. Put $b = b(\bar{c})$. Now, $K_c \cap g^b$ is compact and nonempty. Put $p = \Lambda'\Gamma\text{cat}(K_c \cap g^b)$, and choose a closed neighborhood $N \supset K_c \cap g^b$ such that $\Lambda'\Gamma\text{cat}(N) = p$, $\bar{\epsilon} = \bar{c} - c$, $\epsilon < \bar{\epsilon}$ and η, as in the Deformation Theorem. For every $m \ge 1$, there exists a compact set A_m such that $\Lambda'\Gamma\text{cat}(A_m) \ge 2m$, and $A_m \subset f^{c+\epsilon}$. Furthermore,

$$A_m \subset (A_m \cap g^b - N) \cup (A_m \cap g_b) \cup N,$$

and $A_m \cap g_b \subset f^{c+\epsilon} \cap g_b \subset f^{\bar{c}} \cap g_{b(\bar{c})}$. Hence, $\Lambda'\Gamma\text{cat}(A_m \cap g_b) \le 2$. Therefore,

$$\Lambda'\Gamma\text{cat}((A_m \cap g^b) - N) \ge 2m - (p+2)$$

for all m. Applying the deformation η, we obtain

$$\eta(((A_m \cap g^b) - N), 1) \subset f^{c-\epsilon} \cap g^b \subset f^{c-\epsilon} \cup (f^{\bar{c}} \cap g_b),$$

and $\Lambda'\Gamma\text{cat}(\eta((A_m \cap g^b) - N, 1)) \ge 2m - (p+2)$ for all m. Fix m_0 so that $c - \epsilon < c_{m_0} < c$. This contradicts the definition of c_{m_0}. □

Thus, using Λ'-category, we have established the last step in the proof of Theorem 3.2 above.

4 Historical Remarks

Note that, in the proof of Proposition 3.5, if $u \in K_c \cap g^{b(\bar{c})}$, then $g(u) \le b(\bar{c})$ for some choice of $\bar{c}$ such that $f(u) = c < \bar{c}$. Thus, $g(u) \le b(\bar{c})$ is an a priori bound for u. In [75, Majer-Terracini] one finds an explicit formula for a suitable function $b(c)$.

References

1. Adams, J.F.: On the cobar construction. Colloque de Topologie Algebrique. Louvain (1958).
2. Adams, J.F.: On the non-existence of elements of Hopf invariant one. Ann. of Math. **72** (1960) 20-104.
3. Adams, J.F.: Vector fields on spheres. Ann. of Math. **75** (1962) 603-32.
4. Adams, J.F.: Vector fields on spheres. Topology **1**(1962) 63-65.
5. Adams, J.F., and Hilton, P.J.: On the chain algebra of a loop space. Comm. Math. Helv. **30** (1956), 305-30.
6. Alexander, J.W.: Deformations of an n-cell. Proc. Nat. Acad. Sci. U.S.A.. **9** (1923) 406-407.
7. Artin, E.: Theory of Braids. Ann. of Math. **48** (1947) 101-26.
8. Bahri, A., and Rabinowitz, P.H.: Periodic solutions of Hamiltonian systems of 3-body type. Ann. Inst. H. Poincaré, Anal. Nonlin. **8** (1991) 561-49.
9. Birman, J.S.: Braids, links, and mapping class groups. Ann. of Math. Studies **82**. Princeton University Press (1984).
10. Bödigheimer, F.-C., Cohen, F.R., and Milgram, R.J.: Truncated symmetric products and configuration spaces. Math. Z. **214** (1994) 179-216.
11. Bott, R.: Configuration spaces and imbedding invariants. Turk. J. Math. **20** (1996) 1-17.
12. Bott, R., and Samelson, H.: On the Pontryagin product in spaces of paths. Comment. Math. Helv. **27** (1953) 320-37.
13. Bott, R., and Taubes, C.: On the self-linking of knots. J. Math. Phys. **35** (1994) 5247-87.
14. Bourbaki, N.: Groupes et algèbres de Lie, chap. 2-3. Hermann, Paris (1972).
15. Brown, R.F., and White, J.H.: Homology and Morse theory of third configuration spaces. Indiana U. Math. J. **30** (1981) 501-12.
16. Chow, W-L.: On the algebraic braid group. Ann. of Math. **49** (1948), 654-58.
17. Cohen, F.R.: The homology of $\mathcal{C}_{n+1}$-spaces, $n \geq 0$. The homology of iterated loop spaces, LNM **533** 207-351. Springer-Verlag, New York (1976).
18. Cohen, F.R., and Gitler, S.: On loop spaces of configuration spaces. Preliminary report.
19. Cohen, F.R., and Taylor, L.R.: Configuration spaces: applications to Gelfand-Fuks cohomology. Bull. Amer. Math. Soc. **84** (1978) 134-36.
20. Cohen, F.R., and Taylor, L.R.: Computations of Gelfand-Fuks cohomology, the cohomology of function spaces, and the cohomology of configuration spaces. Geometric applications of homotopy theory I, LNM **657** 106-43. Springer-Verlag, New York (1978).
21. Cohen, R.L.: A model for the free loop space of a suspension. LNM **1286** 193-207. Springer-Verlag, New York (1987).
22. Crabb, M.C., and James, I.M.: Fiberwise configuration spaces. Bol. de la Soc. Mat. Mex. **37** (1992) 83-97.

23. Crowell, R.H., and Fox, R.H.: Introduction to knot theory. GTM **57**. Springer-Verlag, New York (1963).
24. Dold, A.: Partitions of unity in the theory of fibrations. Ann. of Math., 2d ser. **78** (1963) 223-55.
25. Dold, A.: Lectures on Algebraic Topology. Springer-Verlag, New York (1972).
26. Dold, A., and Lashof, R.: Principal quasi-fibrations and fibre homotopy equivalence of bundles. Ill. J. Math. **3** (1959) 285-305.
27. Eilenberg, S. and Mac Lane, S.: On the groups $H(\Pi, n)$, I. Ann. of Math. **58** (1953) 55-106.
28. Eilenberg, S., and Moore, J.C.: Homology and fibrations, I. Colagebras, cotensor product and its derived functors. Comm. Math. Helv. **40** (1966) 199-236.
29. Fadell, E.R.: Homotopy groups of configuration spaces and the string problem of Dirac. Duke Math. J. **29** (1962) 119-26.
30. Fadell, E.: Generalized normal bundles for locally-flat imbeddings. Trans. Amer. Soc. **114** (1965) 488-513.
31. Fadell, E.: Lectures on cohomological index theories of G-spaces with applications to critical point theory. Raccolta di Sem. Univ. della Calabria (1985).
32. Fadell, E.R., and Hurewicz, W.: On the structure of higher differential operators in spectral sequences. Ann. of Math., 2d ser. **68** (1958) 314-47.
33. Fadell, E., and Husseini, S.: Relative cohomological index theories. Advances in Math. **64** (1987) 1-31.
34. Fadell, E., and Husseini, S.: An ideal-valued cohomological index theory with applications to Borsuk-Ulam and Bourgin-Yang theorems. Ergod. Th. and Dynam. Sys. **8**, Charles Conley Memorial Issue (1988) 73-85.
35. Fadell, E., and Husseini, S.: A note on the category of the free loop space Proc. Amer. Math. Soc. **107** (1985) 527-36.
36. Fadell, E., and Husseini, S.: Category of loop spaces of open subsets in Euclidean space. J. Nonlin. Anal. **17** (1991) 1153-61.
37. Fadell, E., and Husseini, S.: Infinite cup length in free loop spaces with an application to a problem of N-body type. Ann. Inst. H. Poincaré, Anal. Nonlin. **9** (1992) 305-19.
38. Fadell, E. and Husseini, S.: Relative category, products and coproducts. Rendiconti del Sem. Mat. e Fis. di Milano bf 64 (1994) 99-115.
39. Fadell, E., and Husseini, S.: On the growth properties of the homology of free loop spaces of configuration spaces. Proc. of the First World Congress of Nonlin. Anal., pp. 2127-34. Walter de Gruyter, Berlin and New York (1996).
40. Fadell, E., and Husseini, S.: The space of loops on configuration spaces and the Majer-Terracini index. Toplogical Methods in Nonlinear Analysis, J. of the Schauder Center **11** (1998) 249-71.
41. Fadell, E., and Husseini, S.: When are configuration bundles fiber homotopically trivial? Bull. des Sc. Math. Quebec, **22** (1998) 149-59.
42. Fadell, E., Husseini, S., and Rabinowitz, P.: Borsuk-Ulam Theorems for arbitrary S^1 actions and applications. Trans. Amer. Math. Soc. **274** (1982) 345-60.
43. Fadell, E., and Neuwirth, L.: Configuration spaces. Math. Scand. **10** (1962) 111-18.
44. Fadell, E., and Van Buskirk, J.: The Braid groups of E^2 and S^2. Duke Math. J. **29** (1962) 243-57.
45. Falk, M., and Randell, R. : Lower central series of a fiber-type arrangement. Inventiones Math. **82** (1985) 77-88.
46. Feichtner, E.M., and Ziegler, G.M.: The integral cohomology of ordered configuration spaces of spheres. Preprint (1996).
47. Fournier, G., and Willem, M.: Multiple solutions of the forced pendulum equation. Ann. Inst. H. Poincarè, Anal. Nonlin. **6** suppl. (1989) 259-81.

48. Fulton, W., and MacPherson, R.: A compactification of configuration spaces. Ann. of Math. **139** (1994) 187-229.
49. Gillette, R., and Van Buskirk, J.: The word problem and its consequences for the braid group and mapping class groups of the 2-sphere. Trans. Amer. Math. Soc. **131** (1968) 277-96.
50. Gottlieb, D.H.: A certain subgroup of the fundamental group. Amer. J. Math. **87** (1965) 840-56.
51. Haefliger, A.: Plongements différentiables de variétés dans variétés. Comm. Math. Helv. **36** (1961) 47-82.
52. Haefliger, A.: Plongements différentiables dans le domaine stable. Comm. Math. Helv. **37** (1962) 155-76.
53. Hall, M. Jr.: The Theory of Groups. The Macmillan Company, New York (1959).
54. Hansen, V.L.: Braids and coverings: selected topics. Lond. Math. Soc. Student Texts **18** (1989). Cambridge Univ. Press.
55. Hilton, P.J.: On the homotopy groups of the union of spheres. J. London Math. Soc. **30** (1955) 154-72.
56. Hopf, H.: Über die Abbildungen von Sphären auf Sphären niedregen Dimension. Fund. Math. **5** (1935) 427-40.
57. Husseini, S.Y.: Constructions of the reduced product type. Toplogy **2** (1963) 213-37.
58. Husseini, S.Y.: Constructions of the reduced product type, II. Topology **3** (1965) 59-79.
59. Husseini, S.Y.: When is a complex fibered by a subcomplex? Trans. Amer. Math. Soc. **124** (1966) 249-91.
60. Husseini, S.Y.: Topology of the classical groups. Gordon and Breach, N.Y. (1969)
61. Husseini, S.Y.: Generalized Lefshetz Numbers. Trans. Amer. Math. Soc. **272** (1982) 247-74.
62. Jambu, M., and Papadima, S.: A generaliztion of fiber-type arrangements and a new deformation method. Topology **37** (1998) 1135-64.
63. James, I.M.: Reduced product spaces. Ann. of Math. **62** (1955) 170-97.
64. James, I.M.: The topology of Stiefel manifolds. London Math. Soc. Lecture Notes **24** (1976). Cambridge University Press.
65. James, I.M., and Whitehead, J.H.C.: On the homotopy theory of sphere-bundles over spheres, I, II. Proc. London Math. Soc., **4**(1954) 196-218, **5** (1955) 148-66.
66. Kallel, S.: Brace products, homology operations and the structure of free loop spaces. (To appear).
67. Kohno, T.: Monodromy representations of braid groups and Yang-Baxter equations. Annales Inst. Fourier, **37** (1987) 139-60.
68. Kohno, T.: Linear representations of braid groups and classical Yang-Baxter equations. Contemp. Math. **78** (1988) 339-63.
69. Kohno, T.: Vassiliev Invariants and the de Rham Complex on the space of knots. Symplectic Geometry and Quantization, Contemp. Math. **179** (1994) 123-38.
70. Loday, J.-L., and Quillen, D.: Cyclic homology and the Lie algebra homology of matrices. Comm. Math. Helv. **59** (1984) 565-91.
71. Lyndon, R.C., and Schupp, P.E.: Combinatorial group theory. Springer-Verlag, New York (1977).
72. Mac Lane, S.: Homology. Springer-Verlag, New York (1963).
73. Magnus, W., Karrass, A., and Solitar, D.: Combinatorial troup theory. Wiley, New York (1966).

74. Majer, P.:Two variational methods on manifolds with boundary. Toplogy **34** (1994), 1-12.
75. Majer, P., and Terracini, S.: Periodic solutions to some problems of n-body type. Arch. Rational Anal. **124** (1993) 381-404.
76. Majer, P., and Terracini, S.: Periodic solutions to some N-body type problems: the fixed energy case. Duke Math. J. **63** (1993) 683-97.
77. Majer, P., and Terracini, S.: Multiple periodic solutions to some n-body type problems via a collision index. Variational Methods in Nonlinear Analyses, Arice, 1992, pp. 245-62. Gordon and Breach, Basel (1995).
78. Majer, P., and Terracini, S.: On the existence of infinitely many periodic solutions to some problems of n-body type. Comm. Pure Appl. Math. **48** (1995) 449-70.
79. Massey, W.S.: The homotopy type of certain configuration spaces. Bol. de la Soc. Mat. Mex. **37** (1992) 355-65.
80. Mathews, H.V.: Cellular twisted products. Ph.D. Thesis, University of Wisconsin-Madison (1993).
81. Mawhin, J., and Willem, M.: Critical Point Theory and Hamiltonian Systems. Applied Mathematical Sciences **74** (1989). Springer-Verlag, New York.
82. Milnor, J.W.: Construction of universal bundles I, II. Ann. of Math. **63** (1956) 272-84, 430-36.
83. Milnor, J.W., and Moore, J.C.: On the structure of Hopf algebras. Ann. of Math. **81** (1965) 211-64.
84. Moore, J.C.: Le Theorème de Freudenthal, la suite exacte de James at l'invariant de Hopf gènèralisè. Exposè 22, Séminaire H. Cartan, E.N.S. (1954/55).
85. Moulton, V. L.: Vector braids. J. Pure App. Algebra **131** (1998) 245-96.
86. Orlik, P., and Terao, H.: Arrangements of hyperplanes. Grundlehren **300** (1992), Springer-Verlag, New York.
87. Palais, R. S.: Local triviality of the restriction map for embeddings. Comm. Math. Helv. **34** (1960) 305-12.
88. Palais, R. S.: Homotopy theory of infinite dimensional manifolds. Topology **5** (1966) 1-16.
89. Ramsey, J. R.: Extensions of Lusternik-Schnirelman Category theory to relative, equivariant and invariant Theories. Ph.D. Thesis, University of Wisconsin-Madison (1989).
90. Reeken, M.: Stability of critical points under small perturbations, Part I. Manus. Math. **7** (1972) 387-411.
91. Riahi, H.: Periodic orbits of n-body type problems. Ph.D. Thesis, Rutgers University, New Brunswick, N.J. (1992).
92. Riahi, H.: Periodic orbits of n-body type problems: the fixed period case. Trans. Amer. Math. Soc. **347** (1995) 4663-85.
93. Riahi, H.: Study of the critical points at infinity arising from the failure of the Palais-Smale condition for n-body type problems. Amer. Math. Soc. Memoirs **138**, No. 658 (1999).
94. Roos, J.E.: Homology of free loop spaces, cyclic homology, and non-rational Poicarè series in commutative algebra. LNM **1352** (1988) 173-89. Springer-Verlag, New York.
95. Rothenberg, M., and Steenrod, N.E.: The cohomolgy of classifying spaces of H-spaces. Bull. Amer. Math. Soc. **58** (1965) 872-75.
96. Samelson, H.: A connection between the Whitehead and the Pontryagin product. Amer. J. Math. **75** (1953) 744-52.
97. Serre, J-P.: Homologie singulière des espaces fibré. Applications. Ann. of Math. **54** (1951) 425-505.

98. Serre, J-P.: Lie algebras and Lie groups. 1964 Lectures at Harvard University. 2d ed. LNM **1500** (1992). Springer-Verlag, New York.
99. Shapiro, A.: Obstructions to the imbedding of a complex in a Euclidean space, I. The first obstruction. Ann. of Math. **66** (1957) 256-69.
100. Shnider, S., and Sternberg, S.: Quantum groups. From coalgebra to Drinfel'd algebras. A guided tour. Graduate Texts in Mathematical Physics, II. International Press, Cambridge, MA (1993)
101. Spanier, E. H.: Algebraic Topology. Springer-Verlag, New York (1966).
102. Stallings, J.: Centerless groups–an algebraic approach to Gottlieb's theorem. Topology **4** (1965) 129-34.
103. Steenrod, N.E.: The topology of fibre bundles. Princeton University Press, Princeton, NJ (1951).
104. Steenrod, N.E., and Whitehead, J.H.C.: Vector fields on the n-sphere. Proc. Nat. Acad. Sci. **37** (1951) 58-63.
105. Sullivan, D., and Vigué-Poirrier, M.: The homology theory of the closed geodesic problem. J. Diff. Geom. **11** (1976) 633-44.
106. Szulkin, A: A relative category and applications to critical point theorey for strongly indefinite functionals. J. Nonlin. Anal. **15** 725-39.
107. Thom, R.: La classification des immersions. Séminaire Bourbaki, 10e année: 1957-1958. Textes de Conférences, Exposé **157** (1957).
108. Vasiliev, V.A.: Complements of discriminants of smooth maps: topology and applications. Trans. of Math. Monographs **98**. Rev. ed. Amer. Math. Soc.(1994)
109. Vigué-Poirrier, M.: Homotopie rationnelle et croissance du nombre de géodésiques fermées. Ann. scient. Éc. norm. sup., 4e série, t. **17** (1984) 413-31.
110. Whitehead, G.W.: Elements of homotopy theory. Springer-Verlag, New York (1978).
111. Wu, W.-T.: Sur les classescaractéristiques des structures fibrés sphériques. Actualités Sci. Indust., **1183** 1-89. Hermann, Paris (1952).
112. Wu, W.-T.: On the imbedding of manifolds in a Euclidean space. Sci. Sinica **13** (1964) 682-83.
113. Xicoténcatl Merino, M. (1997): Orbit configuration spaces, infinitesmal braid relations in homology and equivariant loop spaces. Ph.D. Thesis, University of Rochester, Rochester, NY (1997).

Index

GPSR Compliance

The European Union's (EU) General Product Safety Regulation (GPSR) is a set of rules that requires consumer products to be safe and our obligations to ensure this.

If you have any concerns about our products, you can contact us on ProductSafety@springernature.com

In case Publisher is established outside the EU, the EU authorized representative is:

Springer Nature Customer Service Center GmbH
Europaplatz 3
69115 Heidelberg, Germany

Batch number: 10584501

Printed by Printforce, the Netherlands